Undocumented Secrets of
MATLAB®-Java Programming

Undocumented Secrets of MATLAB®-Java Programming

Yair Altman

CRC Press
Taylor & Francis Group
Boca Raton London New York

CRC Press is an imprint of the
Taylor & Francis Group, an **informa** business
A CHAPMAN & HALL BOOK

CRC Press
Taylor & Francis Group
6000 Broken Sound Parkway NW, Suite 300
Boca Raton, FL 33487-2742

© 2012 by Taylor & Francis Group, LLC
CRC Press is an imprint of Taylor & Francis Group, an Informa business

No claim to original U.S. Government works

Printed in the United States of America on acid-free paper
Version Date: 20111004

International Standard Book Number: 978-1-4398-6903-1 (Hardback)

Library of Congress Cataloging-in-Publication Data

Altman, Yair M.
   Undocumented secrets of MATLAB-Java programming / Yair M. Altman.
      p. cm.
   Includes bibliographical references and index.
   ISBN 978-1-4398-6903-1 (hardback)
   1. MATLAB. 2. Numerical analysis--Data processing. 3. Java (Computer program language) I. Title.

QA297.A544 2011
518.0285'53--dc23

2011036044

Visit the Taylor & Francis Web site at
http://www.taylorandfrancis.com

and the CRC Press Web site at
http://www.crcpress.com

To my family, who make it all worthwhile, and without whose support this herculean effort would never have seen the light of day.

Tovi — An excellent wife, who can find? For her worth is far above jewels . . . Many daughters have done virtuously, but thou excellest them all.

(Proverbs 31)

Contents at a Glance

Contents

Preface

The MATLAB[†] programming environment uses Java[‡] for numerous tasks, including net-working, data-processing algorithms, and graphical user interface (GUI). MATLAB's internal Java classes can often be easily accessed and used by MATLAB users. MATLAB also enables easy access to external Java functionality, either third-party or user-created. Using Java, we can extensively customize the MATLAB environment and application GUI, enabling the creation of very esthetically pleasing applications.

Unlike MATLAB's interface with other programming languages, the internal Java classes and the MATLAB–Java interface were never fully documented by The MathWorks (TMW), the company that manufactures the MATLAB product.

This is really quite unfortunate: Java is one of the most widely used programming languages, having many times as many programmers as MATLAB.[§] Using this huge pool of knowledge and components can significantly improve MATLAB applications.

As a consultant, I often hear clients claim that MATLAB is a fine programming platform for prototyping but not suitable for real-world modern-looking applications.

This book aims to correct this misconception. It shows how using Java can significantly improve MATLAB program appearance and functionality and that this can be done easily and even **without any prior Java knowledge**. In fact, many basic programming requirements can-not be achieved (or are difficult) in pure MATLAB but are very easy in Java. As a simple example, maximizing and minimizing windows is not possible in pure MATLAB but is a trivial one-liner using the underlying Java code:[¶]

```
>> set(get(handle(gcf),'JavaFrame'), 'Maximized',true);
```

Integrating Java in MATLAB is easy and extremely beneficial. By adhering to a few simple programming rules, many potential pitfalls of using MATLAB–Java can be avoided while gaining access to a vast variety of benefits.

Moreover, while most of this book relies on undocumented functionality that is not officially supported by TMW, in reality much of it is actually well documented and supported in the Java

---

[†] MATLAB is a registered trademark of The MathWorks (www.mathworks.com).

[‡] Java is a registered trademark of Javasoft, now an Oracle company (http://java.sun.com/).

[§] MathWorks advertises that there are "over one million" MATLAB programmers worldwide (http://bit.ly/hWzWSk). This number compares with 6–10 million Java programmers (e.g., http://bit.ly/gPXVTK).

[¶] See Chapter 7 for additional details.

world. Once we understand how to use the undocumented MATLAB–Java interface, we can easily use well-documented Java code to our advantage.

MATLAB's internal Java classes, although fully undocumented, are often based on existing (documented) Java classes and have remained relatively stable over the past MATLAB releases. So if we exercise caution, we can make very good use of them.

Very little information was ever published on these topics, most of it on the MATLAB forum (comp.soft-sys.matlab newsgroup, affectionately called CSSM).[1] Some notable CSSM posters in this regard include Amir Ben-Dor, Peter Boettcher, Steve Eddins, Arwel Hughes, Joshua Kaplan, Malcolm Lidierth, Eric Ludlam, Brad Phelan, Brett Shoelson, Donn Shull, Matthew Whitaker, Ed Yu, and Urs Schwarz (aka "us") who had posted a humorous comment about the source of *his* knowledge a few years ago.[2]

Unfortunately, only a few dozen CSSM posts deal directly with undocumented MATLAB–Java features in over 300,000 total threads at last count. Outside CSSM, there are even fewer public references to undocumented MATLAB; there are only a few other MATLAB forums[3] or blogs[4] with original material (not CSSM mirrors), and they are far outshined by CSSM in terms of traffic and content volumes.

Over the years, I have painstakingly researched the MATLAB code group (available in the MATLAB installation) and performed extensive trials and errors. I published many of my findings in CSSM in response to user queries. In later years, I also started to publish related articles on a dedicated blog: http://www.UndocumentedMatlab.com.

This accumulated knowledge about MATLAB–Java is now seeing light for the first time in this book, which includes all previously published work on the topic (as far as I know) as well as much more that was never published.

During the past few years, TMW released product updates of MATLAB twice annually, in March and September. Each of these versions deletes or modifies some hidden elements and adds a few. It is a real challenge to keep up with this rate of change, and I am certain that quite a few hidden niches and errors escaped my notice due to this.

I, therefore, gladly welcome your feedback on the book's website: http://www.Undocumented-Matlab.com/books/Java/. On this site, I will post the book's errata list and changes in future MATLAB releases. I presume these to be of particular importance, as unavoidable changes will occur in future MATLAB releases.

Book Organization

This book is organized in chapters grouped in related functionality/usage and ordered from easiest (novice Java use) to advanced. It is NOT necessary to read the book in order; the chapters and sections are mostly independent and can stand alone. You can safely skip almost any section that you find difficult or uninteresting.

Chapter 1 (Introduction to Java in MATLAB) provides a description of the internal Java engine shipped in MATLAB. Chapters 2 (Using Non-GUI Java Libraries in MATLAB) and 3

(Rich GUI Using Java Swing) describe how this internal Java engine can be used to extend MATLAB, both programmatically (Chapter 2) and visually (Chapter 3).

Chapter 4 (Uitools) describes a set of undocumented built-in MATLAB user-interface functions that use Java components. Many of these functions are MATLAB wrappers for Java components presented in Chapter 3. Using these tools and some simple customizations, we can significantly improve our MATLAB programs' usability.

The notion of customizing our MATLAB user-interface using Java is expanded in Chapters 5 (Built-in MATLAB Widget and Java Classes), 6 (Customizing MATLAB Controls) and 7 (The Java Frame). Chapter 8 (The MATLAB Desktop) discusses customization of the MATLAB environment rather than that of a MATLAB application.

Chapter 9 (Using MATLAB from within Java) discusses the other side of the coin, namely, how to call the MATLAB engine from within a Java program. Unlike calling Java in MATLAB, and unlike calling MATLAB from C/C++/VB, the Java-to-MATLAB interis entirely undocumented and unsupported — Chapter 9 fills this gap.

This book concludes with Chapter 10 (Putting It All Together), which describes a utid an application that tie together many issues presented in this book.

From a supportability viewpoint, progressively advanced chapters of this book are in undocumented territory, are less supported, have fewer online references, and are incy prone to change or malfunction in some future MATLAB release.

No prior Java knowledge is required. All code snippets and examples f-contained and can generally be used as-is. Advanced Java concepts are sometim, but understanding them is not required to run the code. Java-savvy readers wit easy to tailor code samples for their particular needs; for Java newcomers, an intn to Java (Appendix A) and numerous references to online resources will help to e learning curve.

To reduce irrelevant clutter, only bare-bones code snippets are presented within t. Only the concluding chapter contains a full listing. Users wishing to utilize the codets should include exception handling, edge-case handling, additional comments, etny real-world code, as the full listing hopefully shows.

Throughout this book, numerous references are provided that point to online res. Readers can use these resources to expand their knowledge and gain a deeper insigh is possible to achieve within this book. The resources often contain the poster's email, er-ested readers can follow-up directly with the poster.

No toolbox, Simulink or Stateflow is necessary for using this book — only core MATLAB product. These extra tools indeed contain many other Java-based aspects, they are not covered in this book. Perhaps a future book will describe them.

This book shows readers how to use and discover the described components, using ning but MATLAB itself as the discovery tool. In no case is illegal hacking implied or necess for the discovery or usage of anything presented in this book. As far as I know, everything this book is legal and within the bounds of the MATLAB license agreement. However, I n an engineer, not a lawyer, so this is by no means an official legal opinion. If you have any doubt, please contact MathWorks for a formal answer.

A Quick Q&A

I don't know Java—is this book for me? Absolutely yes. This book is intended for MATLAB programmers and users, and no Java knowledge is assumed. Java-savvy programmers will feel find it easier to use and extend some of the more advanced topics. However, even programmers with absolutely no Java experience can still use most of this book as-is. I hope the presentation will suit both audiences equally well.

Is it legal? Yes. I am an engineer, not a lawyer, but as far as I can tell, everything presented in this book is perfectly legal to use as long as one has access to a legal version of MATLAB. Still, if one has any specific concern about a particular aspect, MathWorks will gladly answer the question.

Does MathWorks endorse this book? Unfortunately not. This book often relies on undocumented and unsupported features. MathWorks allows us to use these features but does not officially endorse or support them.

How can I help to promote this work? One can help by sending feedback and by promoting the book and the website to colleagues.

Conventions Used in This Book

The following special text formatting conventions are used within this book:

Fixed-width font is used for MATLAB or Java code segments and for Java package and class names. MATLAB code segments will not normally be given with the Command-Line's prompt (>>), except where this would help distinguish between user-entered text and MATLAB's response:

```
>> version -java
ans =
Java 1.1.8 from Sun Microsystems Inc.
```

Regular bold is used for handle/object property names, which are often camel-cased (e.g., **JavaFrame**), as well as for occasional emphasis.

Bold italic is used for MATLAB function names (e.g., ***gcf*** or ***uicontrol***).

Regular italic is used for Java function names (e.g., *setBorder()*), MATLAB function arguments (e.g., *InputColor*), file names (e.g., *classpath.txt*), as well as for occasional emphasis and the introduction of new terms.

Disclaimer and Warning

Do not use any undocumented feature or function unless you are fully aware of the possible consequences: such features are generally unsupported by TMW; may break in future MATLAB versions without prior notice; may behave differently on different platforms or systems; may have undiscovered undesirable side effects and may even cause MATLAB to crash or hang (become unresponsive).

Much effort was invested to ensure the correctness and accuracy of the presented information. In addition, I tried to highlight potential pitfalls and the ways to avoid them. I also included a dedicated section (Section 1.5) about practices for safe programming.

However, due to its mostly undocumented nature, there is no guarantee that the information is complete or error-free. Quite the contrary: it should be assumed to be incomplete and inaccurate. The author and the publisher of this book cannot take any responsibility for possible consequences due to this book.

When using suggestions, sample code, or ideas from this book, readers *must* therefore take extreme care and should either use them at their own risk or not at all.

References

1. Available, for example, on Google groups http://groups.google.com/group/comp.soft-sys.matlab, on MathWorks's website (http://www.mathworks.com/matlabcentral/newsreader), via a personal newsreader (news:comp.soft-sys.matlab), or archived on http://mathforum.org/kb/forum.jspa?forumID=80 (long URL references such as this will often be accompanies by shortened equivalents; in this case, http://bit.ly/cXqBNr).
2. http://www.mathworks.com/matlabcentral/newsreader/view_thread/115423#292260 (or http://bit.ly/bfcs2S).
3. For example, see Kluid at http://www.kluid.com/mlib/index.php?c=1 (or http://bit.ly/9TNMbZ), DSP-Related at http://www.dsprelated.com/groups/matlab/1.php (or http://bit.ly/9l8oPa), or Stack Overflow (http://stackoverflow.com/).
4. The official MATLAB blog (http://blogs.mathworks.com) contains the most original content; outside MathWorks, the now-defunct BlinkDagger.com is well known and respected and there are several others mentioned here: http://www.mathworks.com/matlabcentral/newsreader/view_thread/251652 (or http://bit.ly/bemxVR) and http://www.mathworks.com/matlabcentral/answers/1822-foreign-matlab-forums (or http://bit.ly/mYCxhT).

MATLAB and Simulink are registered trademarks of The MathWorks, Inc. For product information, please contact:

The MathWorks, Inc.
3 Apple Hill Drive
Natick, MA 01760-2098 USA
Tel: 508-647-7000
Fax: 508-647-7001
E-mail: info@mathworks.com
Web: www.mathworks.com

Chapter 1

Introduction to Java in MATLAB®

Java is a programming language introduced in 1995. Its main strength when compared with other object-oriented languages of its time (C++ being the most important) was its portability: Java was designed to be architecture-neutral, so that Java programs written on a Mac,[†] for example, would behave exactly in the same way as on Windows[‡] and Linux machines, and in fact on any platform that supports Java. This design, coupled with built-in security measures, modern object-oriented programming features, and easily accessible graphical user interface (GUI) and I/O, significantly reduced development work and made Java a favorite among programmers worldwide. A basic introduction to Java is presented in Appendix A, and readers who are unfamiliar with Java are advised to read this appendix first.

Java integration has been available in MATLAB since Release 12 (MATLAB 6.0). Since R12, MATLAB has always been shipped with a bundled Java engine (*Java Virtual Machine* or *JVM*) on all supported MATLAB platforms except Mac OS X, in which MATLAB uses the Mac OS's JVM[1] (note that Apple plans to discontinue its internally-ported Java runtime engine (JRE) in future Mac OS releases[2]). This pre-bundled Java engine has the benefit of hiding the nuts and bolts from the user while gaining access to a full-fledged Java engine.

One important point should be made from the onset: *it is not necessary to have any Java experience when using Java in MATLAB.* MathWorks has done a pretty good job in integrating Java. The end result is that Java objects appear as simple MATLAB objects that can be accessed using regular MATLAB means, familiar even to novice MATLAB users. Therefore, *no prior Java knowledge is assumed in this book.*

MATLAB can access Java objects without any additional tool or knowledge other than simple MATLAB. In fact, *most of this book can be used without ever needing a Java editor or a compiler.* You would indeed need such tools to create custom Java classes, but this is unnecessary for the vast majority of customizations presented in this book.

While some Java-related MATLAB functions are documented, many remain undocumented to this day (2011). MathWorks even declared some of the documented functions as "unsupported", resulting in queries not being answered and bugs not being fixed. While understandable for a major new release (as R12 was in 2000), it is difficult to understand so many releases later.

The following chapters attempt to bridge this undocumented gap. The reader is cautioned that being undocumented (and more important, unsupported), features discussed may be modified or removed without prior notice in some future MATLAB release. As long as we continue to use our current MATLAB version, this will not be a problem — we only need to retest our application when upgrading MATLAB releases.

Having said that, in practice, only a small fraction of unsupported features actually change, and fewer still become unavailable, between MATLAB releases. MATLAB itself relies on many undocumented features for its own fully documented supported functions, so modifying or discontinuing these undocumented features would require a significant effort by MathWorks to redevelop much of its existing code base for no tangible benefit. We may imagine that this is

---

[†] Mac, Macintosh, and Apple are trademarks of Apple, Inc.

[‡] Windows, ActiveX, and COM are trademarks of Microsoft, Inc.

not something easily done. As we dig deeper into the undocumented territory, changes in future versions become more likely (for an apt example with the MATLAB Desktop, see Chapter 8).

It is not always possible to predict which functions are more prone to change than others. Some obvious cases are mentioned as such in this book, but it should be understood that these mentions are pure speculation and should be taken with proper precaution. There is always an inherent uncertainty about the supportability of undocumented functionality in future MATLAB versions, and we should take this into consideration whenever we depend on such functionality in our application.

Java-related issues in MATLAB may be divided into the following main categories, covered in detail in the following chapters:

- Chapter 2: Using Non-GUI Java Libraries in MATLAB
- Chapter 3: Rich GUI Using Java Swing
- Chapter 4: Uitools
- Chapter 5: Built-In MATLAB Widgets and Java Classes
- Chapter 6: Customizing MATLAB Controls
- Chapter 7: The Java Frame
- Chapter 8: The MATLAB Desktop
- Chapter 9: Using MATLAB from within Java

Much of the "plumbing" part of the MATLAB-to-Java interface is documented and supported by MATLAB. It is described below in this chapter, providing a comprehensive introductory picture of the MATLAB–Java interface. Most of the focus is given to the issue of using Java as a useful extension of MATLAB. On the other hand, Chapters 5 through 9 are deeply undocumented and quite prone to change in future MATLAB releases.

All code snippets and examples are self-contained and can be run as-is from the MATLAB Command Window or any MATLAB script/function. MATLAB users familiar with Java will find it easier to expand the provided examples for their particular needs. For the rest, references to online resources are provided to ease the learning curve.

1.1 Creating Java Objects

1.1.1 The Basics

Before using Java in our MATLAB application, we first need to ensure that Java is supported on our specific MATLAB installation (which it should, normally). This check is done using MATLAB's built-in functions *usejava* or *javachk* (which is just a wrapper for *usejava*).[†] There are four documented levels of Java that can be checked ("jvm", "awt", "swing", and "desktop"[3]);

[†] *usejava* itself wraps a *system_dependent* call — an undocumented precursor for the similarly undocumented *feature* function.

an additional undocumented level ("mwt") checks for MathWorks Swing extensions used by custom MATLAB controls (see Chapter 5):

```
if ~usejava('jvm')
    error('this feature requires Java');
end
error(javachk('jvm','this feature requires Java'));   % equivalent
```

Java code is based on *classes* that group Java functionality (functions or *methods*) and properties. Methods and properties, as well as the classes themselves, have *access modifiers*, which determine whether they can be accessed outside the class's methods. MATLAB can only access Java elements that are declared to have a *public* access modifier. Classes are grouped into *packages* and can be *extended* by new classes that add new functionality or modify (*override*) existing ones.

To use Java classes, we create (*instantiate*) objects of their type, and then use their classes' *public*-access methods. Instantiation can only be done if the class has a *constructor* method, which is basically just a method named exactly as its class's name, and declared without any return value (unlike all other class methods). Alternately, we can use the classes' static methods without needing any object. MATLAB's class object system (MCOS)[4] uses many of these concepts, although the actual implementation of classes and modifiers looks different than in Java.

The simplest way to integrate a standard Java class in MATLAB is to directly invoke its constructor method, getting an object reference in return:

```
>> java.awt.Dimension                    % default constructor (no args)
ans =
java.awt.Dimension[width = 0,height = 0]

>> dim = java.awt.Dimension(12,25)       % non-default constructor
dim =
java.awt.Dimension[width = 12,height = 25]

>> java.lang.Thread                      % a different example
ans =
Thread[Thread-161,5,main]    <= note the different representation format
```

MATLAB can only access publicly accessible methods and classes. Due to this, MATLAB cannot instantiate objects from Java classes that do not have a publicly accessible constructor. Forgetting to declare both the class and its constructor(s) as public is a common cause of error when trying to use Java classes in MATLAB.

Static public methods can be invoked using the class name, even when the class does not have a public-access constructor and no objects can be instantiated from it:

```
>> % MJUtilities has no public constructor so it can't be instantiated
>> object = com.mathworks.mwswing.MJUtilities;
??? No constructor 'com.mathworks.mwswing.MJUtilities' with matching signature
found.

>> % This uses the MJUtilities class's static public beep method
>> com.mathworks.mwswing.MJUtilities.beep;    % no error (beep sound)
```

Some Java classes have both public constructor(s) and static methods. In such cases, we can instantiate class objects and can call the static methods either via the class object or directly via the class name:

```
% Three different ways to instantiate a Color object:
>> color = java.awt.Color(1,0.7,0.5)        % use a regular constructor
color =
java.awt.Color[r = 255,g = 179,b = 128]

>> color = java.awt.Color.cyan              % use a static property
color =
java.awt.Color[r = 0,g = 255,b = 255]

>> color = java.awt.Color.decode('0xff00ff')  % use a static method
c =
java.awt.Color[r = 255,g = 0,b = 255]

>> % Instantiate a Java Frame object and use it to list active frames
>> jFrame = java.awt.Frame
jFrame =
java.awt.Frame[frame0,0,0,0x0,invalid,hidden,layout = java.awt.
BorderLayout,title = ,resizable,normal]

>> jFrame.getFrames        % or: getFrames(jFrame)
ans =
java.awt.Frame[]:
    [com.mathworks.mde.desk.MLMainFrame          ]
    [com.mathworks.mde.desk.MLMultipleClientFrame]
    [com.mathworks.mwswing.MJFrame               ]
    [javax.swing.SwingUtilities$SharedOwnerFrame ]
    [com.mathworks.mwswing.MJFrame               ]
    [java.awt.Frame                              ]
>> % getFrames() is a public static method, therefore we don't need
>> % to instantiate an object in order to use it in MATLAB:
>> java.awt.Frame.getFrames
ans =
java.awt.Frame[]:
    [com.mathworks.mde.desk.MLMainFrame          ]
    [com.mathworks.mde.desk.MLMultipleClientFrame]
    . . .
```

Java classes can also be created using MATLAB's ***javaObject*** function, which is useful when the fully qualified class name (FQCN) is long or when the class name is stored in a program variable. Such cases are usually rare.

A more common need is to create instances of Java *nested-classes*. Nested classes are classes that are defined within the context of a parent class. In Java source code, they are referred to using dot-notation (e.g., `Package.Name.ParentClassName.NestedName`), although their class file name uses "$" (*ParentClassName$NestedName.class*). This confuses MATLAB into thinking that the nested class is called `ParentClassName$NestedName` (with a $).

Unfortunately, since "$" is not a valid MATLAB character, we cannot directly create any instance of such a class. The solution is to use *javaObject*:[5]

```
>> jObject = javax.swing.ScrollPaneLayout.UIResource⁶
??? No appropriate method, property, or field UIResource for class javax.swing.
ScrollPaneLayout.

>> jObject = javax.swing.ScrollPaneLayout$UIResource
??? jObject = javax.swing.ScrollPaneLayout$UIResource
                                          |
Error: The input character is not valid in MATLAB statements or expressions.

>> jObject = javaObject('javax.swing.ScrollPaneLayout$UIResource')
jObject =
javax.swing.ScrollPaneLayout$UIResource@695c5b
```

Yet another form of Java object creation is using the unsupported semidocumented ***awtcreate*** function. The main difference between ***javaObject*** and ***awtcreate*** is that ***javaObject*** is executed immediately, in the main MATLAB computational thread, whereas ***awtcreate*** is placed on the Java AWT[†] *Event Dispatch Thread* (EDT; described in Section 3.2) and executed only after all the other pending AWT events have ended. ***awtcreate*** is typically needed when constructing Java GUI components that rely on other GUI commands to finish first. In all other cases, use of ***awtcreate*** is discouraged — its cumbersome JNI notation[7] argument format ensures that programmers would normally and rightly not use this function.

MATLAB release R2008a (7.6) added an important addition solving much of the ***awtcreate*** and ***awtinvoke*** (see below) frustration: ***javaObjectEDT*** (and the corresponding ***javaMethodEDT***) behaves just like ***javaObject*** (and ***javaMethod***) except that it runs on the EDT, without any of the cumbersomeness of ***awtcreate*** (and ***awtinvoke***). Moreover, ***javaObjectEDT*** accepts any reference of an existing Java object and ensures that all method invocations on this reference object from that point onward will automatically be dispatched on the EDT without any code change. Unfortunately, I have found that relying on R2008a (7.6)'s version of ***javaObjectEDT*** sometimes causes MATLAB to hang. As far as I could test, this was corrected in MATLAB release R2008b (7.7), and since then it is most advisable to use ***javaObjectEDT*** for all GUI-related Java components (subclasses of `java.awt.*` or `javax.swing.*`).[‡]

1.1.2 Accessing Java Objects

Regardless of how a Java object is created, the object reference is a reference in the Java sense, which is different from MATLAB's normal objects. Modifying a copy of such a Java reference modifies both the copy and the original, unlike the case for MATLAB objects (we shall see later on in this book that there are many cases of undocumented MATLAB objects that are also references).

---

† AWT is the *Abstract Windowing Toolkit* — Java's very basic graphics framework.
‡ See Section 3.2 for additional details.

Objects can be checked in runtime as to whether they are a Java reference or not, using MATLAB's built-in *isjava* function. Java objects, like MATLAB objects, can be tested for class membership (directly or via inheritance) by using MATLAB's built-in *isa* function:

```
>> isjava(dim)
ans =
     1      <= logical true
>> isa(dim,'java.awt.geom.Dimension2D')
ans =
     1      <= logical true (Dimension inherits from Dimension2D)
>> isa(dim,'java.lang.NoSuchClassName')
ans =
     0      <= logical false
```

A class's fully-qualified package name need not be repeatedly specified: MATLAB's *import* function can simplify our code, as the following MATLAB code snippet shows:

```
import java.util.* java.lang.*
hash1 = java.util.Hashtable;      % long, fully-qualified format
hash2 = Hashtable;                % short, more readable format
stack = Stack;        % another short-format, now for java.util.Stack
```

Note: Do not import any MathWorks-derived (`com.mathworks.*`) packages on the same line as other packages on MATLAB 7.5 (R2007b) and older because this **crashes MATLAB R2007b.**[†] This bug was solved in MATLAB 7.6 (R2008a), but to support older releases, separate such imports into different lines:

```
% This crashes MATLAB 7.5 (R2007b); OK on 7.6 (R2008a) and newer
import javax.swing.* com.mathworks.mwswing.*

% This is ok in all MATLAB releases
import javax.swing.*
import com.mathworks.mwswing.*
```

Imported packages apparently use lazy loading. This means that classes are loaded into memory only when actually needed. Therefore, `import java.util.*` does NOT load the entire util package into memory — it only facilitates code readability. Imported packages/classes can be unimported using MATLAB's `clear import` command (available at the Command Window only, not from within a running function).

Note that using the `import XXX.*` syntax may fail in compiled (deployed) applications. The solution is always to use fully-qualified class-names (*FQCN*) to access Java objects.[8]

User-defined Java classes can be used in MATLAB by adding their containing folder (or *JAR* file) to the Java *classpath*, dynamically using MATLAB's *javaclasspath* or *javaaddpath* function, or statically by adding them to the top of the *classpath.txt* file (*edit('classpath.txt')*).[‡]

[†] At least on MATLAB 7.5 (R2007b) running JVM 1.6 on a Windows XP PC.

[‡] See related bugs in MATLAB versions prior to R2006a (MATLAB 7.2): http://www.mathworks.com/support/solutions/en/data/1-1Y9R5V (or http://tinyurl.com/notqx5). While writing, this page has been removed by MathWorks from their support website. A workaround is to add your folders/JARs to the beginning of the *classpath.txt* file rather than to the end of this file.

We can also add Java classes to one of the existing folders on the dynamic or static classpath (e.g., %matlabroot%/java/patch/).

In general, using the static classpath solves many of the problems that occur on the dynamic classpath, possibly due to the different classloaders used in both cases.[9] Quite a few Java-related problems are solved by simply placing the classes in the static classpath.

The default *classpath.txt* file is located in %matlabroot%/toolbox/local/, but a custom editable copy can also be placed in MATLAB's startup folder (in which case, it will have precedence over the default file).[10] Using a local *classpath.txt* is important for distributions where we do not have administrator modification privileges for MATLAB's default *classpath.txt*, or if we do not wish to modify the root installation.

If we compile (deploy) our MATLAB application, we need to manually add the Java class/ JAR files and *classpath.txt/librarypath.txt* files to the build process using MATLAB's ***deploytool***. This is necessary since MATLAB's compiler is not smart enough to automatically integrate ***javaaddpath*** code directives in ***deploytool***. Alternately, manually place these files in the application's deployment (distribution) folder — this is generally safe to do in any case. In fact, I have seen a case where adding *classpath.txt/librarypath.txt* in ***deploytool*** actually resulted in errors (this may be related to mapped network paths)[†] — the solution was to remove these file references from ***deploytool*** and simply add the files to the deployment folder. It is the deployed application's equivalent of placing these files in MATLAB's startup folder.[‡]

classpath.txt entries should be separated by whitespaces, typically newlines. Comments start with the hash (#) sign. Entries may be a folder name or JAR file name — in both cases, the full path name should be specified, either directly (e.g., C:/MyClasses/New) or relative to the ***matlabroot*** (e.g., $*matlabroot/java/jar/util.jar*). Folders must be separated with a forward slash (/), not a backwards slash (\), even for Windows systems. This follows the Java/JVM convention. Classpath folders (not JAR files) should also end with a trailing "/" character.[§] Entries may also be preceded with target platforms, and the JVM will only load these specified classpath entries on these platforms (e.g., mac=$*matlabroot/java/jarext/aquaDecorations.jar*).[¶] The classpath order is important: classes which appear earlier in the classpath will have precedence over other versions of these classes which appear lower in the classpath.

[†] Debugging such problems is extremely difficult. The first step is to run the deployed application from an operating system command prompt (rather than a desktop icon), in order to see warning and error messages.

[‡] Note that if you deploy a COM (DLL) component, the *librarypath.txt* file needs to be placed in the DLL's actual startup folder, which is typically **not** the same as the folder in which the DLL is located. I recently spent a few hours trying to understand why a compiled DLL loaded in Excel croaks, although its MATLAB and EXE counterparts worked flawlessly. Unlike them, DLLs have no console onto which we can spill debug messages and error traces, making debugging extremely difficult. In the end it turned out that Excel loaded the DLL with a startup folder of the user's Windows home directory. Go figure...

[§] On relatively old MATLAB releases only — new releases also accept classpath folders without trailing slashes: http://www.mathworks.com/support/solutions/en/data/1-1AJWE (or http://bit.ly/abEcRD).

[¶] Available platforms on MATLAB 7.1: alpha,glnx86,sol2,unix,win32,mac. On 7.5 also glnxa64,sol64,win64,maci but without alpha (MATLAB support discontinued). See the main comment of *classpath.txt* for an up-to-date list for your MATLAB version.

By default, the first entry in *classpath.txt* is $matlabroot/java/patch. This means that Java classes in this folder override similarly named classes taken from other folders or JARs. This is used to post updated Java classes (aka *patches*) in MATLAB releases.[11]

We can also use this patch mechanism to override MATLAB's default classes with our own version. If we do this, then it will only work on the platform on which we have copied the corrected class to the patch folder. This causes a distribution headache, since the patch folder is in the program installation tree (matlabroot) and not on the application tree. To facilitate distribution, we can add a code segment at the beginning of our application, which copies a version of the class from the current (application) path to the patches folder if it is not detected there already. Then simply distribute our patch class file together with our application code. When the application first loads, it will automatically try to install (copy to the patch folder) the patch file, so that it could be used later by that same application. In cases of copy failure (probably due to permission constraints on MATLAB's installation tree), a popup warning will notify users to request their system admin to install the file for them:

```
className = 'MyClass.class';
patchName = fullfile(matlabroot,'/java/patch/',className);
if ~exist(patchName,'file')
    [successFlag,msg] = copyfile(className, patchName);
    if ~successFlag
        msg = ['Could not copy ' fullfile(pwd,className) ' to ' ...
               patchName ': ' msg '. Please ask sysadmin to patch it'];
        msgbox([msg, [className ' patch'], 'warn');
    end  % if failed to copy patch class file
end  % if patch class file not detected in patch folder
```

Unfortunately, MATLAB does not automatically create the patch/folder when installing MATLAB. It also does not copy any existing patch files from earlier installations (it copies *prefdir* folder and other settings, but not patch/) — we need to do this manually.

MATLAB R2008a (7.6) and earlier had a limited fixed number of classpath characters — a longer classpath would **crash MATLAB**. This was reportedly fixed in R2008b (7.7).[12]

Related MATLAB function *javarmpath* removes a folder/JAR from the dynamic classpath. Once a Java class is recognized on the Java classpath, MATLAB's built-in *exist* function returns the value 8 when the class name is passed as an input argument.

Determining which Java classes are currently loaded in memory is done via *inmem*:

```
>> [mFunctions, mexFunctions, javaClasses] = inmem
...
javaClasses =
    'schema.package'
    'schema.class'
    'handle.listener'
    ...
    'uitools.uimodemanager'
    'java.util.Hashtable'
    'com.mathworks.mlwidgets.workspace.WhosInformation'
```

Java classes can be cleared from memory using MATLAB's **clear java** function, which was added following user requests on CSSM.[13] Note that this function fails if there are any references of instantiated objects of the cleared class still in memory.[14]

```
>> clear java
Warning: Objects of org/jfree/chart/ChartPanel class exist - not clearing java
```

A nasty side effect of MATLAB's **javaaddpath** and **javaclasspath** commands is that they clear all MATLAB global variables. This nasty side effect of clearing globals has existed at least since MATLAB 7.2 (possibly earlier), and to this day (MATLAB 7.12, R2011a).[15] In MATLAB 7.4 (R2007a) and earlier releases, all Java object references in the MATLAB Desktop Workspace were also cleared.[16]

Serializable Java objects (those implementing the java.io.Serializable interface) can be saved and loaded from *mat* files via MATLAB's built-in **save**, **load** functions.[17]

Arrays behave slightly differently between Java and MATLAB: Java array indices start with zero, whereas MATLAB indices start with 1 — this is a common pitfall when using Java in MATLAB. Multidimensional arrays are also handled differently by the two programming languages. Java object references may be concatenated in a MATLAB array as long as all the references have exactly the same type or some common superclass (java.lang.Object being a last-resort superclass).

The MATLAB functions **javaArray** and **javaArray2cells** are fully documented and supported functions that handle Java arrays. Refer to the official documentation[18] for additional details. Note that **javaArray** and **javaArray2cells** only support named classes, so arrays of Java primitives such as int[] or byte[] cannot be created.[19]

All the Java classes presented in this book are precompiled and are either directly available in the MATLAB installation or downloadable (I will present the relevant URLs). We can also use our own Java classes (see Section 1.6 for details).

As a final note, when Java objects are displayed in MATLAB's Command Window, the object's *toString()* method is automatically invoked to generate a displayable string. Different Java objects have different implementations of *toString()*. If a Java object's class does not have an internal *toString()* method, a default *toString()* method is invoked, displaying the object's FQCN (xxx.yyy.classname) and hashCode value in hex format. All Java objects ultimately extend java.lang.Object, so they all have a hashCode and at least this base Object's default *toString()* method:

```
>> color = java.awt.Color.cyan
color =
java.awt.Color[r = 0,g = 255,b = 255]   <= non-default toString() method

>> jObject = java.lang.Object
jObject =
java.lang.Object@2342a6      <= default representation format class@hash

>> hashCode = dec2hex(jObject.hashCode)
hashCode =
2342A6
```

1.1.3 Memory Usage

Java references appear as zero-byte objects in the workspace. Unfortunately, there is no simple workaround for this:[20]

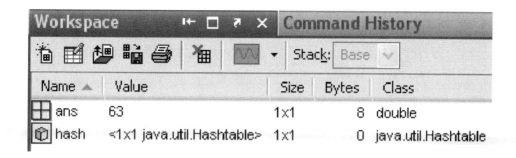

```
>> whos
Name        Size    Bytes      Class              Attributes
ans         1x1         8      double
dim         1x1                java.awt.Dimension
```

One (imperfect) solution for reporting Java object size in MATLAB is to use utilities such as Classmexer[21] that rely on JVM Instrumentation (available since JVM 1.5).[22] To use Classmexer, download its JAR file, add it to the Java classpath (statically or dynamically), and then add the following line to our *java.opts* file (see Section 1.9):

```
-javaagent:classmexer.jar
```

Now, after starting MATLAB, we are able to query the Java references size in runtime. For simple objects such as `java.awt.Dimension`, a simple "shallow" query of the base size is sufficient, using *memoryUsageOf()*:

```
>> com.javamex.classmexer.MemoryUtil.memoryUsageOf(dim)
ans =
    16                <= memory usage in Bytes
```

However, for more complex classes, which hold private references to other classes, we need to use the *deepMemoryUsageOf()* method:[23]

```
>> % Prepare the hashtable:
>> hash = java.util.Hashtable;
>> hash.put('dim',   java.awt.Dimension(12,25));
>> hash.put('color', java.awt.Color.red);

>> % The base size reports a misleadingly-small memory footprint
>> disp(com.javamex.classmexer.MemoryUtil.memoryUsageOf(hash))
    40
```

```
>> % So we must include private reference objects to get the full size
>> disp(com.javamex.classmexer.MemoryUtil.deepMemoryUsageOf(hash))
   296
```

1.2 Java Object Properties

Once we have a reference to the new Java object, we can inspect (using *fieldnames*, *get*, or *inspect*), read (*get*), and update (*set*) its properties.[†] Note the default properties assigned by MATLAB to all Java classes, and the different equivalent invocation forms:

```
>> dim = java.awt.Dimension(12,25);
>> fieldnames(dim)
ans =
     'width'
     'height'
>> ans = dim.fieldnames;              % equivalent invocation form

>> fieldnames(dim,'-full')           % some extra information...
ans =
     'int width'
     'int height'
>> ans = dim.fieldnames('-full');     % equivalent form

>> dim.get
```

```
class          ⎧ Class = [ (1 by 1) java.lang.Class array]
               ⎪ Height = [25]
properties     ⎨ Size = [ (1 by 1) java.awt.Dimension array]
               ⎩ Width = [12]
```

```
               ⎧ BeingDeleted = off      <= start of default MATLAB properties
               ⎪ ButtonDownFcn =
               ⎪ Children = []
               ⎪ Clipping = on
               ⎪ CreateFcn =
               ⎪ DeleteFcn =
    HG         ⎪ BusyAction = queue
  (Handle      ⎨ HandleVisibility = on
 Graphics)     ⎪ HitTest = on
 properties    ⎪ Interruptible = on
               ⎪ Parent = []
               ⎪ Selected = off
               ⎪ SelectionHighlight = on
               ⎪ Tag =
               ⎪ Type = java.awt.Dimension
               ⎪ UIContextMenu = []
               ⎪ UserData = []
               ⎩ Visible = on
```

```
>> ans = dim.get;                     % equivalent form
% Note: getting partial case-insensitive property names is
% ^^^^ supported (as in MATLAB), as long as they are unique:
```

[†] Use of *get/set* is undocumented and unsupported since it causes a memory leak — refer to Section 3.4 for details.

```
>> height = get(dim,'h')
??? Error using = = > get
Ambiguous java.awt.Dimension property: 'h'.
>> height = get(dim,'he')
height =
     25
>> height = dim.getHeight;          % equivalent form, case-sensitive
>> height = dim.height;             % equivalent form, case-sensitive
% Note: this latter obj.propName form is only available for public
% ^^^^ properties - those returned by the fieldnames function
>> size = dim.getSize   % non-primitive data can also be retrieved
size =
java.awt.Dimension[width = 12,height = 25]

>> data = get(dim,{'wid','hei'})     % get multiple prop values
data =
    [12]       [25]
>> set(dim)
```

class properties ⎰ Class / Height / Size / Width

HG (Handle Graphics) properties:
```
    ButtonDownFcn: string -or- function handle -or- cell array
    Children
    Clipping: [ {on} | off ]
    CreateFcn: string -or- function handle -or- cell array
    DeleteFcn: string -or- function handle -or- cell array
    BusyAction: [ {queue} | cancel ]
    HandleVisibility: [ {on} | callback | off ]
    HitTest: [ {on} | off ]
    Interruptible: [ {on} | off ]
    Parent
    Selected: [ on | off ]
    SelectionHighlight: [ {on} | off ]
    Tag
    UIContextMenu
    UserData
    Visible: [ {on} | off ]
```
```
>> ans = dim.set;                    % equivalent form
% Note: the following is lexically ok, but in this specific case
% ^^^^ Width & Height must be set using the Size property
>> set(dim,'wid',20,'hei',30);          % lexically ok, but no effect
>> set(dim,'size',java.awt.Dimension(20,30));
>> dim.setSize(java.awt.Dimension(20,30));   % equivalent form
>> setSize(dim,java.awt.Dimension(20,30));   % equivalent form
```

In R2010b (MATLAB 7.11), when using *set* on one of the Handle-Graphics (HG) properties, we get a warning asking us not to use *set* but rather to use its equivalent Java form:

```
Warning: Possible deprecated use of set('Visible','off') on Java boolean
property: use jobj.setVisible(false) instead.
(Type "warning off MATLAB:hg:JavaSetHGProperty" to suppress this warning.)
```

1.3 Java Object Methods and Actions

Once we have a reference to the new Java object, we can also inspect (using ***methods*** or ***methodsview***) its publicly accessible Java methods. In the following code snippets, note the different equivalent invocation forms:

```
>> dim.methods
Methods for class java.awt.Dimension:
Dimension   equals      getHeight   getWidth    notify      setSize     wait
clone       getClass    getSize     hashCode    notifyAll   toString
>> methods(dim);   % equivalent form

>> dim.methods('-full')                  % some extra information...
Methods for class java.awt.Dimension:
Dimension(java.awt.Dimension)
Dimension(int,int)
Dimension()
java.lang.Object clone()     % Inherited from java.awt.geom.Dimension2D
boolean equals(java.lang.Object)
java.lang.Class getClass()  % Inherited from java.lang.Object
double getHeight()
java.awt.Dimension getSize()
double getWidth()
int hashCode()
void notify()                % Inherited from java.lang.Object
void notifyAll()             % Inherited from java.lang.Object
void setSize(int,int)
void setSize(double,double)
void setSize(java.awt.Dimension)
void setSize(java.awt.geom.Dimension2D) % Inherited from ...
java.lang.String toString()
void wait() throws java.lang.InterruptedException % Inherited ...
void wait(long,int) throws java.lang.InterruptedException % ...
void wait(long) throws java.lang.InterruptedException % ...

>> methods(dim,'-full');   % equivalent form

>> methodsview(dim);
>> dim.methodsview;        % equivalent form
```

The ***methods*** and ***methodsview*** functions accept either a Java object reference or a class name as input argument. The window presented by ***methodsview*** displays up to six columns (columns having no data are hidden from view): Quantifiers (*synchronized, abstract, static*), Return Type, method Name, method Arguments, Other (thrown Exception), and Inheritance parent (inheritance info was removed by R2010a). The data is presented in a Java Frame (not a regular MATLAB figure) window and can also be gotten by the [unused,data] = methods(dim, '-full') command.

Return Type	Name	Arguments	Other	Inherited From
	Dimension	(java.awt.Dimension)		
	Dimension	(int,int)		
	Dimension	()		
java.lang.Object	clone	()		java.awt.geom.Dimension2D
boolean	equals	(java.lang.Object)		
java.lang.Class	getClass	()		java.lang.Object
double	getHeight	()		
java.awt.Dimension	getSize	()		
double	getWidth	()		
int	hashCode	()		
void	notify	()		java.lang.Object
void	notifyAll	()		java.lang.Object
void	setSize	(int,int)		
void	setSize	(double,double)		
void	setSize	(java.awt.Dimension)		
void	setSize	(java.awt.geom.Dimension2D)		java.awt.geom.Dimension2D
java.lang.String	toString	()		
void	wait	()	throws java.lang.InterruptedException	java.lang.Object
void	wait	(long,int)	throws java.lang.InterruptedException	java.lang.Object
void	wait	(long)	throws java.lang.InterruptedException	java.lang.Object

methodsview sample screenshot (R2008a)

Java methods can also be found using MATLAB's built-in **which** function, which searches the specified input argument in all Java classes loaded in memory:

```
>> which setSize
setSize is a Java method          % javax.swing.JPanel method
>> which setSize -all
setSize is a Java method          % javax.swing.JPanel method
setSize is a Java method          % javax.swing.JComponent method
setSize is a Java method          % java.awt.Dimension method
setSize is a Java method          % java.awt.geom.Dimension2D method
setSize is a Java method          % com.mathworks.mwt.MWFrame method
setSize is a Java method          % com.mathworks.mwt.MWListbox method
setSize is a Java method          % javax.swing.Box$Filler method
setSize is a built-in method      % javahandle.java.awt.Dimension method
setSize is a built-in method      %
            javahandle_withcallbacks.java.awt.Dimension method
. . .
```

A MATLAB utility that I have uploaded to the File Exchange,[24] called UIINSPECT, incorporates all the information available in the *inspect* and *methodsview* functions, as well as other information that is normally well hidden in undocumented MATLAB territory (and will be described later in this book). Here is a preview of ***uiinspect*** for the `Dimension` object that was created above:

```
>> uiinspect(dim);
>> dim.uiinspect;    % equivalent form
```

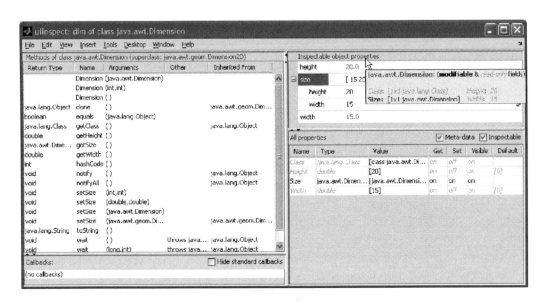

***uiinspect* sample screenshot**

Accessible object methods can be invoked using either direct method invocation (in the form `object.`*method*`(...)`), ***javaMethod***, or the unsupported semidocumented ***awtinvoke*** (that is described in more detail below). Direct method invocation is the preferred invocation form in most cases. The alternative forms, using ***javaMethod*** or ***awtinvoke***, would normally be used only in one of the following cases:

- When the method name is very long (beyond MATLAB's ***namelengthmax*** value)
- When the method name itself is stored in a program variable
- When the method modifies some Java GUI and should be synchronized with the Java Swing event queue (more on this in Section 3.2), where ***javaMethodEDT*** or ***awtinvoke*** should normally be used

awtinvoke, like ***awtcreate***, uses the cumbersome JNI notation to specify argument types. This by itself ensures that only the most persistent programmers, in the rare cases that actually require sequential asynchronous invocation on the Java EDT, would use ***awtinvoke*** instead of using ***javaMethod*** or, easier still, the direct invocation syntax. This said, beware of potential problems when NOT using the EDT. As long as no GUI update is involved, using ***javaMethod*** or direct invocation syntax is safe. If the EDT must be used, ensure to use ***awtinvoke*** or, since

R2008b (7.7), *javaMethodEDT* (or *javaObjectEDT* on its object prior to method invocation). Refer to Section 3.2 for additional information and details about the EDT and its usage in MATLAB.

In MATLAB 7.7 (R2008b) and later releases, we can use *javaMethodEDT* instead of *awtinvoke. javaMethodEDT* is normally much easier to use (no need for the awkward JNI syntax). There are two specific use-cases, however, in which we would choose to use *awtinvoke*:

- If our code should be backward-compatible and should also work on pre-R2008b MATLAB releases.
- If we want to use *awtinvoke's* semi-documented feature of enabling setting a delayed-action callback function that is triggered when the Java method has completed. This functionality, which is unavailable in *javaMethodEDT*, is only documented within the awtinvoke.m function, and not in its main help section (which itself is not normally visible before editing this file). The syntax of this *awtinvoke* functionality is simply: *awtinvoke(javaObj, jniMethodStr, param1, ... , paramN, @cbFunction, cbParam1, ... , cbParamM)*.

In the following code snippet, all invocation forms of the *setSize()* method are equivalent. Note that invocations using *awtinvoke* queue their execution on the Java EDT, while the other forms execute *setSize()* immediately, in the main MATLAB thread:

```
>> newDim = java.awt.Dimension(20,30);
>> dim.setSize(newDim);
>> setSize(dim,newDim);
>> javaMethod('setSize', dim, newDim);
>> javaMethodEDT('setSize', dim, newDim); % use EDT, MATLAB 7.6 +

>> % awtinvoke JNI interface for all MATLAB versions not just 7.6 +
>> awtinvoke(dim,'setSize(Ljava.awt.Dimension;)',newDim); % use EDT
>> dim.awtinvoke('setSize(Ljava.awt.Dimension;)',newDim); % use EDT
```

Several common generic Java converter methods are used in MATLAB in a very intuitive manner: Java's *equals()* method is invoked when MATLAB needs to use *isequal*; *toString()* is invoked whenever MATLAB needs to use *disp* to display a class:

```
>> url = java.net.URL('http://www.cnn.com')
url =
http://www.cnn.com        <= note the use of toString() here
```

All primitive Java types have corresponding MATLAB types and the conversion is usually transparent and intuitive to the user.[25] The *toChar()* and *toDouble()* Java methods are invoked when MATLAB's *char* and *double* type-casting functions are used.

Note that conversions that seem trivial are sometimes not automatically done by MATLAB. For example, if a Java method expects a java.lang.Comparable, an interface which java. lang.String honors, MATLAB still fails to send a MATLAB string ('abcd') to that function by automatic type casting — we need to help MATLAB with an explicit type cast: `java.lang.String('abcd')`. A real-life example is presented in Section 3.5.1.

It should also be noted that automatic-type conversion occurs only when passing data from MATLAB to Java, but only partially on the reverse path. On the reverse path (from Java to MATLAB), scalar built-in (primitive) numeric values are converted into MATLAB doubles, and Java `booleans` into MATLAB logical values. It is only object types which are not automatically converted. The full list of automatic and nonautomatic conversions is documented in MATLAB's External Interfaces/Java section.

In the code snippet above, 'http://www.cnn.com' is a MATLAB char string which is automatically converted into the `java.lang.String` object that is expected by the `java.net.URL` constructor. On the reverse path, the output of the *toString()* method is not converted into a MATLAB char string, and the result remains a `java.lang.String` object. It is left up to the MATLAB programmer to convert this into a MATLAB char string:

```
>> urlName = url.toString
urlName =
http://www.cnn.com

>> urlName2 = char(url.toString);

>> whos urlName*
  Name       Size     Bytes    Class                 Attributes
  urlName    1x1               java.lang.String
  urlName2   1x18      36      char
```

Another important difference between passing data to and from Java is that all data passed from MATLAB to Java except for object references are passed *by value*, whereas all objects (nonprimitive types) returned from Java are passed *by reference*. This means that arrays/structs passed from MATLAB cannot be modified by Java methods. Conversely, modifications by MATLAB of a returned (referenced) object will affect the original Java object. Therefore, if you wish the Java method to modify MATLAB data, you need to encase this data in a Java reference, using MATLAB's *javaArray* function.

Note: Try to minimize the number or frequency of invoked Java method calls, because MATLAB creates a large number of temporary objects upon each call, and this can bog down the JVM with frequent garbage collection. Whenever possible, try to use Java methods that group internal Java methods.[26]

Java exceptions are converted into MATLAB errors that can be caught and handled using MATLAB's built-in *try-catch* mechanism, just like regular MATLAB errors:

```
>> hash = java.util.Hashtable;
>> iter = hash.keys;
>> iter.nextElement
??? Java exception occurred:
java.util.NoSuchElementException: Hashtable Enumerator
    at java.util.Hashtable$EmptyEnumerator.nextElement(Unknown...)
```

```
>> try iter.nextElement, catch err = lasterror, end
err =
        message: [1x150 char]
     identifier: 'MATLAB:Java:GenericException'
          stack: [0x1 struct]
>> disp(err.message)
Java exception occurred:
java.util.NoSuchElementException: Hashtable Enumerator
    at java.util.Hashtable$EmptyEnumerator.nextElement(...)
```

Note that MATLAB only catches errors/exceptions thrown on the main MATLAB thread. Exceptions thrown asynchronously, by the AWT (Java GUI) event thread (EDT), are displayed on the Command Window and to the best of my knowledge cannot be caught or suppressed by MATLAB as of release 7.12 (R2011a).[†]

Java GUI components often use an `ActionMap`[27] to store runnable `Actions`[28] that are invoked by listeners on mouse, keyboard, property, or container events.[29] Unlike object methods, Actions cannot be directly invoked by MATLAB. However, we can use the following workaround:

```
import java.awt.event.ActionEvent;
action = hObject.getActionMap.get(actionName);
actionEvent = ActionEvent(hObject, ActionEvent.ACTION_PERFORMED, '');
   % Note: we can use either '' or actionName when creating ActionEvent
action.actionPerformed(actionEvent);

% Or, for an EDT-safe action invocation:
awtinvoke(action,'actionPerformed(Ljava.awt.event.ActionEvent;)', ...
          actionEvent);
```

This workaround can be used to force display of a GUI component's tooltip (in this case, actionName is simply 'postTip').[30]

1.4 Java Events and MATLAB Callbacks[31]

When loading Java classes into MATLAB, MATLAB callbacks are automatically assigned to their corresponding Java events. Refer to the following Java class example:[32]

```
public class EventTest
{
    private java.util.Vector data = new java.util.Vector();
    public synchronized void addMyTestListener(MyTestListener lis) {
        data.addElement(lis);
    }
```

[†] There is an internal class called `com.mathworks.mwswing.EdtExceptionHandler` with a single property named **SuppressingHandledExceptions**, which appears by name to control exactly this, but unfortunately does not.

```
    public synchronized void removeMyTestListener(MyTestListener lis) {
        data.removeElement(lis);
    }
    public interface MyTestListener extends java.util.EventListener {
        void testEvent(MyTestEvent event);
    }

    public class MyTestEvent extends java.util.EventObject {
        private static final long serialVersionUID = 1L;
        public float oldValue,newValue;
        MyTestEvent(Object obj, float oldValue, float newValue) {
            super(obj);
            this.oldValue = oldValue;
            this.newValue = newValue;
        }
    }

    public void notifyMyTest() {
        java.util.Vector dataCopy;
        synchronized(this) {
            dataCopy = (java.util.Vector)data.clone();
        }

        for (int i = 0; i<dataCopy.size(); i++) {
            MyTestEvent event = new MyTestEvent(this, 0, 1);
            ((MyTestListener)dataCopy.elementAt(i)).testEvent(event);
        }
    }
}
```

When compiling *EventTest.java*, three class files are created: *EventTest.class*, *EventTest$MyTestEvent.class*, and *EventTest$MyTestListener.class*. After placing them on MATLAB's Java classpath (see Section 1.1), they can be accessed as described above

```
>> which EventTest
EventTest is a Java method % EventTest constructor

>> evt = EventTest
evt =
EventTest@16166fc

>> evt.get
        Class = [ (1 by 1) java.lang.Class array]
        TestEventCallback =
        TestEventCallbackData = []

        BeingDeleted = off
        ButtonDownFcn =
        Children = []
        Clipping = on
        CreateFcn =
        DeleteFcn =
        BusyAction = queue
        HandleVisibility = on
```

```
            HitTest = on
            Interruptible = on
            Parent = []
            Selected = off
            SelectionHighlight = on
            Tag =
            Type = EventTest
            UIContextMenu = []
            UserData = []
            Visible = on
>> set(evt)
            Class
            TestEventCallback: string -or- function handle -or- cell array
            . . .

>> set(evt,'TestEventCallback',@(h,e)disp(h))

>> get(evt)
            Class = [ (1 by 1) java.lang.Class array]
            TestEventCallback = [ (1 by 1) function_handle array]    <= ok
            TestEventCallbackData = []
            . . .

>> evt.notifyMyTest   % invoke Java event
                0.0009765625   % <= MATLAB callback
```

Note how MATLAB automatically converted the Java event `testEvent`, declared in `interface MyTestListener`, into a MATLAB callback **TestEventCallback** (the first character is always capitalized). All Java events are automatically converted in this fashion,[†] by appending a "Callback" suffix. Here is a code snippet from R2008a's \toolbox\matlab\uitools@ opaque\addlistener.m that shows this (slightly edited):

```
hSrc = handle(jobj,'callbackproperties');
allfields = sortrows(fields(set(hSrc)));
for i = 1:length(allfields)
   fn = allfields{i};
   if ~isempty(findstr('Callback',fn))
      disp(strrep(fn,'Callback',''));
   end
end

callback = @(o,e) cbBridge(o,e,response);
hdl = handle.listener (handle(jobj), eventName, callback);
function cbBridge(o,e,response)
   hgfeval(response, java(o), e.JavaEvent)
end
```

Note that ***hgfeval***, which is used within the `cbBridge` callback function, is a semi-documented pure-MATLAB built-in function.[33]

[†] This is not always true for non-Java objects.

Also note that there is no specific need in this particular case to use ***handle***(evt, 'CallbackProperties'), although it also works and is generally advisable, as will be explained in Section 3.4:

```
>> hevt = handle(evt,'CallbackProperties')
hevt =
javahandle_withcallbacks.EventTest

>> set(hevt,'TestEventCallback',@(h,e)disp(h))

>> hevt.get
                    Class: [1x1 java.lang.Class]
        TestEventCallback: @(h,e)disp(h)
    TestEventCallbackData: []
```

If several events have the same case-insensitive name, then the additional callbacks will have an appended underscore character (e.g., "TestEventCallback_"):[†]

```
// In the Java class:
public interface MyTestListener extends java.util.EventListener {
    void testEvent(MyTestEvent e);
    void testevent(TestEvent2 e);
}

% ...and back in MATLAB:
>> evt = EventTest; evt.get
        Class = [ (1 by 1) java.lang.Class array]
        TestEventCallback =
        TestEventCallbackData = []
        TestEventCallback_ =
        TestEventCallback_Data = []
        ...
```

If there are several such additional callbacks, they will all have the same name, causing run-time errors and a locking of the object, preventing property value update:

```
// In the Java class:
public interface MyTestListener extends java.util.EventListener {
    void testEvent(MyTestEvent e);
    void TestEvent(MyTestEvent e);
    void testevent(TestEvent2 e);
}

% ...and back in MATLAB:
>> evt = EventTest; evt.get
        Class = [ (1 by 1) java.lang.Class array]
        TestEventCallback =
        TestEventCallbackData = []
        TestEventCallback_ =
```

[†] Unfortunately, the ***classhandle*** events are not differentiated in a similar manner — in this case, only a single event is created, named *Testevent*. ***classhandle*** events, and their relationship to the preceding discussion, are described later in this section.

```
              TestEventCallback_Data = []
              TestEventCallback_ =
              TestEventCallback_Data = []
              . . .
>> set(evt,'TestEventCallback','123')
??? Error using = = > set
Attempt to write a locked object.
```

To complete this discussion, it should be noted that MATLAB also automatically defines corresponding Events in the Java object's ***classhandle*** (see Appendix B for details):

```
>> ch = classhandle(handle(evt))
ch =
      schema.class
>> get(ch)
                 Name: 'EventTest'
              Package: [1x1 schema.package]
          Description: ''
          AccessFlags: {0x1 cell}
               Global: 'off'
               Handle: 'on'
        Superclasses: [0x1 handle]
      SuperiorClasses: {0x1 cell}
     InferiorClasses: {0x1 cell}
              Methods: [11x1 schema.method]
           Properties: [3x1 schema.prop]
               Events: [1x1 schema.event]
        JavaInterfaces: {0x1 cell}
>> get(ch.Properties,'Name')
ans =
    'Class'
    'TestEventCallback'
    'TestEventCallbackData'
>> get(ch.Events)
                      Name: 'TestEvent'
     EventDataDescription: [1x142 char]
>> ch.Events.EventDataDescription
ans =
JavaEventData:
   Source:    the Java object initiating the event
   Type:      the name of the Java event
   JavaEvent: the Java event object
```

Unfortunately, differently capitalized ***classhandle*** events are not differentiated as Java events. For the Java events TestEvent, testEvent, and testevent, only a single ***classhandle*** event is created, named *Testevent*. This is the name of the last declared event with a capitalized leading character, which is the ***classhandle*** standard.

In some cases, users of Java classes encountered problems when trying to use Java Events as MATLAB callbacks. Loading the Java class using the static rather than the dynamic classpath (see Section 1.1) has reportedly solved this problem.[34]

Finally, note an alternative to using callbacks on Java events, as shown above: we could also use undocumented ***handle.listener***s for the same effect (see Appendix B):

```
hListener = handle.listener(handle(evt), 'TestEvent', callback);
```

Java-based events can be used to easily transfer execution to MATLAB code, without needing complex JMI mechanisms (see Chapter 9). In one specific case, a CSSM reader wanted to use Java threads to periodically invoke MATLAB code.[35] Instead of using JMI, the Java thread could simply notify an object's event, where that event is already listened-to within MATLAB. Then, when the thread evokes the event, the MATLAB callback is invoked and execution transfers to the MATLAB code.

It is sometimes necessary to trigger an object event programmatically (synchronously), at a specific predetermined MATLAB program execution point. This can be done using the undocumented built-in ***send*** function, as follows:

```
send(handle(evt),'TestEvent');
```

Custom EventData can be set for the raised event by specifying an optional third argument of type ***handle.event***.

Note that events that were invoked using ***send*** will only trigger callbacks that were set using the ***handle.listener*** function — regular callbacks that were set by setting the object's corresponding callback property will not be triggered.

1.5 Safe Java Programming in MATLAB — A How-to Guide

As already mentioned, there are several potential pitfalls when programming using MATLAB–Java. This section provides a suggested general programming guide for this environment. Programmers who abide by these suggestions should hopefully avoid most of the major pitfalls.

In this book's code snippets, I have often neglected to use these rules. This was done merely for book space considerations, but care should be taken while writing real code.

Rule #1: Program Defensively

Never assume that the code will execute as expected: it may behave differently on different computers; it may depend on some hidden timing order; it may depend on a particular JVM major or minor version or for that matter on a specific MATLAB release, it may depend on a specific user action sequence, and so on.

In short, the code may depend on things that we have never suspected and which may not even be under our control. When any of these dependencies fail, so will our code.

While this is possible also in fully documented/supported code, it is especially important in undocumented/unsupported code such as the MATLAB–Java interface.

The solution is simple: program defensively. Lavish the code with ***try-catch*** blocks to catch and handle run-time exceptions, and always test return values for illegal or unexpected values. In cases where you run some Java code on which you depend, proactively test the code's successful completion (e.g., test the new GUI component's visibility flag).

```
try                                    % Execute doSomething(), no ret val
    % Execute doSomething()            javaObject.doSomething();
    value = javaObject.doSomething();
                                       % continue normally without checking
    % Check the return value
    if (value ~= 0)
        % error handling
    else
        % ok - continue normally
    end
catch
    % exception handling
end
```
                Do   Don't

Rule #2: Never Forget EDT

Whenever you program anything in Java that is displayed onscreen (as opposed to doing some non-GUI computational task), always remember to use the Event Dispatch Thread (EDT, explained in detail in Section 3.2).

Neglecting to use EDT is very tempting: the code is simpler and it will even work most of the time. But every now and then, our GUI (and MATLAB itself) will hang, crash, or behave in entirely unexpected ways. Ouch …!

In practice, especially for MATLAB versions since R2008b (7.7), using EDT is relatively painless: simply lavish the code with ***javaObjectEDT*** function calls whenever you create or use any Java reference for the first time. MATLAB's internal EDT-auto-delegation will then take over all the dirty work. Unfortunately, there is no such magic wand for MATLAB versions earlier than R2008b.

```
% Create a JButton on the EDT          % Create a JButton on Main Thread
jButton = javax.swing.JButton('OK');   jButton = javax.swing.JButton('OK');
jButton = javaObjectEDT(jButton);
                                        % Not ok - might cause problems...
% R2008b+: rely on EDT auto-delegation jButton.setLabel('Not OK');
jButton.setLabel('Also OK');

% R2008a-: use awtinvoke()
awtinvoke(jButton, ...
          'setLabel(java.lang.String)', ...
          'Also OK');
```
                Do   Don't

Rule #3: Use *handle*(…,'CallbackProperties')

Whenever you plan to use an event callback on a Java reference, never set the callback on the naked (un-*handle*d) Java reference, but always on the *handle*d reference, as explained in Section 3.4. Failing to do so will result in memory leaks and occasional run-time errors, whereas using *handle*d appears to carry no penalty other than slight extra code complication.

```
% Create a handled JButton reference
jButton = javax.swing.JButton('OK');
hButton = handle(jButton, . . .
                'CallbackProperties');
% Set the requested callback
set(hButton,'MouseClickedCallback',...)
                    Do
```
```
% Create a naked JButton reference
jButton = javax.swing.JButton('OK');



% Set the requested callback
set(jButton,'MouseClickedCallback',...)
                    Don't
```

Rule #4: Use Java Property Accessor Methods

MATLAB has a very convenient way to access Java objects' property values: we can use the built-in *get* and *set* functions to, respectively, retrieve and modify the specified object's property values.

Unfortunately, using *get* and *set* on naked Java references causes memory leaks and should be avoided. In R2010b, a warning is even displayed in the Command Window whenever we attempt to do so.

Instead, either use *get* and *set* on the *handle*d Java reference (see Rule #3), or better still use the object's natively supported property accessor methods, which are typically named *getPropertyName* or *isPropertyName* (*is* for Boolean [logical flag] properties; *get* for all the others), and *setPropertyName*.

```
% Use set on handled Java reference
set(hButton,'Label','This is OK');

% Use the native Java accessor method
jButton.setLabel('Even better');
                    Do
```
```
% Use set on naked Java reference
set(jButton,'Label','Not OK')




                    Don't
```

Rule #5: Concentrate Undocumented Code

Concentrate as much of the code that uses undocumented features in a single location — a single function or m-file. This way, if and when our code breaks under a specific set of circumstances (e.g., new MATLAB release or a different running platform), it will be easier to diagnose and possibly fix the affected code. If the undocumented-features-dependent code is scattered throughout the entire program, then we would effectively need to diagnose and debug the entire program. Ouch, not again ...!

Rule #6: Test Backward Compatibility

As a corollary to Rule #1, try to test the code, especially sections that rely on undocumented features, on earlier MATLAB releases. In some cases, MATLAB has modified the interface and/or behavior of its internal Java classes across MATLAB releases. Since we often cannot know in advance on which MATLAB release our code will run, it makes sense to test our code on at least one old MATLAB release.

An example of using this rule (together with Rule #1) is given in Section 8.3:

```
try
    cmdWinFrame = cmdWin.getTopLevelAncestor;     % MATLAB 7
catch
    cmdWinFrame = cmdWin.getTopLevelWindow;       % MATLAB 6
end
```

1.6 Compiling and Debugging User-Created Java Classes in MATLAB

MATLAB comes pre-bundled with a vast number of Java classes that are available for immediate use in any MATLAB program, as we have seen above. However, it is sometimes useful to have a user-created Java class. Such classes may be an extension of an existing class with some functionality, or perhaps an entirely new class.

MATLAB runs directly from the source m-files, creating the executable code on-the-fly. Java, on the other hand, needs a compiled (*class*) file to run and cannot run directly from the source (*java*) file. When we use any of the MATLAB-bundled Java classes mentioned above, we are actually using a precompiled class file, the source code for which is not available. For our user-created classes, we need to compile our source files into similar class files.

We have two options for compiling user classes: we can either use a standalone Java compiler or use an integrated Java development environment (IDE).

There are many Java compilers, but the simplest is probably the *javac* compiler, which is part of the official Java Development Kit (JDK), which can be downloaded from http://java.sun.com/javase/downloads/previous.jsp and used freely. While doing so, ensure to use a JVM version not newer than the one used by the MATLAB version, as reported by the "`version -java`" command in the MATLAB desktop (see details in Section 1.8). For example, if our application is meant to target MATLAB R2007a (7.4), which uses JVM 1.5, then we should be careful to compile using JDK 1.5 rather than the latest version (1.6).[36] In practice, JVM

1.5 provides almost all the functionalities of 1.6, while enabling backward compatibility with all MATLAB releases since 2005 (R14 SP2 a.k.a 7.0.4). To provide earlier backward compatibility, you will need to use JDK 1.4.2 or even 1.3, but these JDKs lack important Java functionality (e.g., Generics). In summary, I suggest to normally compile with JVM 1.5, except in rare cases.

To use the javac compiler, simply run "`javac MyJavaClass.java`" from the operating system's command line, replacing "`MyJavaClass`" with the actual name of the class. If the class uses some other classes, indicate their classpath location using the –cp command-line switch. If the source code has errors, they will be reported. Otherwise, we will get a file called *MyJavaClass.class* which should now be placed in MATLAB's dynamic or static Java classpath (see Section 1.1).

Note: Java classes must be placed in files that are named exactly like their contained class. So, class `MyJavaClass` must be placed in *MyJavaClass.java*.

A much preferred alternative to javac is to use a Java IDE. There are several excellent free and commercial IDEs. Two of the best free IDEs are Eclipse[37] and NetBeans[38] (some other popular Java IDEs are JBuilder, IntelliJ, and JDeveloper, but I will only discuss Eclipse and NetBeans below). There is an active "religious war" among developers regarding which of these IDEs is better .... In all IDEs, compilation is done on the fly, and errors are visually displayed next to their offending source. It is also easy to modify the compiler to use an earlier JDK, a need that was explained above.[†]

Once the class file is generated, we need to place it in MATLAB's Java classpath, as explained in Section 1.1. This can be done either dynamically (using *javaaddpath* or *javaclasspath*), statically by adding the class path-name to the *classpath.txt* file (*which('classpath.txt')*), or by adding the class file to one of the classpath folders.

Java classes are not reloaded automatically by MATLAB, when recompiled outside MATLAB. To reload a modified Java class, we need to restart the JVM by restarting MATLAB. For classes placed on the dynamic classpath, you can try MATLAB's *clear('java')* command, while remembering its side effect of clearing all globals. However, this does not always work (e.g., if the class signature has changed).

Expert Java programmers can try to use Paul Milenkovic's suggestion[39] for a proxy class-loader, as an alternative to restarting MATLAB or clearing Java. As Dan Spielman explains,[40] "the rough idea is that you create a classloader for your class, and then access it through the

---

† An Eclipse plugin for MATLAB called Matclipse (http://itp.tugraz.at/wiki/index.php/Matclipse or http://bit.ly/huUIXZ) was developed, but never released. A user request to have an official Eclipse MATLAB plugin has generated some interest from Scott Hirsch, MATLAB's product manager (http://linkd.in/dUh2g7), so perhaps this will be added in some future MATLAB release.

classloader. After you recompile, you kill the classloader and then create a new instance of it, which then reads the recompiled class".[†]

In practice, I suggest restarting MATLAB after Java classes are recompiled, even when this is not strictly necessary. It may save a lot of frustrating debugging and chasing down errors that only happen because MATLAB keeps an old class in memory.

This brings us to the issue of debugging. When using user-created Java classes in MATLAB, we often need to debug the Java code. There are many online resources for debugging Java code.[41] A general advice when creating such Java classes is to remember to set `public` visibility to all Java elements (fields/methods/classes) that need to be visible in MATLAB, otherwise we will encounter hard-to debug problems.[42]

Unfortunately, MATLAB's built-in debugger can only debug the MATLAB code, stepping over Java invocations. There are two basic approaches for debugging the Java code:

The simplest way is to add debug printouts within the Java code. These printouts will be displayed on the Java console, which in MATLAB JVM's case is redirected to the MATLAB Command Window. It is ugly but it works:

```
public int myMethod(int row, int column, double value) {
    System.out.println(row + "," + column + "=> value: " + value);
    // do something useful here...
}
```

The down-side to this approach is that whenever printouts should be modified or removed, the Java code needs to be recompiled.

A slight improvement of this stone-age debugging is to include a debug flag that can be turned on/off at will. The drawback is that, while preventing the need for recompilation, the compiled class remains unnecessarily bloated even after all the development debugging has ended and the debugging framework is no longer needed:

```
public class MyClass
{
    private boolean _debug = false;
    public void setDebug(boolean flag) {
        _debug = flag;
    }
    public int myMethod(int row, int column, double value) {
        if (_debug) {
            System.out.println(row + "," + column + "=> " + value);
        }
```

[†] Java classloader issues are a sore point in advanced MATLAB–Java integration. Another such example is presented in Section 2.2.2 (Connecting to a Database). This has also been referenced in several CSSM and StackOverflow threads (e.g., http://bit.ly/hEaCWD, which sheds some light on the internal mechanism).

```
            // do something useful here...
        }
    }
```

A significantly more powerful method of debugging entails step-by-step debugging using a Java integrated debugger/editor (IDE). A very simple and useful trick is to add a `public static main` method to the Java class, thereby enabling it to run (and be debugged) as a standalone Java application in the Java IDE. Once you have fully debugged the class in the Java IDE, it can be used in MATLAB.

To connect the Java IDE to a running MATLAB, simply start MATLAB with the jdb (Java debugger) flag and then connect the Java debugger to the MATLAB process via default port 4444, as described by Brad Phelan:[43]

```
matlab -jdb
```

MATLAB will now display the following message when it starts:

```
JVM is being started with debugging enabled.
Use "jdb -connect com.sun.jdi.SocketAttach:port = 4444" to attach debugger.
```

Note that the –jdb flag is incompatible with some other startup flags (e.g., -nojvm and -nosplash). MATLAB may fail to start if incompatible startup flags are specified.

Also note that this works only on MATLAB installations having JVM 1.5 or higher (i.e., MATLAB release 7.0.4 (R14 SP2) and higher), which supports remote debugging.[44] For earlier releases, add –Xdebug and –Xrunjdwp options to the *java.opts* file (see Section 1.9). Each option must be placed on a separate line in the *java.opts* file:

```
-Xdebug
```

and one of the following (the second is for Windows platforms only):[45]

```
-Xrunjdwp:transport = dt_socket,address = 4444,server = y,suspend = n
-Xrunjdwp:transport = dt_shmem,address = matlab,server = y,suspend = n
```

These options are incompatible with some MATLAB versions/platforms (particularly modern MATLAB releases), causing MATLAB to fail to start without any generated error message explaining what happened. In some cases, MATLAB may fail to start after setting the *java.opts* options due to a missing *jdwp.dll* library file required by the Java debugger, as explained in MathWorks solution 1-OVU1L.[46] Also, *java.opts* options apply to ALL the MATLAB sessions, whereas the –jdb startup flag can be specified only for the rare sessions in which MATLAB–Java debugging is required. This is important since the remote Java debugging entails some overhead on the MATLAB/Java side, which, in most cases, when debugging is unneeded, is undesirable.

Here is the complete MATLAB–Java debugging process, adapted from a post by Ed Yu on CSSM.[47] This applies to Eclipse IDE, but can easily be adapted for other IDEs:

1. Create an Eclipse project (we must use JDK 1.5 or above); to include the source code of the Java classes, remember the folder where you put the output classes when you defined the project.
2. Start MATLAB with the –jdb command-line startup option (or on pre-R14 MATLAB releases, use the *java.opts* modification that was explained above).
3. In MATLAB, add the folder of the output classes from step 1 into MATLAB's java-classpath. We can use either static or dynamic classpath here.
4. In Eclipse, add a "Remote Java Application" debug configuration: Set Connection Type "Standard (Socket Attach)", Host "localhost", Port 4444.

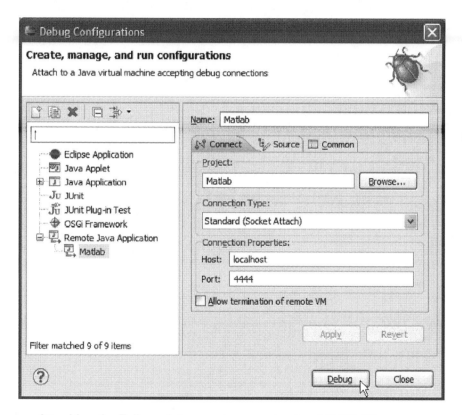

Attaching the Eclipse debugger to a running local MATLAB process

5. Now click the "Debug" button to debug the "Remote Java Application" in Eclipse. If Eclipse fails to connect to MATLAB, you will see an error message; otherwise, you will see no message and will be able to proceed onward.
6. Place a breakpoint in the Java code within Eclipse.
7. Back in MATLAB, instantiate or execute the Java code. You should see the breakpoint popping up in Eclipse when MATLAB calls the Java code and reaches the first breakpoint.

The steps for using NetBeans IDE are very similar: Instead of steps #4 and 5, simply select Debug/Attach Debugger from the main menu or toolbar, and then select SocketAttach to port 4444 or ProcessAttach to the MATLAB process ID (which is actually the JVM's process ID). Socket attachment is useful for remote (intercomputer) debugging, while process attachment is useful for local (same computer) debugging. NetBeans will automatically attempt the debug connection upon pressing the OK button, and will then display MATLAB's threads in a new threads panel, or report an error if the connection failed for any reason:

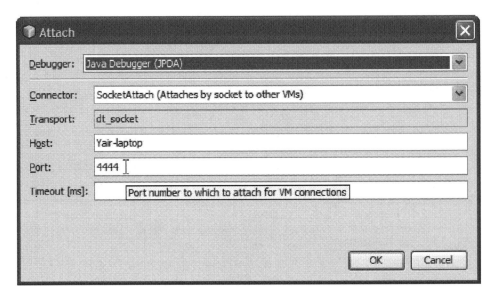

Attaching NetBeans debugger to a specific socket (useful for remote debugging)

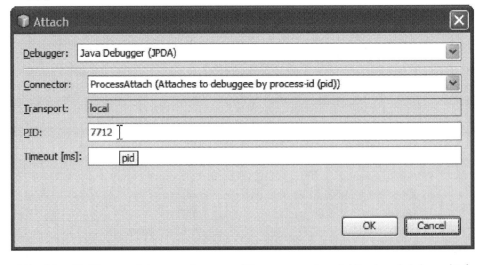

Attaching NetBeans debugger to a specific process (useful for local debugging)

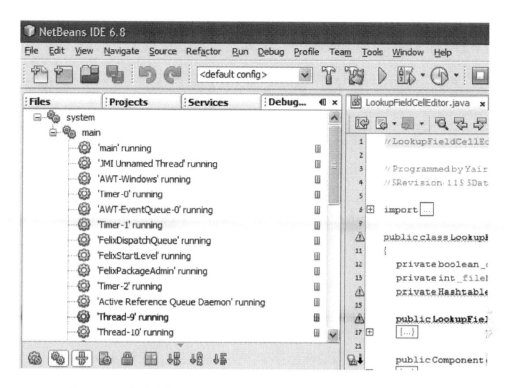

NetBeans IDE debugger attached to a running MATLAB Process

With NetBeans (and possibly also Eclipse), there is a reported problem that MATLAB locks the generated JAR files, preventing recompilation in the IDE. A CSSM-suggested workaround is to point MATLAB's javaclasspath to the Java project's build/classes folder, rather than to the JAR file.[48] When Java files are modified and saved in the IDE, they are automatically recompiled and reloaded by MATLAB.

Profiling memory and CPU usage of Java classes can be done in several ways: we can of course use the facilities available in our chosen Java IDE. Both Eclipse[49] and NetBeans[50] have extensive support for this, with many important plugins for both IDEs to provide a clear picture. At a very basic level (outside the IDE), we can turn on instruction (-Xt) and method (-Xtm) tracing via *java.opts*.[51] We can even generate an hprof text file of called methods and timing, similar to MATLAB's profiler.[52]

Ed Yu has posted a detailed post about debugging a Java class called from within MATLAB.[53] He has also posted detailed accounts of his experience, first using a JBuilder/Eclipse-integrated OptimizeIt profiler,[54] then using an external JProfiler application,[55] and finally using an external YourKit Java profiler.[56]

MATLAB's built-in profiler can also be used. It does not profile the internal steps within the Java methods and reports the Java profile data at the class (not method) level. This is much less useful than the other profilers, but may still be sufficient for simple MATLAB–Java integration tasks. And there is no beating the ease-of-use of using MATLAB's integrated profiler, detailing both MATLAB and Java calls at the same place:

Function Name	Calls	Total Time	Self Time*	Allocated Memory	Freed Memory	Self Memory	Peak Memory	Total Time Plot (dark band = self time)
te42	1	1.250 s	0.516 s	332.00 Kb	0.00 Kb	328.00 Kb	240.00 Kb	
createTable	2	0.718 s	-0.000 s	4.00 Kb	0.00 Kb	0.00 Kb	4.00 Kb	
createTable>processParams	2	0.625 s	0.625 s	4.00 Kb	0.00 Kb	4.00 Kb	4.00 Kb	
createTable>DNDCallback	15	0.311 s	0.311 s	0.00 Kb	0.00 Kb	0.00 Kb	0.00 Kb	
uitable	2	0.046 s	0.000 s	0.00 Kb	0.00 Kb	0.00 Kb	0.00 Kb	
uitools\private\uitable_deprecated	2	0.046 s	0.016 s	0.00 Kb	0.00 Kb	0.00 Kb	0.00 Kb	
mdbfileonpath	1	0.032 s	0.016 s	0.00 Kb	0.00 Kb	0.00 Kb	0.00 Kb	
...peer.UitablePeer$PeerSpreadsheetTable (Java-method)	97	0.032 s	0.032 s	0.00 Kb	0.00 Kb	0.00 Kb	0.00 Kb	

MATLAB's integrated Profiler displaying information about Java methods (see bottom row)

Memory profiling is normally part of the profiling tools, but also has some dedicated tools. First, note there are several JVM switches that can be turned on in *java.opts*: -Xrunhprof[:help]| [:option = value,...], -Xprof, -Xrunprof, -XX: +PrintClassHistogram and so on.[57] There are several memory-monitoring ("*heap-walking*") tools: the standard JDK jconsole, jmap, jhat, and jvisualvm (with its useful plugins) provide good basic coverage. MathWorks has posted a tutorial on using jconsole with MATLAB.[58] There are a number of other third-party tools such as JMP (for JVMs 1.5 and earlier)[59] or TIJMP (for JVM 1.6).[60] Within MATLAB, we can use utilities such as Classmexer[61] to estimate a particular object's size or use java.lang. Runtime.*getRuntime()*'s methods[62] (*maxMemory()*, *freeMemory()* and *totalMemory()*) to monitor overall Java memory.

Starting in MATLAB 7.13 (R2011b), we can use the built-in Java memory monitor:

```
com.mathworks.xwidgets.JavaMemoryMonitor.invoke
```

MATLAB 7.13 (R2011b)'s new Java memory monitor

To complete the picture, note that there are several dedicated tools for Java code analysis that can be used for static or dynamic code coverage and analysis reports.[63]

1.7 Compatibility Issues

While supporting a large subset of the core Java functionality, some advanced Java constructs are not accessible via MATLAB.[†] Use of Generics,[64] introduced in J2SE 5 (MATLAB 7.0.4, R14 SP2), is an example of one such unsupported Java functionality:

```
>> jCollection = javaObject('java.util.Collection<java.lang.String>');
??? Error using ==>javaObject
No class java.util.Collection<java.lang.String> can be located on Java
classpath
```

Also, MATLAB cannot invoke Java code that is not class-wrapped. Always remember that the JVM shipped with MATLAB is only used to enable MATLAB to access Java classes, not to act as an internal Java interpreter. For similar reasons, Java interfaces, annotations, and anonymous classes and methods (typically used by event callback actions) cannot be directly accessed from MATLAB. However, this is not a big limitation, since the MATLAB programming language is very powerful. Java code can often be ported into MATLAB, or used as-is via class or object reference.

Another limitation is that inner classes and enumerations cannot be accessed directly: Inner classes can only be accessed via *javaObject/javaMethod* using 'ParentClass$InnerClass' $-notation, but not via the expected standard ParentClass.InnerClass dot-notation. Accessing enumerations is even more awkward, and using $-notation is not sufficient.

For example, JVM 1.6 (in MATLAB 7.5 R2007b onward) enabled access to the new TrayIcon functionalities (see Section 3.6). One of its functionalities is displaying a message next to the tray icon, using java.awt.TrayIcon.*displayMessage()*. This method expects an object of type java.awt.TrayIcon.MessageType,[65] which is an enumeration within the TrayIcon class. However, MATLAB's dot-notation does not recognize what should have been the following correct notation, so we need to resort to Java reflection:

```
>> trayIcon.displayMessage('title', 'info msg', ...
                          TrayIcon.MessageType.INFO);
??? No appropriate method or public field MessageType for class java.awt.
TrayIcon

>> trayIconClasses = trayIcon.getClass.getClasses;
>> trayIconClasses(1)
ans =
class java.awt.TrayIcon$MessageType          <=hurray!!!
```

[†] These constructs may be interesting only to readers with advanced Java knowledge; other readers may safely skip this section.

```
>> MessageTypes = trayIconClasses(1).getEnumConstants
MessageTypes =
java.awt.TrayIcon$MessageType[]:
     [java.awt.TrayIcon$MessageType]          <= 1: ERROR
     [java.awt.TrayIcon$MessageType]          <= 2: WARNING
     [java.awt.TrayIcon$MessageType]          <= 3: INFO
     [java.awt.TrayIcon$MessageType]          <= 4: NONE
>> trayIcon.displayMessage('title','info msg',MessageTypes(3));
```

We can also access Java enums using their built-in *values()* and *valueOf()* methods:

```
>>
msgType = javaMethod('valueOf','java.awt.TrayIcon$MessageType','I
NFO')
msgType =
INFO          <= a java.awt.TrayIcon$MessageType object

>> enums =
cell(javaMethod('values','java.awt.TrayIcon$MessageType'));
>> msgType = enums{3};   % alternative way to find the INFO
enum value

>> cellfun(@(c)c.toString.char, enums, 'uniform',false)'
ans =
     'ERROR'     'WARNING'     'INFO'     'NONE'
```

Inner classes can also be accessed using the Java classloader, although this is more cumbersome.[66]

Trying to access internal static fields in java.nio.channels.FileChannel. MapMode,[67] in order to use them to create a memory-mapped file using FileChannel. *map(...)*, is similarly problematic.[68] Luckily, in this case, we have the built-in *memmapfile* MATLAB function as a much simpler alternative, but for the record, we could do this:

```
>> channel = java.io.FileInputStream('234.jpg').getChannel
channel =
sun.nio.ch.FileChannelImpl@1c7a5d3     <= which extends FileChannel

>> innerClasses = ch.getClass.getSuperclass.getDeclaredClasses;
>> innerClasses(1)
ans =
class java.nio.channels.FileChannel$MapMode

>> read_only_const = innerClasses(1).getField('READ_ONLY').get(1)
read_only_const =
READ_ONLY     <= a java.nio.channels.FileChannel$MapMode object
```

```
>> fields = innerClasses(1).getFields;
>> read_only_const = fields(1).get(1);    % an alternative

>> buffer = channel.map(read_only_const, 0, channel.size);
```

1.8 Java Versions in MATLAB

1.8.1 Pre-Bundled JVM Versions

As noted above, since R12 MATLAB has always shipped with a bundled Java engine (JVM), on all supported MATLAB platforms except Mac OS X (where MATLAB uses the Mac OS's JVM;[69] note that Apple will reportedly discontinue its internally ported JRE in future Mac OS releases). This pre-bundled Java engine has the benefit of hiding the nuts and bolts from the user, while gaining access to a full-fledged Java engine.

There are drawbacks to MATLAB's use of a pre-bundled JVM: there is a notable increase in the installer and installed disk size, run-time memory, and CPU usage. Also, the pre-bundled Java engine is installed and used by MATLAB, even if the user already has another JVM installed, and even if MATLAB's JVM version is older than the user's version (Section 1.8.2 explains how to bypass this). A bundled JVM also means that Java applications called from within MATLAB (more on this later) may inadvertently close the calling MATLAB applications, unless we are careful.[70]

Different MATLAB releases have integrated progressively advanced JVM engines. The following list is for Windows platforms — other operating systems may vary slightly:[71]

MATLAB Release	MATLAB Version	JVM Version
R12	6.0	1.1.8
R12.1	6.1	1.3.0
R13 [72]	6.5	1.3.1
R14	7.0	1.4.2
R14 SP2	7.0.4	1.5.0 (J2SE 5)
R14 SP3	7.1	1.5.0
R2006a	7.2	1.5.0
R2006b	7.3	1.5.0
R2007a	7.4	1.5.0_07
R2007b	7.5	1.6.0 (Java SE 6)
R2008a	7.6	1.6.0
R2008b	7.7	1.6.0_04
R2009a	7.8	1.6.0_04
R2009b	7.9	1.6.0_12
R2010a	7.10	1.6.0_12
R2010b	7.11	1.6.0_17
R2011a	7.12	1.6.0_17
R2011b	7.13	1.6.0_17

Each MATLAB release, in addition to its internal MATLAB-based enhancements, thus included all the fixes and improvements included in the new Java engine. Java 1.5, for example, improved GUI performance and smoothness.

Some MATLAB versions shipped with different releases of the same JVM version. For example, R2008b (7.7) shipped with JVM 1.6 Update 4 (aka Java 1.6.0_04) for all platforms except the Apple Macintosh, whereas R2007b (7.5) and R2008a (7.6) shipped with the first 1.6 release (update 0, so to speak).[†]

On Macs, even R2009a still uses Java 1.5.0 update 16 (=1.5.0_16). MathWorks explains that Apple's Java 1.6 was developed for Intel 64-bit Macs only, not for 32-bit (maci) applications such as MATLAB up to version 7.7 (R2008b).[73] R2009a is a 64-bit application, but still uses the older Java 1.5 for some reason (possibly due to unsolved compatibility issues); however, R2009a can be modified to use Java 1.6 by setting the JAVA_JVM_VERSION environment variable, as explained below. The 32- versus 64-bit issue is also to be blamed for the activation problem of R2009a on Snow Leopard — in this case, we actually need to switch Snow Leopard's from using the default 64-bit Java 1.6 (Java SE 6) to its nondefault 32-bit Java version. This is done using Snow Leopard's Java Preferences utility from the /Application/Utilities folder.[74] After activation, switch back to the 64-bit Java 1.6 (Java SE 6) or Java 1.5 (J2SE 5).

The MATLAB installer itself uses Java and may require installation of the bundled JVM even if a newer JVM is already installed on the computer. We can safely approve the MATLAB-bundled Java installation, since it will install Java (or rather, JRE) in a subfolder of MATLAB (sys\java\jre) and will not affect the pre-existing JVM. After MATLAB is installed, we can then instruct MATLAB to use the newer JVM, as explained below.

Some MATLAB releases use Java features available only on the latest prebundled JVM. This causes some errors on platforms that bundle with an older JVM. For example, MATLAB 6.5 (R13) shipped with JVM 1.3.1 on Windows and Linux, but only 1.1.8 on Alpha and SGI. This caused the ***colormapeditor*** MATLAB function to fail on Alpha & SGI, since it requires JVM 1.3 and above.

A similar source of errors happens when trying to use Java classes compiled with a newer JVM version than the one MATLAB uses. In this case, the classes' code may be incompatible with MATLAB's JVM, causing a run-time error.

In some cases, the bundled JVM fails to install, thereby preventing the main MATLAB installation. In such cases, search the Internet for the correct JVM version and install it separately, then rerun the MATLAB installer. If all goes well, then the MATLAB installer will detect the JVM and continue with the main MATLAB installation process.[75]

To see MATLAB's JVM version, type the following at MATLAB's command prompt:

```
>> version -java
ans =
Java 1.1.8 from Sun Microsystems Inc.
```

[†] R2009a did not change the JVM (1.6.0_04); R2009b and R2010a shipped with update 12 (=1.6.0_12); R2010b uses 1.6.0_17.

1.8.2 Configuring MATLAB to Use a Different JVM

We may configure MATLAB to use a JVM version different from the pre-bundled version, by following the steps outlined here (taken from the MathWorks Support site).[76] In all cases, the purpose is to point the MATLAB_JAVA environment variable to the path of the JVM (i.e., the folder that contains the *lib/rt.jar* file):

- ■ Windows NT/2000/XP:
1. Click on Settings in the Start Menu.
2. Choose Control Panel.
3. Click on System.
4. Choose the "Environment" tab (NT) or the "Advanced" tab and the "Environment Variables …" button.
5. Set/add a MATLAB_JAVA system environment variable to your JVM path.

- ■ Windows 95/98/ME:
1. Open the C:\AUTOEXEC.BAT file in any text editor.
2. Add a line that points the MATLAB_JAVA environment variable to your JVM path. For example,

    ```
    SET MATLAB_JAVA = D:\jre
    or:
    SET MATLAB_JAVA = "D:\Program Files\Java 1.3"
    ```

- ■ Unix/Linux:
Set a MATLAB_JAVA environment variable to your JVM path. For example,

    ```
    setenv MATLAB_JAVA /usr/jre                    (csh shell)
    or
    set MATLAB_JAVA = /usr/jre; export MATLAB_JAVA (sh shell)
    or
    export MATLAB_JAVA = /usr/jre                  (bash shell)
    ```

- ■ Macintosh:
MATLAB always uses the Mac's default JVM, not its prebundled version.[77] To modify the JVM in use, modify the JAVA_JVM_VERSION environment variable to the version number (not path!). For example,

    ```
    setenv JAVA_JVM_VERSION 1.5         (tcsh shell)
    or
    export JAVA_JVM_VERSION = 1.5       (bash shell)
    ```

In rare cases, MathWorks itself advises to change the pre-bundled JVM to a newer[78] or older[79] sub-version that fixes some specific error(s). Some users found this fixes other problems.[80] For example, installing Java 1.6 solves the problem of running MATLAB 7.4 (R2007a) and earlier on Windows 7,[81] without requiring running in Vista SP2 compatibility mode with Visual Themes/Desktop disabled.[82]

In some cases, the MATLAB_JAVA environment variable may be missing from our specific system. To solve the resulting Java-related problems, simply add the missing variable to the system's environment.[83]

An alternative to modifying the MATLAB_JAVA environment variable was suggested on CSSM: replace the entire JRE subfolder beneath the MATLAB root folder with an upgraded JRE version, *without* changing the JRE subfolder name.[84]

Upgrading the JVM version has the potential added benefit of bug fixes and added functionality. For example, the useful java.lang.String regexps and java.net.URLConnection. *setReadTimeout()* only appeared in JVM 1.5.[85] As another example, JVM 1.6 fixed many JVM 1.5 bugs, improved performance, added some important Swing functionality and Collection classes[86] (see Section 2.1 for details), enabled using the system tray,[87] and many others.[88] A detailed list of changes in each JVM release/update is available online.[89] For example, http://www.oracle.com/technetwork/java/javase/6u5-135446.html describes JVM 1.6 update 5, while http://www.oracle.com/technetwork/java/javase/6u26releasenotes-401875.html describes update 26, the latest update as of this writing (July 2011).

JVM 1.7 (Java SE 7) is a new major Java version, which replaced Java SE 6 in July 2011. I expect MATLAB to start using JVM 1.7 in R2012a or R2012b. Until then, I advise not to try MATLAB with JVM 1.7 — very weird things might happen .... It would be safer to upgrade to the latest Java update that corresponds to your major JVM version. For example, if your MATLAB has JVM 1.6.0_12, then it would probably be safer to upgrade to 1.6.0_26 rather than to 1.7.

You can download the latest JVM from http://www.java.com/en/download/manual.jsp or http://java.sun.com/javase/downloads/index.jsp. Previous JVM versions can be downloaded from http://java.sun.com/javase/downloads/previous.jsp or http://java.sun.com/products/archive/ (which even lets us select the requested update number). Development versions, which usually fix reported bugs, are described and can be downloaded from https://jdk6.dev.java.net/. Be sure to use a JVM version that is compatible with your system: Ed Yu has reported problems trying to run a 32-bit Java with a 64-bit MATLAB.[90]

Using a non-bundled JVM has a risk of causing many built-in MATLAB functions to fail due to JVM incompatibilities, especially when using a JVM version lower than the bundled one. JVMs are generally, although not always, backward-compatible. This means that, in general, we can use a more advanced JVM than the pre-bundled one.

In some cases, upgrading the JVM actually introduced new errors into the MATLAB environment.[91] Therefore, this is mainly useful if we need to use a specific Java feature that is only available in a more advanced JVM than the pre-bundled one. One extreme example is a user report of successfully upgrading R13 (MATLAB 6.5, JVM 1.3.1) to JVM 1.6 in order to use java.net.URLConnection.*setReadTimeout()*.[92]

JVM upgrades sometimes occur automatically or semi-automatically for minor JVM or OS updates. This too may introduce new bugs.[†] The probability of encountering errors increases

---

[†] For example, on Mac OS X 10.5 (Leopard), when upgrading to Update 4: http://www.mathworks.com/support/bugreports/547093 (or http://tinyurl.com/qby5cw).

significantly when upgrading to a major OS version for which our specific MATLAB release was not qualified.

MATLAB on Mac OS, in particular, is plagued by many JVM-related issues, especially with the Mac OS X 10.6 (Snow Leopard),[†] but also on older versions such as Mac OS X 10.5 (Leopard).[‡] A CSSM user suggested that this is due to Mac OS's default installation of JVM 1.6, which is incompatible with MATLAB that requires an older JVM (remember that MATLAB uses Mac's JVM by default, not a pre-bundled JVM). He suggested to actually **downgrade** the JVM by installing version 1.5.[93] A different approach is to reverse the automatic Java upgrade done by OS updates.[94]

Another advice is to install minor JVM updates, which often fixes quirks in earlier JVM versions. In particular, several of the Mac OS X 10.6 (Snow Leopard) issues are apparently fixed by Apple Java 10.6 update 1.[95] Note that not all Mac-MATLAB issues are Java-related: there are also problems with X11 and docking.[96]

A related Mac OS problem is due to 32- versus 64-bit JVM incompatibilities. The suggested fix is to temporarily set the Mac OS's default JVM to the 32-bit version.[97]

Even when using the bundled JVM, there is always some risk that incompatibilities between JVM versions will cause different MATLAB versions to behave differently when running Java functions.[98]

In some cases, due to incompatibilities between the new JVM and MATLAB, pointing MATLAB to use the new JVM may have adverse effects on MATLAB, to a point where MATLAB may occasionally crash or even fail to start.[99] In such cases, we can simply point MATLAB back to its internal prebundled JVM to start afresh.[100]

Warning: if you plan to deploy our MATLAB application to other computers, note that using a non-bundled JVM may cause the MATLAB application to break if it is ever ran on a system other than the one on which the JVM was modified. If the application depends on a non-bundled JVM, then we must therefore include a Java version check in the code and act accordingly. Such version checks should normally check for a minimal version rather than a specific version, since versions are usually forward-compatible and we just need to check for the earliest version supporting some feature:

```
% Two ways to retrieve the JVM version in MATLAB:
javaVersion = char(java.lang.System.getProperty('java.version'));
javaVersion = strtok(strrep(version('-java'),'Java','')); %alternative
⇒      '1.6.0' (or a similar string value)

% Check for a specific JVM version (less frequent usage)
```

[†] Here is an example out of many that flood the CSSM newsgroup and TMW blogs: MATLAB R2007a on Macs was qualified for Mac OS X 10.5 (Leopard) with Java 1.5, but not for Mac OS X 10.6 (Snow Leopard) nor Java 1.6 — retrofitting the newer Java 6 and/or Snow Leopard on Macs running R2007a causes several errors, and only some of which have a patch from MathWorks support: http://www.mathworks.com/matlabcentral/newsreader/view_thread/265588 (or http://bit.ly/ap0vCK).

[‡] http://www.mathworks.com/matlabcentral/newsreader/view_thread/158424 (or http://bit.ly/dhFrYb). On Leopard, some MATLAB releases will not start until libtiff*. dylib is removed from bin/mac, so that the system version of libTIFF is loaded.

```
if ~strncmp(javaVersion, '1.3',3)
    error(['Invalid Java version: ' javaVersion ' - must be 1.3!']);
end

% Check for a minimal JVM version (more frequent usage)
javaVerNum = str2num(regexprep(javaVersion,'(\d + \.\d+)\.?(.*)?','$1'));
if javaVerNum < 1.3
    error(['Invalid Java version: ' javaVersion ' - must be >= 1.3!']);
end
```

If you wish to check for specific platforms/JVMs, MATLAB's built-in com.mathworks. util.PlatformInfo class provides handy convenience methods:

```
>> com.mathworks.util.PlatformInfo.isVersion16    % = JVM version 1.6
ans =
    1
```

MATLAB starts its JVM in client mode by default. This was a conscious decision made by MathWorks, since Java code in MATLAB is mostly used for GUI rather than for computational processing. In this respect, MATLAB resembles a client Java application more than a server one, so the tradeoffs between the server and client JVMs favor the client. Interested readers, and perhaps users who use MATLAB to launch computational-heavy Java code, can modify this setting.[101] Note that the jvm.cfg file mentioned in the referenced CSSM thread actually resides in %matlabroot%/sys/java/jre/win3/jre/lib/i386/ (at least on Windows), and not as reported in the thread. A better way would be to add JVM startup options in the *java.opts* file, as described in the following section. This has the benefit of being cross-platform and session-specific, and the ability to be modified by users with no write access to MATLAB's installation folder.

1.9 *Java.opts*

MATLAB enables users to customize the behavior of its bundled JVM. This is done by placing JVM startup options within the *java.opts* file, which is typically located within the %matlabroot%/bin/%arch% folder (e.g., [matlabroot, '\bin\win32'] on Windows).[102] It may also be placed in MATLAB's startup folder (the current directory when invoking MATLAB). For compiled applications, it should be placed in the compiled application's startup folder.[103] For X11 terminals on Macs, place this file in the Applications/MATLAB/ folder and a copy of this file in the Applications/MATLAB/bin/maci/ folder (or in Applications/MATLAB/ bin/mac/ for PowerPC).[104] On Unix, it should be placed in the folder where MATLAB is started.

When starting MATLAB, the *java.opts* file is read and its contained options are passed to the JVM as command-line options, overriding the default option values.[105] Because of this, MATLAB needs to be restarted for *java.opts* modifications to take effect. Since the JVM is started when MATLAB starts, there is no alternative to *java.opts* for passing startup options to the JVM.[106]

Note a possible pitfall when creating the *java.opts* file on Windows:[107] when saving this file via Notepad or another text editor, the editor may automatically append a .txt suffix to the *java. opts* name, thereby creating a *java.opts.txt* file. Since by default Windows does not display the .txt file extension in the File Explorer, the user may be misled into thinking that their file was created successfully (as *java.opts*), when in fact it was not. MATLAB ignores *java.opts.txt* and displays no warning if the *java.opts* file is not found. One way to prevent this problem is to enclose the file name in quotes ("java.opts"), which tells Notepad to use this name without appending .txt.

All the options in the *java.opts* file must be placed on separate lines. Note that these options are JVM-dependent. Therefore, options that work for JVM 1.6 (R2007b) may not work with JVM 1.1.8 (shipped with MATLAB R12 (6.0)). Refer to the JVM documentation for JVM-specific information and additional details.[108] Also note that MATLAB R2008a (7.6) and earlier had a limited fixed number of options it could pass on to the JVM engine — if the *java.opts* file had more options, MATLAB would unexpectedly **crash**. This issue was reportedly fixed in R2008b (7.7).[109]

JVM 1.1.8, used by MATLAB R12 (6.0), uses an "option=value" format to set the JVM options; JVM 1.3 onward, used by MATLAB R13 (6.5), use a more compact "-option" format. Refer to the JVM documentation for additional details.

To confirm that the *java.opts* changes are in effect, type the following in the MATLAB Command Window and verify that the result is as expected:

```
java.lang.System.getProperty('apple.awt.graphics.UseQuartz');
```

(or whatever other property we were trying to set). Alternately, run the following MATLAB code segment to list all the current Java system properties:[†]

```
propValues = java.lang.System.getProperties.elements;
propKeys   = java.lang.System.getProperties.keys;
while propKeys.hasMoreElements
   disp([propKeys.nextElement ' = ' propValues.nextElement]);'
end
        java.runtime.name = Java(TM) SE Runtime Environment
        java.protocol.handler.pkgs = null|com.mathworks.util.jarloader
        java.vm.version = 1.6.0-b105
        sun.awt.nopixfmt = true
        java.vm.vendor = Sun Microsystems Inc.
        java.vendor.url = http://java.sun.com/
        os.name = Windows XP
        path.separator = ;
        . . .
```

[†] One of the useful Java system properties is "line.separator" which returns the newline string on the current platform (=char(10) on Unix, char(10) + char(13) on windows and char(13) on Macs): double(java.lang.System.getProperty('line. separator').char).

[‡] These properties can easily be placed in a MATLAB struct by using the following within the while loop:`propsStruct. (strrep(propKeys.nextElement,'.','_'))=propValues.nextElement;`

If you see an unexpected property value, ensure that the *java.opts* file is in the correct location and then restart MATLAB.

When MATLAB starts, several default properties (all of which can be updated in *java.opts*) are set by default, as JVM startup parameters. Here is the list of JVM startup arguments for MATLAB R2007b (7.5) on a Windows XP platform:

- -Xms64m
- -Xmx128m
- -Xss512k
- -Xshare:off
- -XX:PermSize=32M
- -XX:MaxPermSize=64M
- -XX:MaxDirectMemorySize=1200000000
- -XX:NewRatio=3
- -Dsun.java2d.noddraw=true
- -Dsun.awt.nopixfmt=true
- -Djava.library.path=C:\Program Files\MATLAB\R2007b\bin\win32
- vfprintf abort

These JVM parameters can be determined in run-time by running the JConsole or JVisualVM utilities (part of the JDK distribution), or using the following code snippet in the MATLAB Command Window (note that my JVM uses a few non-standard parameters, which shall be explained below, and in Sections 1.1 and 1.6):[110]

```
>> mXBean = java.lang.management.ManagementFactory.getRuntimeMXBean;
>> inputArgs = cell(mXBean.getInputArguments.toArray);
inputArgs =
    '-Xss512k'
    '-XX:PermSize = 32m'
    '-Xms64m'
    '-XX:NewRatio = 3'
    '-XX:MaxPermSize = 64m'
    '-Xmx128m'
    '-XX:MaxDirectMemorySize = 1200000000'
    '-Dsun.java2d.noddraw = true'
    '-Dsun.awt.nopixfmt = true'
    '-Xshare:off'
    '-Xrs'
    '-Dsun.awt.disableMixing = true'
    '-Xmx256m'
    '-Xdebug'
    '-javaagent:classmexer.jar'
    '-Djava.library.path = C:\Program Files\MATLAB\R2010b\bin\win32'
    'vfprintf'
    'abort'
```

The following *java.opts* options (in JVM 1.3+ compact format) were reported in reference to MATLAB and may be of interest to MATLAB users (some esoteric mentions omitted). Options starting with a –X are non-standard and subject to change in future JVMs; options starting with –XX are experimental and even more likely to change in future JVMs.[111] The reader is referred to the JVM documentation for additional options and details.[112] Like everything in Java, these options are case-sensitive:

```
-Xdebug
-Xrunjdwp:transport = dt_socket,address = 1044,server = y,suspend = n[113]
```

These options enable debugging Java from within MATLAB (see details in Chapter 2)

```
-Xms128m[114]
```

Increases the minimum memory limit allocated to Java (heap space) to 128 MB.[†] Other settings are possible: –Xms256m, –Xms262144k, or –Xms268435456 all set the limit to 256 MB. Note that MATLAB 7.10 (R2010a) added a preference option to set the heap size and suggests using either this new preference option or the -Xms *java.opts* option, but not both.[115]

```
-Xmx128m[116]
```

Increases the maximum memory limit allocated to Java (heap space), from the default 64 MB to 128 MB.[‡] Other settings are possible, just like for the minimum limit. Do not set this limit to more than 66% of the available physical RAM, nor to more than 256 MB, as this may cause a **MATLAB crash**.[117] Some related options are –XX:PermSize, –XX:MaxPermSize, and –XX:ThreadStackSize which are described below.

```
-XX:PermSize = 128m
-XX:MaxPermSize = 128m[118]
```

This option increases the space allotted by the JVM for permanent class (code) memory storage. MATLAB alone uses most of the default 32 MB[§] of permanent class memory, so increasing this size solves out-of-memory problems for large-code applications. Note that the XX: prefix indicates experimental JVM options which are prone to change between JVM (and therefore also MATLAB) versions.

```
-XX:-UseGCOverheadLimit[119]
```

[†] In R2007b (7.5), the default limit was 64 MB; note the following whitepaper for limit considerations: http://java.sun.com/performance/reference/whitepapers/tuning.html#section4.1.2 (or http://tinyurl.com/lfzgvf).

[‡] Since R2007b (7.5), the default limit was increased to 128 MB.

[§] In R2007a (7.4), MathWorks promised (http://www.mathworks.com/support/solutions/en/data/1-4HCPJ8/ or http://bit.ly/aQjS9j) to increase this to 64 MB and indeed XX:MaxPerSize increased to 64 MB in R2007b (7.5) although XX:PermSize remained 32 MB.

This is another experimental JVM option, which is reported to solve some memory-related MATLAB crashes by preventing the JVM from committing "suicide" if it detects a long memory-hogging operation.

```
-XX:ThreadStackSize = 8192
```

Yet another memory-related experimental JVM option, solving MATLAB crashes[120] and startup failures.[121] The setting above allots 8 MB stack size per thread. Note that this may cause unexpected problems with the Help Browser on MATLAB R14SP1.[122]

```
-Xrs
```

This option, which prevents the JVM from responding to external signals, has been found to solve several problems[123] with MATLAB components shutting down when another component exits.

```
-Djava.compiler = NONE[124]
```

This option disables the just-in-time compiler (JITC) for Java. Note that it has no effect on the MATLAB interpreter JITC. It has a similar effect to running MATLAB with the "–nojvm" command-line option. Note that this prevents many of MATLAB's GUI capabilities.[125] Unfortunately, in some cases there is no alternative. For example, when running on a remote console or when running pre-2007 MATLAB releases on Intel-based Macs.[126] In such cases, using the undocumented "-noawt" command-line option, which enables the JVM yet prevents JAVA GUI, is a suggested compromise.[127] Disabling JIT also solves a bug that may cause MATLAB to crash when debugging structs or MCOS objects.[128] Note that JITC can be dynamically turned on/off during MATLAB operation, using one of the following commands:

```
feature('accel','on/off');  % several online references by MathWorks[129]
feature('jitallow','structs','on/off');
feature('scopedAccelEnablement','on/off');

-Djava.library.path = /usr/lib/jni/[130]
-Djava.library.path = /Applications/MATLAB_R2009a.app/bin/maci/[131]
```

This option enables accessing native libraries (DLLs) by the JVM (see Section 9.5).

```
-Djava.awt.headless = true[132]
```

This option solves some problems with the Distributed Computing Toolbox (DCT) running on Macintosh machines on MATLAB R14 (7.0). Setting this property enables using the Java AWT toolkit in headless (i.e., no display device) mode.[133]

```
-Djava.net.preferIPv4Stack = true[134]
```

This is another property that solves a Macintosh R14 DCT problem.

```
-Dswing.noxp = true[135]
```

This option sets a flag that prevents Java Swing from rendering UI controls such as buttons with Windows XP styling.

```
-Dswing.aatext = true136
```

This option sets automatic antialiasing of text wherever possible.

```
-Dsun.java2d.noddraw = true137
```

Setting this option fixes a figure drawing bug in MATLAB 7.0 (that was fixed in 7.0.1). This is now the default setting in MATLAB and so need not be directly specified.

```
-Dsun.java2d.pmoffscreen = false138
```

Setting this option fixes a problem of extreme GUI slowness when launching MATLAB on a remote Linux/Solaris computer.

```
-Dmathworks.WarnIfWrongThread = false139
```

This option disables warnings about misappropriate use of Java Swing calls (not via the EDT). See additional details in Chapter 3, "Rich GUI Using Java Swing".

```
-Dmatlab.desktop.disableVirtualScreenBounds = true140
```

This property resolves a problem that crashes MATLAB versions R2006b (7.3) through R2008a (7.6) when Microsoft NetMeeting is running on a Windows system. Separate patches on the MathWorks bug page solve the problem for earlier MATLAB releases. Note: using the patch/property causes side effects on dual-monitor configurations.

```
-Dapple.laf.useScreenMenuBar = true141
```

This option fixes a Macintosh problem that displays MATLAB menus in an unexpected position, but may cause menu flickering, memory overflows, and even MATLAB crashes.

```
-Dapple.awt.graphics.UseQuartz = true142
```

This option is a workaround for GUI slowness on Macintosh OS X 10.5 (Leopard) with MATLAB R2007a (7.4) onward, also discussed on CSSM[143] and elsewhere.[144] This option is pre-included in MATLAB R2008a (7.6) onward.

```
-Dapple.awt.textantialiasing = off
-Dawt.useSystemAAFontSettings = false145
```

These options, respectively, for Apple Macintosh and Microsoft Windows, disable font anti-aliasing. This is particularly useful on small-resolution screens, large font sizes, some mono-spaced fonts, and/or large color contrast (or extremely sharp eyes, etc.).

```
-Dorg.apache.commons.logging.Log = org.apache.commons.logging.impl.SimpleLog¹⁴⁶
-Dorg.apache.commons.logging.simplelog.log.org.apache.commons.httpclient = error
```

These options solve a network connection error message in MATLAB R2008a (7.6).

```
-Dgnu.io.rxtx.SerialPorts = /dev/rfcomm0:/dev/ttySa0 (Linux)
-Dgnu.io.rxtx.SerialPorts = COM1004:COM1005:COM1006¹⁴⁷ (Windows)
```

These options enable accessing serial ports above 256 that are otherwise inaccessible.

```
-Duser.home=C:\Documents and Settings\Yair Altman\¹⁴⁸
```

This option solves a problem in the integrated MATLAB Editor on MATLAB R14 (7.0). The specified property value should be the path of the application data, which is the leading part of the path returned by MATLAB's ***prefdir*** function.

References

1. http://www.mathworks.com/help/techdoc/matlab_external/f98533.html#f122001 (or http://bit.ly/5lklN1).
2. http://developer.apple.com/library/mac/#releasenotes/Java/JavaSnowLeopardUpdate3Leopard Update8RN/NewandNoteworthy/NewandNoteworthy.html (or: http://bit.ly/aUEmlo); http://www.math works.com/matlabcentral/newsreader/view_thread/294570 (or: http://bit.ly/cVIbkY).
3. http://www.mathworks.com/matlabcentral/newsreader/view_thread/284363 (or http://bit.ly/c2662Q) used in ***winmenu.m.*** Note the equivalent functions com.mathworks.jmi.Support.*useXXX()* and son on. (see Section 9.2.2).
4. http://www.mathworks.com/help/techdoc/matlab_oop/ug_intropage.html (or http://bit.ly/bc75Cb).
5. http://www.mathworks.com/matlabcentral/answers/1082-jfreechart-pointer-for-dialdemo (or http://bit.ly/fMIZMa).
6. http://download.oracle.com/javase/6/docs/api/javax/swing/ScrollPaneLayout.UIResource.html (or http://bit.ly/gLKfLx).
7. http://java.sun.com/docs/books/jni/ or http://java.sun.com/j2se/1.5.0/docs/guide/jni/spec/types.html#wp276 (http://bit.ly/cuQAiV).
8. http://www.mathworks.com/matlabcentral/newsreader/view_thread/285933#759307 (or http://bit.ly/9FuVEo).
9. http://www.mathworks.com/matlabcentral/newsreader/view_thread/298304 (or http://bit.ly/eN1Cvi) or the related http://stackoverflow.com/questions/4376565/java-jpa-class-for-matlab (or http://bit.ly/dHPpIG) and many others.
10. http://stackoverflow.com/questions/4376565/java-jpa-class-for-matlab#comment-4771530 (or http://bit.ly/exlWkR); https://www.kitware.com/InfovisWiki/index.php/MATLAB_Titan_Toolbox#matlab_configuration (or http://bit.ly/dCC2eh).
11. For example, see http://www.mathworks.com/support/bugreports/100183 (or http://tinyurl.com/n5hr4f); http://www.mathworks.com/support/bugreports/303235 (or http://tinyurl.com/nvpjj6).
12. http://www.mathworks.com/support/bugreports/452487 (or http://tinyurl.com/m3ejan).
13. http://www.mathworks.com/matlabcentral/newsreader/view_thread/59047 (or http://tinyurl.com/2e6cst).
14. http://www.mathworks.com/matlabcentral/newsreader/view_thread/149079#375133 (or http://tinyurl.com/yo4ur5); http://www.mathworks.com/matlabcentral/newsreader/view_thread/101092 (or http://tinyurl.com/d5g394).

15. See http://www.mathworks.com/Matlabcentral/newsreader/view_thread/287891 (or http://bit.ly/bWXYLJ); also see related website: http://www.mathworks.com/Matlabcentral/newsreader/view_thread/300963 (or http://bit.ly/f28NdL).

16. http://www.mathworks.com/matlabcentral/newsreader/view_thread/163362(orhttp://tinyurl.com/29rsu7); this side effect of clearing Java references in the workspace was fixed in MATLAB 7.5 (R2007b).

17. http://www.mathworks.com/help/techdoc/matlab_external/f4873.html#f46890 (or http://tinyurl.com/2xgslo).

18. http://www.mathworks.com/help/techdoc/matlab_external/f15351.html (or http://tinyurl.com/ypffkf).

19. http://iheartmatlab.blogspot.com/2009/09/tcpip-socket-communications-in-matlab.html (or http://bit.ly/elZM8D).

20. http://blogs.mathworks.com/desktop/2010/04/26/controlling-the-java-heap-size/#comment-6991 (or http://bit.ly/9QecaC); http://stackoverflow.com/questions/2388409/how-can-i-tell-how-much-memory-a-handle-object-uses-in-matlab (or http://bit.ly/c7lFjg); http://www.javaworld.com/javaworld/javaqa/2003-12/02-qa-1226-sizeof.html (or http://bit.ly/aumv0S); http://blogs.mathworks.com/desktop/2010/04/26/controlling-the-java-heap-size/#comment-6991 (or http://bit.ly/9QecaC).

21. http://www.javamex.com/classmexer/ (or http://bit.ly/bE5zpx).

22. http://java.sun.com/j2se/1.5.0/docs/api/java/lang/instrument/package-summary.html (or http://bit.ly/bvR2xk).

23. http://www.javamex.com/tutorials/memory/instrumentation.shtml (or http://bit.ly/bldcNU).

24. http://www.mathworks.com/matlabcentral/fileexchange/17935 (or http://tinyurl.com/ytn33w).

25. http://www.mathworks.com/help/techdoc/Matlab_external/f6425.html (or http://tinyurl.com/26t7ye) and http://www.mathworks.com/help/techdoc/Matlab_external/f6671.html (or http://tinyurl.com/2yxtvs).

26. http://www.mathworks.com/matlabcentral/newsreader/view_thread/246361 (or http://bit.ly/carKgd); http://www.mathworks.com/matlabcentral/newsreader/view_thread/281221 (or http://bit.ly/dsynlK).

27. http://java.sun.com/javase/6/docs/api/javax/swing/ActionMap.html (or http://bit.ly/9xgDib).

28. http://java.sun.com/javase/6/docs/api/javax/swing/Action.html (or http://bit.ly/3DhUbx).

29. http://www.javalobby.org/java/forums/t19402.html (or http://bit.ly/c605Ng); http://java.sun.com/docs/books/tutorial/uiswing/misc/action.html (or http://bit.ly/3MXZgi); http://java.sun.com/docs/books/tutorial/uiswing/misc/keybinding.html (or http://bit.ly/agtTK3).

30. http://UndocumentedMatlab.com/blog/spicing-up-matlab-uicontrol-tooltips/#comment-1173 (or http://bit.ly/5oFn8M).

31. http://UndocumentedMatlab.com/blog/matlab-callbacks-for-java-events/ (or http://bit.ly/hlaBBH).

32. http://www.mathworks.com/matlabcentral/newsreader/view_thread/272321 (or http://bit.ly/9bHagh); download *EventTest.zip*, which includes source and class files, from http://UndocumentedMatlab.com/files/EventTest.zip (or http://bit.ly/bKWLtw).

33. http://UndocumentedMatlab.com/blog/hgfeval/ (or http://bit.ly/aIgaOa).

34. Donn Shull, private communication.

35. http://www.mathworks.com/matlabcentral/newsreader/view_thread/274961#722084 (or http://bit.ly/bVHqoL).

36. http://java.sun.com/javase/downloads/widget/jdk6.jsp (http://bit.ly/cqFMkW), useful on MATLAB R2007b (7.5) onward.

37. http://eclipse.org/downloads/ (or http://bit.ly/95wdV4).

38. http://netbeans.org/downloads/ (or http://bit.ly/cISU1K).

39. http://www.medsch.wisc.edu/~milenkvc/pdf/javaproxy.htm (or http://tinyurl.com/d44qpq).

40. http://www-math.mit.edu/~spielman/ECC/javaMatlab.html (or http://tinyurl.com/ccvwya).

41. An excellent place to start is http://java.sun.com/javase/6/webnotes/trouble/ (or http://bit.ly/97CA0G).

42. http://www.mathworks.com/matlabcentral/newsreader/view_thread/240212 (or http://tinyurl.com/dca5wu).

43. http://xtargets.com/snippets/posts/show/48; Brad's website is currently offline, but cached (archived) versions of this Web page can be found online, for example, http://bit.ly/9taZaZ.

44. http://www.mathworks.com/matlabcentral/newsreader/view_thread/155261#633890 (or http://bit.ly/71fDNa).

45. JDWP, which is part of the Java Platform Debugging Architecture (JPDA) framework, is described here: http://java.sun.com/javase/technologies/core/toolsapis/jpda/ (or http://bit.ly/akdNIA); the relevant -Xrunjdwp options are described here: http://java.sun.com/javase/6/docs/technotes/guides/jpda/conninv.html#jdwpoptions (or http://bit.ly/cihXXj).

46. http://www.mathworks.com/support/solutions/en/data/1-OVU1L (or http://tinyurl.com/nvzhms).

47. http://www.mathworks.com/matlabcentral/newsreader/view_thread/155261#404006(orhttp://bit.ly/4EL9Re), edited.
48. http://www.mathworks.com/matlabcentral/newsreader/view_thread/262737#697025 (or http://bit.ly/cCLKrA).
49. http://www.eclipse.org/tptp/home/documents/tutorials/profilingtool/profilingexample_32.html (or http://bit.ly/dmPKcT).
50. http://profiler.netbeans.org/ (or http://bit.ly/cztYQz).
51. http://blogs.sun.com/watt/resource/jvm-options-list.html (or http://bit.ly/bsWNcV).
52. http://prefetch.net/blog/index.php/2008/02/02/profiling-java-methods-with-the-heap-profiling-agent/ (or http://bit.ly/bLZ2l1).
53. http://www.mathworks.com/matlabcentral/newsreader/view_thread/246361 (or http://bit.ly/carKgd).
54. http://www.mathworks.com/matlabcentral/fileexchange/23272 (or http://bit.ly/ceE7x2).
55. http://www.mathworks.com/matlabcentral/fileexchange/23275 (or http://bit.ly/cuo851); JPRofiler is also used by MathWorks: http://blogs.mathworks.com/desktop/2009/08/17/calling-java-from-matlab/#comment-7107 (or http://bit.ly/bpjCGC).
56. http://www.mathworks.com/matlabcentral/fileexchange/23513 (or http://bit.ly/9gp2OT).
57. http://blogs.sun.com/watt/resource/jvm-options-list.html (or http://bit.ly/bsWNcV); http://java.sun.com/javase/technologies/hotspot/vmoptions.jsp (or http://bit.ly/bMiAAm); http://docs.sun.com/app/docs/doc/806-7930/6jgp65iki?a=view (or http://bit.ly/aMAKh0).
58. http://www.mathworks.com/support/solutions/en/data/1-3L4JU7 (or http://bit.ly/bpCztL).
59. http://www.khelekore.org/jmp/ (or http://bit.ly/coj2xm).
60. http://www.khelekore.org/jmp/tijmp/ (or http://bit.ly/dkrUCi).
61. http://www.javamex.com/classmexer/ (or http://bit.ly/bE5zpx).
62. http://www.mathworks.com/matlabcentral/newsreader/view_thread/296813#797410 (or http://bit.ly/cJ8m7c).
63. For example, the open-source FindBugs utility: http://findbugs.sourceforge.net/ (or http://bit.ly/agOjYw).
64. http://java.sun.com/j2se/1.5.0/docs/guide/language/generics.html (or http://tinyurl.com/6nfhp).
65. http://java.sun.com/javase/6/docs/api/java/awt/TrayIcon.MessageType.html (or http://tinyurl.com/cumm3h).
66. http://www.mathworks.com/matlabcentral/answers/15711-how-do-i-access-a-java-inner-class-from-matlab (or http://bit.ly/nX1adr).
67. http://java.sun.com/javase/6/docs/api/java/nio/channels/FileChannel.MapMode.html (or http://bit.ly/chLZYD).
68. http://www.mathworks.com/matlabcentral/newsreader/view_thread/286165#759993 (or http://bit.ly/dAq8KX).
69. http://www.mathworks.com/help/techdoc/matlab_external/f98533.html#f122001 (or http://bit.ly/5lklN1).
70. http://www.mathworks.com/matlabcentral/newsreader/view_thread/44363 (or http://bit.ly/97GcJH).
71. http://www.mathworks.com/matlabcentral/newsreader/view_thread/288627 (or http://bit.ly/9ro58r).
72. http://www.mathworks.com/support/solutions/en/data/1-1A75W (or http://bit.ly/bjXNno); similarly for other releases.
73. http://www.mathworks.com/support/solutions/en/data/1-78HLNI/ (or http://bit.ly/cb8OYl). MathWorks' Brian Arnold explained the details in http://www.mathworks.com/matlabcentral/newsreader/view_thread/161842#408891 (or http://bit.ly/bbdNdP).
74. http://robert.scullin.name/blog/index.php/2009/09/09/install-activate-matlab-in-snow-leopard/ (or http://bit.ly/aYrquA); http://www.mathworks.com/support/solutions/en/data/1-8GS5S1/ (or http://bit.ly/aia2v6).
75. http://www.mathworks.com/matlabcentral/newsreader/view_thread/266889 (or http://bit.ly/5dqhpG); http://www.mathworks.com/matlabcentral/newsreader/view_thread/258941 (or http://bit.ly/cXoqYp).
76. http://www.mathworks.com/support/solutions/en/data/1-1812J (or http://bit.ly/dbFVqh).
77. http://www.mathworks.com/support/solutions/en/data/1-78HLNI/ (or http://bit.ly/cb8OYl).
78. http://www.mathworks.com/support/solutions/en/data/1-1A2L7 (or http://tinyurl.com/nowqmc); http://www.mathworks.com/support/solutions/en/data/1-1881O (or http://tinyurl.com/m5uv4f); http://blogs.mathworks.com/desktop/2009/08/31/pouncing-on-snow-leopard/#comment-6710 (or http://bit.ly/4FvyLD); http://www.mathworks.com/support/bugreports/452486 (or http://bit.ly/d2OYIy) and more than a dozen others: http://www.mathworks.com/support/bugreports/search_results?search_executed=1&keyword=JRE&release=0 (or http://bit.ly/9LGyf2).
79. http://www.mathworks.com/support/solutions/en/data/1-1A2HO (or http://tinyurl.com/kqvgkq).

80. http://www.mathworks.com/matlabcentral/newsreader/view_thread/157118#395347 (or http://tinyurl.com/3c3sb8); http://www.mathworks.com/matlabcentral/newsreader/view_thread/242556#630627 (or http://tinyurl.com/c3ym6m); http://www.mathworks.com/matlabcentral/newsreader/view_thread/161248# 651824 (or http://tinyurl.com/pwosf8); http://www.mathworks.com/matlabcentral/newsreader/view_ thread/242556#658743 (or http://tinyurl.com/nevkx2).

81. http://social.technet.microsoft.com/Forums/en-US/w7itproappcompat/thread/4dba5d57-3127-48f3-9461-eb1ef5d7c70e (or http://bit.ly/czpDyM); http://recluze.wordpress.com/2009/07/30/matlab-7-under-windows-7/ (or http://bit.ly/diqHvm); http://www.mathworks.com/matlabcentral/newsreader/ view_thread/272598#763011 (or http://bit.ly/aZuf4S); http://www.mathworks.com/matlabcentral/ newsreader/view_thread/295889#799946 (or http://bit.ly/fbecCw).

82. http://www.mathworks.com/matlabcentral/newsreader/view_thread/266480 (or http://tinyurl.com/yhp6erz).

83. http://www.mathworks.com/matlabcentral/newsreader/view_thread/244727#773600 (or http://bit. ly/983O1l).

84. http://www.mathworks.com/matlabcentral/newsreader/view_thread/161248 (or http://bit.ly/bcSrlN).

85. http://www.mathworks.com/matlabcentral/newsreader/view_thread/268395#702321 (or http://bit. ly/8JBBVM); http://www.mathworks.com/matlabcentral/newsreader/view_thread/164917#599155 (or http://bit.ly/bKrl3E).

86. http://java.sun.com/javase/6/docs/technotes/guides/collections/changes6.html (or http://tinyurl.com/5ea3b5).

87. http://java.sun.com/developer/technicalArticles/J2SE/Desktop/javase6/systemtray (or http://tinyurl.com/ cbup8c).

88. http://www.devx.com/Java/Article/30722 or http://java.sun.com/javase/6/webnotes/features.html or http://java.sun.com/javase/6/features.jsp

89. http://java.sun.com/javase/6/webnotes/ReleaseNotes.html (or http://bit.ly/bKTibY).

90. http://www.mathworks.com/matlabcentral/newsreader/view_thread/155261#717300 (http://bit.ly/9tJ9Ht).

91. http://www.mathworks.com/support/bugreports/495091 (or http://tinyurl.com/mnsdra); http://www. mathworks.com/support/bugreports/495103 (or http://tinyurl.com/ndupwx); http://www.mathworks.com/ matlabcentral/newsreader/view_thread/245448 (or http://tinyurl.com/c2j24n); http://www.mathworks.com/ matlabcentral/newsreader/view_thread/254319 (or http://tinyurl.com/n5od2n); http://www.mathworks.com/ matlabcentral/newsreader/view_thread/272205 (or http://bit.ly/cw4Kqm) and several others.

92. http://www.mathworks.com/matlabcentral/newsreader/view_thread/268395 (or http://bit.ly/4zQgpY).

93. http://www.mathworks.com/matlabcentral/newsreader/view_thread/260767#716249 (or http://bit.ly/ aw6lgl); also see http://www.mathworks.com/matlabcentral/newsreader/view_thread/263909#718604 (or http://bit.ly/cZOQvh).

94. http://www.mathworks.com/matlabcentral/newsreader/view_thread/309912 (or: http://bit.ly/n2uPCx).

95. For example, http://support.apple.com/kb/DL972 (or http://bit.ly/63elim); http://blogs.mathworks.com/ desktop/2009/08/31/pouncing-on-snow-leopard/#comment-6710 (or http://bit.ly/4FvyLD).

96. http://www.mathworks.com/matlabcentral/newsreader/view_thread/162064 (or http://bit.ly/9rmR5h); http:// www.mathworks.com/matlabcentral/newsreader/view_thread/167685 (or http://bit.ly/bMfNrV); http://www. mathworks.com/support/solutions/en/data/1-5XUP9M/ (or http://bit.ly/a6G0H6); http://www.mathworks. com/support/solutions/en/data/1-9A6FYK/ (or http://bit.ly/aEqzPU) and the very informative http://blogs. mathworks.com/desktop/2010/02/08/starting-matlab-from-the-os-x-dock/ (http://bit.ly/bZxV4m). Also see the following undocumented hint regarding removal of a useless MATLAB icon from the Mac Dock bar: http://www.macosxhints.com/article.php?story=20080212162806562 (or http://bit.ly/9lPJDc).

97. http://www.mathworks.com/matlabcentral/newsreader/view_thread/290935#778666 (or http://bit.ly/ c3UR7f).

98. http://www.mathworks.com/matlabcentral/newsreader/view_thread/160076 (or http://tinyurl. com/27tmxy).

99. http://www.mathworks.com/matlabcentral/newsreader/view_thread/271127 (or http://bit.ly/93SIzC).

100. For example, http://www.mathworks.com/matlabcentral/newsreader/view_thread/269553 (or http://bit. ly/72K4M3).

101. http://www.mathworks.com/matlabcentral/newsreader/view_thread/69629 (or http://bit.ly/co0QrW).

102. http://www.mathworks.com/support/solutions/en/data/1-18I2C/ (or http://bit.ly/86OW7J).

103. http://www.mathworks.com/support/solutions/en/data/1-1KY3U1 (or http://tinyurl.com/mfge8w).

104. http://www.mathworks.com/support/solutions/en/data/1-72H2IS (or http://tinyurl.com/lzlm2n).

105. http://blogs.mathworks.com/desktop/2009/07/06/calling-java-from-matlab/#comment-6609 (or http://tinyurl.com/yg9kucw).
106. For example, http://www.mathworks.com/matlabcentral/newsreader/view_thread/266104 (or http://tinyurl.com/y8aztll); http://blogs.mathworks.com/desktop/2009/07/06/calling-java-from-matlab/#comment-6604 (or http://tinyurl.com/ydhapy9).
107. http://www.mathworks.com/matlabcentral/newsreader/view_thread/154954 (or http://tinyurl.com/2fn2ys).
108. http://blogs.sun.com/watt/resource/jvm-options-list.html (or http://tinyurl.com/2r2hff). See http://java.sun.com/javase/6/docs/technotes/guides/2d/flags.html for a specific list of Java2D options on JVM 1.6 (there are similar pages on the java.sun.com website for earlier JVM releases).
109. http://www.mathworks.com/support/bugreports/452487 (or http://tinyurl.com/m3ejan).
110. http://technology.amis.nl/blog/4214/accessing-jvm-arguments-from-java-to-determine-if-jvm-is-running-in-debug-mode (or http://bit.ly/9GMKsB).
111. http://java.sun.com/javase/technologies/hotspot/vmoptions.jsp (or http://tinyurl.com/2dx4mz).
112. http://java.sun.com/javase/6/docs/technotes/tools/windows/java.html (or http://tinyurl.com/32yvlk) — similar Web pages are also available for non-Windows and earlier JVM versions.
113. http://www.mathworks.com/matlabcentral/newsreader/view_thread/155261#404006 (or http://tinyurl.com/yq9g2h).
114. http://www.mathworks.com/support/solutions/en/data/1-18I2C (or http://tinyurl.com/nsjbce); http://www.mathworks.com/matlabcentral/newsreader/view_thread/252354 (or http://bit.ly/dDhasJ); http://www.mathworks.com/matlabcentral/newsreader/view_thread/275348 (or http://bit.ly/bvR7lT).
115. http://www.mathworks.com/help/techdoc/rn/bsct_ou-1.html#bseyszd-1 (or http://bit.ly/9XcSPJ).
116. http://www.mathworks.com/support/solutions/en/data/1-11D2U1 (or http://tinyurl.com/n74v64); http://www.mathworks.com/support/solutions/en/data/1-6K8OPU (or http://tinyurl.com/kupgmq); http://www.mathworks.com/support/bugreports/273783 (or http://tinyurl.com/mb2uo2).
117. http://www.mathworks.com/support/bugreports/398525 (or http://tinyurl.com/n76wj4).
118. http://www.mathworks.com/support/solutions/en/data/1-20FV2Z (or http://tinyurl.com/noyw8j); http://www.mathworks.com/support/solutions/en/data/1-4HCPJ8 (or http://tinyurl.com/m7an94).
119. http://www.mathworks.com/support/solutions/en/data/1-8G7XG1 (or http://tinyurl.com/na2jnu).
120. http://www.mathworks.com/support/solutions/en/data/1-19Z14 (or http://tinyurl.com/kpdk7m).
121. http://www.mathworks.com/support/bugreports/251291 (or http://tinyurl.com/m5vawn).
122. http://www.mathworks.com/support/solutions/en/data/1-13HCX7 (or http://tinyurl.com/lx7ln3).
123. For example, http://www.mathworks.com/support/solutions/en/data/1-30EYL4 (or http://tinyurl.com/m3cadf); http://www.mathworks.com/support/solutions/en/data/1-XW8A0 (or http://tinyurl.com/n3l4ow).
124. http://www.mathworks.com/support/solutions/en/data/1-1IMUDO (or http://tinyurl.com/nzeswo).
125. http://www.mathworks.com/help/techdoc/rn/bropbi9-1.html#brubkzc-1 (or http://bit.ly/c3jo3A).
126. http://www.mathworks.com/support/faq/macintel.html (or http://bit.ly/cWFqnX).
127. For example (Macs again), http://www.mathworks.com/matlabcentral/newsreader/view_thread/145515#373700 (or http://bit.ly/9VuSJm); http://www.mathworks.com.au/matlabcentral/newsreader/view_thread/239001#610205 (or http://bit.ly/aBHLeQ).
128. http://www.mathworks.com/support/bugreports/595677 (or http://bit.ly/cmsPHG).
129. http://www.mathworks.com/matlabcentral/newsreader/view_thread/76874#195833 (or http://bit.ly/cseyjE); http://www.mathworks.com/support/solutions/en/data/1-21E4Y8/ (or http://bit.ly/cseyjE).
130. http://www.mathworks.com/matlabcentral/newsreader/view_thread/247047#636162 (or http://tinyurl.com/cdkdhf); http://www.mathworks.com/matlabcentral/newsreader/view_thread/246481#636161 (or http://tinyurl.com/cajgq7).
131. http://www.mathworks.com/matlabcentral/newsreader/view_thread/247097 (or http://tinyurl.com/c9t74t).
132. http://www.mathworks.com/support/bugreports/249088 (or http://tinyurl.com/mryvoz); http://www.mathworks.com/support/bugreports/240836 (or http://tinyurl.com/otmx6a).
133. http://java.sun.com/developer/technicalArticles/J2SE/Desktop/headless/ (or http://tinyurl.com/3qwch6); http://java.sun.com/j2se/1.4.2/docs/guide/awt/AWTChanges.html#headless (or http://tinyurl.com/freza).
134. http://www.mathworks.com/support/bugreports/249097 (or http://tinyurl.com/lmew8q).
135. http://www.mathworks.com/support/bugreports/194025 (or http://tinyurl.com/mjoatj).
136. http://www.javalobby.org/forums/thread.jspa?forumID=61&threadID=14179 (or http://bit.ly/4CEWTC).
137. http://www.mathworks.com/support/solutions/archived/1-1BYDY.html (or http://tinyurl.com/yvsho2).
138. http://www.mathworks.com/matlabcentral/newsreader/view_thread/160387 (or http://tinyurl.com/bxfj4p).

139. http://www.mathworks.com/matlabcentral/newsreader/view_thread/156388#400906 (or http://tinyurl.com/2yqyzn).
140. http://www.mathworks.com/support/bugreports/303235 (or http://tinyurl.com/nvpjj6).
141. http://www.mathworks.com/support/bugreports/230485 (or http://tinyurl.com/lzhgg3).
142. http://www.mathworks.com/support/bugreports/412219 (or http://tinyurl.com/qvx5wb); http://www.mathworks.com/support/solutions/en/data/1-31CIOM (or http://tinyurl.com/kpz5mh); http://blogs.mathworks.com/loren/2007/02/08/string-annotations-for-plots/#comment-32288 (or http://bit.ly/jTpTak); http://www.mathworks.com/matlabcentral/newsreader/view_thread/309912#847852 (or http://bit.ly/qTxMDl).
143. http://www.mathworks.com/matlabcentral/newsreader/view_thread/160534 (or http://tinyurl.com/2b4nxv), http://www.mathworks.com/matlabcentral/newsreader/view_thread/144169#398840 (or http://tinyurl.com/yueky5), and so on.
144. http://www.macosxhints.com/article.php?story=2007112503413229 (or http://tinyurl.com/4b97rv); http://blogs.mathworks.com/desktop/2009/01/05/happy-new-year-2/#comment-6340 (or http://tinyurl.com/neuf5z).
145. http://blogs.mathworks.com/desktop/2007/03/16/introducing-the-desktop-blog/#comment-2135 (or http://bit.ly/b5SrYQ); http://www.mathworks.com/support/bugreports/404319 (or http://tinyurl.com/272bf2) — this page was removed by MathWorks; http://www.mathworks.com/support/solutions/en/data/1-5AZS7W (or http://tinyurl.com/n8x6w6).
146. http://www.mathworks.com/support/bugreports/450681 (or http://tinyurl.com/lehofw).
147. http://www.mathworks.com/support/solutions/en/data/1-6LEG2H (or http://tinyurl.com/n7r9ct); http://www.mathworks.com/support/solutions/en/data/1-3ZT8GP (or http://tinyurl.com/km94xk).
148. http://www.mathworks.com/support/bugreports/234906 (or http://tinyurl.com/lfsgqh).

Chapter 2

Using Non-GUI Java Libraries in MATLAB®

As noted earlier, Java has been available in MATLAB since R12 (MATLAB 6.0). Integration of non-GUI Java classes within MATLAB has always been relatively easy to use. Support for integration of Java GUI components within MATLAB figures has been added in R14 SP3 (MATLAB 7.1).

This chapter covers the programmatic (non-GUI) aspects of these integrations, using some useful Java libraries in our MATLAB code. Chapter 3 (Rich GUI Using Java Swing) shows how we can use the Java GUI integration to enrich plain-vanilla MATLAB GUI.

It is interesting to note that non-GUI integration of Java classes is fully documented and supported in MATLAB,[1] whereas the newer GUI integration is still officially unsupported. Possible reasons for this discrepancy will be presented in Chapter 3.

While being fully documented, in general, even non-GUI Java integration sometimes exhibits undocumented aspects, as will be seen in Section 2.2 (Database Connectivity).

MATLAB itself uses undocumented features, for example, to implement its Database Toolbox. However, whenever any MATLAB release changes break such features, MathWorks is careful to distribute Toolbox releases that work well with the new behavior. When designing MATLAB applications that similarly rely on such undocumented aspects, care should be taken to retest the application with new MATLAB releases, and release application updates, just like MathWork's toolboxes.

2.1 Complex Data Structures

This section and the following one (Database Connectivity) provide detailed examples showing how built-in Java functionality can greatly increase MATLAB programming productivity and the resulting program power. These are simply two examples of the enormous potential Java has for MATLAB applications. While not really "undocumented" in the true sense (most of the material is actually documented separately in MATLAB and Java), in practice, such Java gems are unfortunately often overlooked by MATLAB programmers. It is hoped that Chapter 1 has made us feel at ease with the MATLAB–Java interface, enough to take advantage of the available power of Java.

2.1.1 Java Collections

Java contains a wide variety of predefined data structures (specifically Collections and Maps), which can easily be adapted to fit most programming tasks. It is unfortunate that the MATLAB programming language does not contain similar predefined collection types, apart from its basic cell, array and struct elements. MATLAB R2008b (7.7) also added ***containers.Map***, which is a much-scaled-down MATLAB version of the `java.util.Map` interface, but is a step in the right direction. Some MATLAB programmers prepared their own implementations of data structures, which can be found on the File Exchange.[2]

However, this limitation of MATLAB can easily be overcome with Java's out-of-the-box set of predefined classes, as described below. Java collections have many advantages over hand-coded MATLAB equivalents, in addition to the obvious time saving: Java's classes are extensively debugged and performance-tuned, which is especially important when searching large collections. Also, these classes provide a consistent interface, are highly configurable and extendable, enable easy cross-type interoperability, and generally give MATLAB programmers the full power of Java's collections without needing to program the nuts-and-bolts.

To start using Java collections, readers should first be familiar with them. These classes are part of the core Java language and are explained in any standard Java programming textbook. The official java.sun.com website provides a detailed online tutorial[3] about these classes, their usage and differences, in addition to a detailed reference of these classes. JVM 1.6, included in MATLAB since the R2007b (7.5) release, significantly enlarged the Collection Framework compared to JVM 1.5 (used by MATLAB releases 7.0.4–7.4). It is therefore important to choose the reference version that is relevant for your particular JVM.[4] A detailed list of enhancements added to the different JVM versions can also be found on the java.sun.com website.[5]

Java Collections include interfaces and implementation classes. As of MATLAB R2011a (7.12), Java interfaces cannot be used directly — only the implementation classes. Of the many Collection classes, the following are perhaps most useful (all classes belong to the `java.util` package, unless otherwise noted):

- `Set`:[6] an interface that is implemented by classes characterized by their prevention of duplicate elements. Some notable implementation classes:
  - `EnumSet`:[†] stores same-type enumerated values; this is the best-performing `Set`.
  - `HashSet`: stores elements in a hash table and is the second fastest `Set`.
  - `LinkedHashSet`: stores elements in a hash table whose elements are linked based on insertion order; slightly slower than `HashSet`.
  - `TreeSet`: stores elements in a value-ordered balanced (red-black) tree; much slower than `HashSet`.

- `List`:[7] an interface that is implemented by classes characterized by ordered elements (aka *sequences*), which may be duplicates of each other and accessed based on their exact position within the `List`. Specially optimized internal algorithms[8] enable sorting, shuffling, reversing, rotating, and other modifications of the `List`. Some notable implementation classes are as follows:
  - `Vector`: stores elements in a growable `Set`.
  - `Stack`: a `Vector` specialization subclass, which implements last-in-first-out (LIFO) behavior — compare to the `Queue` class below.[9]
  - `LinkedList`: stores elements in a list whose elements are also linked, enabling usage as a stack, queue, or double-ended queue (`Deque`).[10]

- `Queue`:[11] an interface which is implemented by classes designed for holding elements prior to processing, in an ordered list accessible only at one (=*head*) or two (*head* and *tail*) positions. All classes include specialized insertion, extraction, and inspection methods. Some notable implementation classes:

[†] Available only since JVM 1.5 (i.e., MATLAB 7.0.4, R14SP2).

- LinkedList: a List which is also a Stack, Queue, and Deque.
- ArrayDeque:[†] stores elements in a resizable array which is accessible in either LIFO (=stack) or as FIFO (=queue) mode, hence its name (DEQUE=**D**ouble-**E**nded **QUE**ue).
- PriorityQueue:[‡] a queue whose elements are dynamically sorted based on their priority value.
- java.util.concurrent.ArrayBlockingQueue:[§] stores elements in a fixed-capacity FIFO queue.
- java.util.concurrent.SynchronousQueue:[¶] enables synchronization between producer and consumer threads (=rendezvous mechanism).
- java.util.concurrent.DelayQueue:[††] a queue whose elements are stored for a minimal time duration before being accessible for extraction.

- Map:[12] an interface which is implemented by classes characterized by elements of unique keys paired with associated values. Early Java versions used the java.util. Dictionary abstract superclass, but this was subsequently replaced by the java. util.Map interface class. Maps contain specialized algorithms for fast retrieval based on a supplied key. Some of the notable implementation classes:
  - EnumMap:[‡‡] stores same-type enumerated keys, similar to EnumSet.
  - HashMap: stores elements in unsorted hash table, similar to HashSet.
  - Hashtable: synchronized (thread-safe) null-key-enabled HashMap.
  - TreeMap: stores elements in a value-ordered balanced (red-black) tree, similar to TreeSet.
  - LinkedHashMap: stores elements in a hash table whose elements are linked based on insertion order, similar to LinkedHashSet.

It should be noted that MATLAB R2008b (7.7)'s new ***containers.Map*** class is a scaled-down MATLAB version of the java.util.Map interface. It has the added benefit of seamless integration with all MATLAB types (unlike Java Collections — see below), as well as the ability since MATLAB 7.10 (R2010a) to specify data types.[13] Serious MATLAB implementations requiring key-value maps/dictionaries should still use Java's Map classes to gain access to their larger functionality if not performance. MATLAB versions earlier than R2008b have no real alternative in any case and must use the Java classes. The reader may also be interested to examine pure-MATLAB object-oriented (class-based) Hashtable implementations, which is available on the File Exchange.[14]

[†] Available only since JVM 1.6 (i.e., MATLAB 7.5, R2007b).
[‡] Available only since JVM 1.5 (i.e., MATLAB 7.0.4, R14SP2).
[§] Available only since JVM 1.5 (i.e., MATLAB 7.0.4, R14SP2).
[¶] Available only since JVM 1.5 (i.e., MATLAB 7.0.4, R14SP2).
[††] Available only since JVM 1.5 (i.e., MATLAB 7.0.4, R14SP2).
[‡‡] Available only since JVM 1.5 (i.e., MATLAB 7.0.4, R14SP2).

A potential limitation of using Java Collections is their inability to contain nonprimitive MATLAB types such as structs.[15] To overcome this, either down-convert the types to some simpler type (using ***struct2cell*** or programmatically) or create a separate Java object that will hold the information and store this object in the Collection.

Many additional Collection classes offer implementation of specialized needs. For example, `java.util.concurrent.LinkedBlockingDeque` implements a `Queue`, which is also a `LinkedList`, is a double-ended queue (`Deque`), and is blocking (meaning that extraction operations will block until at least one element is extractable).

All the Java Collections have intentionally similar interfaces, with additional methods specific to each implementation class based on its use and intent. Most Collections implement the following common self-explanatory methods (simplified interface):

```
int size()
int hashCode()
boolean isEmpty()
boolean contains(Object element)
boolean containsAll(Collection c)
Iterator iterator()
boolean add(Object element)
boolean remove(Object element)
boolean addAll(Collection c)
boolean removeAll(Collection c)
boolean retainAll(Collection c)
void clear()    % no return value
Object clone()
Object[] toArray()
String toString()
```

The full list of supported methods in a specific Collection class can, as any other Java object/ class, be inspected using MATLAB's ***methods*** or ***methodsview*** functions:

```
>> methods('java.util.Hashtable')
Methods for class java.util.Hashtable:
Hashtable      containsKey      equals      isEmpty     notifyAll      size
clear          containsValue    get         keyset      put            toString
clone          elements         getClass    keys        putAll         values
contains       entrySet         hashCode    notify      remove         wait
```

2.1.2 Collections Example: Hashtable

A detailed MATLAB example that utilizes Hashtable[16] for a phonebook application is detailed in MATLAB's External Interface/Java section.[†] The following code snippet complements that example by displaying some common characteristics of Collections:

```
>> hash = java.util.Hashtable;
>> hash.put('key #1','myStr');
```

[†] Actually, `java.util.Properties`, which is a subclass of `java.util.Hashtable`.

```
>> hash.put('2nd key',magic(3));
>> disp(hash)   % same as: hash.toString
{2nd key=[[D@59da0f, key #1 = myStr}
>> disp(hash.containsKey('2nd key'))
      1                          % = true
>> disp(hash.size)
      2


>> disp(hash.get('key #2'))    % key not found
      []


>> disp(hash.get('key #1'))    % key found and value retrieved
myStr

>> disp(hash.entrySet) % java.util.Collections$SynchronizedSet object
[2nd key =[[D@192094b, key #1 = myStr]
>> entries = hash.entrySet.toArray
entries =
java.lang.Object[]:
    [java.util.Hashtable$Entry]
    [java.util.Hashtable$Entry]
>> disp(entries(1))
2nd key=[[D@192094b
>> disp(entries(2))
key #1 = myStr


>> hash.values  % a java.util.Collections$SynchronizedCollection
ans =
[[[D@59da0f, myStr]
>> vals = hash.values.toArray
vals =
java.lang.Object[]:
    [3 x 3 double]
    'myStr'
>> vals(1)
ans =
      8      1      6
      3      5      7
      4      9      2
>> vals(2)
ans =
myStr
```

2.1.3 *Enumerators*

Note that Java *Iterators* (aka *Enumerators*), such as those returned by the *hash.keys()* method, are temporary memory constructs. A common pitfall is to directly chain such constructs. While legal from a syntax viewpoint, this would produce results that are repetitive and probably unintended, as the following code snippet shows:

```
>> hash.keys
ans =
java.util.Hashtable$Enumerator@7b1d52    <= enumerator reference
>> hash.keys
ans =
java.util.Hashtable$Enumerator@127d1b4   <= new reference object

>> disp(hash.keys.nextElement)
2nd key    <= 1st key enumerated in the hash
>> disp(hash.keys.nextElement)
2nd key    <= same key returned, because of the new enumeration obj
>> % Wrong way: causes an endless loop since hash.keys regenerates
>> % ^^^^^^^^^^ so hash.keys.hasMoreElements is always true
>> while hash.keys.hasMoreElements, doAbc();  end   % endless loop

>> % Correct way: store the enumerator in a temporary variable
>> hashKeys = hash.keys;
>> while hashKeys.hasMoreElements,  doAbc();  end

>> hash.keys.methods
Methods for class java.util.Hashtable$Enumerator:
equals             hasNext        nextElement     remove
getClass           hashCode       notify          toString
hasMoreElements    next           notifyAll       wait

>> % And similarly for ArrayList iterators:
>> jList = java.util.ArrayList;
>> jList.add(pi); jList.add('text'); jList.add(magic(3)); disp(jList)
[3.141592653589793, text, [[D@1c8f959]
>> iterator = jList.iterator
iterator =
java.util.AbstractList$Itr@1ab3929

>> disp(iterator.next); fprintf('hasNext: %d\n',iterator.hasNext)
         3.14159265358979
hasNext: 1
>> disp(iterator.next); fprintf('hasNext: %d\n',iterator.hasNext)
text
hasNext: 1

>> disp(iterator.next); fprintf('hasNext: %d\n',iterator.hasNext)
     8     1     6
     3     5     7
     4     9     2
hasNext: 0
```

2.2 Database Connectivity

Interfacing MATLAB to a database can be done in many different manners: we can purchase and use MathWorks' Database Toolbox,[17] use a third-party solution (many of which are free),[18] use one of numerous utilities on the MATLAB File Exchange,[19] use an ActiveX COM server,[20] use a proprietary ODBC driver, or use Java DataBase Connectivity[21] (JDBC).[22] Many databases offer Java connectors or drivers that easily integrate with MATLAB.[23] The MATLAB File Exchange also contains some interesting tutorials and demos for accessing databases from MATLAB.[24]

Any of these methods may be used: it is specifically NOT necessary to purchase the expensive Database Toolbox to connect databases to MATLAB. This section will only detail the Java-based JDBC approach.

2.2.1 *Using Java Database Connectivity (JDBC) in MATLAB*

JDBC is a Java-accessed driver with a standard interface and functionality.[25] The basic concept is that any Java program can access any data storage (database, spreadsheet, data file, etc.) which has a JDBC driver, in exactly the same manner (*Interface*), using exactly the same code, as any other data storage.

Since MATLAB has access to Java functionality, it is therefore very easy to connect any MATLAB program to any database which has a JDBC driver. Nowadays, it is very difficult to find a database which does NOT have a JDBC driver,[26] and by extension which *cannot* be accessed from MATLAB.

In practice, even if we cannot find a JDBC driver, we can always use the database's ODBC driver with Java's built-in JDBC–ODBC bridge, which wraps the ODBC driver in a JDBC interface.[27]

MATLAB's Database Toolbox, as well as some of the relevant submissions on the File Exchange, use JDBC at its core, and wrap the basic JDBC calls with lots of code that handles error checking, parameter passing, function wrapping, and other such programming tasks that may save the programming time of the user.

MATLAB users who are not proficient in JDBC, or who are programming a large database-driven application, may find the extra cost of the Database Toolbox cost-effective compared with the time it would take to wrap all the basic JDBC calls in MATLAB.

It should be noted that while database access is possible and relatively easy using the JDBC interface described below, there are many nooks and crannies which are beyond the scope of this book. Interested readers are referred to the official JDBC tutorial,[28] JDBC books,[29] or to other online resources.[30]

As noted previously, users are advised to use the reference version suitable for their MATLAB release. Most online references given in this book are for the JVM 1.6 version (suitable for MATLAB release R2007b (7.5) onward). References for other JVMs can easily be inferred by a simple URL modification, described in this[31] reference note.

Let us now examine the basic tasks of connecting to an existing database, reading information, and updating data. For more advanced database functionality, the reader is referred to the additional resources mentioned above.

2.2.2 Initializing the JDBC Driver

Many databases provide a readymade Java connector that simplifies the task of connecting to the database. For example, MySQL provides a list of such connectors,[32] and a recently submitted utility on the MATLAB File Exchange uses one of these (Connector/J) for creating a MATLAB connector.[33] Whenever possible, I advise using such readymade connectors rather than programming the database connectivity from scratch. While programming is fun and informative, it may take some time to get all the pieces and edge-cases working properly.

In general, database tasks are composed of four JDBC subtasks:

- Initializing the JDBC driver (a one-time process)
- Setting up the application's *Connection* to the database (a one-time process)
- Sending processing queries (*Statements*) to the database
- Process information (*ResultSets*) retrieved from the database

Initializing the JDBC driver requires[†] updating Java's `jdbc.drivers` system property with the driver's URL prior to MATLAB/JVM launch, or loading the driver into memory and registering it in the built-in `java.sql.DriverManager`.[34] The latter is done using the very simple code snippet below. This particular example is for Apache's Derby database[35] JDBC driver — other databases will have similar JDBC driver class names which should be used instead. A list for the most common databases can be found here,[36] and a more complete (although some years out-of-date) list here.[37] The built-in JDBC–ODBC bridge[38] is called "sun.jdbc.odbc.JdbcOdbcDriver"; MySQL's[39] is called "com.mysql.jdbc.Driver".[40]

Unfortunately, only drivers that appear directly on MATLAB Java's static classpath can be easily loaded. Theoretically, we could add the driver class or JAR file(s) to the Java classpath in run-time (using *javaaddpath*).[41] Unfortunately, this fails to properly register the `Driver` in MATLAB, since MATLAB's default classloader does not access the dynamic classpath for some reason.[42] MathWorks says that it is not a bug, but an expected behavior.[43]

The solution is to add the driver's class or JAR file(s) to MATLAB's static classpath in *%MATLABroot%/toolbox/local/classpath.txt* and it should work after restarting MATLAB.[44] We can then use MATLAB's classloader (or direct class invocation) to load the driver onto memory and finally register the driver class with the `DriverManager`:[‡]

```
try
    driverClassName = 'org.apache.derby.jdbc.EmbeddedDriver';
```

[†] Strictly speaking, there is an alternative way of connecting to databases, using Java Naming and Directory Interface (JNDI) and `javax.sql.DataSource`. This is an advanced topic well outside the scope of this book. Refer to Java JDBC tutorials or references for further information.

[‡] This latter part is unnecessary for JVM 1.6 (MATLAB R2007b) and upward, but is suggested for backward compatibility.

```
try
    % This works when the class/JAR is on the static Java classpath
    % Note: driver automatically registers with DriverManager
    java.lang.Class.forName(driverClassName);
catch
    try
        % Try loading from the dynamic Java path
        classLoader =
            com.mathworks.jmi.ClassLoaderManager.getClassLoaderManager;
        driverClass = classLoader.loadClass(driverClassName);
    catch
        try
            % One more attempt, using the system class-loader⁴⁵
            classLoader = java.lang.ClassLoader.getSystemClassLoader;
            % An alternative, using the MATLAB Main Thread's context CL
            %classLoader =
            %   java.lang.Thread.currentThread.getContextClassLoader;
            driverClass = classLoader.loadClass(driverClassName);
        catch
            % One final attempt - load directly, like this:
            driverClass = eval(driverClassName);
            % Or like this (if the driver name is known in advance):
            driverClass = com.mysql.jdbc.Driver;
        end
    end
    % Now manually register the driver with DriverManager
    % Note: silently fails if driver is not in static classpath⁴⁶
    DriverManager.registerDriver(driverClass.newInstance);
end

% continue with database processing
catch
    error(['JDBC driver' driverClassName 'not found!']);
    % do some failover activity
end
```

Unfortunately, there seems to be a bug in MATLAB releases R2009a–R2010a (7.8–7.10) which causes MATLAB startup errors when some JAR files (MySQL's JDBC driver included) are added to the static Java classpath.[†] The recommended solution in such cases is to try to add the driver using the dynamic classpath (via *javaaddpath*). We can add this command to the startup.m script in order to avoid the necessity of repeating the command in each new MATLAB session:

```
javaaddpath('C:\mysql-connector-java-5.1.12-bin.jar');
```

As an alternative to using *javaaddpath*, Ed Yu (whom we recall from Section 1.6) has suggested using *javaclasspath* that reportedly works better for JDBC.[47]

[†] http://www.mathworks.com/support/bugreports/624963 (or http://bit.ly/d1Gc5x); the bug was apparently fixed in R2010b (7.11).

2.2.3 *Connecting to a Database*

In order to connect to the database, we need to know its connection string. This has a slightly different version for each database type. The general format is 'jdbc:<dbType>:<dbName> <extraProperties>'. Refer to the online documentation[48] and the database's documentation for details about the database's specific connection string.

Sample connection strings are:

- Oracle MySQL: 'jdbc:mysql://localhost:3306/testDB'
- Oracle Database: 'jdbc:oracle:thin:@localhost:1521:xe'
- Apache Derby: 'jdbc:derby:sampleDB;create=true;'
- Microsoft Access: 'jdbc:odbc:Driver={Microsoft Access Driver (*.mdb)}; DBQ=c:/Yair/test.mdb;DriverID=22;READONLY=true}'

This connection string can now be used for the actual connections, in either secure (username/password) or non-secure mode:[†]

```
% Non-secured (no username/password required for database access)
connStr = 'jdbc:derby:sampleDB';
con = java.sql.DriverManager.getConnection(connStr);
% Secure login (username/password required)
import java.sql.*
con = DriverManager.getConnection(connStr,'username','password');
```

Some databases (e.g., SQLServer) enable defining a *Trusted Connection*, which enables connecting to the DB without requiring authentication. For example,[49]

```
connStr = 'Provider = sqloledb;Data Source = oledbserver;Initial
          Catalog = oledbcat;Trusted_Connection = yes;';
con = java.sql.DriverManager.getConnection(connStr);
```

Microsoft Windows enables setting up an ODBC connection to data sources and readers might be tempted to use this for their database connection. My advice is to try to use JDBC directly, rather than the ODBC bridge. The reason is that in many cases the ODBC driver is simply using JDBC under the hood, thereby adding another point of possible failure, a tie-in to a specific computer configuration, and worse performance than a direct JDBC connection.

DriverManager attempts to wait forever for the connection to succeed or fail. This can be limited by setting a timeout value in seconds prior to the connection attempt:

```
DriverManager.setLoginTimeout(3); % wait 3 seconds max
DriverManager.getConnection(connStr,'username','password');
```

[†] Of course, real-world MATLAB implementations should heavily use try-catch blocks on all DB-related actions such as these.

A database connection is successfully established when a valid `java.sql.Connection` object[50] is returned by the call to `DriverManager.getConnection(connStr)`.

2.2.4 Sending SQL Requests

Once a database connection is successfully established (a valid `java.sql.Connection` object is gotten), it can process SQL *Statements* (`java.sql.Statement` objects[51]). The version of SQL accepted is determined by your specific DB — each DB vendor and version supports a slightly different variety of SQL functions and syntax. However, if we adhere to ANSI-standard SQL-92,[52] then we should be safe in general.

There are several types of Statements:

- Data retrieval (SQL "select") queries
- Data manipulation (DML) queries — insert/update/delete
- Data definition (DDL) queries — adding/modifying tables/indices/and so on
- Stored procedure invocation queries — uses the `CallableStatement` subclass[53]

Here are some examples (remember to add ***try-catch*** blocks in your program!):

```
% Create the SQL Statement object
stmt = con.createStatement();

% SQL Select query example
sqlQueryStr = 'SELECT * FROM TableName WHERE x = 2';
resultSet = stmt.executeQuery(sqlQueryStr);

% SQL Update query example
sqlQueryStr = 'UPDATE TableName SET x = 1 WHERE y = 2';
stmt.executeQuery(sqlQueryStr);
numRowsUpdated = stmt.getUpdateCount;

% SQL Insert query example
sqlQueryStr = 'INSERT INTO TableName VALUES(''str'',2.4)';
stmt.executeQuery(sqlQueryStr);
numRowsInserted = stmt.getUpdateCount;

% DDL query example
sqlQueryStr = 'CREATE TABLE NewTable(name varchar, price float)';
stmt.executeQuery(sqlQueryStr);   % no return value

% Stored-procedure invocation example
callableStmt = con.prepareCall('{CALL Show_Suppliers}');
resultSet = callableStmt.executeQuery();
```

Note that none of these SQL statements requires appending the statement terminator at the end of the `sqlQueryStr`. It is fortunate that the JDBC driver does this automatically for us, since different databases have different terminators (e.g., ';' or 'GO'). This enables database-independent SQL query strings and code.

SQL keywords (such as SELECT) are case-insensitive. I normally use uppercase, but this is merely a readability convention — we can use lowercase if we prefer. This is sometimes not the case with identifier names such as table names and field names, which in some databases are case-sensitive. Also note that different databases may have separate behavior for identifier names. In addition to case sensitivity, names may also be limited in their length (e.g., 16 characters) or content (e.g., only a–z, A–Z, 0–9, and '_'). Names containing special characters (e.g., space) may be allowed or not, but when allowed may require enclosures of a specific format (e.g., '[My field name]'). In short, refer to the database documentation for specific information.

These queries may also be passed via the PreparedStatement[54] interface, which enables faster response times and easier invocation when processing parameterized queries. In the following example, a single generic SQL query is precompiled once and then reused multiple times with different parameter values (*Bound Variables*). The alternative would be to prepare a different standard Statement for each loop element and this would be much less efficient for a loop over many elements:

```
salesForWeek = [175, 150, 60, 155];
products = {'Colombian', 'French RFRoast', 'Espresso', 'Decaf'};
sqlQueryStr = 'UPDATE sales SET amount = ? WHERE product like ?';
stmt = con.prepareStatement(sqlQueryStr);
for productIdx = 1 : length(products)
    stmt.setInt(1, salesForWeek(productIdx));
    stmt.setString(2, products{productIdx});
    numRowsUpdated = stmt.executeUpdate();   % note the return value
end   % for productIdx loop
```

Finally, note that we can also use SQL to insert binary data (such as images, etc.) into our DB.[55] This will, of course, depend on our DB's capabilities in storing binary data. For example, Oracle and MySQL use *BLOBs* (Binary Large Objects)[56] to store images, whereas Microsoft Access uses an *OLE Object*.[57] If we use BLOBs whose original data is not binary (e.g., a large set of double-precision values), then we will need to cast the data into an ***int8*** array before inserting into the DB, as explained in the following section.

A related question is how to access *CLOB* (Character Large Object) data in an Oracle DB. It appears that MATLAB treats such a CLOB as a string when inserting or passing to a stored procedure.[58] I have only seen online references to problems with Oracle's CLOB, although other databases (e.g., Apache Derby) also support CLOBs.

2.2.5 *Handling SQL Result Sets*

ResultSets[59] are the standard container for SQL query results. ResultSets are returned for SQL retrieval (SELECT) and Stored-Procedure invocation queries. The other SQL query Statement types return no output, and attempting to assign their output to a variable will cause an error. ResultSets may be viewed as a virtual data table whose rows can be read sequentially. Each row contains data fields based on the invoked SQL query results and can be read using dedicated type-specific getter methods (*getInt()*, *getString()*, *getBoolean()*, *getDouble()*, *getBlob()*, and other more-specific types).

Each getter method has two variants, accepting either a column name or a column index (which starts at 1, despite the fact that Java indexing normally starts at 0). Using column index is faster but may cause problems if the Stored-Procedure or View modified the result columns; column indices are also less readable and maintainable than column names.

Here is an example that reads selective table data and stores it in a **MATLAB** cell matrix:

```
% Prepare and execute a data-retrieval (select) query
sql1Str = 'SELECT [Customer ID],[Customer Name],[Last Year''s Sales]';
sql2Str = 'FROM Customer';
sql3Str = 'WHERE [Customer ID] > 100';
sql4Str = 'ORDER BY [Customer ID] DESCENDING';
sqlQueryStr = [sql1Str sql2Str sql3Str sql4Str];
resultSet = stmt.executeQuery(sqlQueryStr);

% Process the ResultSet data and store in a Matlab cell matrix
% resultSet rows are processed one at a time, until their end
matlabData = {};
while (resultSet.next())
    id = resultSet.getInt('Customer ID');  % or: resultSet.getInt(1)
    name = char(resultSet.getString(2));  % Java String => Matlab char
    sales = resultSet.getDouble(3);
    matlabData(resultSet.getRow,:) = {id, name, sales};
end  % while resultSet

% Display the resulting matlabData (a 169x3 cell array)
disp(matlabData)
    [270]    'Mountain View Sport'    [  65231.7]
    [269]    'Hikers and Bikers'      [   8300.8]
    [268]    'Coastal Line Bikes'     [   1130.4]
    [267]    'Tom's Place for Bikes'  [ 36400.67]
    ...
```

Note that data retrieval attempts to automatically convert the original data type to the requested getter method data type. Therefore, if data = 65231.7 (a double), then we could get it using *getInt()* (getting 65231) or *getString()* (getting "65231.7000" after the conversion from java.lang.String to **MATLAB** string using **MATLAB**'s ***char*** function).

ResultSet row data can only be read once. After a certain field data was read, an error will be thrown if we try to reread it. Other errors will be thrown if a supplied column name is missing in the ResultSet, or when trying to read data before the initial resultSet.*next()* or after the final data row (when resultSet.*next()* returns false).

Database NULL values are special and care should be taken in our **MATLAB** program to account for their possible existence in the data. The *wasNull()* method reports whether the last value read was NULL or not. This can also be done by testing the *getString()* result for [] (using: if isnumeric(val) since non-NULL results return character strings). Unfortunately, a similar testing for *getInt()* or *getDouble()* fails because these methods return a zero (0) value for

NULLs, which cannot be differentiated from a normal non-NULL zero result — these methods require the *wasNull()* method for validity checks:

```
>> rs = stmt.executeQuery('SELECT ''xyz'',null FROM Customer');
>> rs.next;            % go to first data row
>> rs.getString(1)
ans =
xyz
>> rs.getString(2)
ans =
     []
>> rs.wasNull
ans =
     0

>> rs.next;            % go to next row to try to re-read the NULL
>> rs.getDouble(3)
ans =
     0
>> rs.getDouble(2)
ans =
     0              <= note the same result as non-NULL value
>> rs.wasNull
ans =
     0
```

ResultSets are forward-scrolling by default (TYPE_FORWARD_ONLY), meaning data rows can only be read in sequence, from beginning to end.[60] ResultSets are also read-only by default (CONCUR_READ_ONLY), meaning their data cannot be modified. These parameters cannot be modified once the Statement was created, but may be set during Statement construction.[61] For example,

```
stmt = con.createStatement(ResultSet.TYPE_SCROLL_INSENSITIVE,
                           ResultSet.CONCUR_UPDATABLE);
sqlQueryStr = 'SELECT * FROM TableName WHERE x = 2';
resultSet = stmt.executeQuery(sqlQueryStr);
resultSet.absolute(5);  % move to the fifth row of resultSet
resultSet.updateString(2,'xyz');  % update data in 2nd field
resultSet.updateRow();  % update the relevant row in the database
```

Note that the originating database table was only updated when resultSet.updateRow was called. This should be done separately for each modified ResultSet row.

When backward scrolling is enabled (TYPE_SCROLL_INSENSITIVE or TYPE_SCROLL_ SENSITIVE),[†] moving the ResultSet row cursor (which can be envisioned as a pointer to

---

[†] TYPE_SCROLL_INSENSITIVE is insensitive to changes done to the database while the ResultSet is open, whereas TYPE_SCROLL_SENSITIVE propages database changes to the open ResultSet.

the current data row) can be done either to a specific (absolute) row using `resultSet`.*absolute()* or relatively to the current row using `resultSet`.*relative()*. In both cases, the supplied parameter may be positive or negative, affecting the movement direction. For example, `resultSet`.*relative(-5)* moves the `ResultSet` cursor five rows back (if available), whereas `resultSet`.*absolute(-5)* moves to the fifth row before the end.

Several additional `ResultSet` cursor-movement methods are available:

- `previous` — opposite of *next()*, same as *relative(-1)*;
- `first` — moves to first row, same as *absolute(1)*;
- `last` — moves to last row, same as *absolute(-1)*;
- `beforeFirst` — moves to zeroth (invalid) row, just before the first data row;
- `afterLast` — moves to *N* + 1th (invalid) row, just after the last data row;
- `isBeforeFirst` — checks if the cursor is currently on the zeroth (invalid) row;
- `isFirst` — checks if the cursor is currently on the first data row;
- `isLast` — checks if the cursor is currently on the last data row;
- `isAfterLast` — checks if the cursor is currently on the *N* + 1th (invalid) row;
- and a few others.

After we have finished processing a `ResultSet`, it is a good practice to call `resultSet`.*close()* in order to free database, JDBC, and MATLAB resources. Similarly, call `stmt`.*close()* when done with the query. When a database connection is no longer needed, call `con`.*closeAllStatements()* and `con`.*close()*. Care should be taken to free these resources also when an exception occurs, by placing the call to *close()* following the ***try-catch*** block, instead of just within the ***try*** sub-block:

```
% Bad practice (simple retrieval from a ResultSet):
rs = stmt.executeQuery('SELECT ''xyz'', null FROM Customer');
text = rs.next.getString(1);

% A better practice (close everything; extensive exception handling):
try
    rs = stmt.executeQuery('SELECT ''xyz'', null FROM Customer');
    text = rs.next.getString(1);
catch
    msgbox(['Error occurred:' lasterr]);
end
try rs.close();                    catch, end
try stmt.close();                  catch, end
try con.closeAllStatements();      catch, end
try con.close();                   catch, end
```

For additional performance and resource considerations in applications that heavily utilize a database, consider using *Data Source* connections, *Connection Pooling*, and other advanced features available in the database and database driver.

Binary data (images, etc.) can also be retrieved, depending on DB support.[62] A particular problem with Oracle BLOBs is that Oracle returns a `oracle.sql.BLOB` Java object.[63] The BLOB's data can be accessed in MATLAB using the Java object's *getBytes(position,length)* method, which returns an array of byte (MATLAB *int8*) values. If the original BLOB data was not binary (e.g., a large set of double-precision values), then we will need to cast the data before DB inserting (into an *int8* array) and following retrieval (back from *int8*):[64]

```
blobDataForDB = typecast(originalData(:),'int8');      % into DB
originalData = typecast(blobDataFromDB,'int8');         % from DB
   % Note: you may need to reshape() originalData appropriately
```

Similarly, Oracle returns an `oracle.sql.CLOB` Java object when retrieving CLOB data. To convert this into MATLAB, use the Java object's *getSubString(position,length)* method:[65]

```
text = char(clob.getSubString(1,clob.length()));
```

2.3 Miscellaneous Other Uses

Java classes and packages can be used for numerous other uses. Java is such a well-established and popular programming language, that it is very probable that any conceivable programming block has a Java version posted somewhere online, often in multiple variations and styles.

Here are several examples that were posted online of Java classes that have been used in MATLAB applications. These encompass a wide array of programming fields and illustrate Java's versatility and adaptability for MATLAB users:

■ Use Java networking[66] for client–server connectivity between MATLAB and external applications. Quite a few CSSM posts[67] and utilities[68] were posted, and even a detailed tutorial.[69] One power-user, Dirk-Jan Kroon, has even created a Java-based MATLAB webserver.[70]

Note that using networking as a server (as opposed to the much simpler client side) normally requires setting up separate I/O and processing threads. However, as Walter Roberson has pointed out (also see Section 3.2):[71]

> *"… Matlab normally runs as a single Java thread; you can use Java thread creation methods to create additional Java threads, but they will not have access to the Matlab workspaces except by communicating with the single main Matlab thread."*

Still, Andrzej Karbowski's jPar utility[72] on the MATLAB File Exchange was created to connect remote MATLAB sessions using Java RMI (Remote-Method Invocation — a subset of Java networking).

■ Retrieve URLs that require authentication.[73]

■ Send emails with authentication or behind a proxy.[74]

■ Use Java I/O for communication with the serial/parallel/USB ports[75] and for a multitude of file-processing functionalities.[76]

- Use Java XML parsers to process XML data files.[77]
- Communicate with external websites, for example, to get an updated list of stock and options data.[78]
- Compress/decompress data (MATLAB's *zip* function only works with files).[79]
- Use `java.text.DecimalFormat`[80] to format displayed numbers in a Locale-specific manner that is far more customizable than MATLAB's standard built-in functions allow.[81]
- Use `java.security.MessageDigest`[82] to hash, encrypt, and decrypt data.[83]
- Use `java.nio.channels.FileLock`[84] to solve problems with network files being concurrently accessed by several MATLAB processes.[85]
- Automate testing and GUI control using `java.awt.Robot`'s[86] ability to programmatically move the mouse cursor,[87] simulate mouse clicks and key presses,[88] and take partial or full screenshots.[89]

  I have used `java.awt.Robot` for exactly this purpose in my ***ScreenCapture*** utility. This is a relatively small and highly documented utility — the reader may download it free of cost and inspect its source code.[90] Another, Java-based alternative is to use the component's *printAll()* method.[91]

- Use the computer name and the process-id as a random-number seed.[92]
- Use the Java jMEF library for combining Gaussian distribution functions.[93]
- Use the Java sound API to control the computer's speakers.[94]
- Use specific algorithms in graph-theory.[95]
- Use specialized data structures such as *decision trees*.[96]
- Run Javascript code without the overhead of a browser GUI, using a Java-based Javascript processing engine.[97] This could be used, for example, to integrate Google Maps in a MATLAB mapping application.[98]

... and a few more ideas that can be implemented in Java but which I have not seen online (which is not to say they have not been done for Matlab — only that I have personally not come across them):

- Encryption utilities, networking operations that are trivial in Java, are next to impossible in standard MATLAB.
- Serialization of Java objects is currently not supported in MATLAB,[99] but can be done using an external Java class.
- Use `java.math.BigIntger` and `BigDecimal` classes for arbitrary numerical precision.[100]
- Many other interesting ideas in Brian Eubanks' *Wicked Cool Java* book.[101]

2.4 A Short Pause for Reflection

This chapter described how using Java can benefit MATLAB programmers. Much of the MATLAB–Java interface explored in this chapter is well documented, with only a few caveats, bugs, and undocumented behaviors. Nevertheless, Java remains under-utilized in MATLAB applications, perhaps since most MATLAB engineers have no Java knowledge and are therefore unaware of its possible usefulness. It is hoped that after reading this chapter (and the online resources referenced), MATLAB programmers will be much less hesitant to explore and integrate Java classes into their MATLAB applications.

Java has an extensive set of packages and classes dealing with I/O to external files, processes, systems, and hardware. It has extensive support for networking, from low-level TCP/IP to webpage URLs and XML documents. Java contains classes and libraries that provide wider functionality and more granular control than their MATLAB equivalents that (it must be admitted) provide adequate basic support for most needs.

Java's millions of programmers worldwide far outnumber MATLAB programmers. There are also many active Java forums,[102] blogs,[103] articles,[104] tutorials,[105] and source code repositories,[107] with traffic and content far beyond those available in CSSM and the MathWorks' site. Therefore, there is a good likelihood that for any programming task, algorithm, or problem in a MATLAB application, somebody somewhere has already posted a Java solution which can be integrated into our MATLAB program.

Beyond saving hefty toolboxes and algorithm development cost, this book aims to show that MATLAB developers have a real development alternative. The choice is ours based on the application's requirements and our programming capabilities.

This chapter has focused on non-GUI Java components and programming. Using Java GUI (*Swing*) components can significantly improve a MATLAB application's usability. However, this topic is more complex than those presented so far and contains more undocumented niches and pitfalls for the MATLAB programmer. Readers will note that the further we advance in this book, the deeper we get into undocumented areas.

With our improved confidence in MATLAB–Java integration, let us now start *Swing*ing.

References

1. http://www.mathworks.com/help/techdoc/matlab_external/f44062.html (or http://tinyurl.com/2zdvha).
2. For example, see http://www.mathworks.com/matlabcentral/fileexchange/26778 (or http://bit.ly/bagjrG).
3. http://java.sun.com/docs/books/tutorial/collections/ (or http://tinyurl.com/6y9ob).
4. 1.4.2: http://java.sun.com/j2se/1.4.2/docs/guide/collections/ (or http://tinyurl.com/az492n);
   1.5.0: http://java.sun.com/j2se/1.5.0/docs/guide/collections/ (or http://tinyurl.com/bj5chs);
   1.6.0: http://java.sun.com/javase/6/docs/technotes/guides/collections/ (or http://tinyurl.com/bckwmy).
5. 1.6 compared to 1.5: http://java.sun.com/javase/6/docs/technotes/guides/collections/changes6.html (or http://bit.ly/dfhWwA); 1.5 compared to 1.4.2: http://java.sun.com/javase/6/docs/technotes/guides/collections/changes5.html (or http://bit.ly/d8TOCv).
6. http://java.sun.com/javase/6/docs/api/java/util/Set.html (or http://tinyurl.com/2huafe).
7. http://java.sun.com/javase/6/docs/api/java/util/List.html (or http://tinyurl.com/325uax).
8. http://java.sun.com/docs/books/tutorial/collections/algorithms (or http://tinyurl.com/chmfs4).
9. http://stackoverflow.com/questions/4163920/matlab-stack-data-structure (or http://bit.ly/bezONj).
10. http://stackoverflow.com/questions/4142190/is-there-a-queue-in-matlab (or http://bit.ly/aNf6Ek) shows a MATLAB example.
11. http://java.sun.com/javase/6/docs/api/java/util/Queue.html (or http://tinyurl.com/2vrhwd).
12. http://java.sun.com/javase/6/docs/api/java/util/Map.html (or http://tinyurl.com/2sfddr).

13. http://www.mathworks.com/help/techdoc/ref/containers_map.html (or http://bit.ly/aEkKOw); http://stackoverflow.com/questions/3591942/hash-tables-in-matlab (or http://bit.ly/dwpJvI).

14. http://www.mathworks.com/matlabcentral/fileexchange/26778 (or http://bit.ly/bagjrG); http://www.mathworks.com/matlabcentral/fileexchange/28586 (or http://bit.ly/c0XeFl).

15. http://stackoverflow.com/questions/436852/storing-Matlab-structs-in-java-objects (or http://bit.ly/9ToxIj).

16. http://www.mathworks.com/help/techdoc/matlab_external/f18070.html (or http://tinyurl.com/becpyj).

17. http://www.mathworks.com/products/database/ (or http://bit.ly/aanwJK).

18. For example, http://sourceforge.net/projects/mym (or http://bit.ly/dgGp2W) for MySQL; http://www.dertech.com/pgmex/pgmex.html (or http://bit.ly/azLSys) for PostgreSQL; or http://energy.51.net/dbtool/ for any ODBC (shareware with a free demo mode).

19. http://www.mathworks.com/matlabcentral/fileexchange/?term=database (or http://bit.ly/d3YH8K) provides close to 100 utilities, for example, http://www.mathworks.com/matlabcentral/fileexchange/4045-cse-sql-database-library (or http://bit.ly/aanwJK); http://www.mathworks.com/matlabcentral/fileexchange/8385-database-connection-mfiles (or http://bit.ly/bzNyeB); http://www.mathworks.com/matlabcentral/fileexchange/8663-mysql-database-connector (or http://bit.ly/cnde6T); http://www.mathworks.com/matlabcentral/fileexchange/9549-myblob (or http://bit.ly/bVlFsj); http://www.mathworks.com/matlabcentral/fileexchange/18834-myblobtestdb (or http://bit.ly/d89pjv); http://www.mathworks.com/matlabcentral/fileexchange/13621-ado-ole-database-connection (or http://bit.ly/bCFdCM).

20. For example, see the code within http://www.mathworks.com/matlabcentral/fileexchange/13621 (or http://bit.ly/cJ7nk7) or http://www.mathworks.com/matlabcentral/fileexchange/8385 (or http://bit.ly/aXrA0c); or the explanations in http://www.mathworks.com/matlabcentral/newsreader/view_thread/81979#210417 (or http://bit.ly/dzvT8y), http://stackoverflow.com/questions/3100998/getting-names-of-access-database-tables-with-matlab (or http://bit.ly/aULlln), and http://www.toomre.com/Writing_BLOBs_With_MATLAB_ActiveX (or http://bit.ly/d2JTNZ).

21. http://java.sun.com/javase/technologies/database (or http://bit.ly/bBIYaS); http://java.sun.com/docs/books/tutorial/jdbc/ (or http://bit.ly/cerwII).

22. As in http://www.mathworks.com/MATLABcentral/fileexchange/25577 (or http://bit.ly/5oIFtj).

23. For example, MongoDB: http://github.com/mongodb/mongo-java-driver/ (or http://bit.ly/cU7vll); http://stackoverflow.com/questions/3886461/connecting-to-mongodb-from-MATLAB (or http://bit.ly/bbwEtZ).

24. http://www.mathworks.com/matlabcentral/fileexchange/12027 (or http://bit.ly/9AayAJ); http://www.mathworks.com/matlabcentral/fileexchange/13069 (or http://bit.ly/9e1h9F); http://www.mathworks.com/matlabcentral/fileexchange/17897 (or http://bit.ly/9FGla0); http://www.mathworks.com/matlabcentral/fileexchange/28237 (or http://bit.ly/9phRvc) and several others by Dimitry Shvorob: http://www.mathworks.com/matlabcentral/fileexchange/?term=authorid:17777 (or http://bit.ly/dc08Al).

25. http://java.sun.com/products/jdbc

26. http://developers.sun.com/product/jdbc/drivers (or http://tinyurl.com/2bt7f7).

27. http://java.sun.com/j2se/1.4.2/docs/guide/jdbc/bridge.html (or http://bit.ly/c7W4mA). For example, connecting to a Microsoft Access DB: http://blog.taragana.com/index.php/archive/access-microsoft-access-database-from-java-using-jdbc-odbc-bridge-sample-code/ (or http://bit.ly/a5oOSM).

28. http://java.sun.com/docs/books/tutorial/jdbc/index.html (or http://tinyurl.com/2o6jy).

29. http://java.sun.com/products/jdbc/community/books/index.html (or http://bit.ly/a84BwL), or this relatively old comparative review: http://www.javaworld.com/javaworld/jw-01-1998/jw-01-bookreview.html (or http://bit.ly/cZHht3).

30. Such as www.java2s.com's excellent website http://www.java2s.com/Tutorial/Java/0340_Database/Catalog0340_Database.htm (or http://tinyurl.com/aawpe7).

31. For example, here are the different reference sections for the ResultSet class:
    1.4.2: http://java.sun.com/j2se/1.4.2/docs/api/java/sql/ResultSet.html (or http://tinyurl.com/7xm6w);
    1.5.0: http://java.sun.com/j2se/1.5.0/docs/api/java/sql/ResultSet.html (or http://tinyurl.com/zvbr3);
    1.6.0: http://java.sun.com/javase/6/docs/api/java/sql/ResultSet.html (or http://tinyurl.com/2ukwuf).

32. http://www.mysql.com/products/connector/ (or http://bit.ly/aILHB8).

33. http://www.mathworks.com/matlabcentral/fileexchange/28237-queryMySQL (or http://bit.ly/9phRvc).

34. http://java.sun.com/javase/6/docs/api/java/sql/DriverManager.html (or http://tinyurl.com/7y24l5), or http://java.sun.com/j2se/1.5.0/docs/guide/jdbc/getstart/drivermanager.html (or http://tinyurl.com/7y24l5).

35. http://db.apache.org/derby/

36. http://www.java2s.com/Tutorial/Java/0340_Database/AListofJDBCDriversconnectionstringdrivername.htm (or http://bit.ly/b7fkCQ).

37. http://developers.sun.com/product/jdbc/drivers (or http://tinyurl.com/2bt7f7).

38. http://java.sun.com/javase/6/docs/technotes/guides/jdbc/bridge.html (or http://bit.ly/dqv8JY).

39. http://dev.mysql.com/downloads/connector/j/ (or http://bit.ly/9ZSPfq); for a detailed description of the JDBC-MySQL connector, see http://www.developer.com/java/data/article.php/3417381 (or http://bit.ly/9AQk5c).

40. http://stackoverflow.com/questions/2138530/how-can-i-remotely-connect-odbc-using-java-in-windows-xp/2139141#2139141 (or http://bit.ly/9uNWWS).

41. http://www.mathworks.com/MATLABcentral/newsreader/view_thread/146801#369440 (or http://tinyurl.com/cmeqea).

42. http://forums.java.net/jive/message.jspa?messageID=111283 (or http://tinyurl.com/cz6n7x); or http://www.mathworks.com/matlabcentral/newsreader/view_thread/161242#409349 (or http://tinyurl.com/ddjl2u); or http://www.mathworks.com/matlabcentral/newsreader/view_thread/259499 (or http://tinyurl.com/l2bj7f).

43. http://www.mathworks.com/support/solutions/en/data/1-1YFUFB (or http://tinyurl.com/nxm42o); unfortunately, this Web page was removed by MathWorks and can no longer be seen. See a related comment on this issue: http://blogs.mathworks.com/desktop/2009/07/06/calling-java-from-matlab/#comment-6862 (or http://bit.ly/e0dJ4h).

44. http://www.mathworks.com/MATLABcentral/newsreader/view_thread/242091 (or http://tinyurl.com/cy9nh8).

45. http://www.mathworks.com/MATLABcentral/newsreader/view_thread/37146 (or http://bit.ly/9TksuX).

46. There are several CSSM posts about this issue. For example, see http://www.mathworks.com/matlabcentral/newsreader/view_thread/58397 (or http://tinyurl.com/99qv3k).

47. http://www.mathworks.com/matlabcentral/newsreader/view_thread/242091#636516 (or http://bit.ly/cwJYJ5).

48. http://www.java2s.com/Tutorial/Java/0340__Database/AListofJDBCDriversconnection stringdrivername.htm (or http://bit.ly/b7fkCQ); http://java.sun.com/docs/books/tutorial/jdbc/basics/connecting.html (or http://tinyurl.com/2u6uu8).

49. http://www.toomre.com/Writing_BLOBs_With_MATLAB_ActiveX (or http://bit.ly/d2JTNZ).

50. http://java.sun.com/javase/6/docs/api/java/sql/Connection.html (or http://tinyurl.com/2g6wbd).

51. http://java.sun.com/javase/6/docs/api/java/sql/Statement.html (or http://tinyurl.com/3a859v).

52. http://www.w3schools.com/sql/default.asp or http://en.wikipedia.org/wiki/SQL

53. http://java.sun.com/javase/6/docs/api/java/sql/CallableStatement.html (or http://tinyurl.com/65dt5w).

54. http://java.sun.com/javase/6/docs/api/java/sql/PreparedStatement.html (or http://tinyurl.com/375pes).

55. http://www.mathworks.com/matlabcentral/newsreader/view_thread/276333#730751 (or http://bit.ly/b4aRHZ). The extremely detailed solution in this thread is mySql-specific, but can be adapted to other DBs relatively easily. For solution to the Microsoft Access DB, see http://www.mathworks.com/support/solutions/en/data/1-90B0EB/ (or http://bit.ly/aGo-DCt) and for SQLServer, see http://www.toomre.com/Writing_BL OBs_With_MATL AB_ActiveX (or http://bit.ly/d2JTNZ).

56. http://download.oracle.com/otn_hosted_doc/jdeveloper/1012/jdbc-javadoc/oracle/sql/BLOB.html (or http://bit.ly/heFMmY).

57. http://support.microsoft.com/?scid=kb%3Ben-us%3B824263&x=13&y=5 (or http://bit.ly/b2oHAy).

58. http://www.mathworks.com/matlabcentral/newsreader/view_thread/278869 (or http://bit.ly/bOdJXt).

59. http://java.sun.com/javase/6/docs/api/java/sql/ResultSet.html (or http://tinyurl.com/2ukwuf).

60. http://www.careerride.com/JDBC-resultset-types.aspx (or http://bit.ly/cFokTR).

61. http://java.sun.com/docs/books/tutorial/jdbc/basics/retrieving.html (or http://tinyurl.com/2ng4f).

62. http://stackoverflow.com/questions/2647621/retrieve-blob-field-from-mysql-database-with-matlab (or http://bit.ly/bIEKRB); http://www.mathworks.com/matlabcentral/newsreader/view_thread/12454 (or http://bit.ly/9BxHp1).

63. http://www.mathworks.com/matlabcentral/newsreader/view_thread/257024 (or http://bit.ly/aTiqxi).

64. http://www.mathworks.com/support/solutions/en/data/1-4A456Y/ (or http://bit.ly/drVrtC).

65. http://www.mathworks.com/support/solutions/en/data/1-90T4HK/ (or http://bit.ly/drkMXe). The link to Oracle's javadoc of CLOB provided in MathWork's solution page is now defunct. Instead, use this link: http://web.archive.org/web/20050818194023/http://download-uk.oracle.com/otn_hosted_doc/jdeveloper/904preview/jdbc-javadoc/oracle/sql/CLOB.html (or http://bit.ly/942tdo).

66. http://java.sun.com/docs/books/tutorial/networking/index.html (or http://bit.ly/b1hXSl).

67. http://www.mathworks.com/matlabcentral/newsreader/view_thread/244801 (or http://bit.ly/ccKz3q); http://www.mathworks.com/matlabcentral/newsreader/view_thread/258876 (or http://bit.ly/aa9N1A).

68. http://www.mathworks.com/matlabcentral/fileexchange/23728-tcpip-distributed-waitbar (or http://bit.ly/aB0LeF); http://www.mathworks.com/matlabcentral/fileexchange/24524-tcpip-communications-in-matlab (or http://bit.ly/bkqfyo); http://www.mathworks.com/matlabcentral/fileexchange/24525-a-simple-udp-communications (or http://bit.ly/9hZnj6); http://www.mathworks.com/matlabcentral/fileexchange/25249-tcpip-socket-communication (or http://bit.ly/hlwWxd); http://www.mathworks.com/matlabcentral/fileexchange/27999-ssh-from-matlab-updated (or http://bit.ly/9w3CiS). Also see MEX-based http://www.mathworks.com/matlabcentral/fileexchange/345-tcpudpip-toolbox-2-0-6 (or http://bit.ly/ceazff).

69. http://iheartmatlab.blogspot.com/2008/08/tcpip-socket-communications-in-matlab.html (or http://bit.ly/ba32Wq) continued on http://iheartmatlab.blogspot.com/2009/09/tcpip-socket-communications-in-matlab.html (or http://bit.ly/ehVcdV).

70. http://www.mathworks.com/matlabcentral/fileexchange/29027-web-server (or http://bit.ly/alRWpZ).

71. http://www.mathworks.com/matlabcentral/newsreader/view_thread/289271#771282 (or http://bit.ly/ds2C0S).

72. http://www.mathworks.com/matlabcentral/fileexchange/24924-jpar-parallelizing-matlab (or http://bit.ly/cOMVSP); described in this paper: http://www.ia.pw.edu.pl/~karbowsk/jpar/jpar-para08-abstract.pdf (or http://bit.ly/atiWw7).

73. http://www.mathworks.com/support/solutions/en/data/1-4EO8VK/ (or http://bit.ly/bEhDrP); this may appear to be a pure ma tlab solution but in fact the entire implementation of **urlread** and **urlwrite** is Java-based and can easily be customized, extended, and configured.

74. http://www.mathworks.com/matlabcentral/newsreader/view_thread/254080#658549 (or http://bit.ly/dv2rZb); like **urlread** and **urlwrite**, *sendmail* also uses Java functionality, which can easily be customized, extended, and configured.

75. http://www.mathworks.com/matlabcentral/newsreader/view_thread/246850 (or http://bit.ly/aJnA4w); also look athttp://www.mathworks.com/matlabcentral/fileexchange/25478 (or http://bit.ly/byFbRd) for a MEX solution.

76. http://www.mathworks.com/matlabcentral/newsreader/view_thread/253537#799898 (or http://bit.ly/eUlvRD).

77. http://blogs.mathworks.com/desktop/2010/11/01/xml-and-matlab-navigating-a-tree/ (or http://bit.ly/ 9wg9Zp).

78. http://www.mathworks.com/matlabcentral/fileexchange/28071 (or http://bit.ly/ahgr9T).

79. http://www.mathworks.com/matlabcentral/fileexchange/25656-compression-routines (or http://bit.ly/bE473p); http://www.mathworks.com/matlabcentral/newsreader/view_thread/245803 (or http://bit.ly/952Pfo); http://www.mathworks.com/matlabcentral/newsreader/view_thread/289360 (or http://bit.ly/c5VSuQ); MATLAB's entire set of compression functionality (*zip*, *unzip*, *tar*, *untar*, *gunzip*) is Java-based. Users can easily adapt their code for their needs.

80. http://download.oracle.com/javase/6/docs/api/java/text/DecimalFormat.html (or http://bit.ly/nJmlml).

81. http://UndocumentedMatlab.com/blog/formatting-numbers/ (or http://bit.ly/qSvXy4).

82. http://java.sun.com/javase/6/docs/api/java/security/MessageDigest.html (or http://bit.ly/9dN43m).

83. http://www.mathworks.com/matlabcentral/newsreader/view_thread/282298#759462 (or http://bit.ly/bGPP4q).

84. http://java.sun.com/j2se/1.4.2/docs/api/java/nio/channels/FileLock.html (or http://bit.ly/91qgfW); http://www.exampledepot.com/taxonomy/term/194 (or http://bit.ly/9wK5tM).

85. http://www.mathworks.com/matlabcentral/newsreader/view_thread/279510 (or http://bit.ly/95nV5w); http://stackoverflow.com/questions/3451343/atomically-creating-a-file-lock-in-matlab-file-mutex (or http://bit.ly/bCYexx).

86. http://java.sun.com/j2se/1.5.0/docs/api/java/awt/Robot.html (or http://bit.ly/bOBRXh).

87. http://UndocumentedMatlab.com/blog/controlling-mouse-programmatically/ (or http://bit.ly/cqrOJK).

88. http://www.mathworks.com/matlabcentral/newsreader/view_thread/308658#839089 (or http://bit.ly/mcmmwZ).

89. http://www.mathworks.com/support/solutions/en/data/1-2X10AT/ (or http://bit.ly/duWbUP); http://www.mathworks.com/matlabcentral/newsreader/view_thread/254587 (or http://bit.ly/b2Er37); http://www.mathworks.com/matlabcentral/newsreader/view_thread/235825#668975 (orhttp://bit.ly/cGgVHb);

http://www.mathworks.com/matlabcentral/newsreader/view_thread/243113 (or http://bit.ly/9SXnl0);
http://www.mathworks.com/matlabcentral/newsreader/view_thread/262754 (or http://bit.ly/du378H);
http://www.mathworks.com/matlabcentral/newsreader/view_thread/269154 (or http://bit.ly/d6LYfR);
http://www.mathworks.com/matlabcentral/newsreader/view_thread/284445#753949 (or http://bit.ly/9jzvCi).

90. http://www.mathworks.com/matlabcentral/fileexchange/24323-screencapture (or http://bit.ly/czfUSE).

91. http://www.mathworks.com/matlabcentral/newsreader/view_thread/246001 (or http://bit.ly/c7MnHt).

92. http://www.mathworks.com/matlabcentral/newsreader/view_thread/246706#635636 (or http://tinyurl.com/ch8dbd).

93. http://www.mathworks.com/matlabcentral/fileexchange/26079 (or http://bit.ly/c5SLE6); http://www.lix.polytechnique.fr/~nielsen/MEF/ (or http://bit.ly/btUYhf).

94. http://UndocumentedMatlab.com/blog/updating-speaker-sound-volume/ (or http://bit.ly/aYSs0W); http://www.mathworks.com/matlabcentral/fileexchange/28394-jAudio (or http://bit.ly/fLoQ5n); http://www.mathworks.com/matlabcentral/fileexchange/25584-SoundVolume (or http://bit.ly/dSFGjb).

95. http://www.mathworks.com/matlabcentral/newsreader/view_thread/285315#757436 (or http://bit.ly/biQS3c).

96. http://en.wikipedia.org/wiki/Alternating_decision_tree (or http://bit.ly/lh2Ylz), implemented in the open-source JBoost package (http://jboost.sourceforge.net/ or http://bit.ly/iOVmYT), which can easily be integrated in MATLAB (http://www.mathworks.com/matlabcentral/newsreader/view_thread/308764 or http://bit.ly/iGkNN1; http://UndocumentedMatlab.com/blog/jboost-integrating-an-external-java-library-in-matlab/ or http://bit.ly/nteRXx).

97. http://weblog.raganwald.com/2007/07/javascript-on-jvm-in-fifteen-minutes.html (or http://bit.ly/dOLrfw).

98. http://stackoverflow.com/questions/4778852/how-to-embed-google-map-api-in-matlab (or http://bit.ly/hdRUX9); also see http://www.mathworks.com/matlabcentral/fileexchange/?term=tag%3Agoogle+map (or http://bit.ly/i73mCB), http://blogs.mathworks.com/loren/2010/05/06/oilslick/ (or http://bit.ly/eSBkEk).

99. See related Web pages: http://www.mathworks.com/matlabcentral/newsreader/view_thread/289283 (or http://bit.ly/gZhnVY) andhttp://www.mathworks.com/matlabcentral/fileexchange/12063-serialize (or http://bit.ly/fxX5qr).

100. Note John D'Errico's related File-Exchange utility at http://www.mathworks.com/matlabcentral/fileexchange/22725-variable-precision-integer-arithmetic (or http://bit.ly/epQwnj).

101. http://my.safaribooksonline.com/1593270615?tocview=true (or http://bit.ly/bM7l67).

102. For example, the entire comp.lang.java.* Usenet tree; several dedicated forums on groups.google.com; forums.sun.com; www.java-forums.org; forums.java.net; www.javakb.com; http://forums.devshed.com/java-help-9/; http://www.codeguru.com/forum/forumdisplay.php?f=5; http://www.ibm.com/developerworks/forums/dw_jforums.jspa, and so on.

103. blogsearch.google.com/blogsearch?q=java+code; ww.onjava.com/onjava and many others.

104. For example, see java.sun.com; www.devx.com/Java; www.developer.com/java; www.javaworld.com/features; today.java.net/pub/q/articles; www.java2s.com/Article/Java/CatalogJava.htm; oreilly.com/pub/q/all_onjava_articles, and others.

105. For example, www.freejavaguide.com; www.java2s.com/Tutorial/Java/CatalogJava.htm; java.sun.com/docs/books/tutorial; math.hws.edu/javanotes; www.javacoffeebreak.com/tutorials, and others.

106. For example, www.javadb.com; www.java2s.com/Open-Source/Java/CatalogJava.htm; wikis.sun.com/display/code/Home; javaboutique.internet.com/javasource; www.thefreecountry.com/sourcecode/java.shtml, and others.

Chapter 3

Rich GUI Using Java Swing

MATLAB provides programmers with extensive GUI functionality, including a visual graphic designer (**guide**), and programmatic access to component properties and callback events. However, the level of customizability and extensibility of the basic MATLAB components is limited. MATLAB provides a wide set of basic building blocks, and if we wish something looking slightly more stylish or modern, then we have a problem. Well, that is the official documented version. Unofficially, programmers have access to almost all the modern GUI facilities that Java provides through its Swing class library (*toolkit*)[1] — a toolkit on which the MATLAB GUI itself is based.

As a consultant who has competed for large-scale MATLAB projects, it was frustrating at times to lose contracts to non-MATLAB implementations just on the grounds that MATLAB GUI looks childlike and outdated. It was like competing with a Ferrari engine encased in a 20-year-old pickup frame. This has also been a frustrating experience to other MATLAB users.[2] Using Swing now enables your programs to also *look* like a Ferrari. Sometimes, a single image is worth a thousand words: take a look at the screenshots on the following page, which compare MATLAB's Swing-based Preference panel with a corresponding panel created with pure-MATLAB controls using MATLAB's built-in GUIDE.[3] In this simple example, we note several visual differences:

- Collapsible tree items in the main listbox
- Central vertical alignment of the panel title ("Color Preferences")
- Natural resizing behavior (components keep their size, only positions change)
- Complex color combo-box controls (not just simple buttons or popup menus)
- Syntax-highlighted multiline editbox without the annoying vertical scrollbar
- Panel-wide scrollbar that hides some controls at the bottom

There are numerous books, tutorials, and online resources about Java/Swing programming. MATLAB programmers who have gotten used to relying only on the internal documentation, and possibly also on very few websites, will be overwhelmed with thousands of websites dedicated to Swing programming (and Java in general).[4] This book is too small to include more than short code snippets. Readers are encouraged to continue exploring online, based on the keywords and ideas presented here.

3.1 Adding Java Swing Components to MATLAB Figures

3.1.1 *The javacomponent Function*[5]

Swing components can be added very easily onto a MATLAB figure window, using the undocumented *javacomponent* function, available since MATLAB R14 (7.0).[6] Programmers can use *usejavacomponent* (or the equivalent *usejava('awt')* or *javachk* as described at the beginning of Chapter 2) to programmatically check whether or not their system supports *javacomponent*:

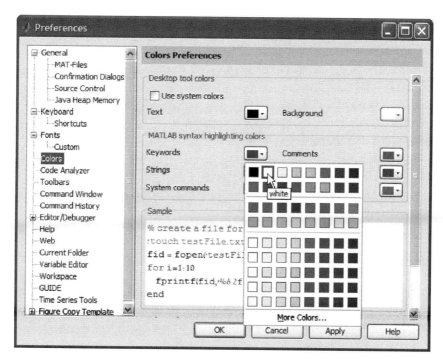

Stylish window using Swing controls (**See color insert.**)

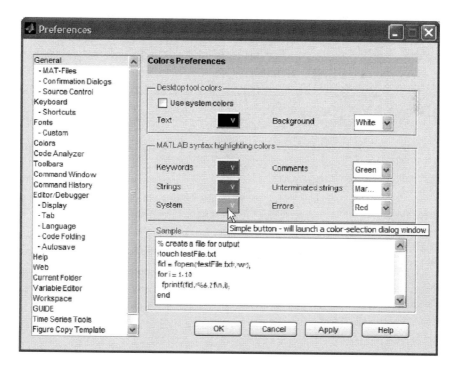

Simulation of the same dialog window, using standard MATLAB uicontrols (**See color insert.**)

```
if ~usejava('swing')  % or the equivalent: if ~usejavacomponent()
    error('this platform does not support Java Swing');
end
error(javachk('swing','this platform'));    % equivalent
```

javacomponent accepts a component class name (a string) or a reference to a previously created component object, an optional pixel position parameter (default = [20,20,60,20], just like ***uicontrol***; it may also contain the strings 'North', 'South', 'East', or 'West'[†]) and an optional parent container handle (defaults to the current figure). ***javacomponent*** then adds the requested component as a child of the requested parent container and wraps it in a MATLAB Handle-Graphics (HG) container. ***javacomponent*** returns two handles: the MATLAB HG container handle and a reference (handle) to the Java component:

```
>> [jButton, hButton] = javacomponent('javax.swing.JButton')
hButton =
        javahandle_withcallbacks.javax.swing.JButton
jButton =
            158.002197265625
>> javacomponent('javax.swing.JButton','North');
>> javacomponent(javax.swing.JButton('Click me!'),[50,40,80,30]);
>> javacomponent(javax.swing.JButton('Click me!'),'East',hFig);
```

Note the difference between Java object creation and ***javacomponent***: a pre-created Java object only resides in JVM memory, not onscreen, until ***javacomponent*** is called to display it. ***javacomponent*** only creates objects, as a convenient service to programmers, when a class name (string) parameter is passed to it. In practice, it is better to separate these two actions: create the Java object separately and then pass the object's reference handle to ***javacomponent*** for display. This enables easier error-trapping if the Java object cannot be created or fails to initialize, before attempting to display the object:

```
% Create and initialize a JScrollBar object
try
    jScrollbar = javaObjectEDT('javax.swing.JScrollBar');
    jScrollbar.setOrientation(jScrollbar.HORIZONTAL);
catch
    error('Cannot create Java-based scroll-bar!');
end

% Display the object onscreen
try
    javacomponent(jScrollbar,'South');
catch
    error('Cannot display Java-based scroll-bar!');
end
```

[†] Based on the platform's java.awt.BorderLayout.NORTH, and so on. This string is case-sensitive (i.e., 'NORTH' or 'north' will fail). Also, note that 'Center' is not supported for some unknown reason.

javacomponent accepts parent handles that are figures, toolbars,[†] ***uipanel***s,[‡] or ***uicontain-ers***.[7] Unfortunately, frames are not ***uicontainer***s and, therefore, cannot be used as ***javacomponent*** parents.

 Note: Due to a bug in R2007a, ***javacomponent***s cannot be added to ***uicontainer***s, since javacomponent.m checks if `isa(hParent,'uicontainer')` (and similarly for `'uiflowcontainer'`, `'uigridcontainer'`) instead of `isa(hParent,'hg.uicontainer')` (and similarly for the others). If we modify javacomponent.m accordingly (add "hg." in lines 98–100), this bug will be fixed. Since R2007b, `isa(..., 'hg.uicontainer')` is equivalent to `isa(..., 'uicontainer')`, so this fix is unnecessary.

Once the component has been created, even before it has been placed onscreen, it can be manipulated just like any other Java object. For example,[§]

```
jButton.setText('do not click!');  % or: set(jButton,'text','...')
```

Some manipulations obviously require the component to be visible, for example,

```
jButton.requestFocus();
jButton.hide();  % or: jButton.setVisible(0), set(jButton,'Visible',0)
```

Note that it is not assured that an exception will be thrown if an object manipulation that requires a visible object is requested when the object is hidden. Swing objects very often simply ignore the requested action in such cases. This often happens when MATLAB programmers place Java code in their GUIDE-created *_OpeningFcn() function,[¶] since_OpeningFcn() is called before the figure window becomes visible.

The component can also be manipulated to some extent via its HG container, which is of a special MATLAB type (class) called hgjavacomponent. This includes getting/setting the component position, position units, visibility, resizing callback, tag, UserData, and so on:

```
set(hButton,'units','norm','position',[.2,.3,.1,.05]);
set(hButton,'visible','off'); %note: on/off, not true/false as in Java
set(hButton,'ResizeFcn',{@resizeCallbackFunc,param1,param2});
```

When adding Java components which are container classes (descendants of `java.awt.Container`), it is important to remember that only other Java components can be added to

---

[†] The toolbar option as a possible ***javacomponent*** parent has been around from the beginning, and yet until MATLAB 7.6 (R2008a), it remained entirely undocumented and not just semi-documented like the rest of the function. Since R2008a, parents of type ***uisplittool*** and ***uitogglesplittool*** can also be used (see Section 4.5.4).

[‡] Old MATLAB releases failed to mention ***uipanel*** as a possible parent in ***javacomponent***'s help comment (this documentation flaw was fixed in new releases).

[§] In these examples, remember to either use EDT or to place a short pause beforehand (see Sections 1.5 and 3.2).

[¶] If a figure is called MyFig, GUIDE will automatically create MyFig.m with an internal function called MyFig_OpeningFcn().

these containers. MATLAB objects such as axes (for plots or images) and ***uicontrol***s cannot be added, since they do not have a `Container` wrapper.† Therefore, feel free to use these Java containers as long as their contained GUI is limited to Java components (`JButton`, `JComboBox`, etc.). This limitation is very annoying — it would be very useful to be able to place MATLAB axes or ***uicontrol***s within a `JTabbedPane` or `JSplitPane`. Instead, we need to rely on MATLAB-based workarounds (***uitab*** and ***uisplitpane,*** see Section 4.3 and Chapter 10, respectively), which are cumbersome compared with their Java counterparts.

Swing components can also be added directly to the figure frame's peer (`jFrame.addchild(component)`), but this has numerous drawbacks and is not advisable. MATLAB automatically creates an HG container for the component and sets the default position ([20,20,60,20] pixels). In addition, ***javacomponent*** adds some important event listeners and checks, and so it is advisable to use ***javacomponent*** rather than directly adding peer components.

Note that when directly adding figure peers, as opposed to using ***javacomponent***, updating the component's size or position must be done after a corresponding change to the component's parent; otherwise, the change will have no visual effect:[8]

```
jButton.getParent.setSize(100,30);
jButton.setSize(100,30);
```

A better way to modify the component's size/position is to use the HG container's *setPosition()* method. This updates both container and component at the same time:

```
jButton.getParent.setPosition(50,100,70,40);
```

Also, note that the HG container created this way fails to register properly as a figure child and, therefore, cannot be found or accessed via the HG hierarchy (using ***findall***, etc.), and it will not be cleared with the ***clf*** command and other similar side effects. It is actually not a regular (double-value) HG handle at all, but rather a Java object of class `com.mathworks.hg.peer.HGPanel`. The bottom line is that this method of adding components is nice as a one-time exercise, but it should not be used in practice.‡

If we set the component's pixel position ourselves (rather than using the container's layout manager to determine the position), note that Java positions start at (0,0) in the **top**-left corner of the content pane (which includes the figure's toolbars but not its menubar), increasing **downwards** toward the lower right. This is contrary to the MATLAB positions, which start at (1,1) in the **bottom**-left corner, increasing **upwards** toward the upper right, exclusive of the toolbar area. This difference may cause confusion when placing the component and also when trying to compare the MATLAB handle's **PixelPosition** property (via ***getpixelposition***) and the Java component's position (via its *getBounds()* method).

---

† There actually *is* a very awkward, undocumented, unsupported, and very problematic method to add MATLAb components to Java Swing containers (see Section 3.8).

‡ Other bugs prevent using hgjavacontainers as container parents: ***uicontrol***s are not displayed at all; ***javacomponent***s are badly positioned: http://www.mathworks.com/matlabcentral/newsreader/view_thread/295520#804720 (or http://bit.ly/eEFuxU).

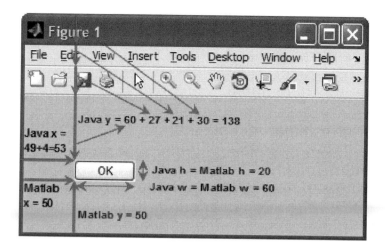

Java versus MATLAB pixel positions (See color insert.)

Note that each HG object has an undocumented **PixelBounds** property — be careful NOT to use it: it sometimes reports incorrect values and is unreliable and misleading.[†]

The exact (pixel) positions can be used to take screenshots of individual components or even the entire window. This can be done using java.awt.Robot's *createScreenCapture(...)* method[9] or using the component's *printAll()* method.[10]

Swing components can also be added directly to a figure window's content pane:

```
jFrame = get(handle(gcf),'JavaFrame');
jButton = javax.swing.JButton('Click me!');
jPane = jFrame.fHG1.getContentPane;‡
jPane.add(jButton, java.awt.BorderLayout.NORTH);
jPane.revalidate;           % repaint jPane with the added JButton
```

Note the mandatory call to *revalidate()*; otherwise, the figure window will not repaint the content area and the added component will not become visible.[11] Also, note that Java components added directly in this fashion will have no HG container (not even a com.mathworks. hg.peer.HGPanel) and will thus be inaccessible via the HG hierarchy.

 Note: When using *javacomponent* to add Java components to MATLAB figures, we should be aware of several additional important issues:

- The default background color of *javacomponent*s is a slightly different shade of gray than the default *uicontrol* background color.

[†] For example, sometimes it reports the 5-pixel margin surrounding uicontrols and sometimes not; when the uicontrols moves or resizes, the PixelBounds value sometimes fails to be updated, and so on.

[‡] In R2007b and earlier, use fFigureClient rather than fHG1Client.

- Java components are not automatically hidden with their ancestor container panel. This is also the root cause of the failure of Java components to disappear when switching tabs in a **uitab** (see Section 4.3).
- Java components are slightly mis-aligned vertically with **uicontrol**s, even when positioned using the same Y position.
- Unlike **uicontrol**s, Java callbacks are activated even when the affected value does not change. Therefore, setting a value in the component's callback could well cause an infinite loop of invoked callbacks.

For a detailed description and workarounds for these and other issues with **javacomponent**, please refer to the online reference that is provided at the end of this sentence.[12]

3.1.2 *The Swing Component Library*

Java's Swing toolkit contains many controls (a.k.a. *widgets*) and containers. The official online tutorial provides an excellent, although somewhat disorderly, introduction to Swing features[13] and components,[14] and how to utilize these within GUIs.[15]

Most Swing controls have MATLAB extensions. For example, `com.mathworks.mwswing.MJButton` extends `javax.swing.JButton`. These extensions are detailed in Chapter 5. All the standard MATLAB **uicontrol**s have Swing equivalents. This is not surprising considering the fact that internally MATLAB controls actually use the MATLAB extensions. Thus, a MATLAB button **uicontrol** is basically a `com.mathworks.hg.peer.PushButtonPeer` object that extends `javax.swing.JButton` via the intermediate `com.mathworks.mwswing.MJButton`. Chapter 6 will detail the objects that underlie the standard MATLAB **uicontrol**s.

Swing Component	Type	MATLAB Equivalent	MATLAB Doc
JButton	Basic control	*uicontrol('style', 'pushbutton')*	Full
JCheckBox	Basic control	*uicontrol('style', 'checkbox')*	Full
JComboBox	Basic control	*uicontrol('style', 'popupmenu')*	Full
JList	Basic control	*uicontrol('style', 'listbox')*	Full
JMenu	Basic control	*uimenu, uicontextmenu*	Full
JPopupMenu	Basic control	*uimenu*	Full
JCheckBoxMenuItem	Basic control	*uimenu('checked', 'on')*	Full
JRadioButton	Basic control	*uicontrol('style', 'radiobutton')*	Full
JSlider	Basic control	—	—
JSpinner	Basic control	—	—

(continued)

Swing Component	Type	MATLAB Equivalent	MATLAB Doc
JScrollBar	Basic control	*uicontrol('style', 'slider')*†	Full
JTextField	Basic control	*uicontrol('style', 'edit')*	Full
JFormattedTextField	Basic control	—	—
JPasswordField	Basic control	—	—
JColorChooser	Complex control	*uisetcolor*	Full
JEditorPane	Complex control	—	—
JTextPane	Complex control	—	—
JTextArea	Complex control	*uicontrol('style', 'edit')*	Full
JFileChooser	Complex control	*uigetfile, uiputfile*	Full
JTable	Complex control	*uitable*	Full since R2008a
JTree	Complex control	*uitree*	Semi
JLabel	Noninteractive	*uicontrol('style', 'text')*	Full
JProgressBar	Noninteractive	– (simulate with *plot* or *waitbar*)	—
JSeparator	Noninteractive	*uimenu('separator', 'on')*	Full
JToolTip	Noninteractive	*set(hControl, 'tooltipString', '...')*	Full
JApplet	Top-level container	—	—
JDialog	Top-level container	*dialog*	Full
JFrame	Top-level container	*figure*	Full
JWindow	Top-level container	—	—
JDesktopPane	Top-level container	–	—
JInternalFrame	Complex container	—	—
JLayeredPane	Complex container	– (use *uipanel* + *uistack*)‡	Full
JRootPane	Complex container	—	—
JMenuBar	Basic container	*uimenu*	Full
JPanel	Basic container	*uipanel*	Full
JScrollPane	Basic container	–	—
JSplitPane	Basic container	- (*UISplitPane* in Chapter 10)	—
JTabbedPane	Basic container	*uitab, uitabgroup*	Semi
JToolBar	Basic container	*uitab\bar*	Full
Box	Basic container		—

---

† Note that *uicontrol('style', 'slider')* actually produces a scroll–bar and not a slider. See the discussion later in this section.

‡ To a limited extent only. For example, plot axes are always displayed beneath uipanels, which are always displayed beneath javacomponents, which are themselves always displayed beneath uicontrols. The built-in MATLAB function *uistack* only affects the z-ordering within subgroups.

Here is the full list of the standard Swing components, with their MATLAB counterparts. A visual list can be found in the official online tutorial.[16] All Swing components are classes in the `javax.swing` package (e.g., `javax.swing.JButton`):

Here is a visual presentation of some Swing components and effects which are missing in MATLAB, as displayed using ***javacomponent***.[†]

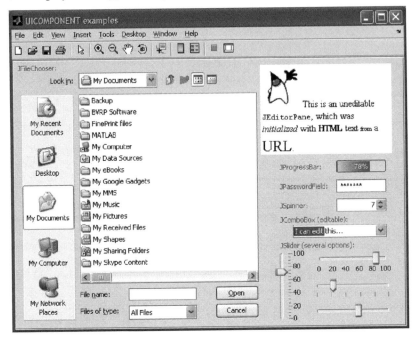

Sample Swing components integrated in a MATLAB figure window (See color insert.)

Some Swing components can be mimicked with pure-MATLAB code, but very inefficiently. For example, `JPasswordField` can be simulated by trapping keyboard clicks, adding '*' characters to a text ***uicontrol*** and saving typed characters as the password; `JProgressBar` can be programmed as an axes with no tick marks or tick labels, a ***patch*** plot object that fills the left part of the axes up to the required percentage point, and a ***text*** object ('78%') on top. Some effects, like the non-**StringPainted** variation of `JProgressBar` (▓▓▓▓▓▓▓), can be very difficult to simulate. In all cases, lots of MATLAB code is required to get the same look-and-feel that can be achieved with a short ***javacomponent*** command.

Having said this, readers should always suit the solution to the problem — sometimes using Java is simply overkill for simple GUI tasks. A standard ***waitbar*** in a modal popup window may be simpler than integrating a `JProgressBar`.[17]

Using standard Swing (or Swing-derived) components can often improve MATLAB GUI effectiveness, responsivity, and visual appearance. Consider the following figure window, containing both a `javax.swing.JScrollBar` and a MATLAB ***uicontrol***(*'style'*,*'slider'*):

[†] This screenshot was taken to illustrate ***uicomponent*** (see Section 3.1.4), but ***uicomponent*** uses ***javacomponent***.

```
jScrollbar = javax.swing.JScrollBar;
jScrollbar.setOrientation(jScrollbar.HORIZONTAL);
javacomponent(jScrollbar, [10,40,200,20]);
uicontrol('style','slider', 'position',[10,10,200,20]);
```

A Swing `JScrollBar` component (top) and a MATLAB slider *uicontrol* (bottom)
(See color insert.)

The topmost component (a `JScrollBar` object) not only looks visually more appealing than the MATLAB *uicontrol* beneath it (which looks so Windows95 1990s style in my opinion), but also has a more consistent look-and-feel with other MATLAB *uicontrol*s, since they are all based more closely on Swing.[†]

`JScrollBar` can also be customized in appearance and behavior to a higher degree than possible with the properties exposed by MATLAB in the slider *uicontrol*. This is typical of all Java components. Section 3.3 describes customization of Java controls.

`JScrollBar` has a very important functionality, which the slider *uicontrol* lacks: the ability to set a continuous-motion callback. The *uicontrol* only has a single action callback property (simply called **Callback**), which is fired only when one of the arrow buttons or unoccupied internal scrollbar area is clicked, or <u>after</u> dragging the scrollbar gripper (handle), that is, when the mouse button is released. Applications often need the callback to fire continuously when the gripper is dragged, but this is not the case for the slider *uicontrol*.[18]

This missing functionality can be easily achieved by using the `JScrollBar` object (or the *uicontrol*'s underlying Swing object, as described below). `JScrollBar` has 29 callbacks which can be set, ranging from focus gain/loss (enabling mouse hover effects), mouse wheel, mouse buttons, and keyboard actions (clicked/released), and also the specific callback that we need in this case: **AdjustmentValueChangedCallback** fires continuously, whenever the slider position value changes, by whichever means (by interactively moving the gripper with the mouse, by a keyboard event, or programmatically).[‡] Section 3.4 provides more details about Swing callbacks.

[†] Also note a slight mis-documentation error for the slider uicontrol: http://www.mathworks.com/matlabcentral/newsreader/view_thread/118214#298452 (or http://tinyurl.com/azlyez).

[‡] http://www.mathworks.com/matlabcentral/newsreader/view_thread/144980 (or http://tinyurl.com/b99jnb); http://www.mathworks.com/matlabcentral/newsreader/view_thread/272224 (or http://bit.ly/9LNCzZ). Note that the continuous-callback issue can be solved in several other ways, using the undocumented ***handle.listener*** function as described in Appendix A and in http://UndocumentedMatlab.com/blog/continuous-slider-callback/ (or http://bit.ly/bexwI9).

The lesson of this short discussion of JScrollBar, which is representative of other controls, is that Java Swing components generally provide all the functionalities of the standard MATLAB controls, with additional benefits in appearance, behavior, and customizability. This topic shall be further explored in Chapter 6.

3.1.3 Displaying Swing-Derived Components

javacomponent can be used to place not only Swing components, but also Swing-extended components onscreen. MATLAB itself almost never uses Swing components as is, instead preferring to use MathWorks-derived extensions of these components, generally in the com. mathworks.mwswing or com.mathworks.widgets packages. These packages and their classes, described in Chapter 5, are all in the static Java classpath and are, therefore, automatically available for use by MATLAB programmers.

It sometimes makes sense to use the MATLAB components instead of the generic Swing counterparts. For example, com.mathworks.mwswing.MJButton extends the standard javax.swing.JButton and normally looks exactly like a JButton. However, when calling MJButton.*setFlyOverAppearance(1)*, which is a new method added by MJButton, the button's appearance changes to look flat with a special effect upon mouse hover. This new feature is handy for toolbar buttons, but it can also be used anywhere in our GUI. I will show a real-life use for this in Section 10.2.

Another internal MATLAB class, com.mathworks.mwt.MWCheckbox, enables a tri-state (yes/no/maybe) checkbox, unlike the standard javax.swing.JCheckBox (yes/no). There are many other tri-state checkbox alternatives available.[19] The point here is only that sometimes it makes sense to use MATLAB's internal derived-classes rather than the original Swing components, but, of course, nothing prevents us from using a better component class that we can find elsewhere. Internal MATLAB classes do have the advantage of being inherently accessible on all platforms of the same MATLAB release, whereas if we use non-MATLAB components, we must include their Java class files in our deployment package.

Note that in some cases, derived classes may remove functionality available in the original Swing class. For example, MATLAB's *uitree*, derived from Swing's JTree, prevents setting tooltips, which is very easy in JTree.[20] The internal MATLAB classes shall be detailed later, in Chapter 5.

Just like MATLAB components, *javacomponent* can also display third-party or our own Swing-derived components. There are quite a few online sources[21] for Swing components that can easily be incorporated into our MATLAB application. Simply download the relevant class files, add them to the static (via *classpath.txt*) or dynamic (via *javaaddpath*) Java classpath, use *javacomponent* to display them, and then use their reference handle to manipulate their appearance and behavior.

3.1.4 *UIComponent and JControl*

javacomponent, useful as it is, has several limitations. In its string variant (class name), it requires a fully qualified classname (FQCN) that is not inferred automatically.† It also has a different parameters format than *uicontrol*, which may confuse users. *javacomponent* also cannot display `java.awt.Window` components. Finally, it returns two handles — one is a *handle()* reference of the Java object and the other is an HG handle (a double numeric value) of the automatically created HG container — users are often confused as to which property should be set on which of these handles.[22]

To overcome these limitations, I created UIComponent — a utility that merges *uicontrol* and *javacomponent*. *uicomponent* is available for free download on the File Exchange.[23] It accepts all *uicontrol* parameters and styles, as well as any other displayable Java (Swing/AWT) class. *uicontrol*'s calling syntax was preserved for full backwards compatibility. *uicomponent* uses the built-in *uicontrol* whenever possible (for standard MATLAB styles) and *javacomponent* for all other Java classes.

uicomponent supports the entire Swing/AWT and any user-defined class, in several equivalent case-insensitive formats. For example, 'Spinner', 'spiNNer', 'JSpinner', and 'javax.swing.jspinner' are all recognized. Notable new styles that are now available (unavailable in the built-in *uicontrol*): spinner, slider, editable comboBox, passwordField, tree, table, fileChooser, colorChooser, and progressBar. But there is much more: whatever is available in Java is seamlessly available in *uicomponent*.

uicomponent returns the same two handles that *javacomponent* returns (namely a Java reference handle and a numeric HG handle), modified to include each other's properties and handles (using *schema.prop* and *handle.listener* — see Appendix B). Here are some examples (more can be found in *uicomponent*'s help comment):

```
uicomponent('style','edit', 'String','hello');   % a regular uicontrol
uicomponent(hFig, 'style','edit', 'String','hello'); % specify parent
uicomponent('style','jspinner','value',7);
uicomponent('style','javax.swing.jslider','tag','myObj');
uicomponent('style','JComboBox',{1,pi,'text'},'editable',true);
```

Another File Exchange submission that aims to tackle some of *javacomponent*'s limitations is Malcolm Lidierth's JCONTROL.[24] *jcontrol* uses MATLAB's new object-oriented class approach and has the benefit of returning just a single handle object, which aggregates the handles for both HG container and the contained Java object.

3.2 MATLAB's Main Thread and the Event Dispatch Thread (EDT)

All Java UI components should run on the Event Dispatch Thread (EDT) or risk ill-effects ranging from intermittent event processing and errors to MATLAB crashes and/or hangs.

---

† Using MATLAB's *import* statement — see Section 1.1.2.

The reason for this is not due to MATLAB but due to the way Swing (or actually, AWT) works.[25]

All graphic events, ranging from component creation through rendering and callback event processing, should funnel through a single processing thread called the EDT. Any action done on another thread (MATLAB's main processing thread in our case) risks a race condition or deadlock with the EDT, which could (and often does) result in weird, non-deterministic, and non-repetitive behavior, all of which should be avoided in any application which should behave in a precisely deterministic manner.

Creating and processing UI components on the EDT are done using MATLAB's ***awtcreate, awtinvoke, javaObjectEDT***, and ***javaMethodEDT*** functions introduced in Section 1.1. Non-EDT usage should be reserved only for very simple GUIs and for non-GUI processing (e.g., data structures or database access described in Chapter 2).

Multiple EDTs could theoretically be used for different independent GUIs. However, setting this up correctly is not easy. Also, the potential performance benefit is small, if we abide by the recommended programming paradigm to minimize EDT processing, running major calculations on a separate thread — MATLAB's main processing thread.

A very rare CSSM thread[26] dwelt on the EDT issue as it pertains to MATLAB, with some very interesting remarks by MathWorks personnel. It is the most detailed MATLAB-specific online reference that I could find.

MATLAB R2007b (7.5) and earlier versions used ***awtcreate*** within ***javacomponent*** for newly created components, that is, components requested as classname from the ***javacomponent*** function, for example, ***javacomponent('javax.swing.JButton')***. This solves the EDT problem for those invocations of ***javacomponent*** but not for invocations of ***javacomponent*** with pre-created Java references, nor for directly adding Swing components and subcomponents to onscreen containers. In all these cases, special care should be taken to use ***awtcreate***. R2008a (and better yet, R2008b) only extended the EDT solution for all ***javacomponent*** invocations, and not for all other cases.

Prior to the introduction of ***javaObjectEDT*** and ***javaMethodEDT***, MATLAB programmers had to use ***awtcreate*** and ***awtinvoke*** in order to create and process components on the EDT. ***awtcreate*** and ***awtinvoke***'s cumbersome JNI notation[27] argument format ensured that programmers rarely used these functions. These functions also had limitations/bugs[†] in their internal (undocumented) function ***parseJavaSignature***,[‡] preventing usage in some cases. In other cases, problems in the Java objects' reflection visibility, which is used by ***awtinvoke***, cause ***awtinvoke*** to fail. In all these cases, we had to resort to using the direct calling syntax:

```
jbutton.setLabel('Close');
```

The problem with this is that it might cause a race condition or deadlock with the EDT, especially following object creation when the EDT has not yet finished rendering the component onscreen. Back then (again, prior to ***javaObjectEDT*** and ***javaMethodEDT***), the workaround

---

† For example, invoking methods that accept a `java.lang.Object`.
‡ %MatlabROOT%/toolbox/matlab/uitools/private/***parseJavaSignature.m***

was to place a time **pause** just before our direct invocation. This had two effects which mitigated the risks: it yielded the CPU to the EDT and gave EDT time to finish its pending actions. The more complex the component (e.g., `JTree` or `JTable`), the longer the required pause. In practice, a 5 or 10 millisecond wait (i.e., **pause(0.01)**) was usually enough.[28] Sometimes, but not always, this **pause** can be replaced with a simple **drawnow**. Normally, such pauses are unnoticeable, but in some cases (e.g., expanding a deeply nested `JTree`), where the pause is repeated dozens or hundreds of times, it may well be annoying.

Starting with MATLAB R2008b (7.7), users should use **javaObjectEDT** and **javaMethodEDT** for all GUI-related Java components (subclasses of `java.awt.*` or `javax.swing.*`), instead of all these crude workarounds, thereby removing the risks altogether as well as all the unnecessary pauses. **javaObjectEDT** (and the corresponding **javaMethodEDT**) behaves just like **javaObject** (and **javaMethod**), except that it runs on the EDT, without any of the cumbersomeness of **awtcreate** (and **awtinvoke**). Moreover, **javaObjectEDT** accepts any reference of an existing Java object and ensures that all the method invocations on this reference object from that point onward will automatically be dispatched on the EDT, without any code change.

If we only use **javacomponent** to place components onscreen, then we only need to worry about using EDT for subcomponents, since **javacomponent** uses **javaObjectEDT** automatically. However, if we set up specialized subcomponents (e.g., CellRenderers and CellEditors described in Section 4.1.1), these are not handled by **javacomponent** and should be handled manually.

 Note: **javaObjectEDT** and **javaMethodEDT** were actually released as undocumented built-in functions in MATLAB 7.6 (R2008a). Unfortunately, I have found that relying on R2008a (7.6)'s version of **javaObjectEDT** sometimes causes MATLAB to hang. As far as I could test, this was corrected in MATLAB release R2008b (7.7).

```
% Try to use the EDT from now on, if available
majorVersion = str2double(regexprep(version,'^(\d+).*','$1'));
minorVersion = str2double(regexprep(version,'^\d+\.(\d+).*','$1'));
if majorVersion >= 7 && minorVersion > 6
   result = javaObjectEDT(jObject);
end

% Alternate method suggested by Jan Simon:29
V = sscanf(version, '%d.', 2);
if V(1) >= 7 && V(2) > 6
   result = javaObjectEDT(jObject);
end
```

Note a common pitfall when checking MATLAB versions: MATLAB's version number is stored as a *major.minor* string format, which is NOT a decimal representation. Therefore, 7.10 is a newer version than 7.9, and is NOT the same as the old MATLAB 7.1. In fact, MATLAB 7.2 is much older than MATLAB 7.10 or 7.11. For this reason, we need to separately test the *major* version number (7) and the *minor* number, as has been done in the code snippet above.[30]

Recent MATLAB releases have added the **verlessThan** built-in function that does this for us, but unfortunately **verLessThan** did not exist in earlier releases and so code relying on this function would not work on these releases. Using the code snippet above solves this problem for all MATLAB 7 releases (earlier releases did not have the **regexprep** function, but this can easily be fixed by other means).

Note that we cannot simply call **javaObjectEDT** within a **try-catch** block, since MATLAB 7.6 (R2008a) does have this function (and will thus not fail), although we should not use it because of its bugs on that version. The code snippet above ensures correct behaviors on all MATLAB versions, both old and new.

---

javaObjectEDT and **javaMethodEDT** were undocumented until R2009a (7.8), when they became officially supported.[31] Similarly, in R2008a onward, there are corresponding **javaObjectMT** and **javaMethodMT** for creating objects and invoking their methods on MATLAB's Main Thread. Since this is **javaObject** and **javaMethod**'s normal behavior and since I cannot see reasons to invoke methods of EDT-created objects on the Main Thread, I see little practical use for **javaObjectMT** and **javaMethodMT**.

 Note: There is also an undocumented built-in **edtObject** function. This function is mentioned in %MATLABroot%/bin/registry/jmi.xml, but I have not seen it being used anywhere in the visible MATLAB code corpus, and I do not know its exact use or purpose. It appears to accept the same input arguments as **javaObjectEDT** (at least, it complains otherwise...), so I assume that it is either a helper function for **javaObjectEDT** or its equivalent. Similarly related yet unfamiliar functions are **java_method**, **java_object**, and **java_array**, which seem to correspond to **javaMethod**, **javaObject**, and **javaArray**.

Extra care should be taken to use **javaObjectEDT** or **awtcreate** when directly adding Java Swing components and subcomponents to onscreen containers. Remember that all Swing components are also potential containers, since `javax.swing.JComponent` extends `java.awt.Container`. Therefore, once a Java component is placed onscreen, nothing prevents a direct addition of subcomponents or invoking their methods outside EDT. Doing so will actually work most of the time, with occasional EDT-related effects.

Wherever possible (R2008b onward), users should always use **javaObjectEDT** instead of **awtcreate**. Not only are **awtcreate**'s limitations solved, but **javaObjectEDT** also marks the object for automatic delegation on the EDT. This means that any future action on this object will automatically occur on the EDT without any need for special programming setup. In contrast, **awtcreate** only uses EDT for the initial object creation, and any future actions on this object need to use **awtinvoke** (or **javaMethodEDT**) in order to use EDT.

Similar use of the EDT (or more precisely, asynchronous GUI operations) should be done in Java code, either standalone or code called from MATLAB. A good example of this is the simple and yet incorrect creation of a Java `JFrame` followed by its immediate population with internal components and its display:

```
import javax.swing.JFrame;
public class Test
{
   public static void main(String[] args) {
     JFrame jframe = new JFrame();
     jframe.setVisible(true);
   }
}
```

While this may work in console mode, it will fail when called from MATLAB, since the Java *main()* code runs on the main MATLAB thread, whereas the `JFrame` creation runs on the EDT. This causes a race condition resulting in an empty-looking `JFrame`. The solution is to use EDT, as shown in the following Java code snippet:[32]

```
import javax.swing.JFrame;
public class Test
{
   JFrame jframe;
   Test() {
     jframe = new JFrame();
   }
   public static void main(String[] args) {
     java.awt.EventQueue.invokeLater(new Runnable()
     {
       public void run() {
         Test test = new Test();
         test.jframe.setVisible(true);
       }
     });
   }
}
```

When debugging using a Java IDE (see Section 1.6), EDT appears in the list of threads as "AWT-EventQueue-0" (look closely at the NetBeans screenshot in Section 1.6).

Sometimes, an EDT-related warning will appear in the Command Window — see *disableThreadSafeGetMethods()* in Section 8.1.1 for a way to stop them. Note that these warnings have a very good reason for existence, and so this should be used with care.

I believe that an explanation of the internal MATLAB interpreter implementation at this point would help our understanding. Note that this explanation may be inaccurate, as I have no access to the internal MATLAB code. However, it fits well-known programming practices and explains observed phenomena, so I suspect that it is correct.

My hunch is that MATLAB's main computational thread is implemented as an endless loop that waits for computational chunks. These chunks may be Command-Window requests, or runnable MATLAB function code, or invoked GUI/timer callbacks, and so on. The chunks wait in a queue to be processed by the main MATLAB thread, which only handles a single chunk at a time. Java to MATLAB calls (using JMI — see Section 9.2) are simply another chunk and need to wait for their processing turn. Therefore, **calling MATLAB from the EDT is a bad idea that could even deadlock MATLAB**. In a multithreaded application, calling **pause** would have enabled JMI code to execute on another thread. However, in MATLAB there is only a single computational thread (the Main Thread), so the MATLAB code has to finish before any JMI request can be handled.

If all this is correct, it means that true multithreaded MATLAB applications cannot be implemented, even when using Java or C++ threads.[33] Although MATLAB applications can have multiple Java or C++ threads,[34] if they need the single-threaded MATLAB core engine, they would simply need to wait in line for their turn. If such a need arises, it is better to handle the processing in the Java/C++ threads and not in the MATLAB thread. Unfortunately, even this does not guarantee "thread-safety" for file I/O.[35]

3.3 Customizing Java Components

3.3.1 *Component Properties and Methods*

Refer again to the screenshot of Swing components shown in Section 3.1.2. The components on the bottom right of this screenshot, namely JSlider and JComboBox, show how slight modifications of the component properties can have a distinct visual effect. JSlider shows distinct variants by changing only one or two properties. For JComboBox, a single property enabled interactive editing of the combo-box's content — something not directly possible when using the MATLAB *uicontrol* equivalent (later in this chapter, we shall see how to achieve this even for MATLAB uicontrols, by modifying properties of the Java components which underlie the MATLAB uicontrols).

Let us take JProgressBar as an illustrative example of customizing Swing controls:

```
% Create a progress-bar with initial value = 57% (type A)
jProgressBar = javax.swing.JProgressBar;
jProgressBar.setValue(57);        % default range is [0-100]
[jhProgressBar,hContainer] = javacomponent(jProgressBar,[20,20,100,40]);
```

```
jProgressBar.setStringPainted(true);    % (type B)
% alternative: set(jProgressBar,'StringPainted','on')   %see note below
```

```
jProgressBar.setString('57%');
```

```
jProgressBar.setIndeterminate(true);   % (type C) 36
```

Animated (indeterminate) progress-bar

Slightly different appearances can be achieved using different Swing look-and-feels (see the discussion in Section 3.3.2). For example, in the Windows Classic L&F:

Progress-bars in the Windows Classic look-and-feel (types A, B, and C)

Another Swing component whose appearance can radically be modified with very few property changes is JSlider, as the following code snippets show:

```
% Create a slider with horizontal orientation (jSlider.HORIZONTAL = 0),
% with no labels nor tick marks, and initial value = 57%
jSlider = javax.swing.JSlider;
jSlider.setValue(57);
[jhSlider,hContainer] = javacomponent(jSlider,[10,10,100,40]);
```

```
set(jSlider, 'Value',84, 'MajorTickSpacing',20, 'PaintLabels',true);
```

 Note: Some Java components, such as JSlider, require Java-style scalar (true/false or 1/0) data for Boolean properties like **PaintLabels**, while other components such as JProgressBar require MATLAB-style 'on'/'off'. Moreover, JSlider appears to have accepted 'on'/'off' until MATLAB release R2008a (7.6) or so . . . . I do not know how to tell in advance whether a particular Java class expects 'on'/'off' or true/false. I guess we need to try each case separately — a run-time error will be thrown if we are wrong:

```
>> set(jSlider, 'PaintLabels','on');
??? Parameter must be scalar.

>> set(jProgressBar, 'StringPainted',true)
??? Error using ==> set
Bad property value found.
Object Name : javax.swing.JProgressBar
Property Name : 'StringPainted'.
```

```
set(jSlider, 'Value',22, 'PaintLabels','off', 'PaintTicks',true);
```

```
jSlider.setPaintLabels(true);        % or: jSlider.setPaintLabels(1);
```

```
set(jSlider, 'Value',72, 'Orientation',jSlider.VERTICAL, ...
             'MinorTickSpacing',5);
set(hContainer,'position',[10,10,40,100]);   %note container size change
```

And now, here is a JSlider that was more extensively configured:[37]

```
rm = javax.swing.DefaultBoundedRangeModel(1,0,1,10); % 1-to-10 range
js = javax.swing.JSlider(rm);
set(js,'Background',[0,0,0],'Foreground',[1,1,1]);   % white on black

tickLabel{1} = javax.swing.JLabel(java.lang.String('1'));
set(tickLabel{1},'Foreground',[0,1,0]);              % green '1'
tickLabel{2} = javax.swing.JLabel(java.lang.String('10'));
set(tickLabel{2},'Foreground',[1,0,0]);              % red '10'

jht = java.util.Hashtable();
jht.put(java.lang.Integer(1), tickLabel{1});
jht.put(java.lang.Integer(10), tickLabel{2});
js.setLabelTable(jht);

js.setMinorTickSpacing(1);
js.setMajorTickSpacing(2);
js.setSnapToTicks(true);                             % snap to integers
js.setPaintTicks(true);
js.setPaintLabels(true);
[jsh, hContainer] = javacomponent(js);               %jsh = handle(js)
set(hContainer,'Position', [100, 100, 200, 40]);
```

In addition to the component and container classes, Swing also contains many useful utility classes. For example, `BorderFactory` facilitates creation of `Border` objects (the outline that surrounds the components, giving a colored outline, 3D appearance, or other visual effects).[38] Here is a sample `JButton` with a few different borders:[39]

```
import javax.swing.* java.awt.*
jb1 = JButton('Click me #1!');
jb2 = JButton('Click me #2!');
jb3 = JButton('Click me #3!');
etchedBorder = javax.swing.border.EtchedBorder.LOWERED;
jb1.setBorder(BorderFactory.createLineBorder(Color.red));
jb1.setBorder(BorderFactory.createEtchedBorder(etchedBorder));
jb3.setBorder(BorderFactory.createRaisedBevelBorder());
javacomponent(jb1,[10,10,100,40]);
javacomponent(jb2,[130,10,100,40]);
javacomponent(jb3,[250,10,100,40]);
```

Buttons with different border types

Each Swing component can have its own specific mouse hover cursor, an object of class `java.awt.Cursor`.[40] There are MATLAB functions (the documented *set*(gcf,'Pointer*', ...) and the semi-documented *setptr*) for setting a figure-wide cursor, but Java is required if we need to set component-specific shapes.

The component's cursor is controlled via the **Cursor** property (and the corresponding *setCursor(), getCursor()* methods). This controls the cursor shape when the mouse pointer is located within the component's bounds. The related method *isCursorSet()* determines whether or not a nondefault cursor was set for this component:

```
>> jb = javax.swing.JButton
jb =
javax.swing.JButton[,0,0,60 x 20,alignmentX = 0.0,alignmentY = 0.5,border = ...]
>> jb.getCursor
ans =
java.awt.Cursor[Default Cursor]
>> jb.isCursorSet
ans =
    0
```

```
>> % Now set a non-default cursor:
>> jb.setCursor(java.awt.Cursor(java.awt.Cursor.HAND_CURSOR));
>> jb.getCursor
ans =
java.awt.Cursor[Hand Cursor]
>> jb.isCursorSet
ans =
     1
```

To restore the default cursor, simply call *setCursor([])*. The different predefined cursor types are listed below (the displayed cursor shapes are from Windows — they look slightly different on other platforms):[41]

Name	Value	Cursor	Name	Value	Cursor
DEFAULT_CURSOR	0		NE_RESIZE_CURSOR	7	
CROSSHAIR_CURSOR	1		N_RESIZE_CURSOR	8	
TEXT_CURSOR	2		S_RESIZE_CURSOR	9	
WAIT_CURSOR	3		W_RESIZE_CURSOR	10	
SW_RESIZE_CURSOR	4		E_RESIZE_CURSOR	11	
SE_RESIZE_CURSOR	5		HAND_CURSOR	12	
NW_RESIZE_CURSOR	6		MOVE_CURSOR	13	

Custom cursor shapes can be set via the java.awt.Toolkit.*createCustomCursor()* method.[42] For example, let us use the MATLAB icon as a custom cursor image:

```
% Create the custom cursor
myIcon = fullfile(MATLABroot,'/toolbox/MATLAB/icons/MATLABicon.gif');
imageToolkit = java.awt.Toolkit.getDefaultToolkit;
iconImg = imageToolkit.createImage(myIcon);
hotSpot = java.awt.Point(20,0);    % =MATLAB icon point (top)
myCursor = imageToolkit.createCustomCursor(iconImg,hotSpot,'My Cursor')
      => sun.awt.windows.WCustomCursor[My Cursor]
```

```
% Now use the new cursor
jb.setCursor(myCursor);
```

Note that Java automatically converts iconImage into the closest supported cursor size (imageToolkit.*getBestCursorSize(iconWidth,iconHeight)* = 32 × 32 pixels on Windows XP) and colormap (imageToolkit.*getMaximumCursorColors()* = 256 colors on Windows XP). Since our original iconImage size was 16 × 16, it was resized to 32 × 32, thereby becoming pixelized, as seen above. Note that because of this resizing, we had to set the hotSpot point to (20,0) rather than to (10,0) on the original iconImage.

Also, note that the JButton methods *setCursor(), getCursor()* surprisingly do <u>not</u> have corresponding properties,[†] so we cannot use the standard ***set/get*** stock functions:

```
>> jb.ismethod('setCursor')
ans =
    1
>> jb.isprop('Cursor')
ans =
    0
>> get(jb,'Cursor')
??? Error using ==> get
Invalid javax.swing.JButton property: 'Cursor'.
```

As another example of the usefulness of customized Java components, let's display a hyperlink in our GUI window using a simple border-less Java Swing JButton:[‡]

```
% Prepare the Java JButton object
str = 'Undocumented Matlab.com';  % split link to display multi-line
link = strrep(str,' ','');  % the actual link should have no spaces...
jButton=javax.swing.JButton(['<html><center><a href="">' str '</a>']);
jButton.setToolTipText(['Visit the ' link ' website']);
jButton.setCursor(java.awt.Cursor(java.awt.Cursor.HAND_CURSOR));
jButton.setVerticalAlignment(javax.swing.SwingUtilities.CENTER);
jButton.setMargin(java.awt.Insets(0,0,0,0));
jButton.setContentAreaFilled(false);
jButton.setBorder([]);

% Assign the action callback and display onscreen
hButton = handle(jButton,'CallbackProperties');
set(hButton,'ActionPerformedCallback',['web(''' link ''');']);
[hjButton, hcontainer] = javacomponent(jButton, pixelPos, hParent);
```

[†] This may be a bug in MATLAB's wrapping of the Java object. There are other such cases, for example, the java.awt. Panel, that encloses a ***uicontrol**('style','slider')*, has the *getX()* & *getY()* methods but no 'X' or 'Y' properties.

[‡] Additional methods of displaying hyperlinks are discussed in Sections 5.5.1, 6.5.2, 6.9, 8.3.1, and 8.3.2.

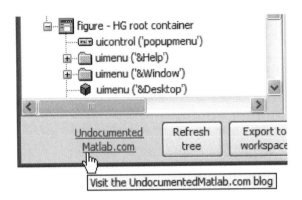

A simple `JButton` appearing as a hyperlink label

3.3.2 Look-and-Feel

One of MATLAB's great advantages is its cross-platform compatibility. Generally speaking, MATLAB applications written on Windows will work as is on Macintosh and Linux. Java has similar cross-platform compatibilities, but it is admittedly slightly easier to verify that two platforms share the same MATLAB version than to verify Java compatibility. Java-based applications might fail on other platforms that have other JVM versions installed, just as MATLAB applications might fail on platforms that have other MATLAB releases. To use Java in a MATLAB application, design for the lowest JVM version it might run on. Also, if the application will run on different platforms, try to use a platform-independent *Look-and-Feel*. This is the topic of this section.[43]

Components can be set to have a different look-and-feel (PLAF or L&F) than their platform's standard.[44] This involves using Swing's `javax.swing.UIManager` class, calling the static `javax.swing.UIManager.setLookAndFeel()` prior to object creation:

```
javax.swing.UIManager.setLookAndFeel('javax.swing.plaf.metal.MetalLookAndFeel')
javacomponent(javax.swing.JSlider);†
```

| Metal L&F | Motif L&F | Windows L&F |

... and similarly for checkboxes, tabs, buttons, and so on (**See color insert.**):

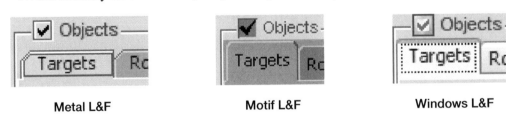

| Metal L&F | Motif L&F | Windows L&F |

---

† And similarly for com.sun.java.swing.plaf.motif.MotifLookAndFeel, com.sun.java.swing.plaf.windows.WindowsLookAndFeel, and com.sun.java.swing.plaf.windows.WindowsClassicLookAndFeel.

The list of available L&Fs can be retrieved using the static method `javax.swing.` UIManager.*getInstalledLookAndFeels()*:

```
>> lafs = javax.swing.UIManager.getInstalledLookAndFeels
lafs =
javax.swing.UIManager$LookAndFeelInfo[]:
    [javax.swing.UIManager$LookAndFeelInfo]      % Metal
    [javax.swing.UIManager$LookAndFeelInfo]      % Nimbus
    [javax.swing.UIManager$LookAndFeelInfo]      % Motif
    [javax.swing.UIManager$LookAndFeelInfo]      % Windows
    [javax.swing.UIManager$LookAndFeelInfo]      % Windows Classic

>> for lafIdx = 1:length(lafs), disp(lafs(lafIdx)); end
javax.swing.UIManager$LookAndFeelInfo[Metal
               javax.swing.plaf.metal.MetalLookAndFeel]
javax.swing.UIManager$LookAndFeelInfo[Nimbus
               com.sun.java.swing.plaf.nimbus.NimbusLookAndFeel]
javax.swing.UIManager$LookAndFeelInfo[CDE/Motif
               com.sun.java.swing.plaf.motif.MotifLookAndFeel]
javax.swing.UIManager$LookAndFeelInfo[Windows
               com.sun.java.swing.plaf.windows.WindowsLookAndFeel]
javax.swing.UIManager$LookAndFeelInfo[Windows Classic
               com.sun.java.swing.plaf.windows.WindowsClassicLookAndFeel]
```

Although not listed in the installed L&Fs, MATLAB also enables access to the third-party Plastic/Plastic 3D L&F by jgoodies.com,[45] which generates a stylish GUI:

```
javax.swing.UIManager.setLookAndFeel(
               'com.jgoodies.looks.plastic.Plastic3DLookAndFeel')
```

Plastic3D L&F

The JIDE class library bundled with MATLAB (see Section 5.7), and specifically the *jide-common.jar* file located in the %MATLABroot%/java/jarext/jide/ folder, contains a separate set of third-party L&Fs: Aqua,[†] Eclipse (Metal[‡] and Windows[§] variants), Office2003,[¶] VSNet

---

[†] `com.jidesoft.plaf.aqua.AquaJideLookAndFeel`. This L&F requires `apple.laf.appleLookAndFeel`, which is normally unavailable on Windows platforms and so cannot be used there. Aqua is the default L&F on Macs. Read here: http://blogs.mathworks.com/desktop/2009/03/23/more-mac-like-tabs/#comment-6180 (or http://tinyurl.com/l9tvox).

[‡] `com.jidesoft.plaf.eclipse.EclipseMetalLookAndFeel`

[§] `com.jidesoft.plaf.eclipse.EclipseWindowsLookAndFeel`

[¶] `com.jidesoft.plaf.office2003.Office2003WindowsLookAndFeel`

(Metal[†] and Windows[‡] variants), and Xerto.[§] As far as I can tell, only EclipseMetal and VsnetMetal have any visible effect, adding a small bluish tint gradient to the Metal L&F (I could find no difference between these two); Office2003 appears similar to the Windows L&F, except that the menus get a bluish-orange gradient tint. I would be happy to hear feedback from readers JIDE's L&Fs.

Starting with the *jide-common.jar* file bundled in MATLAB release R2008b, JIDE stopped including full L&F classes and started using L&F extensions using their com.jidesoft. plaf.LookAndFeelFactory class.[46] I have not been able to use this class effectively, but readers are welcome to try (please let me know if you succeed).

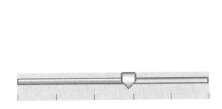

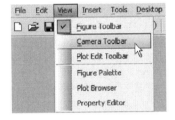

JIDE EclipseMetal L&F   **JIDE Office2003 L&F**

External L&Fs can also be downloaded and then be used in MATLAB.[47]

The current and standard L&Fs can be retrieved using the respective static methods javax.swing.UIManager.*getLookAndFeel()* and *getSystemLookAndFeelClassName()*:

```
>> disp(javax.swing.UIManager.getLookAndFeel)
[The JGoodies Plastic 3D Look and Feel - © 2001-2006 JGoodies Karsten Lentzsch
- com.jgoodies.looks.plastic.Plastic3DLookAndFeel]

>> disp(javax.swing.UIManager.getSystemLookAndFeelClassName)
com.sun.java.swing.plaf.windows.WindowsLookAndFeel
```

MATLAB has a utility class com.mathworks.mwswing.plaf.PlafUtils that contains static methods to query the current L&F: *isPlasticLookAndFeel()*, *isAquaLookAndFeel()*,[¶] *isMetalLookAndFeel()*, *isMotifLookAndFeel()*, and *isWindowsLookAndFeel()*.[††]

Modifying the L&F affects *all* components created from then on. It also automatically updates the MATLAB Desktop and Editor's L&F, which may cause unexpected errors.[‡‡] In some cases, this can at least partially be solved as follows:[48]

[†] com.jidesoft.plaf.vsnet.VsnetMetalLookAndFeel

[‡] com.jidesoft.plaf.vsnet.VsnetWindowsLookAndFeel

[§] com.jidesoft.plaf.xerto.XertoWindowsLookAndFeel

[¶] The Aqua L&F apparently refers to the Apple Macintosh L&F.

[††] There is, unfortunately, no such method for the new Nimbus L&F, which is discussed later in this section.

[‡‡] For example, on a WindowsXP platform, an exception about XP combo-box button when using WindowsClassic or Plastic L&F; also see http://www.mathworks.com/matlabcentral/newsreader/view_thread/257284 (or http://tinyurl.com/mynybr).

```
jFrame = com.mathworks.mde.desk.MLDesktop.getInstance.getMainFrame();
javax.swing.SwingUtilities.updateComponentTreeUI(jFrame);
jFrame.repaint;
```

Or possibly, it can be solved with the following additional settings:

```
jFrame.getRootPane.putClientProperty('defeatSystemEventQueueCheck',1)
jFrame.getRootPane.putClientProperty('ClassLoader',
                              jFrame.getClass.getClassLoader)
```

In any case, my advice is to restore the default L&F immediately after creating any component with a non-standard L&F:

```
originalLnF = javax.swing.UIManager.getLookAndFeel;         % class
newLnFName = 'javax.swing.plaf.metal.MetalLookAndFeel';    % string
javax.swing.UIManager.setLookAndFeel(newLnFName);
jComponent = javacomponent(...);                           % Create GUI
drawnow;    % ensure the controls are displayed before restoring L&F
javax.swing.UIManager.setLookAndFeel(originalLnF);   % Restore L&F
```

Components can update their L&F to the current L&F using their jComponent.*updateUI()* method. Components that are not specifically updated by invoking their *updateUI()* method will retain their existing (original) L&F — the L&F which was active when the components were created or last updated.

The default settings for the L&F can be retrieved using the static method javax.swing.UIManager.*getDefaults()*, which returns an enumeration of the many hundreds of all default settings (1019 on my Windows platform ...):

```
>> defaults = javax.swing.UIManager.getDefaults;
>> propValues = defaults.elements; propKeys = defaults.keys;
>> while propKeys.hasMoreElements
     key = propKeys.nextElement;
     value = propValues.nextElement;
     disp([char(key) ' = ' evalc('disp(value)')]);
   end
SplitPane.dividerSize =        5
DockableFrameTitlePane.stopAutohideIcon = javax.swing.ImageIcon@1f4e4c0
FormattedTextField.caretBlinkRate =      500
Table.gridColor = javax.swing.plaf.ColorUIResource[r = 128,g = 128,b = 128]
... (1015 other property settings)
```

Specific settings can be modified using javax.swing.UIManager.*put(key,newValue)*.

Readers wishing for complete control over the look-and-feel are referred to Swing's Synth L&F.[49] Synth enables customization of practically every aspect of the visual appearance and component behavior, using XML configuration files.

With Java 1.6 update 10 onward (available in MATLAB R2010a (7.10), or as a retrofit as explained in Section 1.8.2), it is possible to use the Nimbus L&F[50] instead of Synth: Nimbus enables better cross-platform vectorized appearances and all the customizability that Synth offers. Nimbus was specifically designed to enable creating personalized skins as a derivative of the basic L&F.[51] However, most designers who target application for a particular platform (Windows or Unix) apparently favor Java's native (a.k.a. "System") L&F for that platform.[52] Also, the Nimbus L&F takes some time to tweak and appears to generate many errors in MATLAB. Still, Nimbus may be useful for cross-platform applications as well as for specific GUI components. Nimbus is preinstalled as a non-default L&F in MATLAB R2010a.

Readers may be interested in Malcolm Lidierth's MUtilities package on MATLAB's File Exchange.[53] MUtilities encapsulated L&F (and other GUI) functionality in MATLAB.

3.3.3 HTML Support

A common feature of Swing components is their acceptance of HTML (and partial CSS[†]) for any of their JLabels.[54] Since all MATLAB *uicontrol*s are based on Swing-derived components, this Swing feature automatically applies to MATLAB *uicontrol*s as well.[‡] Whatever can be formatted in HTML (font, color, size, etc.) is inherently available in MATLAB controls. Note that HTML tags do not need to be closed (<tag> ...</tag>), although it is good practice to close them properly. For example, let us create a multi-colored MATLAB listbox:[55]

```
uicontrol('Style','list', 'Position',[10,10,70,70], 'String', ...
   {'<HTML><FONT color = "red">Hello</Font></html>', 'world', ...
    '<html><div style = "font-family:impact;color:green"><i>What a', ...
    '<Html><Font color = "rgb(0,0,255)" face = "Comic Sans MS">nice day!'});
```

HTML rendering in standard MATLAB uicontrols (See color insert.)

Note the alternative ways of specifying colors in this example: <font color="red">, <font color="rgb(0,0,255)">, and <div style="color:green">. Also note the use of the double quotes (") for the HTML strings: HTML also accepts single quotes ('), but the double quotes do not get mixed-up with MATLAB's string quotes, thereby improving readability (a similar trick is often used for JavaScript code that is inlined in HTML).

The supported HTML subset includes the <img> tag and can, therefore, display images.[§] However, the image src (filename) needs to be formatted in a URL-compliant format such as

---

† For example, the text-align CSS directive appears to be ignored, while font directives (color/size, etc.) are honored.

‡ Note that the text uicontrol is based on a class which overrides the HTML support: http://UndocumentedMatlab.com/blog/html-support-in-matlab-uicomponents/#comment-12 (or http://bit.ly/bujkFq).

§ Chapter 6 will show how the controls can be made to display images using other means.

'http://www.website.com/folder/image.gif' or 'file:/c:/folder/subfolder/img.png.' Warning: if we try to use a non-URL-format filename, the image will not be rendered, only a placeholder box:

```
uicontrol('Position',..., 'String','<Html><img src = "img.gif">'); %bad
uicontrol('Style','list', ... '<Html><img src = "img.gif"/>'}); %bad
```

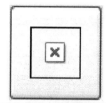 

Ill-specified HTML <img>s in MATLAB uicontrols

```
>> iconsFolder = fullfile(MATLABroot,'/toolbox/MATLAB/icons/');
>> iconUrl = strrep(['file:/' iconsFolder 'MATLABicon.gif'],'\','/');
>> str = ['<Html><img src = "' iconUrl '">']
str =
<Html><img src = "file:/C:/Program
Files/MATLAB/.../icons/MATLABicon.gif" >

>> uicontrol('Position',..., 'String',str);
>> uicontrol('Style','list', ... str});
```

Correctly specified HTML <img>s in MATLAB uicontrols (See color insert.)

HTML can also be used in tooltips. For example, let us place an image directly in the tooltip HTML:

```
filePath = 'C:\Yair\Undocumented MATLAB\Images\table.png';
str = ['<html><center><img src = "' filePath '"><br>' filePath];
set(hButton,'tooltipString',str);
```

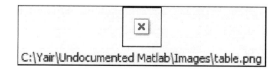

Tooltip with invalid HTML img source URL

If we fix `filePath` to be a valid URL, it now looks as intended:

```
filePath = 'C:\Yair\Undocumented MATLAB\Images\family.jpg;
filePath = strrep(['file:/' filePath],'/','/');
str = ['<html><center><img src = "' filePath '"><br> ' ...
      '<b><font color = "blue">' filePath];
set(hButton,'tooltipString',str);
```

file:/C:/Yair/Undocumented Matlab/Images/family.jpg

Tooltip with a valid HTML image and caption

This tooltip looks enormous (it was actually downsized to fit this page ...), because our HTML <img> size was not limited, so the tooltip was automatically enlarged to contain the full image size. To limit the tooltip size, simply add the Height and Width attributes to the <img> tag, remembering to preserve the original image aspect ratio.

HTML support is very useful when trying to overcome MATLAB's text label *uicontrol*'s limitation of not supporting either Tex or HTML. Instead, we can simply use a standard Java Swing `JLabel`, which does support HTML.[56] For example,

```
%show the 'for all' and 'beta' symbols and other HTML formatting
str = '<html>&#8704;&#946; <b>bold</b><i> <font color = "red">label</html>';
jLabel = javaObjectEDT('javax.swing.JLabel',str);
[hcomponent,hcontainer] = javacomponent(jLabel,[100,100,80,20],gcf);
```

∀β **bold** *label*

HTML-rendered label (See color insert.)

Menus and tooltips can also be HTML-customized in a similar fashion (note: menu customization will be detailed separately in Section 4.6):[57]

```
uicontrol('Style','popup', 'Position',[10,10,150,100], 'String', ...
   {'<HTML><BODY bgcolor="green">green background</BODY></HTML>', ...
    '<HTML><FONT color="red" size = " + 2">Large red font</FONT>', ...
    '<HTML><BODY bgcolor="#FF00FF"><PRE>fixed-width font'});
```

HTML-rendered controls (popup menus and tooltips) (See color insert.)

HTML-rendered menus (See color insert.)

A blog reader has suggested[58] using HTML to display font names in their own font:

```
fontStr = @(font) ['<html><font face = "' font '">' font '</font></
html>'];
htmlStr = cellfun(fontStr,listfonts,'uniform',false);
uicontrol('style','popupmenu', 'string',htmlStr, 'pos',[20,350,60,20]);
```

HTML-rendered popup (combo-box) menu

There are some caveats when using HTML with *uicontrol*s: HTML buttons do not underline the mnemonic character (e.g., <u>C</u>ancel) — we need to underline it ourselves (<u>C</u>ancel). Also, a Java bug[59] causes HTML text to retain its color (not become gray) when the control is disabled.

In some cases, such as multiline editboxes, HTML support is present but cannot be turned on in MATLAB. Sections 6.5.2 and 6.5.3 show how to turn it on using the underlying Java component (`JEditorPane` in the multiline editbox case):

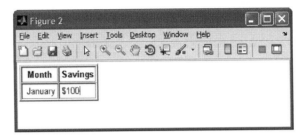

Editable HTML-aware `JEditorPane`

In a related note, a fully capable browser component can easily be included in MATLAB GUI figures to display HTML messages and even entire webpages (local or on the Internet). This is discussed in Section 8.3.2.

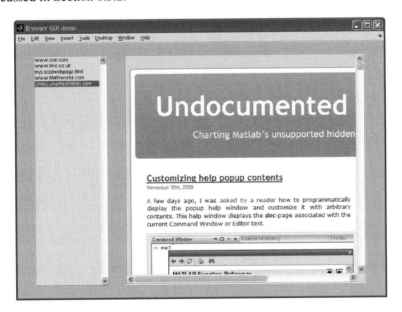

Browser object integrated in a MATLAB GUI

3.3.4 *Focus Traversal*[60]

Java Swing components have several focus-related properties (and accessor methods) that relate to the component's focus cycle, that is, selecting (=setting the focus on) the component using the keyboard.[61] MATLAB documentation calls the focus cycle "*tab-order*", but it only allows selecting the focus cycle order, using the ***uistack*** function — for all the extra functionalities, we need to use these Java properties.

One aspect that is often encountered and easily fixed is MATLAB's default exclusion of all *javacomponent*s from its focus traversal cycle. This means that if we place several *uicontrol*s and *javacomponent*s together onscreen, clicking TAB or Shift-TAB will only move the focus between the regular *uicontrol*s, but none of the *javacomponent*s. The *javacomponent*s can still get the focus, but only programmatically or by mouse click — not by keyboard-clicking TAB or Shift-TAB.

```
h1 = uicontrol('style','edit','position', [10 10 120 20]);
h2 = uicontrol('style','edit','position', [10 40 120 20]);
h3 = javacomponent(javax.swing.JTextField, [10 70 120 20]);
```

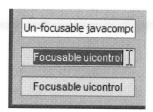

Problem: the *javacomponent* is not TAB-focusable
(TAB only switches focus between bottom uicontrols)

The fix for this problem is immediate:

```
h3.setFocusable(true);
```

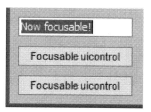

Solution: the *javacomponent* is now TAB focusable as expected
(TAB now cycles between all three controls)

Note that this fixes the problem despite the fact that **Focusable** is already true. The reason is that *setFocusable()* invokes a back-end Java method that resets the component to its standard focus policy, overriding MATLAB's non-standard policy.

3.4 Component Callbacks

Components are useless if they cannot be queried and updated programmatically or if they do not call asynchronous actions (a.k.a. *fire callbacks*) when certain events occur. These callbacks have to be asynchronous or the computer will have to keep polling the controls to query them for changes, instead of doing useful stuff. In short, good callback support and implementation are at the heart of every successful GUI system.

Swing provides a standard set of callbacks for all its components, with a few additional specific callbacks for specific components. All callbacks can be accessed from within MATLAB.[62] Here is a description of the standard callbacks supported by all Swing components.[63] All callbacks fire just once, unless specified as continuous:

- **AncestorMovedCallback** — fired when one of the component's container ancestors has changed its position relative to its container.[64]
- **AncestorAddedCallback** — fired when one of the component's container ancestors has been added to another container.
- **AncestorRemovedCallback** — fired when one of the component's container ancestors has been removed from the component's hierarchy.
- **AncestorResizedCallback** — fired when one of the component's container ancestor has been resized.
- **ComponentAddedCallback** — fired when a subcomponent is added as a direct child to this component.[65] Compare: **ComponentRemovedCallback.**
- **ComponentHiddenCallback**—fired when the component is hidden (*setVisible(false)*).[66] Compare: **ComponentShownCallback.**
- **ComponentMovedCallback** — fired when the component is moved within its container. Since Java components are enclosed in a tight-fitting HG container, this callback never fires for them: it does not fire when the container moves or resizes but only when the component's starting position is moved within it.
- **ComponentRemovedCallback** — fired when a subcomponent is added as a direct child to this component. Compare: **ComponentAddedCallback.**
- **ComponentResizedCallback** — fired when the component is resized, either directly or because its container was resized.
- **ComponentShownCallback**—fired when the component is displayed (*setVisible(true)*). Compare: **ComponentHiddenCallback.**
- **FocusGainedCallback** — fired when the component gains GUI focus,[67] by a mouse click,† <Tab> key click,‡ or calling the component's *requestFocus()* method. Can be used at the figure-level to detect window focus events.[68]
- **FocusLostCallback** — fired when the component loses focus to another component or window. Compare: **FocusGainedCallback**.[69]
- **HierarchyChangedCallback** — fired when the component changes its ancestors (e.g., moved from one panel to another).[70]

---

† Mouse clicks can be either interactive or programmatic, as explained here: http://www.mathworks.com/matlabcentral/newsreader/view_thread/235825#668975 (or http://tinyurl.com/l5b7av).

‡ Like mouse clicks, <tab> clicks can also be invoked programmatically using java.awt.Robot (*keyPress()/keyRelease()* methods).

- **KeyPressedCallback** — fired continuously when a keyboard button (including Shift, Ctrl, etc.) was pressed while the component had focus.† The meta-data contains details about the specific key and modifiers (Alt, Shift, Ctrl, etc.) that were pressed.[71] Compare: **KeyReleasedCallback**, **KeyTypedCallback**.

- **KeyReleasedCallback** — fired when a keyboard button was released while the component had focus. The meta-data contain details about the specific key and modifiers (Alt, Shift, Ctrl, etc.) that were pressed. Compare: **KeyPressedCallback**.

- **KeyTypedCallback** — similar to **KeyPressedCallback**, but only fired (continuously) when an actual printable character is clicked. So, for Shift-A, **KeyPressedCallback** will fire twice (Shift, 'A'), but **KeyTypedCallback** will only fire once. Compare: **KeyPressedCallback**, **KeyReleasedCallback**.

- **MouseClickedCallback** — fired when a mouse button is pressed and then released (=clicked) within the component's bounds. If either the press or the release occurs outside the component's bounds, the event will not fire. The figure's **SelectionType** property‡ will be 'normal', 'extend', or 'alt' depending on which button was pressed. Compare: **MousePressedCallback**, **MouseReleasedCallback**. Also, note an undocumented change in the **SelectionType** property behavior starting in MATLAB 7.6 (R2008a).[72]

- **MouseDraggedCallback** — fired continuously when the mouse is clicked in the component's bounds and then moved while the button is still depressed (i.e., dragged), even beyond the component's bounds.[73] The callback event's meta-data contains the movement's delta-x and delta-y (positive for x-right/y-down; negative for x-left/y-up). Handling drag beyond the component's bounds depends on the component's **Autoscrolls** property. Compare: **MouseMovedCallback**.

- **MouseEnteredCallback** — fired when the mouse is moved (depressed or not) into the component's bounds. Compare: **MouseExitedCallback**.

- **MouseExitedCallback** — fired when the mouse is moved (depressed or not) out of the component's bounds. Compare: **MouseEnteredCallback**.

- **MouseMovedCallback** — fired continuously when the mouse is moved undepressed within the component's bounds. The callback event's meta-data will contain the delta-x and delta-y of the movement (positive for x-right/y-down; negative for x-left/y-up). This can be used to complement MATLAB's **WindowButtonMotionFcn** property when tracking mouse movements over figure components.[74] Compare: **MouseDraggedCallback**.

---

† This can be used to intercept keyboard events in a much more fine-grained manner than that enabled by the pure-MATLAB callbacks. See, for example, http://www.mathworks.com/matlabcentral/newsreader/view_thread/159849 (http://bit.ly/9yYXqf), or http://www.mathworks.com/matlabcentral/newsreader/view_thread/115753#292219 (http://bit.ly/c9PFLy).

‡ This information is part of the Java event's meta-data, but it is not conveyed in the MATLAB meta-data. Luckily, there is this workaround. See http://www.mathworks.com/matlabcentral/newsreader/view_thread/148366 (or http://tinyurl.com/d3h9df), which was a good lesson in humility for me and also shows how sometimes a simple answer lies in unexpected places.

- **MousePressedCallback** — fired when the mouse button is depressed (even before it has been released) within the component's bounds.[75] The callback event meta-data will contain the click location within the component's bounds. Compare: **Mouse-ClickedCallback, MouseReleasedCallback.**

- **MouseReleasedCallback** — fired when the mouse button is released within the component's bounds.[76] The callback event's meta-data will contain the click location within the component's bounds. Compare: **MousePressedCallback.**

- **MouseWheelMovedCallback** — fired when the mouse wheel is turned (even before it has been released) within the component's bounds.[77]

- **PropertyChangeCallback** — fired when one of the component's properties has changed. For example, after setting the component's text, tooltip, or border. Does not fire when modifying the component's callback properties.

- **VetoableChangeCallback** — fired upon a constrained property value change, allowing the callback to intercept and prevent the property change by raising an exception. Of all the Swing components, only `JInternalFrame` actually declares vetoable properties which can be intercepted.[†] We can use this callback for our own components by setting up a `VetoableChangeListener`[78] and calling the component's *addVetoableChangeListener()* method to register it (for a Java callback) or by calling *fireVetoableChange(propertyName, oldValue, newValue)* in our component class's setter methods.[79] A simpler solution involves **schema.prop**, but this is outside the scope of this book.

It should be noted that these callbacks are standard in all Swing GUI controls. Thus, they can be used not just for MATLAB uicontrols' underlying Java objects, but also for any Swing component that you display using MATLAB's *javacomponent* function.

The specific list of callbacks supported by each component depends on component type. As noted above, some components have additional specific callbacks.[80] For example, **ActionPerformedCallback** is fired when a user has performed the main action associated with the control (selecting/clicking, etc.).[81] This is one of the most commonly used callbacks, one of the few exposed by MATLAB HG handles. This callback is implemented by `JButton` and `JCheckBox` (for instance), but not by `JList` or `JMenu`. **CaretUpdateCallback** and **CaretPositionChangedCallback**[82] are only supported by text-entry controls such as `JTextField` or `JEditorPane`, but not by `JSlider` or `JTabbedPane`. Other components have other such specific callbacks. To see the full list of supported callbacks for a particular object, use *uiinspect* (described above) or use the following code snippet:

```
>> props = sort(fieldnames(get(javax.swing.JButton)));
>> callbackNames = props(~cellfun(@isempty,regexp(props,'Callback$')));
callbackNames =
    'ActionPerformedCallback'
    'AncestorAddedCallback'
```

[†] Only four properties are vetoable: Closed, Icon, Maximum, and Selected.

```
'AncestorMovedCallback'
. . .
```

A nice example of using Java callbacks to automatically select (highlight) the content text in a textbox when focus is gained was one of the first online posts in CSSM to use MATLAB 7's new *javacomponent* Swing integration.[83]

To prevent memory leaks in complex GUIs, it is advisable to *get* and *set* callbacks using the *handle()* object, instead of directly using the Java reference.[84] Starting in R2010b, setting callbacks on un-*handle*d Java references evokes a warning message:

```
>> jb = javax.swing.JButton;
>> jbh = handle(jb,'CallbackProperties');
>> set(jbh,'ActionPerformedCallback',@myCallbackFcn) % ok!

>> set(jb, 'ActionPerformedCallback',@myCallbackFcn) % bad! memory leak
Warning: Possible deprecated use of set on a Java callback.
(Type "warning off MATLAB:hg:JavaSetHGProperty" to suppress this warning.)
```

handle() objects implement the *Adapter* design pattern,[85] exposing all properties and methods (but not the public fields) of the original Java reference. The original Java reference can always be accessed via the *handle*'s *java()* method:

```
>> jScrollPane = handle(javax.swing.JScrollPane,'CallbackProperties')
jScrollPane =
    javahandle_withcallbacks.javax.swing.JScrollPane

>> scrollPolicy = jScrollPane.VERTICAL_SCROLLBAR_NEVER
??? No appropriate method, property, or field VERTICAL_SCROLLBAR_NEVER for class
javahandle_withcallbacks.javax.swing.JScrollPane.

>> scrollPolicy = jScrollPane.java.VERTICAL_SCROLLBAR_NEVER
scrollPolicy =
    21
```

Whenever possible, it is good practice to always use *handle* objects instead of directly using the Java reference.[86] Not only does it prevent memory leaks, a good enough reason, but it also enables placing handles of different (even dissimilar) Java objects in a single MATLAB array (which is often useful in GUI programming), provides access to *schema.class*, *schema. prop*, and other similar goodies (see Appendix B), and enables using MATLAB's *getappdata/ setappdata* functions to store control-specific data like standard MATLAB *uicontrol*s:[†]

```
>> setappdata(jbh,'data1',{'cell','array',magic(3)})
>> setappdata(jbh,'data2',gcf)
>> getappdata(jbh)
```

[†] MATLAB R2010b has a bug that sometimes causes an error when using *getappdata* or *setappdata* on Java references in callback functions. To circumvent this, use a java.util.Hashtable instead (see Section 4.1.2 for a working example).

```
ans =
    data1: {'cell' 'array' [3 × 3 double]}
    data2: 1
```

For some unknown reason,[†] unless the initial **handle()** call has been done with the optional 'CallbackProperties' parameter, all subsequent **handle()** calls, even those which explicitly request 'CallbackProperties', will not expose the callback events. This is reflected in the following code snippet that uses **get()**, but also applies to **set()**:

```
>> jb = javax.swing.JButton;
>> jbh = handle(jb)                    % naked handle() call: ill-advised!
jbh =
      javahandle.javax.swing.JButton
>> jbh = handle(jb,'callbackProperties')
jbh =
      javahandle.javax.swing.JButton     <= naked handle is reused
>> get(jb,'ActionPerformedCallback')
??? Error using ==> get
There is no 'ActionPerformedCallback' property in the 'javax.
swing.JButton' class

>> get(jbh,'ActionPerformedCallback')
??? Error using ==> get
There is no 'ActionPerformedCallback' property in the 'javax.
swing.JButton' class

>> jb = javax.swing.JButton;      %recreate or handle will be reused
>> jbh = handle(jb,'CallbackProperties') %Non-naked handle() call: ok
jbh =
      javahandle_withcallbacks.javax.swing.JButton
>> get(jb,'ActionPerformedCallback')
ans =
      ''                                 <= ok!
>> get(jbh,'ActionPerformedCallback')
ans =
      ''                                 <= ok!
```

Oddly enough, if we **get()** or **set()** a Java reference's callback before calling **handle()**, the reference will expose all callbacks, even though its **handle()** might not:

```
>> jb = javax.swing.JButton; % recreate button or handle will be reused
>> get(jb,'ActionPerformedCallback')
ans =
      ''                            <= ok! (only for Java reference)
>> jbh = jb.handle;              % naked handle() call: ill-advised!
      jbh =
```

[†] Even MATLAB's own *javacomponent.m* file has the following comment: "It seems once a java object is cast to a handle, you cannot get another handle with 'callbackproperties'".

```
javahandle.javax.swing.JButton  <= get a naked handle again
```

```
>> get(jbh,'ActionPerformedCallback')
??? Error using ==> get
There is no 'ActionPerformedCallback' property in the 'javax.
swing.JButton' class
```

Luckily, *javacomponent* automatically creates an initial handle with 'CallbackProperties' when the Java object is first placed onscreen. This ensures that any object using *javacomponent*, unless it has called a naked *handle()* first, has access to the Java Object's callback properties. The problem is more significant with subcomponents (e.g., CellRenderer or CellEditor explained in Section 4.1.1) that are not preprocessed by *javacomponent*. It is, therefore, advisable to never use naked *handle()*: always add the 'CallbackProperties' parameter, even when unneeded.

The preceding discussion focused on *handle()* usage for Java components. *handle()* can also be used for MATLAB components. However, while the *java()* function (*java*(jbh) or jbh.*java*) returns the original Java object for Java handles, it merely returns a Java-bean adapter reference for MATLAB handles. These adapter objects should not be confused with the MATLAB components' underlying Java objects. In fact, the adapters are basically just simple automated Java wrappers for the MATLAB objects and do not expose much additional functionality.[87] To get the real underlying Java object, use *findjobj* (see Section 7.2.2).

Note another quirk: GUI components that are loaded from *.*fig* files (created using GUIDE or saved via their figure menu) are loaded without 'CallbackProperties', and it is, therefore, not possible to access their Java callbacks. Only GUI figures and components created in run-time have accessible callbacks. As a workaround, we can use the *handle.listener* approach (see Section 1.4 and Appendix B):[88]

```
hListener = handle.listener(jbh, 'ActionPerformed', @myCallbackFcn);
```

When setting up the callbacks, we may specify additional parameters that will also be passed to the callback when the event is fired:[89]

```
set(jbh, 'ActionPerformedCallback', 'disp(gcbo)');
set(jbh, 'ActionPerformedCallback', @myCallbackFcn);
set(jbh, 'ActionPerformedCallback', {@myCallbackFcn,Param1,Param2});
set(jbh, 'ActionPerformedCallback', @(h,e) myCallbackFcn(extraData));
```

When fired, callback functions receive at least two input parameters: the component reference handle (e.g., a javahandle_withcallbacks.javax.swing.JButton) and a java.util.EventObject[90] reference (typically a more specific class, like java.awt.event.FocusEvent[91] or java.awt.event.MouseEvent[92]) that contains event metadata (type, time, source, details, etc.). If the callback was set using the cell array notation with extra parameters, these will also be passed to the callback function.

Here is a sample **MouseClickedCallback** event that was fired on a JButton:

```
eventData =
java.awt.event.MouseEvent[MOUSE_CLICKED,(50,7),absolute(554,601),butto
n = 1,modifiers = Button1,clickCount = 1] on javax.swing.
JButton[,0,0,60 × 20,...]
```

```
>> get(eventData)
data =
                AltDown: 'off'
           AltGraphDown: 'off'
           BeingDeleted: 'off'
             BusyAction: 'queue'
                 Button: 1
          ButtonDownFcn: ''
               Children: [0 × 1 double]
                  Class: [1 × 1 java.lang.Class]
             ClickCount: 1
               Clipping: 'on'
              Component: [1 × 1 javax.swing.JButton]
               Consumed: 'off'
            ControlDown: 'off'
              CreateFcn: ''
              DeleteFcn: ''
       HandleVisibility: 'on'
                HitTest: 'on'
                     ID: 500
          Interruptible: 'on'
       LocationOnScreen: [554 601]
               MetaDown: 'off'
              Modifiers: 16
            ModifiersEx: 0
                 Parent: []
                  Point: [50 7]
           PopupTrigger: 'off'
               Selected: 'off'
     SelectionHighlight: 'on'
              ShiftDown: 'off'
                 Source: [1 × 1 javax.swing.JButton]
                    Tag: ''
                   Type: 'java.awt.event.MouseEvent'
          UIContextMenu: []
               UserData: []
                Visible: 'on'
                   When: 1236907484155'
                      X: 50
               XOnScreen: 554
                      Y: 7
              YOnScreen: 601
```

---

† Milliseconds since January 1, 1970. Use date = java.util.Date(eventData.getWhen) to convert this number into a timestamp object and char(date) to get a human-readable string (in this case: "Thu Mar 12 20:24:44 EST 2009"). See http://java.sun.com/javase/6/docs/api/java/util/Date.html (or http://bit.ly/7N3B48) for a description of the Date object.

3.5 Using Third-Party Libraries in MATLAB

3.5.1 JFreeChart and Other Charting Libraries[93]

An extremely powerful and widely used Swing-based class library is **JFreeChart** (www. jfree.org), which includes classes for displaying charts, graphs, and gauges in Java panels. JFreeChart solves MATLAB's limitation that plot axes cannot be added to Java containers. JFreeChart is free open-source[94] under the GNU LGPL license. Used by over 40,000 Java developers worldwide[95] (as well as by some MATLAB developers[96]), it is in constant development and improvement.[†]

JFreeChart has some limitations compared with MATLAB plots, but it can do things that are extremely difficult to achieve in MATLAB, as shown in the following screenshots:

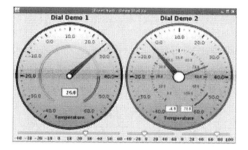
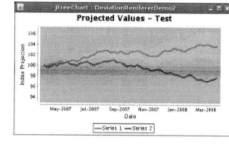

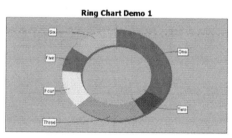
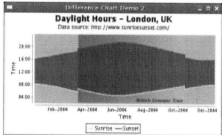

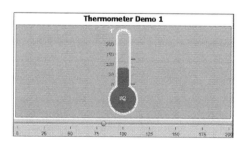
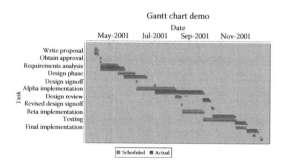

Sample JFreeChart charts, gauges, and plots (See color insert.)

[†] Version 1.0.13 was released in April 2009; by the time you read this, there may be a newer version available.

Let us now integrate a JFreeChart pie chart within a MATLAB figure, as a means of illustrating how to integrate third-party Swing-derived components into MATLAB.

First, download the latest JFreeChart version from its download page on the open-source repository www.sourceforge.net.[97] Next, unzip the downloaded file into some new folder.[†] Now, edit *classpath.txt* (or use ***javaaddpath***) to load *jfreechart-1.0.13.jar* and *jcommon-1.0.16.jar* (which are located in the /lib/ subfolder) to the Java classpath (replace the version numbers as appropriate):[‡]

```
javaaddpath C:/Yair/Utils/JFreeChart/lib/jcommon-1.0.16.jar
javaaddpath C:/Yair/Utils/JFreeChart/lib/jfreechart-1.0.13.jar
```

Within the MATLAB code, load the data into an object that implements the `org.jfree.data.Dataset` interface. There are separate such objects for each specific chart type. For example, in order to display a pie chart, we would use `org.jfree.data.general.DefaultPieDataset`:

```
dataset = org.jfree.data.general.DefaultPieDataset;
dataset.setValue(java.lang.String('C'),       4); §
dataset.setValue(java.lang.String('C++'),     7);
dataset.setValue(java.lang.String('MATLAB'), 52);
dataset.setValue(java.lang.String('Java'),   23);
dataset.setValue(java.lang.String('Other'),  14);
```

Now, prepare an `org.jfree.chart.JFreeChart` object, and update some of its properties:

```
chart3D = org.jfree.chart.ChartFactory.createPieChart3D(...
          'Programming languages', dataset, true, true, false); ¶
plot3D = chart3D.getPlot; % an org.jfree.chart.plot.PiePlot3D obj
plot3D.setForegroundAlpha(0.7); % set transparency level
```

Finally, place the chart in a Swing-compliant panel and display using ***javacomponent***:

```
jPanel = org.jfree.chart.ChartPanel(chart3D);
[jp,hp] = javacomponent(jPanel,[20,20,300,300],gcf);
```

[†] A new folder is advisable, since the zip contents and the javadoc documentation are quite hefty in JFreeChart's case.

[‡] There are several other JAR files in the lib subfolder, but only these two are needed for most cases. The other libraries provide SWT support, development unit tests, servlets for web servers, PDF/RTF/HTML export, and other specialized needs. Read the documentation for details.

[§] Note the explicit casing to `java.lang.String`, since dataset.*setValue()* expects a `java.lang.Comparable` (as seen via ***methodsview*** or the ***uiinspect*** utility) — MATLAB is not smart enough to understand that `java.lang.String` implements the `Comparable` interface so the default type conversion can take place. We, therefore, need to use an explicit type cast.

[¶] Arguments in this case: title string, data set, display legend flag, display tooltips flag, and generate URLs flag.

... and similarly for a 2D exploding pie chart (no need to re-create the panel — simply point it to the new chart using jPanel.*setChart()* and the entire figure is automatically redrawn):

```
chart2D = org.jfree.chart.ChartFactory.createPieChart(...
            'Programming languages', dataset, true, true, false);
plot2D = chart2D.getPlot;      % an org.jfree.chart.plot.PiePlot obj
plot2D.setExplodePercent(0,0.6);'
plot2D.setExplodePercent(3,0.30);'
jPanel.setChart(chart2D);
```

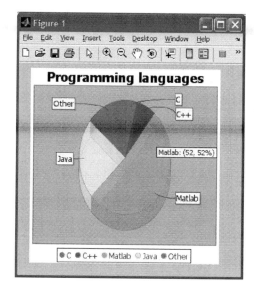
Before: 3D pie chart §

After: 2D pie chart

It is beyond the scope of this book to describe JFreeChart in detail. JFreeChart is indeed a highly customizable and much recommended addition to MATLAB applications. However, exactly due to its extensiveness, we need to read its documentation, available in javadoc format on the download page, together with an installation guide. A detailed developer's guide is also available (at a low cost).

To ease the learning curve, a detailed interactive demo application is available.[98] This demo can also run from the command-line following JFreeChart's installation (from the top-level JFreeChart folder):

```
java -cp lib/jfreechart-1.0.13.jar -cp lib/jcommon-1.0.16.jar -jar jfreechart-
1.0.13-demo.jar
```

---

† Arguments: zeroth index (='C'), 60% outward explosion.

‡ Arguments: third index (='Java', remember Java starts indexing at 0), 30% outward explosion.

§ Note the tooltip when the mouse hovers over the 52% "MATLAB" segment in the three-dimensional pie chart.

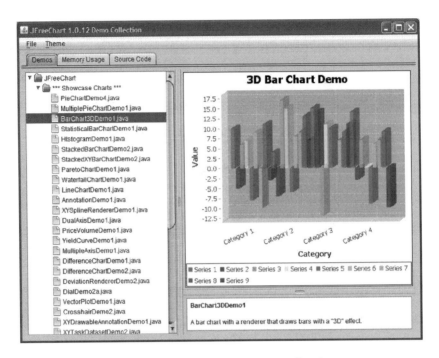

JFreeChart interactive demo application

Other free (yet unofficial) guides for JFreeChart can be found online.[99] Beyond this basic documentation, which should be enough for most programming tasks, there is a very detailed[†] official reference manual available for purchase.[‡]

There are several other Java charting libraries, although JFreeChart is possibly the most widespread. MATLAB users might also be interested in exploring **JMathLib** (www.jmathlib. de), a free MATLAB look-alike written in pure Java, which provides the ability to read MATLAB m-files and present Java-based charts. A MATLAB user has reported using the Java-based Processing charting set (www.processing.org) in MATLAB with mixed success (excellent graphics but also large memory leaks).[100]

3.5.2 JFreeReport and Other Reporting Libraries

Pentaho Reporting (formerly **JFreeReport**)[101] is a Java-based open-source reporting package. It is typically used in conjunction with Pentaho's visual Report Designer (or its now-defunct **JFreeDesigner**[102] predecessor). It has historical ties to JFreeChart, although it was developed independently. These packages create professional-looking reports, invoices, receipts, inventory lists, and so on, that tightly integrate with JFreeChart, enabling easy creation of reports that contain charts and graphs.

---

† 750 pages long at last count ....

‡ http://www.object-refinery.com/jfreechart/guide.html (or http://tinyurl.com/d5xgqm). Do not be confused: JFreeChart is, and plans to remain, entirely free open-source. It is only the book which is not free, as a means to support the JFreeChart project.

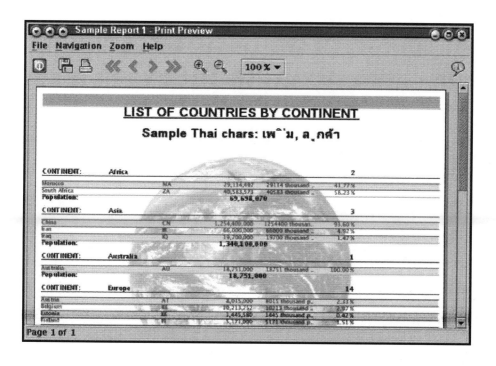

JFreeReport (Pentaho Reporting) sample reports

Another popular open-source reporting package is **JasperReports**[103] (its designer is called **iReport**[104]), winner of the 2007 Java Duke's Choice Open Source Award:[105]

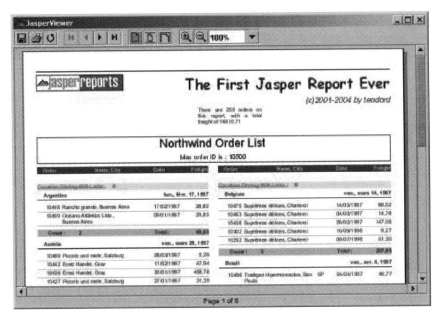

JasperReports sample report

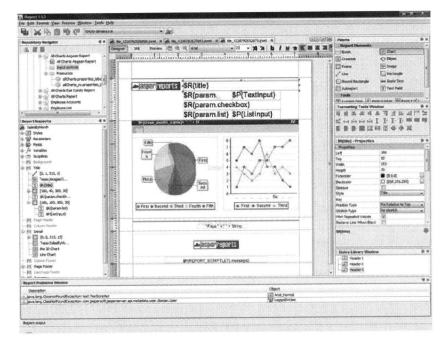

iReports designer for JasperReports

OpenReports[106] is another open-source report package that supports other formats:

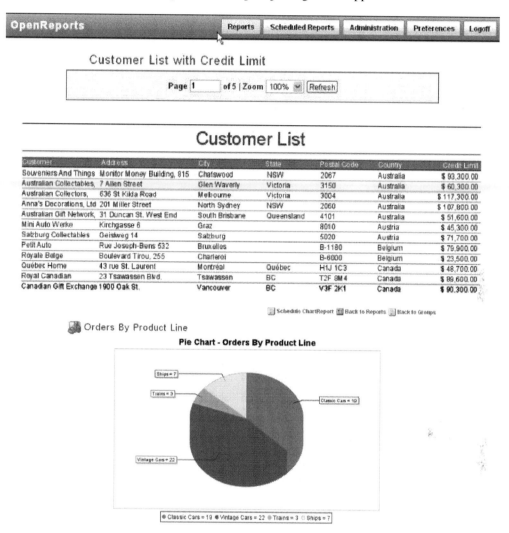

OpenReports sample reports

These reporting packages (and several others) differ not just in their visual capabilities, but also in their acceptance of different data input formats: XML and/or JAVA API and/or database connection, and so on. They all share a feature of being generic in nature, not tailored to any particular field of use.

3.5.3 JGraph and Other Visualization Libraries

An example of a third-party Java charting library integrated in MATLAB is the *JGraphT MATLAB* utility.[107] This utility uses the open-source **JGraphT** Java-based library[108] that provides mathematical graph-theory objects and algorithms, and can be used with the corresponding **JGraph** library[109] for visualization. Apparently, the MATLAB utility is a wrapper for the non-GUI JGraphT, but we can also use JGraph's GUI visualization capabilities, since JGraphT has an extremely simple JGraph adapter.

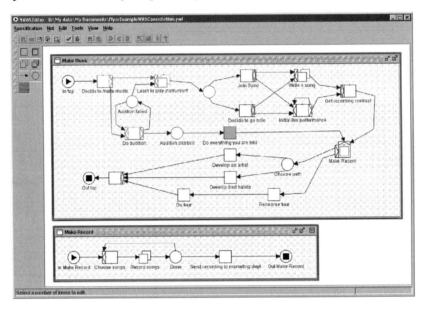

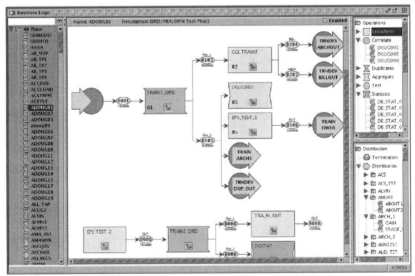

JGraph samples (See color insert.)

To use JGraph, download the JGraphx zip file[110] and then extract the zip file. We will get a jgraphx/ subfolder containing separate subfolders with docs, examples, source code, and a lib/ subfolder with a *jgraphx.jar* file that should be loaded into the classpath:

```
javaaddpath('jgraphx/lib/jgraphx.jar');
```

Here is an example of using JGraph in MATLAB, provided by blog reader Scott Koch:[111]

```
% Make the graph object
graph = com.mxgraph.view.mxGraph;

% Get the parent cell
parent = graph.getDefaultParent();

% Group update
graph.getModel().beginUpdate();

% Add some child cells
v1 = graph.insertVertex(parent, '', 'Hello', 240, 150, 80, 30);
v2 = graph.insertVertex(parent, '', 'World', 20, 20, 80, 30);
graph.insertEdge(parent, '', 'Edge', v1, v2);
graph.getModel().endUpdate();

% Get scrollpane
graphComponent = com.mxgraph.swing.mxGraphComponent(graph);

% Make a figure and stick the component on it
pos = get(gcf,'position');
mypanel = javax.swing.JPanel(java.awt.BorderLayout);
mypanel.add(graphComponent);
[obj, hcontainer] = javacomponent(mypanel, [0,0,pos(3:4)], gcf);
```

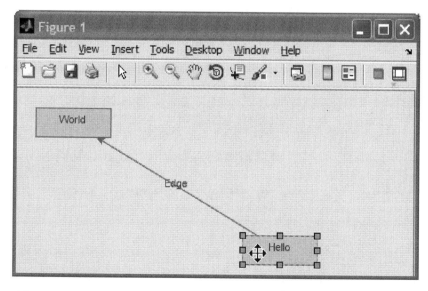

JGraph example within a MATLAB figure (the graph is fully interactive)

The Gephi package,[112] winner of the 2010 Java Duke's Choice Award for Technical Data Visualization,[113] is a free open-source exploration and visualization package, which leaves many costly packages far behind (requires JVM 1.6, MATLAB R2007b+):

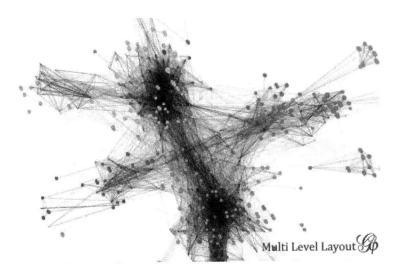

Gephi sample report (See color insert.)

Other packages exist with tailored solutions for specific needs. For example, the Titan/VTK (Visualization Tool-Kit) open-source package[114] targets visualization of informatics data. It even has specific instructions for MATLAB integration.[115] VTK enables display of a complex GUI, which would be extremely difficult (if not impossible) to achieve in pure MATLAB. Here is a sample report from VTK's main wiki page:

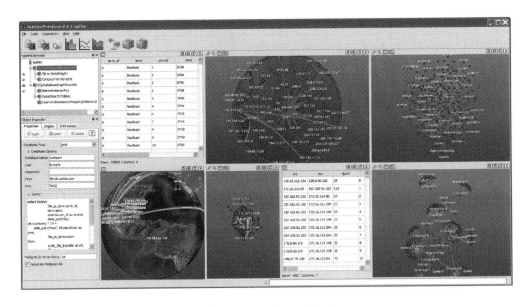

Titan/VTK sample application

3.5.4 ImageJ and Other Image-Processing Libraries

Another open-source Java library, which is extremely popular in the Java world, is ImageJ.[116] This 100,000 source-code-lines fully documented library handles image processing. Whereas one user claims that *"imageJ seems to be doing circles around MATLAB's image processing toolbox"*,[117] well-respected ImageAnalyst disagrees.[118]

ImageJ is under constant development, with new versions being released about twice a month.[119] ImageJ has dozens of plugins contributed by numerous users for different image formats, processing filters, and so on.[120] Tutorial and reference docs are available.[121] Several open-source MATLAB connectors to ImageJ are available online.[122]

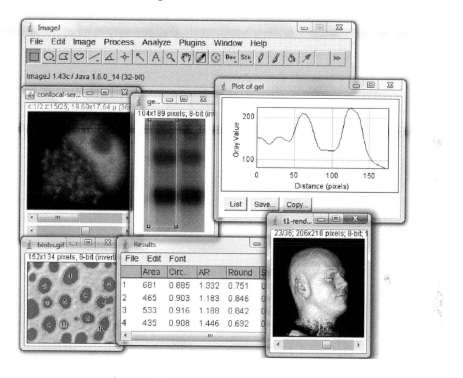

ImageJ image-processing library

ImageJ is by no means the only Java-based image-processing library: there are several other Java-based open-source and commercial libraries available.[123]

3.5.5 Swing Extension Class Libraries

Several open-source class libraries have been developed that enhance and extend the basic Swing. In many cases, these class libraries can be used in MATLAB with as little effort as downloading the relevant JAR file, adding it to the Java (preferably static) classpath, and calling its internal functionality. In some cases, simple Java wrappers need to be coded, which can be called from MATLAB, in order to bypass known limitations (e.g., MATLAB's inability to use Java Generics).

The payoff may well be worth your time to experiment: some of the extensions are simply amazing. A good place to look for such extensions is the official SwingLabs homepage,[124] which holds links to many such extension projects — all of them are open-source. Most projects have online demos (jnlp files) that help illustrate key features.

Some of the interesting projects include:

- SwingX[125] — the semiofficial extension to Swing that includes support for sorting, filtering, and highlighting of tables/trees/lists, auto-completion, tree-tables, collapsible panels, hyperlinks, date/month pickers, animated "busy" indicators, and other similar goodies. Many (but not all) of these are already available in MATLAB via internal classes (refer to Chapter 5 for details).
- JXLayer[126] — enables transformations of the appearance of any displayable control or component: blurring, rotation, stretching, scaling, and so on. JXLayer is widely used and has several active user extensions, blogs, and forum discussions.[127] JXLayer is preinstalled in MATLAB starting in R2011a.
- SwingHelper[128] — a set of swing extension mini-projects.
- Wizard[129] — facilitates creating multipanel wizard GUIs.
- PDF-renderer[130] — viewer and annotator for PDF documents.
- JDIC[131] — integration with and access to computer desktop functionality.

3.5.6 A Note of Caution

A note of caution is due when integrating external Java class libraries in MATLAB applications: not all libraries were developed and debugged with the same amount of attention to detail, quality, and public feedback as the Swing libraries. Some libraries may have internal bugs, inaccuracies, memory leaks, and all sorts of other similar malfunctions. These may affect the stability, behavior, and accuracy of any MATLAB application that integrates these libraries. In some cases, the original library creator may no longer be supporting this package; in other cases, the level of support cannot be guaranteed.

For this reason, I personally prefer, wherever possible, to use an open-source library that has thousands of active users worldwide. This large user base could provide answers, workaround, and/or support in case of need. To me, this appears to be more advisable than using a library from some obscure provider.

While this general caution and advice is true for all integrations of any external software component, it is especially important with the MATLAB–Java interface, which is lacking in documentation, support, and decent debugging tools. Tracing problems across the MATLAB–Java interface can be a painful and frustrating experience indeed.

As a case in point, a CSSM poster asked the MATLAB community's help in solving a memory leak problem in what he thought was the MATLAB–Java interface, but which later turned out to be internal leaks within the external Java library[132] (in this specific case — jPar[133]).

3.6 System-Tray Icons

Java 1.6, available since MATLAB 7.5 (R2007b), enabled programmatic access to system-tray icons on such systems that supported system tray.[134] If the SystemTray[135] object indicates that it *isSupported*, then a TrayIcon[136] can be added, along with an associated tooltip and popup menu (**See color insert.**):[137]

```
sysTray = java.awt.SystemTray.getSystemTray;
if (sysTray.isSupported)
  myIcon = fullfile(MATLABroot,'toolbox/MATLAB/icons/MATLABicon.gif');
  iconImage = java.awt.Toolkit.getDefaultToolkit.createImage(myIcon);
  trayIcon = java.awt.TrayIcon(iconImage, 'initial tooltip');
  trayIcon.setToolTip('click this icon for applicative context menu');
  java.awt.SystemTray.getSystemTray.add(trayIcon); % remove(trayIcon)
end
```

The icon image can be made to automatically resize to the system-tray dimensions, using the trayIcon.*setImageAutoSize(true)* method (by default, the icon image will maintain its original size, getting cropped or appearing small as the case may be).

**Before: small & cropped (large-sized) icons   After: auto-resized icons
(compare the standard-size non-MATLAB shield icon)**

Of course, after initial setup, all the tray icon's properties (icon image, popup, tooltip, etc.) can be modified with convenient *set()* methods (*setImage, setPopupMenu, setTooltip*) or via MATLAB's *set()* function (*set(handle(trayIcon),…, …)*).

Tray icons have several important functionalities, which provide very important visual cues for application users in a very non-obtrusive manner:

- **The displayed icon**. This icon may be modified in run-time, depending on the state of the application. For example, during heavy calculations the icon might be set to a red traffic light, replaced by a green light when all is done. Doing so in run-time is as easy as calling trayIcon.*setImage()* with a modified image.

- **Tooltips**. These are displayed whenever the user hovers the mouse over the icon. The tooltip can contain information which is also updated in run-time, for example, the calculation completion percentage. Multi-line tooltips are supported by inserting new-line (LF = 10) characters, but HTML is not supported. Updating tooltips in run-time is done by trayIcon.*setToolTip()*.

```
trayIcon.setToolTip(sprintf('multi\nline\ncomment')); %or: ['...',10,'...']
```

- **Popup menu.** When a user right-clicks the mouse on the icon, a popup menu will be presented if it has been specified previously. Of course, this too can be customized in run-time, for example, by disabling or enabling some items depending on application state — see below.[†]
- **Double-clicking.** When users double-click on system-tray icons, they usually expect the corresponding application to come into focus in the desktop forefront. This action too is entirely customizable, via the **ActionPerformedCallback** property.
- **Informational messages.** Asynchronous informational messages can be presented next to the system-tray icon in a fashion similar to what we came to expect from modern programs. This could be used to indicate some unexpected event that was detected or the end of a complex calculation phase. The message title, text, and severity icon are all customizable.

Icon popup menus are similar in concept to MATLAB *uicontextmenu*s. Unfortunately, they need to be programmed separately, since Java does not accept *uicontextmenu* handles. This is actually quite easy, as the following code snippet shows:

```
% Prepare the context menu
menuItem1 = java.awt.MenuItem('action #1');
menuItem2 = java.awt.MenuItem('action #2');
menuItem3 = java.awt.MenuItem('action #3');

% Set the menu items' callbacks
set(menuItem1,'ActionPerformedCallback',@myFunc1);
set(menuItem2,'ActionPerformedCallback',{@myfunc2,data1,data2});
set(menuItem3,'ActionPerformedCallback','disp ''action #3...'' ');

% Disable one of the menu items
menuItem2.setEnabled(false);

% Add all menu items to the context menu (with internal separator)
jmenu = java.awt.PopupMenu;
jmenu.add(menuItem1);
jmenu.add(menuItem2);
jmenu.addSeparator;
jmenu.add(menuItem3);

% Finally, attach the context menu to the icon
trayIcon.setPopupMenu(jmenu);
```

Tray icon context (right-click) menu

[†] Popup menus evoke nasty red error messages to the command-line due to an internal MATLAB bug, but they otherwise work ok.

Unfortunately, neither icon tooltip nor its popup menu supports HTML. The reason is that the system-tray functionality resides in the `java.awt` package and does not inherit `javax.swing.JLabel`'s support for HTML. It is actually not part of Swing at all, and the only reason it is included in this chapter is that it complements Swing-based controls in enabling MATLAB applications a richer GUI by use of Java elements.

If we recall the discussion in Section 1.7, we need a simple Java reflection hack to be able to display the informational messages, since the `java.awt.TrayIcon.MessageType` enumeration object cannot be directly accessed:[138]

```
>> trayIconClasses = trayIcon.getClass.getClasses;
>> MessageTypes = trayIconClasses(1).getEnumConstants
MessageTypes =
java.awt.TrayIcon$MessageType[]:
    [java.awt.TrayIcon$MessageType]      <= 1: ERROR
    [java.awt.TrayIcon$MessageType]      <= 2: WARNING
    [java.awt.TrayIcon$MessageType]      <= 3: INFO
    [java.awt.TrayIcon$MessageType]      <= 4: NONE
>> trayIcon.displayMessage('title','info msg',MessageTypes(3));
```

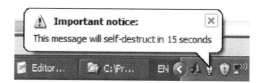 

Sample tray icon message: WARNING...   ...and INFO

Multi-line messages can be created by inserting L&F (10) characters within the string. If the title string is left empty, then neither title nor the severity icon will be displayed. The message can still be manually dismissed by clicking within its boundaries:

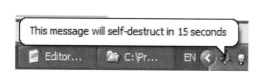 

Tray icon messages without a title (hence also without a severity icon)

The popup messages are automatically positioned by the operating system and are automatically removed after some timeout, if not manually dismissed by the user. The messages replace one another, if a previous message has still not been removed.

I have created a utility function called ***systray***, which facilitates the setup and update of system-tray icons and messages. ***systray*** (with source code) can be downloaded from the MATLAB File Exchange.[139]

3.7 Drag-and-Drop

3.7.1 Data Transfer Mechanism in MATLAB

Drag-and-drop (DND), together with cut-copy-and-paste (CCP), collectively called *Data Transfer* (DT), are standard functionalities in modern GUI applications. Unfortunately, MATLAB's support of these has historically been lacking. Some degree of DT support is available in MATLAB. For example, we can CCP text into a textbox uicontrol and use DND with GUIDE or the MATLAB desktop.[140] However, CCP is often lacking (e.g., we cannot paste into a MATLAB figure or plot) and DND is missing altogether from figure components (axes and controls). This omission was apparently intentional, because MathWorks took the trouble to ensure DND and CCP behavior for some of its internal tools (e.g., in *tstool*, MATLAB's time-series tool[†]). I cannot but wonder why MathWorks chose not to include DT functionality as a generally supported functionality in its HG library.

Over the years, sporadic attempts were made by posters on CSSM and File Exchange to provide DT support. Of these, some were platform-specific (using Windows-specific ActiveX calls[141]), whereas others were source- or target-limited (e.g., limiting their support to axes and figures[142] or to files dropped onto the desktop[143]). An unsuccessful attempt to integrate Java-based DND was reported on CSSM.[144] I am not familiar with any generic cross-application DT solution in MATLAB (e.g., DND or CCP of an image from the computer's browser onto a uicontrol or axes).

This section presents a generic cross-platform solution for incorporating DT in MATLAB. It relies on Java Swing's support for DT:[145] since MATLAB GUI is essentially Swing-based, Swing's DT functionality can be adapted to all MATLAB GUI elements.

DND configuration requires three distinct steps:

- Define draggable source components and allowed actions (copy/move/link).
- Define droppable target components and their internal drop location.
- Define callback actions to invoke when a source is dropped onto a target.

Not all of these steps are mandatory. For example, a component may only be droppable (=drag target) but not draggable (=drag source). Some Java Swing components are not draggable- or droppable-enabled, although, in general, any component that we would expect to be draggable is already pre-enabled as such, and similarly for droppable.[146] To be precise, this does not mean that these components *are* draggable/droppable by default — only that they *can* be set as draggable/droppable if the programmer so chooses, without requiring any complex programming. In cases where DND support is not predefined (either as draggable or droppable or both), full DND support can be added to our custom Swing-extended class using customized programming.[147]

To set a Java Swing component as draggable (=possible drag source), and assuming that the component is draggable-enabled, use its *setDragEnabled(true)* method (the method may be called

---

[†] *tstool*'s DND behavior is set up in %matlabroot%/toolbox/matlab/timeseries/@tsexplorer/@TreeManager/TreeManager.m and dropCallback.m.

setDndEnabled() in some cases). This also applies to all MATLAB uicontrols, since they use underlying Swing components. Non-MATLAB components (e.g., text strings) cannot be set draggable-enabled, and rely on their internal (native) setting to determine whether or not they are draggable.

The way a drop location is displayed is affected by the component's *setDropMode()* method, which accepts `javax.swing.DropMode.USE_SELECTION` (the default value), `DropMode.ON`, `DropMode.INSERT`, or `DropMode.ON_OR_INSERT`.[†] `DropMode.ON` mode enables dropping on top of (replacing) existing elements without modifying the current selection, as `USE_SELECTION` does; `DropMode.INSERT` enables dropping between (i.e., inserting) elements; `DropMode.ON_OR_INSERT` enables a combination of both, depending on the actual cursor pixel position. Note that setting the drop mode is only possible since JVM 1.6, which is available in MATLAB R2007b (7.5) onward.[‡]

DROPMODE.USE_SELECTION

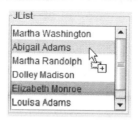
DropMode.ON

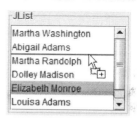
DropMode.INSERT

Unfortunately, enabling DND is not nearly as easy as setting the **DropMode**. Things are even more complicated in MATLAB, which actually prevents regular DND from working within regular MATLAB figure windows. We shall see later how to bypass this limitation, but for the moment let us discuss the dropped-item side of the process.

To set a component as droppable, we need to assign it a `java.awt.dnd.DropTarget`.[148] Most MATLAB components return null ([]) in response to *getDropTarget()*. In such cases, we need to assign a new `DropTarget`. For example, let us set the axis canvas as droppable:

```
dnd = handle(java.awt.dnd.DropTarget(),'CallbackProperties');
jFrame = get(handle(gcf),'JavaFrame');
jAxis = jFrame.getAxisComponent;
jAxis.setDropTarget(dnd);
```

The `DropTarget` object contains the following properties:

```
>> dnd.get
                Active: 1
                 Class: [1 x 1 java.lang.Class]
             Component: []
        DefaultActions: 3
```

[†] http://java.sun.com/docs/books/tutorial/uiswing/dnd/dropmodes.html (or http://bit.ly/bxntHG). It is advised to use one of the nondefault drop modes, since the default was only set for backward compatibility. Some components (e.g., JTable) support additional drop modes: INSERT_ROWS, INSERT_COLS, and the corresponding ON_OR_INSERT_ROWS/COLS.
[‡] Or as a retrofit of JVM 1.6 to older Matlab releases, as explained in Section 1.8.2.

```
        DropTargetContext: [1 x 1 java.awt.dnd.DropTargetContext]
                FlavorMap: [1 x 1 java.awt.datatransfer.SystemFlavorMap]
          DragEnterCallback: []
      DragEnterCallbackData: []
           DragExitCallback: []
       DragExitCallbackData: []
           DragOverCallback: []
       DragOverCallbackData: []
               DropCallback: []
           DropCallbackData: []
    DropActionChangedCallback: []
DropActionChangedCallbackData: []
```

DropTargetContext is the component context to which the dnd object is attached and the **Component** property is a reference to the component itself. **Active** enables easy temporary enabling/disabling of DND behavior for this DropTarget. **DefaultActions** is an enumerator indicating which DND actions are supported by this DropTarget (3 = java.awt.dnd. DnDConstants.ACTION_COPY_OR_MOVE).[149] **FlavorMap** is a hash-map of all supported java.awt.datatransfer.DataFlavor[150] types accepted for drop events. The default map is java.awt.datatransfer.SystemFlavorMap,[151] which is instantiated by the contents of the %matlabroot%/sys/java/jre/.../lib/flavormap.properties file.† It basically includes a list of supported data types and the corresponding Java encapsulation class.

As can be seen, dnd includes five callbacks: **DragEnter, DragOver,** and **DragExit** are fired when a dragged object enters, passes over, and exits the bounds of the component for which dnd was set; **DropCallback** is fired when an object is dropped within the component's bounds; and **DropActionChanged** is fired when the requested **DropAction** changes (e.g., by clicking <Ctrl> in mid-drag to change the action from MOVE to COPY or vice versa). In most cases, **DragEnter, DragOver,** and **DragExit** are simple, at most modifying the cursor icon and/or modifying the component's appearance. Most of the real action occurs in **DropCallback**, which handles the actual processing of the dropped (*imported*) object.

A MATLAB example of setting **DropCallback** for a *uitree* whose nodes are dropped onto other containers can be seen in the Time-Series Tool (*tstool*).‡ Like other MATLAB callbacks, **DropCallback** accepts the Source (a java.awt.dnd.DropTarget) and EventData (a java. awt.dnd.DropTargetDropEvent) standard input arguments. Additional user-defined arguments can be specified when setting the callback.

```
>> eventData.get
        BeingDeleted: 'off'
          BusyAction: 'queue'
        ButtonDownFcn: ''
```

---

† The actual folder location depends on platform (win32/linux, etc.) and JVM version (which depends on the MATLAB release).

‡ *tstool*'s DND behavior is set up in %matlabroot%/toolbox/matlab/timeseries/@tsexplorer/@TreeManager/TreeManager.m and dropCallback.m.

```
           Children: [0 × 1 double]
              Class: [1 × 1 java.lang.Class]
           Clipping: 'on'
          CreateFcn: ''
 CurrentDataFlavors: [1 × 1 java.awt.datatransfer.DataFlavor[]]
CurrentDataFlavorsAsList: [Error]
          DeleteFcn: ''
         DropAction: 2
  DropTargetContext: [1 × 1 java.awt.dnd.DropTargetContext]
   HandleVisibility: 'on'
            HitTest: 'on'
       Interruptible: 'on'
      LocalTransfer: 'off'
           Location: [530 289]
             Parent: []
           Selected: 'off'
   SelectionHighlight: 'on'
             Source: [1x1 java.awt.dnd.DropTarget]
      SourceActions: 1073741827
                Tag: ''
       Transferable: [1x1 java.awt.dnd.DropTargetContext$TransferableProxy]
               Type: 'java.awt.dnd.DropTargetDropEvent'
       UIContextMenu: []
           UserData: []
            Visible: 'on'
```

The properties of interest in `eventData` are **CurrentDataFlavors**, which returns a list of separate **DataFlavors** by which the dropped object can be understood; **DropAction** is the type of drop action[†] (1 = Copy, 2 = Move, 1073741824 = Link/Shortcut); **SourceDropAction** is a corresponding property of the drag source (1073741824 means that all the three actions mentioned above were made possible by the source); **DropTargetContext** is the component context onto which the drop occurred; **LocalTransfer** is a flag indicating whether the DND source was another component within MATLAB (or actually, in the same JVM), or from an external application; **Location** is the dropped pixel position within the component (some controls like `JList` and `JTable` have automatic methods translating this location into a list/row index); **Source** is the component's `DropTarget` object.

`eventData`'s most important property is **Transferable**, which returns an object[‡] that contains the dropped data. **Transferable** is only readable when at least one of the `eventData`'s **CurrentDataFlavors** returns true for `eventData.isDataFlavorSupported (currentData Flavors(id))`. Otherwise, an exception will be thrown:[152]

```
Error using ==> get
Java exception occurred: java.awt.dnd.InvalidDnDOperationException:
The operation requested cannot be performed by the DnD system since it is not in
the appropriate state
```

---

† See DndConstants references above.

‡ http://java.sun.com/javase/6/docs/api/java/awt/dnd/DropTargetContext.TransferableProxy.html (or http://bit.ly/drsfFL).
Note that since this is an inner class, it appears in MATLAB as 'java.awt.dnd.DropTargetContext$TransferableProxy'.

If the dropped data flavor is supported, then the data can be retrieved by using the Transferable object's *getTransferData()* method as follows:

```
>> th = eventData.getTransferable;
>> tdf = th.getTransferDataFlavors
tdf =
java.awt.datatransfer.DataFlavor[]:
    [java.awt.datatransfer.DataFlavor]
    [java.awt.datatransfer.DataFlavor]
    [java.awt.datatransfer.DataFlavor]

>> tdf(1)
ans =
java.awt.datatransfer.
DataFlavor[mimetype = application/x-java-serialized-
object;representationclass = java.lang.String]

>> tdf(2)
ans =
java.awt.datatransfer.DataFlavor[mimetype = text/plain;representationclass = java.
io.InputStream;charset = unicode]

>> tdf(3)
ans =
java.awt.datatransfer.DataFlavor[mimetype = text/rtf;representationclass = java.
io.InputStream]

>> th.getTransferData(tdf(1))
ans =
Rich-text data string

>> th.getTransferData(tdf(2))
ans =
java.io.StringReader@4d3130

>> th.getTransferData(tdf(3))
ans =
java.io.ByteArrayInputStream@19dfe98
```

3.7.2 A Sample MATLAB Application That Supports DND

Let us implement a simple DND GUI, in which a listbox contains some plotting commands, and the user can drag any of them onto one of the two axes. For this example, we will use a standard MATLAB listbox *uicontrol*, demonstrating how DND behavior can be retrofitted to existing MATLAB GUIs, without any need to switch to Java components:

```
% First prepare the figure
hFig = figure('name','DND example','numbertitle','off');

% Add a listbox with several plotting commands
plotNames = {'surfc(peaks)','contour(peaks)','contourf(peaks)', ...
             'surf(membrane)','contour(membrane)','contourf(membrane)'};
```

```matlab
% (rather than use a MATLAB listbox uicontrol, directly use JList)
%hListbox = uicontrol('style','listbox', 'units','norm', ...
%                      'position',[.05,.05,.3,.6], 'string',plotNames);
%jListbox = findjobj(hListbox,'nomenu');
%jListbox = jListbox.getViewport.getView;  % in a scrollpane
jListbox = javax.swing.JList(plotNames);
[hjList,hcList] = javacomponent(jListbox,[10,10,100,200],hFig);
set(hcList,'units','norm','position',[.05,.05,.3,.6]);

% Add some plot axes
hAx1 = axes('position',[.5,.1,.45,.35]);
hAx2 = axes('position',[.5,.55,.45,.35]);

% Enable dragging from the listbox
jListbox.setDragEnabled(true);

% Enable drop on the figure axes
dnd = handle(java.awt.dnd.DropTarget(),'CallbackProperties');
jFrame = get(handle(hFig),'JavaFrame');
jAxis = jFrame.getAxisComponent;
jAxis.setDropTarget(dnd);
set(dnd,'DropCallback',{@dropCallbackFcn,hFig});
set(dnd,'DragOverCallback',@dragCallbackFcn);

% Define the drag movement callback function
function dragCallbackFcn(src,eventData)
   try
      % Enable drop only on top of actual axes - not the entire figure
      hAxes = overobj('axes');
      if isempty(hAxes)
         eventData.rejectDrag;
      else
         eventData.acceptDrag(eventData.getDropAction);
      end
   catch
      % never mind...
   end
end

% Define the drop callback function
function dropCallbackFcn(src,eventData)
   try
      hFig = varargin{3};
      transferable=eventData.getTransferable;  % will crash if invalid
      dataFlavorStr = 'text/plain; class=java.lang.String';
      dataFlavor = java.awt.datatransfer.DataFlavor(dataFlavorStr);
      dataStr = transferable.getTransferData(dataFlavor);
      hAxes = overobj('axes');
      if ~isempty(hAxes)
         axes(hAxes(1));
         eval(dataStr);
      end
      eventData.dropComplete(true);
   catch
```

```
            % never mind...
        end
    end
```

In some cases, implementation of DND in MATLAB fails to process drop events correctly. The reason for this is that apparently MATLAB's default `DropTargetListener` seems to pass the `DropTargetDropEvent` (dtde) to MATLAB callback functions without previously accepting the drop, so dtde.*getTransferable()* fails and the drop event is ignored.[153] This is apparently not an issue for internal MATLAB components, which override the standard `DropTarget` class.

Dirk Engel, who discovered this bug, also provided a suggested fix (overriding the `DropTarget` class), which is provided here with some modifications:[154]

```java
import java.awt.dnd.*;
import java.awt.datatransfer.*;
import java.util.List;
import java.io.IOException;

// Modified DropTarget to be used for drag & drop in MATLAB GUI
public class DropTargetList extends DropTarget
{
  private Transferable transferable;
  private DataFlavor acceptedDataFlavor =
                  DataFlavor.javaFileListFlavor;
  private List<?> transferDataList;
  private String  transferDataStr;
  private boolean debugFlag = false;
  public synchronized void drop(DropTargetDropEvent dtde) {
    dtde.acceptDrop(DnDConstants.ACTION_COPY_OR_MOVE);
    super.drop(dtde);
    transferable = dtde.getTransferable();
    try {
      if (transferable.isDataFlavorSupported(
          DataFlavor.javaFileListFlavor)) {
        transferDataList =
          (List<?>)transferable.getTransferData(acceptedDataFlavor);
      } else {
        // try to interpret as a plain text string
        transferDataStr =
          (String)transferable.getTransferData(DataFlavor.stringFlavor);
      }
    } catch (UnsupportedFlavorException e) {
      return;
    } catch (IOException e) {
      return;
    }
  }

  public void setAcceptedDataFlavor(DataFlavor flavor) {
    acceptedDataFlavor = flavor;
  }
```

```
public DataFlavor getAcceptedDataFlavor() {
    return acceptedDataFlavor;
}
public Transferable getTransferable() {
    return transferable;
}
public List<?> getTransferDataList() {
    return transferDataList;
}
public String getTransferDataStr() {
    return transferDataStr;
}
}
```

Use of this class is very simple, as the following MATLAB code snippet demonstrates:

```
fileList = get(jDropTarget,'TransferDataList');          %Java List
files = cellfun(@(c)char(c),fileList.toArray.cell,'un',0); %cell array
```

Here is a larger snippet from an actual program that drag-and-drops data and external files onto a JTable in a MATLAB GUI:

```
% Enable Drag-&-Drop onto a JTable in MATLAB GUI
function enableDND(jtable)
  %dnd = handle(java.awt.dnd.DropTarget(),'CallbackProperties');
  dnd = handle(javaObjectEDT(DropTargetList),'CallbackProperties');
  set(dnd,'DropCallback',{@DNDCallback,jtable});
  set(dnd,'DragEnterCallback',{@DNDCallback,jtable});
  %set(dnd,'DragOverCallback',{@DNDCallback,jtable});
  jtable.setDropTarget(dnd);
end


% Main Drag-&-Drop callback function, for both drag and drop events
function DNDCallback(jDropTarget,jEventData,jtable)
  persistent isFileDND
  try
    % Drop event
    if jEventData.isa('java.awt.dnd.DropTargetDropEvent')  %nargin>2
      try
        pause(0.05); drawnow;
        if isFileDND
          % File dropped
          fileList = get(jDropTarget,'TransferDataList');
          files = cellfun(@(c)char(c),fileList.toArray.cell,'un',0);
          rowData = FileDropFunction(files,jtable);
        else
          % Table row dropped
          dataStr = get(jDropTarget,'TransferDataStr');
          rowData = DataDropFunction(dataStr,jtable);
        end
```

```matlab
        % If any rows are to be added
        if ~isempty(rowData)
          % Get the drop location (row index)
          dropPoint = jEventData.getLocation;

          % Handle possible table scrolling
          dropY = dropPoint.getY - jtable.getTableHeader.getHeight +
                  jtable.getParent.getViewPosition.getY;
          dropRow = fix(dropY / jtable.getRowHeight);
          if dropRow >= jtable.getRowCount
            dropRow = -1;
          end

          % Insert the new row(s) immediately before drop-location's
          % row, and then move the selection to this latest new row
          stopEditing(jtable);
          for rowIdx = size(rowData,1) : -1 : 1
            if dropRow >= 0
              jtable.getModel.insertRow(max(0,dropRow),
                                        rowData(rowIdx,:));
            else
              jtable.getModel.addRow(rowData(rowIdx,:));
            end
          end

          % Move the selection to Column A of this new row
          selectedRow = dropRow + (dropRow<0)*jtable.getRowCount;
          jtable.changeSelection(selectedRow,0,false,false);
          jtable.repaint;
        end
      catch
        lasterr
      end
      jEventData.dropComplete(true);

    % Drag event
    elseif jEventData.isa('java.awt.dnd.DropTargetDragEvent')
      try
        isFileDND = jEventData.isDataFlavorSupported(
                java.awt.datatransfer.DataFlavor.javaFileListFlavor);
      catch
        % never mind - reuse previous data...
        disp(lasterr)
      end
      jEventData.acceptDrag(java.awt.dnd.DnDConstants.ACTION_COPY);
  catch
      disp(lasterr)
  end
end
```

Further reading resources:

- http://weblogs.java.net/blog/shan_man/archive/2006/02/choosing_the_dr.html
- http://weblogs.java.net/blog/2006/09/15/top-level-drop-swing-and-java-se-6

- http://www.mathworks.com/matlabcentral/newsreader/view_thread/154949
- http://www.java2s.com/Tutorial/Java/0240_Swing/1780_Drag-Drop.htm

A suggested exercise for adventurous readers is to implement a MATLAB figure that displays a plot axes and a `JFileChooser` Swing component side by side; image files from the file chooser or from external applications (e.g., Windows Explorer) can be drag-and-dropped onto the axes, thereby displaying their image contents.

3.8 Adding MATLAB Components to Java Swing Containers

In Section 3.1.1, it has been mentioned that MATLAB objects such as axes (for plots or images) and *uicontrol*s cannot be added to Swing containers, since they do not have a `javax.swing.JContainer` wrapper. It was advised to use these Java containers as long as their contained GUI is limited to Java components (`JButtons`, `JComboBox`, etc.). This limitation is very annoying — it would be most helpful to be able to place MATLAB axes or uicontrols within a `JTabbedPane`, a `JSplitPane`, or a pure-Java frame.[155] Instead, we must rely on MATLAB-based workarounds (*uitab* and *uisplitpane* — see Section 4.3 and Chapter 10), which are cumbersome compared with their Java counterparts.

These were indeed the conventional wisdom and documented knowledge, until now.[156]

There actually is an undocumented, unsupported, and quite problematic method of placing MATLAB components in our Java GUI. It is based on the fact that Swing components (and all MATLAB GUI is ultimately that) can be reparented by being added to any Swing container: MATLAB *uicontrol*s are easy to reparent, since they each have separate Swing component peers and associated container (as explained above).

MATLAB axes cannot be separately reparented, since they do not have separate Swing component peers: as shall be explained in Chapter 7, all the figure's axes and plots are really graphic pixels drawn onto a single large `java.awt.Canvas` that spans the entire figure content area. It is this `Canvas` (or one of its hierarchy ancestors, such as the `ContentPane`) which can be reparented onto any Java container.

Here is a simple example to move the `Canvas` onto a Java `JFrame` (we can move it onto any Swing container instead):

```
% Display a simple plot within a MATLAB figure
plot(1:5);

% Get the MATLAB frame's plotting canvas container
jFrame = get(handle(gcf),'JavaFrame');
jCanvas = jFrame.getAxisComponent;
jFrameProxy = jCanvas.getParent.getParent.getTopLevelAncestor;

% Prepare a new pure-Java frame
jf2 = javax.swing.JFrame;
jf2.pack;
jf2.setSize(jCanvas.getSize);
jf2.setTitle('My pure Java Frame');
```

```
% Add the MATLAB plot canvas to the new pure-Java frame
awtinvoke(jf2,'add(Ljava.awt.Component;)',jCanvas);
awtinvoke(jf2,'setVisible',1);        % the plot shows in the Java Frame

% Hide original MATLAB figure - only the Java frame remains visible
set(gcf,'Visible','off');

% Update the MATLAB plot, actually modifying the axes in the JFrame
plot(1:23); % plot in Java window is updated using regular HG MATLAB
awtinvoke(jf2,'show()');               % re-render (repaint) the plot
```

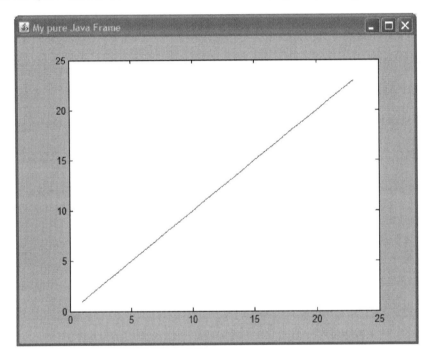

MATLAB axes reparented onto a pure-Java container

Note how we use jf2.*show()* in the last command above: For some unknown reason, *repaint()* and *invalidate()* are not enough to cause a plot re-rendering. Also note that we use **awtinvoke** rather than direct invocation (jf2.*show()*), because this operation needs to be done on the Java EDT thread.

Finally, note that the entire axes content of the MATLAB figure is transferred to the new container, not specific axes or uicontrols. As far as Java is concerned, jCanvas is simply a single graphic image.

The main menu, toolbars, controls, and ActiveXes were not re-parented, since they belong to different containers in the MATLAB figure. If we re-parent the root pane,[†] these components

---

[†] jCanvas.getParent.getParent.getTopLevelAncestor.getComponent(0) or jFrame.fHG1Client. getWindow.getRootPane (jFrame.fFigureClient.getWindow.getRootPane in R2007b and earlier).

are also moved to the new target container, along with the axes in jCanvas. Remember to set the target's size accordingly.

Re-parenting appears to work, sort of: when we close the original MATLAB figure, the Java frame window hangs, so remember to keep them both alive (or closed) together.

Many HG listeners are still connected to the original MATLAB figure, causing the figure to flicker when some HG actions are done. For this reason, keeping the original MATLAB figure alive but hidden is advisable, and this has been done above using *set(gcf,'Visible','off')*. The downside is that when the figure is hidden, several interactive features (e.g., interactive zooming) are disabled.

We also need to handle other listeners (resizing, deletion, etc.) ourselves, since the original component listeners remain attached to the original MATLAB figure, and need to be re-attached to the new Java container.[157]

Such a re-parenting technique can be used to effectively remove the title bar from the figure window, by specifying jf2.*setUndecorated(true)* before jf2 is displayed.[158] Also, see the discussion in Section 7.3.7.

The re-parenting process is deeply undocumented. All in all, using re-parenting is extremely difficult to program correctly, handling all edge cases. For this reason, I reiterate my earlier advice, to use MATLAB figures instead of Java frames; to use Java containers only for pure-Java components and not for MATLAB components; and, in general, to add Java components to MATLAB figures rather than adding MATLAB components to Java frames.

In a related issue, adding Java components to a MATLAB-created frame, from within Java (as opposed to using *javacomponent* from within MATLAB), is also nontrivial and creates problems with repainting.[159]

An altogether different approach was suggested on CSSM: taking the screenshot of the Canvas and placing this as an image in Java.[160] We would not be able to later modify this plot as in the re-parenting approach, but this approach does solve the system-hang issues.

3.9 Alternatives to Swing

This chapter has focused on Swing and its extensive use in MATLAB. However, Swing is by no means the only Java-based GUI framework available for MATLAB programmers. Alternative UI toolkits can be used, such as the popular SWT,[161] the older AWT,[162] or the lesser-known Qt.[163]

MATLAB uses Swing internally, so it is natural to continue developing using Swing, for a consistent look and feel, enabling use of internal MATLAB components and other side benefits. AWT is Swing's predecessor and much outdated, so there is no real reason to use it except for very specific tasks.[†] Interested readers can, however, try to use SWT or Qt, which sometimes provide benefits that Swing lacks.[‡] Some open-source projects[164] attempt to merge Swing and SWT, thus providing the benefits of both.

[†] For example, directly drawing on MATLAB figures' content area (their *Axes Canvas*) using JAWT and JNI.

[‡] For example, whenever tight integration with the native platform (e.g., wrapping ActiveX components with Java code on Windows) is required. See http://www.developer.com/java/other/article.php/10936_2179061_2 (or http://bit.ly/cmWTXT).

Here is a simple "Hello World!" MATLAB–SWT application:[†]

```
import org.eclipse.swt.*;
import org.eclipse.swt.widgets.*;
import org.eclipse.swt.graphics.*;
import org.eclipse.swt.layout.*;

display = Display;
shell = Shell(display);
shell.setLayout(GridLayout);
label = Label(shell, SWT.NONE);
label.setText('Hello, World!');
button = Button(shell, SWT.PUSH);
button.setText('Click me!');
shell.pack;
label.pack;
shell.open;

% Wait for the window to be closed, then dispose (cleanup) components
while (~shell.isDisposed)
    if (display.readAndDispatch)
        display.sleep;
    end
end
display.dispose;
label.dispose;
button.dispose;
```

A simple SWT-based MATLAB GUI

Here are a few standard SWT controls (widgets) — (**See color insert.**):[165]

Button	**Button**	**Button**	**Button**	**Canvas**
(SWT.ARROW)	(SWT.CHECK)	(SWT.PUSH)	(SWT.TOGGLE)	

[†] Adapted from http://www.mathworks.com/matlabcentral/newsreader/view_thread/69793 (or http://bit.ly/cI2kgD). Note that using SWT requires adding the SWT jar files to the static Java classpath, as explained in this CSSM post. The post refers to jar files of a pretty-old Eclipse version (2.1.3), which can be obtained from http://bit.ly/cXMBtj. The latest Eclipse version (and SWT files) can be downloaded from http://www.eclipse.org/downloads/ (or http://bit.ly/95wdV4).

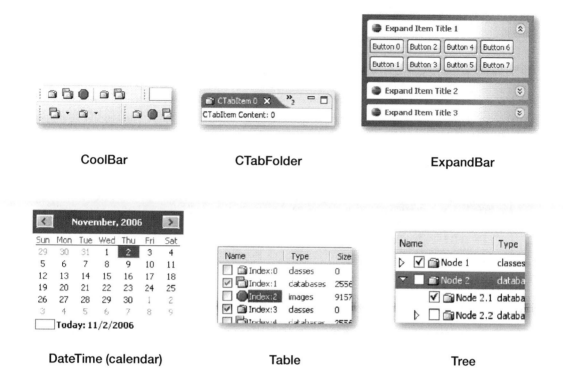

CoolBar	CTabFolder	ExpandBar

DateTime (calendar)	Table	Tree

Note that mixing components (widgets) of different toolkits on the same figure window may cause unexpected behavior. For example, when AWT and Swing components were mixed in JVM versions prior to 1.6 update 12, the AWT components always overlapped the Swing widgets, regardless of component priority. Therefore, because MATLAB uses Swing, if we plan to integrate non-Swing toolkits, we should expect such problems and carefully test our application.

References

1. http://java.sun.com/docs/books/tutorial/ui (or http://tinyurl.com/csnocf).
2. For example, http://www.mathworks.com/matlabcentral/newsreader/view_thread/248979 (or http://tinyurl.com/p758k3).
3. For other great-looking Swing-based applications, using the NetBeans platform, see http://platform.netbeans.org/screenshots.html
4. The most important resource is Sun's official http://java.sun.com/ website, which includes an excellent tutorial (http://java.sun.com/docs/books/tutorial/uiswing) and reference (http://java.sun.com/javase/reference/api.jsp). Some websites dedicated to Java code examples are java2s.com, javadb.com, exampledepot.com, java-tips.org, jexamples.com, javaexamples.com, javalessons.com, and http://www.crionics.com/products/opensource/faq/swing_ex/SwingExamples.html
5. http://UndocumentedMatlab.com/blog/javacomponent/ (or http://bit.ly/8Yxm1K).
6. http://www.mathworks.com/matlabcentral/newsreader/view_thread/79603 (or http://bit.ly/aiifw9); some bugs existed until fixed in R14 SP3: http://www.mathworks.com/matlabcentral/newsreader/view_thread/112313#283635 (or http://bit.ly/bW3fRn).
7. http://UndocumentedMatlab.com/blog/matlab-layout-managers-uicontainer-and-relatives/ (or http://bit.ly/aZ4p7S).

8. http://www.mathworks.com/matlabcentral/newsreader/view_thread/98032 (or http://tinyurl.com/c8y7e8); http://www.mathworks.com/matlabcentral/newsreader/view_thread/97829 (or http://tinyurl.com/cttuq8).

9. Used by my ***ScreenCapture*** utility: http://www.mathworks.com/matlabcentral/fileexchange/24323 (or http://bit.ly/czfUSE).

10. http://www.mathworks.com/matlabcentral/newsreader/view_thread/246001 (or http://bit.ly/c7MnHt).

11. http://www.mathworks.com/matlabcentral/newsreader/view_thread/161098 (or http://tinyurl.com/chugfo).

12. http://UndocumentedMatlab.com/blog/common-javacomponent-problems (or http://bit.ly/q8o2vJ).

13. http://java.sun.com/docs/books/tutorial/ui/features (or http://tinyurl.com/abb7te).

14. http://java.sun.com/docs/books/tutorial/uiswing/components (or http://tinyurl.com/6662x6).

15. http://java.sun.com/docs/books/tutorial/uiswing (or http://tinyurl.com/2wzah).

16. http://java.sun.com/docs/books/tutorial/ui/features/compWin.html (or http://tinyurl.com/awmkty).

17. http://www.mathworks.com/matlabcentral/newsreader/view_thread/246571 (or http://tinyurl.com/bpkm5v).

18. Here is a typical example (of several) of a MATLAB user who asked in vain about this functionality: http://www.mathworks.com/matlabcentral/newsreader/view_thread/163711 (or http://bit.ly/9ZenfE); MathWorks has suggested a workaround using class extension here: http://www.mathworks.com/help/techdoc/matlab_oop/f1-5978.html#f1-20068 (or http://bit.ly/bTtQDU).

19. For example, http://bit.ly/pcZTuP, http://www.javaspecialists.co.za/archive/Issue082.html (or http://bit.ly/p5tLAr); http://stackoverflow.com/questions/1263323/tristate-checkboxes-in-java (or http://bit.ly/o6nd6H).

20. http://www.mathworks.com/matlabcentral/newsreader/view_thread/153690 (or http://tinyurl.com/9amob6).

21. For example, www.javashareware.com, www.swinglabs.org, www.downloadthat.com, www.shareware-connection.com, www.easyfreeware.com, www.l2fprod.com, www.fileheap.com/software/components.html, swing-components.safe-install.com, and many others.

22. http://www.mathworks.com/matlabcentral/newsreader/view_thread/242127 (or http://tinyurl.com/bpyfff); http://www.mathworks.com/matlabcentral/newsreader/view_thread/172873 (or http://tinyurl.com/c6u6hb).

23. http://www.mathworks.com/matlabcentral/fileexchange/14583 (or http://tinyurl.com/cgyz8r).

24. http://www.mathworks.com/matlabcentral/fileexchange/15580 (or http://tinyurl.com/beqxeh).

25. http://java.sun.com/docs/books/tutorial/uiswing/concurrency/dispatch.html (or http://tinyurl.com/2sec5k).

26. http://www.mathworks.com/matlabcentral/newsreader/view_thread/156388 (or http://tinyurl.com/arcqv8).

27. http://java.sun.com/docs/books/jni/ or http://java.sun.com/j2se/1.5.0/docs/guide/jni/spec/types.html#wp276 (http://tinyurl.com/2xcqvk).

28. http://www.mathworks.com/matlabcentral/newsreader/view_thread/160272 (or http://tinyurl.com/b5o2bt).

29. http://UndocumentedMatlab.com/blog/matlab-and-the-event-dispatch-thread-edt/#comment-38092 (or http://bit.ly/faxBmI).

30. See the list of comments here: http://UndocumentedMatlab.com/blog/editormacro-assign-a-keyboard-macro-in-the-matlab-editor/#comments (or http://bit.ly/bZkgkm).

31. http://www.mathworks.com/help/techdoc/rn/brvak9c-1.html#bryg6qx-1 (or http://tinyurl.com/cxz7gt).

32. http://www.mathworks.com/matlabcentral/newsreader/view_thread/259650 (or http://tinyurl.com/npxbrl).

33. http://www.mathworks.com/matlabcentral/newsreader/view_thread/161504 (or http://bit.ly/dabuLI); also see http://www.mathworks.com/support/solutions/en/data/1-V3B5T/ (or http://bit.ly/d5Ugqf) and http://robertoostenveld.ruhosting.nl/index.php/multithreading-matlab-mex/ (or http://bit.ly/9TYAme).

34. For example, http://www.osmanoglu.org/index.php/computing/4-computing/1-matlabjavamultitreading (or http://bit.ly/cbW4Xm).

35. http://www.mathworks.com/matlabcentral/newsreader/view_thread/279510 (or http://bit.ly/95nV5w).

36. Indeterminate Java progress-bars are described here: http://java.sun.com/j2se/1.4.2/docs/guide/swing/1.4/pb.html (or http://bit.ly/bpYXUi); http://java.sun.com/docs/books/tutorial/uiswing/components/progress.html (or http://bit.ly/cFupOM).

37. Based on http://www.mathworks.com/matlabcentral/newsreader/view_thread/238449 (or http://tinyurl.com/c4avbe). Note that contrary to what this news thread says, the posted code does actually work and display the presented result.

38. http://java.sun.com/docs/books/tutorial/uiswing/components/border.html (or http://tinyurl.com/3a2n6e).

39. More border examples can be found here: http://java.sun.com/developer/onlineTraining/GUI/Swing1/shortcourse.html#JFCBorder (or http://tinyurl.com/dxezrm).

40. http://java.sun.com/javase/6/docs/api/java/awt/Cursor.html (or http://tinyurl.com/ck32px).

41. http://java.sun.com/javase/6/docs/api/constant-values.html#java.awt.Cursor.DEFAULT_CURSOR (or http://bit.ly/ceUOn1).

42. http://java.sun.com/javase/6/docs/api/java/awt/Toolkit.html#createCustomCursor(java.awt.Image,%20 java.awt.Point,%20java.lang. String) (or http://tinyurl.com/c5v9zj).

43. http://UndocumentedMatlab.com/blog/modifying-the-look-and-feel/ (or http://bit.ly/dpoZ5i).

44. http://java.sun.com/docs/books/tutorial/uiswing/lookandfeel/plaf.html (or http://tinyurl.com/3az5fu); or the following detailed technical article: http://java.sun.com/products/jfc/tsc/articles/architecture/#pluggable (or http://tinyurl.com/prslvs).

45. http://www.jgoodies.com/freeware/looks/index.html (http://bit.ly/9x0xvz).

46. http://www.jidesoft.com/javadoc/com/jidesoft/plaf/LookAndFeelFactory.html (or http://bit.ly/9ltqrX).

47. For example, Alloy (lookandfeel.incors.com or http://bit.ly/9b9fhD) and Synthetica (www.javasoft.de/ synthetica/ or http://bit.ly/cgVwh6).

48. http://UndocumentedMatlab.com/blog/modifying-matlab-look-and-feel/#comments (or http://bit.ly/dfeNUj).

49. http://java.sun.com/docs/books/tutorial/uiswing/lookandfeel/synth.html (or http://bit.ly/bknOFR).

50. http://java.sun.com/docs/books/tutorial/uiswing/lookandfeel/nimbus.html (or http://bit.ly/bhBnQH).

51. For example, http://www.jasperpotts.com/blog/2008/08/skinning-a-slider-with-nimbus/ (or http://bit.ly/ a2g1Kd).

52. http://www.jasperpotts.com/blog/2009/03/breakdown-of-what-should-be-default-laf-for-java-7/ (or http:// bit.ly/bDsGX0).

53. http://www.mathworks.com/matlabcentral/fileexchange/28326-MUtilities (or http://bit.ly/9sY6Dm).

54. http://java.sun.com/docs/books/tutorial/uiswing/components/html.html (or http://tinyurl.com/5v38m).

55. http://www.mathworks.com/matlabcentral/newsreader/view_thread/164175#416384 (or http://tinyurl. com/b3fqfc).

56. http://www.mathworks.com/matlabcentral/newsreader/view_thread/265569 (or http://tinyurl.com/yjhgcns).

57. http://www.mathworks.com/matlabcentral/newsreader/view_thread/173886#447732 (or http://tinyurl. com/d6ymbt); http://UndocumentedMatlab.com/blog/spicing-up-matlab-uicontrol-tooltips/ (or http:// tinyurl.com/ye9xcp7).

58. http://UndocumentedMatlab.com/blog/html-support-in-matlab-uicomponents/#comment-2187 (or http:// bit.ly/dAW4cU).

59. http://developer.java.sun.com/developer/bugParade/bugs/4783068.html (or http://tinyurl.com/dhcw8b). This webpage contains several workarounds; the full fix is expected in JVM 1.7 (whose MATLAB integration date is currently unknown).

60. http://UndocumentedMatlab.com/blog/fixing-a-java-focus-problem/ (or http://bit.ly/aLB3mP).

61. http://java.sun.com/docs/books/tutorial/uiswing/misc/focus.html (or http://tinyurl.com/5curo); also read the very informative http://java.sun.com/javase/6/docs/api/java/awt/doc-files/FocusSpec.html (or http:// tinyurl.com/cqom4d).

62. http://UndocumentedMatlab.com/blog/uicontrol-callbacks/ (or http://tinyurl.com/yle7okd).

63. http://java.sun.com/docs/books/tutorial/uiswing/events/eventsandcomponents.html (or http://tinyurl. com/6xc38). This page also lists other events (and corresponding callbacks) that are used by only some of the components.

64. http://java.sun.com/javase/6/docs/api/javax/swing/event/AncestorListener.html (or http://tinyurl.com/ab6ghm).

65. http://java.sun.com/docs/books/tutorial/uiswing/events/containerlistener.html (or http://tinyurl.com/bdnsof).

66. http://java.sun.com/docs/books/tutorial/uiswing/events/componentlistener.html (or http://tinyurl.com/2afuln).

67. http://java.sun.com/docs/books/tutorial/uiswing/events/focuslistener.html (or http://tinyurl.com/7rcaj); also read the very informative http://java.sun.com/docs/books/tutorial/uiswing/misc/focus.html (or http:// tinyurl.com/5curo) and http://java.sun.com/javase/6/docs/api/java/awt/doc-files/FocusSpec.html (or http://tinyurl.com/cqom4d).

68. http://UndocumentedMatlab.com/blog/detecting-window-focus-events/ (or http://tinyurl.com/yzf3jcx).

69. http://www.mathworks.com/matlabcentral/newsreader/view_thread/284958#755942 (or http://bit.ly/dBvayJ).

70. http://java.sun.com/javase/6/docs/api/java/awt/event/HierarchyListener.html (or http://tinyurl.com/aduwvu).

71. http://java.sun.com/docs/books/tutorial/uiswing/events/keylistener.html (or http://tinyurl.com/5825h).

72. http://www.mathworks.com/help/techdoc/rn/br3lhn8-1.html#br9d16b (or http://tinyurl.com/p8wk9p); http://www.mathworks.com/help/techdoc/rn/brgysvh-1.html#br9cuil (or http://tinyurl.com/qn7c4f).

73. http://java.sun.com/docs/books/tutorial/uiswing/events/mousemotionlistener.html (or http://tinyurl. com/586jk).

74. http://www.mathworks.com/matlabcentral/newsreader/view_thread/301383 (or http://bit.ly/dLd7AG).

75. http://java.sun.com/docs/books/tutorial/uiswing/events/mouselistener.html (or http://tinyurl.com/3rbt9).

76. Sample usage: http://www.mathworks.com/matlabcentral/newsreader/view_thread/153058 (or http://tinyurl.com/clvfj4).

77. http://java.sun.com/docs/books/tutorial/uiswing/events/mousewheellistener.html (or http://tinyurl.com/dh2hkd); for its use in MATLAB read http://www.mathworks.com/matlabcentral/newsreader/view_thread/104129 (or http://tinyurl.com/cxt28j).

78. http://java.sun.com/javase/6/docs/api/java/beans/VetoableChangeListener.html (or http://tinyurl.com/b6x75l).

79. http://java.sun.com/docs/books/tutorial/javabeans/properties/constrained.html (or http://tinyurl.com/b8dzvy).

80. http://java.sun.com/docs/books/tutorial/uiswing/events/eventsandcomponents.html#many (or http://tinyurl.com/b93c69).

81. http://java.sun.com/docs/books/tutorial/uiswing/events/actionlistener.html (or http://tinyurl.com/2eept).

82. http://java.sun.com/docs/books/tutorial/uiswing/events/caretlistener.html (or http://tinyurl.com/c9vh2b).

83. http://www.mathworks.com/matlabcentral/newsreader/view_thread/80041#204096 (or http://tinyurl.com/cnp2pr).

84. http://mathforum.org/kb/message.jspa?messageID=5950839 (or http://bit.ly/dsjsga), http://www.mathworks.com/matlabcentral/newsreader/view_thread/156388#399260 (or http://bit.ly/aoEmXW) and a few others, including MATLAB's official doc: http://www.mathworks.com/help/techdoc/ref/set.html#f67-433534 (or http://bit.ly/cJe0SP); http://www.mathworks.com/help/techdoc/rn/broifyr-1.html#brrxpv8-1 (or http://bit.ly/9QGAKn).

85. http://en.wikipedia.org/wiki/Adapter_pattern (or http://bit.ly/9yoDeT); I believe that **handle**s are more *Adapter* than *Decorator* objects, but this could also be argued otherwise. The distinction here is merely theoretical and not practical.

86. http://www.mathworks.com/matlabcentral/newsreader/view_thread/246581 (or http://tinyurl.com/c9sobm).

87. http://www.mathworks.com/matlabcentral/newsreader/view_thread/270589#709652 (or http://bit.ly/ 5vVnYt).

88. http://UndocumentedMatlab.com/blog/detecting-window-focus-events/#comment-14472 (or http://bit.ly/b9n0u3).

89. http://www.mathworks.com/help/techdoc/creating_plots/f7-55506.html (or http://tinyurl.com/cmtuot).

90. http://java.sun.com/javase/6/docs/api/java/util/EventObject.html (or http://tinyurl.com/b5l3hp).

91. http://java.sun.com/javase/6/docs/api/java/awt/event/FocusEvent.html (or http://tinyurl.com/3x2dhe), misreported as `java.awt.FocusEvent`.

92. http://java.sun.com/javase/6/docs/api/java/awt/event/MouseEvent.html (or http://tinyurl.com/6mx8x7).

93. http://UndocumentedMatlab.com/blog/jfreechart-graphs-and-gauges/ (or http://bit.ly/dGq4nn).

94. http://sourceforge.net/projects/jfreechart; http://www.jfree.org/jfreechart/

95. As reported by http://www.jfree.org/jfreechart/; also — 400,000 Google references and numerous forum posts.

96. Search CSSM for JFreeChart: http://tinyurl.com/clpzsa

97. http://sourceforge.net/projects/jfreechart/files/ (or http://bit.ly/ehWYc9).

98. http://www.jfree.org/jfreechart/jfreechart-1.0.13-demo.jnlp (or http://bit.ly/gou4Ot).

99. For example, http://www.javaresources.biz/jfreechart_tutorial.jsp (or http://tinyurl.com/d8gqa9); http://www.screaming-penguin.com/node/4005 (or http://tinyurl.com/c79vtq).

100. http://www.mathworks.com/matlabcentral/newsreader/view_thread/269635 (or http://bit.ly/6HBbCO). Installation instructions and sample code: http://home.earthlink.net/~cdunson/ProcessingMatlab.html (or http://bit.ly/6OxqAY).

101. http://sourceforge.net/projects/jfreereport and http://reporting.pentaho.org

102. http://sourceforge.net/projects/jfreedesigner

103. http://sourceforge.net/projects/jasperreports; http://en.wikipedia.org/wiki/JasperReports

104. http://sourceforge.net/projects/ireport

105. http://java.com/en/dukeschoice/07winners.jsp (or http://bit.ly/afhRHF).

106. http://sourceforge.net/projects/oreports

107. http://www.mathworks.com/matlabcentral/fileexchange/27074 (or http://bit.ly/cpgde0).

108. http://jgrapht.sourceforge.net/

109. http://www.jgraph.com/

110. http://www.jgraph.com/jgraphdownload.html (or http://bit.ly/ghr7gr).

111. http://UndocumentedMatlab.com/blog/jfreechart-graphs-and-gauges/#comment-30052 (or http://bit.ly/hI5wBA); http://UndocumentedMatlab.com/blog/jgraph-and-bde/ (or http://bit.ly/hfZUTO).

112. http://gephi.org/

113. http://java.com/en/dukeschoice/ (or http://bit.ly/cBB9Gm).

114. http://vtk.org/, http://en.wikipedia.org/wiki/VTK. VTK was sponsored by SNL (http://titan.sandia.gov/) and Kitware, Inc.

115. https://www.kitware.com/InfovisWiki/index.php/Category:Matlab (or http://bit.ly/f5wbEy); http://www.kitware.com/products/archive/kitware_quarterly1009.pdf (or http://bit.ly/gZ30ae).

116. http://rsbweb.nih.gov/ij/features.html (or http://bit.ly/aylE9m); http://en.wikipedia.org/wiki/ImageJ (or http://bit.ly/bKCx4Q).

117. http://www.mathworks.com/matlabcentral/newsreader/view_thread/281698#745222 (or http://bit.ly/9feEZh).

118. http://www.mathworks.com/matlabcentral/newsreader/view_thread/290176#774399 (or http://bit.ly/9uh26u); also read: http://www.mathworks.com/matlabcentral/newsreader/view_thread/244712#629073 (or http://bit.ly/czXP64).

119. http://rsbweb.nih.gov/ij/notes.html (or http://bit.ly/d2RyTl).

120. http://rsbweb.nih.gov/ij/plugins/index.html (or http://bit.ly/dudT55); http://www.mathworks.com/matlabcentral/newsreader/view_thread/285447 (or http://bit.ly/eCWutl).

121. http://bigwww.epfl.ch/sage/soft/mij/ (or http://bit.ly/iIVuz5); use the static (rather than dynamic) classpath (see Section 1.1.2) to prevent ImageJ plugin issues: http://blog.mostlycurious.com/imagej-within-matlab-finally-working (or http://bit.ly/jzgUFl); Another connector — http://sourceforge.net/projects/imagejmatlab/ (or http://bit.ly/llm2b2); also see http://bit.ly/jV7FzY. Also see http://www.mathworks.com/matlabcentral/fileexchange/32344 (or http://bit.ly/p5gwQb).

122. http://www.imagingbook.com/index.php?id=102 (or http://bit.ly/cgeK6O); http://www.mathworks.com/matlabcentral/newsreader/view_thread/285447#812882 (or http://bit.ly/eSZA9L).

123. http://rsbweb.nih.gov/ij/links.html (or http://bit.ly/d5l94G).

124. https://swinglabs.dev.java.net/ (or http://bit.ly/gwhAop).

125. https://swingx.dev.java.net/ (or http://bit.ly/eQajUo).

126. https://jxlayer.dev.java.net/ (or http://bit.ly/edIeym).

127. For example, http://www.pbjar.org/blogs/jxlayer/jxlayer40/, http://forums.java.net/jive/forum.jspa?forumID=140, and http://forums.java.net/jive/thread.jspa?threadID=21038

128. https://swinghelper.dev.java.net/

129. https://wizard.dev.java.net/ (or http://bit.ly/hpQyrg).

130. https://pdf-renderer.dev.java.net/ (or http://bit.ly/ebv5mt).

131. https://jdic.dev.java.net/ (or http://bit.ly/h4Fozl).

132. http://www.mathworks.com/matlabcentral/newsreader/view_thread/283708#774672 (or http://bit.ly/avTFL1).

133. http://www.mathworks.com/matlabcentral/fileexchange/24924-jpar-parallelizing-matlab (or http://bit.ly/cOMVSP).

134. http://java.sun.com/developer/technicalArticles/J2SE/Desktop/javase6/systemtray (or http://tinyurl.com/y4bq9d); http://blogs.sun.com/CoreJavaTechTips/entry/getting_to_know_system_tray (or http://bit.ly/am5Gy9).

135. http://java.sun.com/javase/6/docs/api/java/awt/SystemTray.html (or http://tinyurl.com/2rld4j).

136. http://java.sun.com/javase/6/docs/api/java/awt/TrayIcon.html (or http://tinyurl.com/cyd4nl).

137. http://UndocumentedMatlab.com/blog/setting-system-tray-icons/ (or http://tinyurl.com/y9pl5y6). Note that `TrayIcon` has some quirks: http://weblogs.java.net/blog/alexfromsun/archive/2008/02/jtrayicon_updat.html (or http://bit.ly/gdvvGb).

138. http://UndocumentedMatlab.com/blog/setting-system-tray-popup-messages/ (or http://tinyurl.com/y8b8cn5).

139. http://www.mathworks.com/matlabcentral/fileexchange/23299 (or http://tinyurl.com/coqhz9).

140. http://blogs.mathworks.com/desktop/2007/10/15/drag-and-drop-data-import/ (or http://tinyurl.com/cpmkla); http://blogs.mathworks.com/desktop/2008/08/18/its-all-about-the-data/ (or http://tinyurl.com/cbwpxu) and others.

141. http://www.mathworks.com/matlabcentral/fileexchange/16312 (or http://tinyurl.com/ccza6s); http://www.mathworks.com/matlabcentral/newsreader/view_thread/155932 (or http://tinyurl.com/cmb52q).

142. http://www.mathworks.com/matlabcentral/fileexchange/4224 (or http://bit.ly/dtfiT0), which is an excellent submission.

143. http://www.mathworks.com/matlabcentral/fileexchange/15294 (or http://tinyurl.com/cezbxn).

144. http://www.mathworks.com/matlabcentral/newsreader/view_thread/154949 (or http://tinyurl.com/chkfyd).

145. http://java.sun.com/docs/books/tutorial/uiswing/dnd/index.html (or http://bit.ly/ahk1HD); http://java.sun.com/j2se/1.4.2/docs/guide/dragndrop/ (or http://bit.ly/aWfPGQ); http://java.sun.com/j2se/1.5.0/docs/guide/dragndrop/ (or http://bit.ly/bfSBlt).

146. http://java.sun.com/docs/books/tutorial/uiswing/dnd/defaultsupport.html (or http://bit.ly/bpMFCA); and more details (yet based on the older JVM 1.4.2): http://java.sun.com/j2se/1.4.2/docs/guide/swing/1.4/dnd.html (or http://bit.ly/aPgN1B).

147. http://java.sun.com/docs/books/tutorial/uiswing/dnd/transferhandler.html (or http://bit.ly/awiQ3d); note that the TransferHandle mechanism changed somewhat between JVMs 1.4, 1.5, and 1.6.

148. http://java.sun.com/javase/6/docs/api/java/awt/dnd/DropTarget.html (or http://tinyurl.com/cuycx7).

149. http://java.sun.com/javase/6/docs/api/java/awt/dnd/DnDConstants.html (or http://bit.ly/a4a35P) and http://java.sun.com/javase/6/docs/api/constant-values.html#java.awt.dnd.DnDConstants.ACTION_NONE (or http://bit.ly/91Rrwz).

150. http://java.sun.com/javase/6/docs/api/java/awt/datatransfer/DataFlavor.html (or http://tinyurl.com/cp3pja).

151. http://java.sun.com/javase/6/docs/api/java/awt/datatransfer/SystemFlavorMap.html (or http://tinyurl.com/cy2kyq).

151. This may be the reason for the following reported problem in CSSM: http://www.mathworks.com/matlabcentral/newsreader/view_thread/154949 (or http://tinyurl.com/chkfyd).

152. http://www.mathworks.com/matlabcentral/newsreader/view_thread/154949#640039 (or http://bit.ly/cSVoGs).

153. Download both Java and class files from http://UndocumentedMatlab.com/files/DropTargetList.zip (or http://bit.ly/d2ik3J).

154. http://www.mathworks.com/matlabcentral/newsreader/view_thread/258956 (or http://tinyurl.com/knqvlz).

155. http://www.mathworks.com/matlabcentral/newsreader/view_thread/154285 (or http://tinyurl.com/cy7a2z); http://www.mathworks.com/matlabcentral/newsreader/view_thread/82411 (or http://tinyurl.com/cqn93t); http://www.mathworks.com/matlabcentral/newsreader/view_thread/148719 (or http://tinyurl.com/cp7j43).

156. See Amir Ben-Dor's comment here: http://www.mathworks.com/matlabcentral/newsreader/view_thread/82411#209607 (or http://tinyurl.com/cqqvh6), where the re-parenting idea was originally suggested. Despite his very few CSSM posts, Amir appears to be one of the earliest and most leading-edge CSSM commenter on the MATLAB–Java integration issue.

157. http://www.mathworks.com/matlabcentral/newsreader/view_thread/284932#755936 (or http://bit.ly/cB9l2d).

158. http://www.mathworks.com/matlabcentral/newsreader/view_thread/167272 (or http://bit.ly/cTkPsP).

159. http://www.mathworks.com/matlabcentral/newsreader/view_thread/285708 (or http://bit.ly/dsDdSl).

160. http://en.wikipedia.org/wiki/Standard_Widget_Toolkit or http://www.eclipse.org/swt. SWT is often used together with the JFace toolkit (http://en.wikipedia.org/wiki/JFace).

161. http://java.sun.com/javase/6/docs/technotes/guides/awt (or http://tinyurl.com/c6oull).

162. http://en.wikipedia.org/wiki/Qt_Jambi or http://www.qtsoftware.com/products

163. http://swingwt.sourceforge.net and http://swtswing.sourceforge.net

164. http://www.eclipse.org/swt/widgets/

Chapter 4

Uitools

Uitools is the name given by MathWorks for a set of user-interface functions. These functions have existed in a pretty stable situation for many MATLAB versions, some as far back as R12 (MATLAB 6.0, released in 2000). Despite this, to this day (MATLAB R2011a), some of these functions are unsupported and only partially documented. Many important uitools are Java based, hence the reason for including this chapter in this book. Here, we will only focus on the Java-based or customizable functions.

The odd thing about uitools is that some of these functions appear in the category help output (result of *help('uitools')*), and in some cases, they may even have a fully visible help section (e.g., *help('setptr')*) but do not have any online help documentation (*docsearch('setptr')* fails and *doc('setptr')* simply displays the readable help text).

In other cases (e.g., *help('uitree')*), the entire readable help section is as follows:

```
WARNING: This feature is not supported in MATLAB
and the API and functionality may change in a future release.
```

In such cases, one has to manually edit the function (*edit('uitree')*) in order to place a leading comment sign (%) at line 4 in order for the help section to become readable in the Command Window:

```
         function [tree, container] = uitree (varargin)
         % WARNING: This feature is not supported in MATLAB
         % and the API and functionality may change in a future release.
fix => %
         % UITREE creates a uitree component with hierarchical data in a
         % figure window.
         % UITREE creates an empty uitree object with default property
         % values in a figure window.
         % ...
```

These two phenomena are called "Semi-Documented Functions" in this book, and many other examples will be presented in this and later chapters. All these functions are officially unsupported by MathWorks, even when having a readable help section. The rule of thumb appears to be that a MATLAB function is supported only if it has online documentation. Note that in some rare cases, a documentation discrepancy may be due to a MathWorks documentation error, and not due to unsupportability.

The reasons for the partial documentation are varied: Some functions (such as *uitable* or *uitree*) are based on internal Java objects that MathWorks may feel are still unstable, buggy, or should be kept under wraps for some reason. Others (such as *moveptr*) were either developed for internal MATLAB development and are not yet ready for public use or (like *uicontainer*) are not well integrated with other MATLAB functionalities.

Whatever be the reasons, the uitools folder, located at %matlabroot%/toolbox/matlab/uitools/, contains a treasure trove well worth exploring. Type *help('uitools')* for an annotated partial list of available uitools and *dir([matlabroot,'\toolbox\matlab\uitools'])* for a full (un-annotated) list.

4.1 Uitable

For many years, the highest-requested MATLAB GUI component was an editable data table/grid. Such a component was missing both from MATLAB's GUI editor (*guide*) and from the supported list of GUI functions until R2008a (aka MATLAB 7.6). This missing component appears, at least according to the number of requests/queries/posts on the MATLAB forums and from the number of submissions/downloads of File Exchange solutions, to be the single most needed missing GUI component in MATLAB.

True, there are many possible methods of creating data tables in MATLAB using basic MATLAB building blocks, and the variety of solutions on the File Exchange reflects this: one can use a two-dimensional matrix of editboxes[1] or labels,[2] a standard listbox,[3] ActiveX components,[4] standard Java Swing widgets,[5] undocumented MathWorks Java classes,[6] and even text labels on a plot axes.[7]

Starting with MATLAB 7 (R14), MathWorks have included the unsupported function *uitable* in the uitools folder. This function uses the internal MATLAB Java widget `com.mathworks.widgets.spreadsheet.SpreadsheetTable`, which derives from the standard Java Swing `JTable`[8] class via `com.mathworks.mwswing.MJTable`. *uitable* became fully supported in R2008a[9] after many years in unsupported mode, and then switched to a `com.jidesoft.grid.SortableTable`-derived table (see Section 5.7).

uitable provides a lightweight and consistent look-and-feel alternative to the non-Java solutions and integrates well with MATLAB GUI. Use of *uitable* has the benefit of employing a scrollable `JTable` without the hassle of setting up a `ScrollableViewport` and other similar nuts and bolts. Also, *uitable* automatically detects Boolean (*islogical*) and Combo-Box (*iscell*) data columns and uses corresponding cell editors for them — checkboxes and dropdowns, respectively.

Since R2008a, MATLAB includes both versions of *uitable*. MATLAB automatically selects the version to use based on the supplied input and output arguments: *uitable* accepts data and parameters in the familiar P–V (property–value) named-pair format, in which case, MATLAB automatically uses the new version of *uitable* (we can force it to use the older version by adding a 'v0' input argument). The old (pre-R2008a) *uitable* also accepts a couple of other formats. In both versions, *uitable* accepts an optional figure handle as the leading (first) argument, which is *gcf* by default.

MATLAB automatically uses the new *uitable* when only a single output argument (or none) is requested and when the properties conform to the P–V pairs of the new *uitable*'s properties. In all other cases, the older *uitable* is used.

Note that if we have not used the 'v0' input argument and MATLAB has determined that it needs to use the old *uitable* version, a warning message will be displayed. This warning can be suppressed using the *warning* function as follows:

```
>> mtable = uitable('Data',magic(3));            % new uitable, no warning
>> [mtable,c] = uitable('v0','Data',magic(3));   % old uitable, no warning
```

```
>> [mtable,c] = uitable('Data',magic(3));              % old uitable + warning
Warning: it appears you are using an obsolete version of uitable.
See the documentation for correct uitable usage:
  help uitable and doc uitable
For more information, click here
> In uitable at 47

>> warning off MATLAB:uitable:OldTableUsage
>> [mtable,c] = uitable('Data',magic(3));              % old uitable, no warning
```

Note: In the following sections, text that deals exclusively with the old *uitable* version is outlined with a left sidebar, as in this paragraph; text that applies only to the new version has a right sidebar; and text that applies to both versions has no sidebar.

Settable properties of the old *uitable* include **CheckBoxEditor, ColumnNames, ColumnWidth, Data, DataChangedCallback, Editable, GridColor, NumColumns, NumRows, Position, Units**, and **Visible**. MATLAB versions prior to 2008 had an undocumented **Parent** property, renamed **UIContainer** in later releases.

An undocumented feature of the older *uitable* is that it also accepts a Java object deriving from javax.swing.table.DefaultTableModel[10] as an optional second argument, following the optional figure handle (which may be skipped, making the user's TableModel the first argument). DefaultTableModel is used as the default TableModel, if none other is specified (this is also standard for Swing's JTable).[11]

The older *uitable* returns two arguments: a handle to the created table (a com.mathworks.hg.peer.UitablePeer Java object wrapped within a MATLAB handle) and an undocumented second optional argument holding a handle to the MATLAB GUI container of the created table. These are exactly the two arguments returned from the *javacomponent* function (see Chapter 3). Use of the first return argument enables the user to specify the table **Units** (e.g., "Normalized"), **Editable** (true/false), **Enabled** (true/false), **Visible** (true/false), and **DataChangedCallback** properties. *uitable*'s help section seems to imply that these properties may be passed directly as P–V pairs, just like **ColumnNames,** etc., but this is, in fact, not so.

Units and **Visibility** may also be set via the second (container) output argument. The container handle also enables changing other standard MATLAB handle properties, such as **Position, UserData, Tag, HandleVisibility etc**. Note that the container handle is a simple MATLAB Handle Graphics (HG) object, whereas the table handle is a Java object wrapped within a MATLAB handle. This means that while we need to pass true/false (or 1/0) to table's **Visible** property, we need to pass 'on'/'off' to the container's:

```
[mtable, container] = uitable(gcf, magic(3), {'A', 'B', 'C'});
set(mtable, 'Visible', true, 'Position',[10,10,280,100]);
set(container, 'Visible', 'on');
```

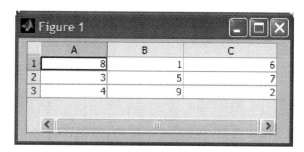

A very simple *uitable* (old version)

The container handle can also be retrieved directly from the table as follows:

```
mtable = uitable(...);
container = mtable.getUIContainer;
container = get(mtable,'UIContainer');         % equivalent method
```

We might expect that setting a table's **UIContainer** property (e.g., to a *uipanel* handle), would automatically move a table into the new container. However, this is not so. Instead, we need to set the container's **Parent** property as follows:

```
set(mtable,'UIContainer',hPanel);                     % doesn't work
set(get(mtable,'UIContainer'),'Parent',hPanel);       % ok
```

Like all containers returned from *javacomponent*, the container handle has some hidden undocumented properties: **JavaPeer** (a handle to the table object), **FigureChild** (same), **PixelBounds, HelpTopicKey, Serializable,** and so on. Note that some properties, while settable, appear to have no effect (the container's **Opaque** and **BackgroundColor** properties, for example). Note that due to a limitation of *javacomponent*, which is used to place the *uitable* onscreen, the *uitable* is always created as a direct child of the container figure.

Once created, *uitable* data can be read by one of the following methods:

```
data = cell(mtable.Data);
data = cell(mtable.getData);
data = cell(get(mtable,'Data'));
```

Note the use of *cell()* casting: this is required, since the returned data is a 2D Java array. Cell casting transforms this Java array into a MATLAB-usable two-dimensional cell array. Similarly, when we set the **Data,** we need to pass a cell array:

```
>> set(mtable,'Data',{1,2,3; 4,'text',true});    % setting mixed data
```

Settable properties for the new *uitable* include **BackgroundColor, CellEditCallback, ColumnFormat, ColumnEditable, ColumnName, ColumnWidth, Data, Enable, FontAngle, FontName, FontSize, FontUnits, FontWeight, ForegroundColor, KeyPressFcn, Position, RearrangeableColumns, RowName, RowStriping, CellSelectionCallback, TooltipString, Units,** and all the standard HG properties (**Parent, Tag, UserData, Visible,** etc.).

The new ***uitable*** returns only a single argument — a MATLAB handle of the created table. The new version of ***uitable*** is created in a similar manner to the old version:

```
mtable = uitable(gcf, 'Data',magic(3), 'ColumnName',{'A', 'B', 'C'});
set(mtable,'Position',[10,10,280,100]);
```

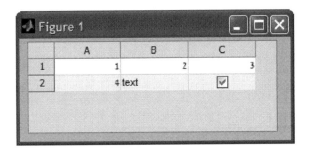

The new version of *uitable*

The new ***uitable***'s data can be retrieved and set much more easily than the old version's by simply using the **Data** property (no ***cell*** casting is necessary):

```
>> data = get(mtable,'Data')                    % getting the data
data =
      8    1    6
      3    5    7
      4    9    2
>> set(mtable,'Data',[1,2,3; 4,5,6]);           % setting numeric data
>> set(mtable,'Data',{1,2,3; 4,'text',true});   % setting mixed data
```

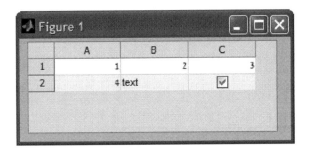

Modified table data with text and logical values

4.1.1 Customizing Uitable

The MATLAB handles returned by ***uitable*** (either the new or the old versions) have some built-in customizations possible. However, using the underlying Java table reference, we can gain access to a much wider range of customizations, which shall be discussed below. Note that the table object returned by ***uitable*** is not actually a Java object but rather a MATLAB handle object. Because of this, I call the returned object `mtable` and the underlying Java object `jtable`.

In the following listing of Java sub-components, we shall use the following annotated screen-shot of a *uitable* (the new version — it is pretty similar to the old version):

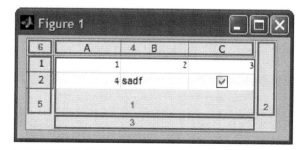

Modified table data with text and logical values, annotated (See color insert.)

Unfortunately, there is no direct way of getting the new *uitable's* underlying jtable reference. However, we can use the *findjobj* utility for this (see Section 7.2.2). The following code snippet shows how we "dig out" jtable (☐1) from its containing scrollpane:

```
>> jscroll = findjobj(mtable)
jscroll =
    javahandle_withcallbacks.com.mathworks.hg.peer.utils.UIScrollPane

>> jscroll.list
com.mathworks.hg.peer.utils.UIScrollPane[,0,0,280x100,...]
 javax.swing.JViewport[,31,19,248x80,...]
1│   com.mathworks.hg.peer.ui.UITablePeer$22[,0,0,225x54,...]
      javax.swing.CellRendererPane[,0,0,0x0,hidden]
   com.mathworks.hg.peer.utils.UIScrollPane$1[,262,9,17x0,hidden,...]
2│    com.sun.java.swing.plaf.windows.WindowsScrollBarUI$WindowsArrowButton
      com.sun.java.swing.plaf.windows.WindowsScrollBarUI$WindowsArrowButton
   com.mathworks.hg.peer.utils.UIScrollPane$2[,31,182,218x17,hidden...]
3│    com.sun.java.swing.plaf.windows.WindowsScrollBarUI$WindowsArrowButton
      com.sun.java.swing.plaf.windows.WindowsScrollBarUI$WindowsArrowButton
   javax.swing.JViewport[,31,1,248x18,...]
4│    com.jidesoft.grid.SortableTable$2[,0,0,248x18,...]
      javax.swing.CellRendererPane[,0,0,0x0,hidden]
   javax.swing.JViewport[,1,19,30x80,...]
5│    com.mathworks.hg.peer.ui.table.RowHeader[,0,0,30x54,...]
      javax.swing.CellRendererPane[,0,0,0x0,hidden]
   javax.swing.JLabel[,1,1,30x18,...]◄───────────  6│

>> jtable = jscroll.getViewport.getView
jtable =
com.mathworks.hg.peer.ui.UITablePeer$22[,0,0,225x54,...]
```

In the old version of *uitable*, get jtable directly from mtable (note that components 4 and 6 are in reversed positions compared with the new *uitable*):

```
[mtable,hcontainer] = uitable (...);
jtable = mtable.getTable;    % or: get(mtable,'table');
>> jscroll = jtable.getParent.getParent; % or: findjobj(mtable.UIContainer)
```

```
jscroll =
  javahandle_withcallbacks.com.mathworks.widgets.spreadsheet.-
SpreadsheetScrollPane

>> jscroll.list
com.mathworks.widgets.spreadsheet.SpreadsheetScrollPane[,0,0,280x100...]
  javax.swing.JViewport[,16,17,263x65,...]
1   com.mathworks.hg.peer.UitablePeer$PeerSpreadsheetTable[,0,0,264x30,...]
      javax.swing.CellRendererPane[,0,0,0x0,hidden]
  javax.swing.JScrollPane$ScrollBar[,0,0,0x0,hidden,...]
2   com.sun.java.swing.plaf.windows.WindowsScrollBarUI$WindowsArrowButton...
    com.sun.java.swing.plaf.windows.WindowsScrollBarUI$WindowsArrowButton...
  javax.swing.JScrollPane$ScrollBar[,16,82,263x17,...]
3   com.sun.java.swing.plaf.windows.WindowsScrollBarUI$WindowsArrowButton...
    com.sun.java.swing.plaf.windows.WindowsScrollBarUI$WindowsArrowButton..
6 javax.swing.table.JTableHeader[,1,1,15x16,...]
    javax.swing.CellRendererPane[,0,0,0x0,hidden]
5 javax.swing.JViewport[,1,17,15x65,...]
    com.mathworks.mwswing.MJTable[,0,0,15x30,...]
      javax.swing.CellRendererPane[,0,0,0x0,hidden]
4 javax.swing.JViewport[,16,1,263x16,...]
    javax.swing.table.JTableHeader[,0,0,264x16,...]
      javax.swing.CellRendererPane[,0,0,0x0,hidden]
```

Setting individual cells is much faster than getting the entire data from the table, modifying it, and then setting the entire data again in the mtable. Just remember that mtable, just as any MATLAB object, uses 1-based indexing, whereas jtable, like any Java object, uses 0-based indexing. So, the first column (or row or anything) is 1 in mtable and 0 in jtable in both versions of **uitable** (both new and old). Forgetting this rule can be an endless source of bugs.

```
% Slow alternative:
data = get(mtable,'Data');    % add cell() for the old uitable version
data{row,col} = newValue;
set(mtable,'Data',data);


% This is much faster:
newValueStr = num2str(newValue);   % uitable cells contain strings
jtable.setValueAt(newValueStr,row-1,col-1); %row,col index start at 0
```

Note that the underlying cell data of jtable is a string. MATLAB automatically converts the numeric data into strings. Therefore, when we use set(mtable,'Data',data), we should use numeric data, whereas if we use jtable.*setValueAt()*, we should use the string representation of the data, otherwise an error will be displayed:

```
>> jtable.setValueAt(345, row-1, col-1)
Warning: Cannot convert logical edit to numeric matrix.
 Please click for more information
(Type "warning off MATLAB:hg:uitable:CellEditWarning" to suppress this warning.)

>> jtable.setValueAt('345', row-1, col-1)   % <= this works ok
```

In the new ***uitable***, if we wish to set a non-numeric string in a specific table cell, the initial data type of the relevant column must be 'char', otherwise the string will display as 'NaN'.

Since the Java object underlying ***uitable*** is basically a Java Swing `JTable`, most `JTable` features and actions are also applicable to the ***uitable*** object. Alternately, if we want to find out how to do something on the ***uitable***, the easiest way is to find out how to do a similar action on a `JTable` and then try to apply this on the ***uitable*** object — it works in almost all cases (some few `JTable` features are inconsistent or unavailable in old Java versions. Type ***version('-java')*** to see your specific JVM version).

Since all Swing objects accept HTML for any of their `JLabels`,[12] this means that ***uitable*** cells, column headers, and tooltips also accept HTML (there is no need to close the tags with `</font>`, `</html>`, etc.). An example (old ***uitable*** version, easily adaptable to the new version) is as follows:[†]

```
mtable = uitable('Data',magic(3),'ColumnNames',{'<html><b>A','B','C'});
jtable = mtable.getTable;
str = '<html><b>line #1</b><br><font color="red">line #2';
jtable.setToolTipText(str);
jtable.getTableHeader.setToolTipText(str);
jtable.setValueAt('<html><i><font size=+1>big italic',row,col);
```

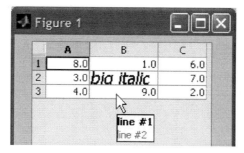

uitable with HTML cell contents and tooltips

Similarly, table cells can display images and icons using the HTML `<img>` tag.[13]

In the preceding discussion, I have shown how column headers can easily be modified and customized. Unfortunately, row headers cannot be changed so easily. While these are configurable in the new ***uitable***, the old version requires access to a separate Java object ([5]) which is used to store these headers.[14]

There have been several dozen references to ***uitable*** in CSSM over the last few years, and readers are referred there for additional references.[15] A CSSM poster once requested to know how to set a specific ***uitable*** column to a specific color.[16] My answer, based on `JTable` usage, was to simply use a nonstandard `CellRenderer`[17] for the requested column as follows (column B):

```
cr = javax.swing.table.DefaultTableCellRenderer;
cr.setForeground(java.awt.Color.red);
```

[†] The following code segment is for the old ***uitable***; for the new version, use ***findjobj*** to get the jtable reference, and also set the initial data to strings via ***arrayfun(@num2str**, magic(3), 'uniform', false).

```
cr.setBackground(java.awt.Color(0.2,0.3,0.8)); %dark-blue
jtable.getColumnModel.getColumn(1).setCellRenderer(cr); %1 = B
jtable.repaint; % repaint the table to use the new CellRenderer
```

uitable with a nonstandard column foreground and background

Separation of Model and View is the basis for Swing's MVC architecture.[18] The basic idea in the MVC design pattern is to separate the data (*Model*) from the presentation format (*View*) and the data-modification functionality (*Controller*).[19] This is the reason for separate cell renderers and cell editors in JTable, JTree, JList, and JComboBox. Cell renderers and editors are *delegate objects*,[20] which are separate helper objects from their target component. They fulfill specific missions of display and editing of their target component's sub-components (cells).

Unfortunately, in the new ***uitable*** design, JIDE and MATLAB have apparently broken this MVC approach by using a DefaultUIStyleTableModel class as the table model. This model not only controls the data but also sets the table's appearance (e.g., row-striping background colors) and disregards column cell renderers. In order for our DefaultTable-CellRenderer to have any effect, we must, therefore, replace MATLAB's standard DefaultUIStyleTableModel with a simple DefaultTableModel:

```
% Replace uitable's standard table model
data = mat2cell(magic(3),[1,1,1],[1,1,1]);       % {8,1,6; 3,5,7; 4,9,2}
cols = {'<html><b>A', 'B', 'C'};
jtable.setModel(javax.swing.table.DefaultTableModel(data,cols));

% Now set the cell-renderers and repaint the table
jtable.getColumnModel.getColumn(1).setCellRenderer(cr);          %1 = B
jtable.repaint;  % repaint the table to use the new CellRenderer
```

Java uses its DefaultTableCellRenderer[21] to decide how to display cell data for the entire table column. If we want to modify cell appearance, we need to modify the specific column's CellRenderer as shown above. This requires programming a Java class. In some cases, smart MATLAB programming can save this Java programming: A CSSM user once requested to know how to set the number of displayed decimal places.[22] The textbook approach would be to create a dedicated CellRenderer. However, in practice, it is easier to preformat the data using ***sprintf*** than to create a separate CellRenderer Java class. Of course, the data will then be stored (and sorted) as strings rather than as numbers. So, the trick was to populate the field with ***str2double***(***sprintf***('%.0f', myData)) or to cast the data as integers (***uint8***(myData)) in order to prevent showing any decimals.

While HTML content provides very easy and simple font customization, greater control can be achieved by using com.jidesoft.grid.StyledTableCellRenderer as the table column's cell renderer (StyledTableCellRenderer extends JIDE's StyledLabel, which is described in Section 5.5.1). There are similar renderers for trees and lists.

A different renderer, which I have called ColoredFieldCellRenderer, can be used to set a specific color (and/or tooltip) for specific table cells.[23]

```
cr = ColoredFieldCellRenderer;
jtable.getColumnModel.getColumn(1).setCellRenderer(cr);
cr.setCellBgColor(0,1,java.awt.Color.cyan);
cr.setCellTooltip(2,1,'cell-specific tooltip');
jtable.repaint;
```

uitable with a nonstandard column cell renderer (See color insert.)

A related common need is to set a nonstandard CellEditor:[24] ***uitable*** automatically detects Boolean (***islogical***) and Combo-Box (***iscell***) data columns and uses corresponding cell editors for them — checkboxes and drop-downs, respectively. For all other data types, ***uitable*** uses a standard text-field editor (javax.swing.DefaultCellEditor[25] initialized with JTextField).

We often need a color editor or some other nonstandard editors. Such nonstandard cell editors cannot be set using the standard MATLAB (mtable) object, but they can relatively easily be set at the Java (jtable) level, using one of several options:

1. Prepare a Java class that extends DefaultTableCellEditor.
2. Prepare a Java class that extends JComboBox, JCheckBox, or JTextField.
3. Use JComboBox or a similar editable Java component.

The usage in MATLAB is very simple. For example, to set the leftmost table column (#0) to use a combo-box (drop-down) editor, we will use a simple JComboBox:

```
comboBox = javax.swing.JComboBox({'First','Last'});
comboBox.setEditable(true);
editor = javax.swing.DefaultCellEditor(comboBox);
jtable.getColumnModel.getColumn(0).setCellEditor(editor);
```

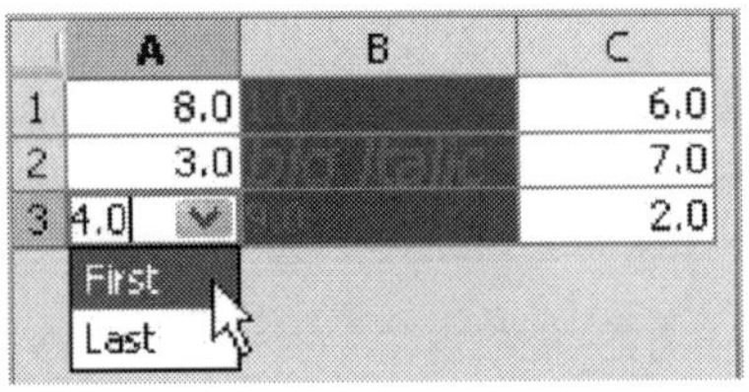

uitable with a nonstandard column cell editor (See color insert.)

Note how the cell editor (in this case, a combo-box) only appears for the currently edited cell — all the other cells appear as defined by their cell renderers (which in this case is a simple label and not a combo-box).

This was not a very powerful example. After all, the drop-down values are static for the entire column. It is useful for cases where the same editor component (e.g., font/color/file selection, etc.) can be used for all data rows.

Let us now demonstrate a more realistic and more powerful example: Very often, the cell drop-down values should depend on its specific row (i.e., not all column cells should have the same drop-down options). For example, a list of selectable dishes in column B might depend on the dish type in column A. Different table rows will have different dish types and thus different selectable dishes:

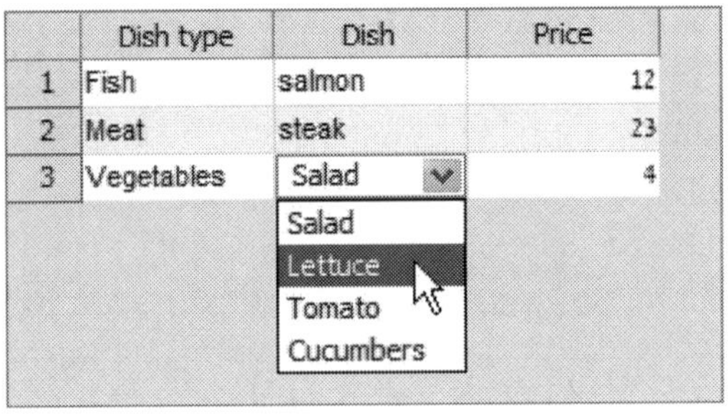

uitable with a more complex nonstandard column cell editor

To implement our dish-type example, we need a custom `CellEditor` Java class. Preparation of a custom `CellEditor` is beyond the scope of this book — users are referred to the official documentation[26] or any good text about Java Swing. Here is a simple skeleton of a sample `DefaultCellEditor`-derived class for our dish-types example: it uses a lookup value in the Dish-type column to set the drop-down values of the Dish column, separately for each row:

```
import java.awt.*;
import java.util.*;
import javax.swing.*;
public class LookupFieldCellEditor extends DefaultCellEditor
{
    private int_lookupColumn = -1;
    private Hashtable_dataVals = null;
```

```
    public LookupFieldCellEditor(Hashtable dataVals, int column) {
        super(new JComboBox());
        _dataVals = dataVals;
        _lookupColumn = column;
    }
    public Component getTableCellEditorComponent(JTable table,
                                                 Object value,
                                                 boolean isSelected,
                                                 int row, int col) {
        JComboBox cell = (JComboBox) super.
            getTableCellEditorComponent(table,value,isSelected,row,col);
        // Modify the selected cell's CellEditor (JComboBox) to display
        // only relevant fields based on this row's lookup column
        if (_lookupColumn >= 0) {
            cell.removeAllItems();
            Object lookupObj = table.getValueAt(row,_lookupColumn);
            if (lookupObj != null) {
                String srcName = (String) lookupObj;
                if (_dataVals.containsKey(srcName)) {
                    String[] vals = (String[])_dataVals.get(srcName);
                    java.util.List <String> dataList = Arrays.asList(vals);
                    java.util.Iterator iter = dataList.iterator();
                    while (iter.hasNext())
                        cell.addItem(iter.next());
                }
            }
        }
        cell.setSelectedItem(value);
        cell.setMaximumRowCount(cell.getItemCount() ==0 ? 1 : 10);
        if (cell.getItemCount() == 0) cell.addItem(' ');
        return cell;
    }
    public void setEditable(boolean flag) {
        ((JComboBox) getComponent()).setEditable(flag);
    }
}
```

Save this code in *LookupFieldCellEditor.java*, compile it, and place the generated class file in your Java classpath (see Section 1.6 for details). Alternately, download *LookupFieldCellEditor.zip*,[27] which includes both source and class files, and add this file to your Java classpath. Usage of this class in MATLAB is very simple:

```
set(mtable,'ColumnEditable',[true,true,false]); %make columns editable
fieldsHashtable = java.util.Hashtable;
fieldsHashtable.put('Meat',        {'steak','veal','beaf',...});
fieldsHashtable.put('Fish',        {'cod','whitefish','salmon',...});
fieldsHashtable.put('Vegetables',  {'Salad','Lettuce','Tomato',...});
ed = LookupFieldCellEditor(fieldsHashtable,0);            % 0 = column A
jtable.getColumnModel.getColumn(1).setCellEditor(ed);     % 1 = column B
```

If we use a custom CellEditor, we may also need a corresponding CellRenderer. For example, if we use a com.jidesoft.combobox.ColorComboBox (see Section 5.4.1) for selecting a color, we would probably wish to have a color label as our `CellRenderer`. I have done just that, and readers can download the relevant renderer and editor classes from the website.[28] Their usage in MATLAB is, again, very simple:

```
data = {'Fish',          'salmon',      '223,34,145'; ...    % pink
        'Meat',          'steak',       '200,0,0'; ...       % dark red
        'Vegetables','Lettuce',         '0,120,0'};          % green
cols = {'Dish type', 'Dish', 'Color'};
mtable = uitable(gcf, 'Data',data, 'ColumnName',cols);
set(mtable,'Position',[10,10,280,150],'ColumnEditable',true(1,3));
jscroll = findjobj(mtable);
jtable = jscroll.getViewport.getView;
jtable.setModel(javax.swing.table.DefaultTableModel(data,cols))
jtable.getColumnModel.getColumn(2).setCellRenderer(ColorCellRenderer);
jtable.getColumnModel.getColumn(2).setCellEditor(ColorCellEditor);
```

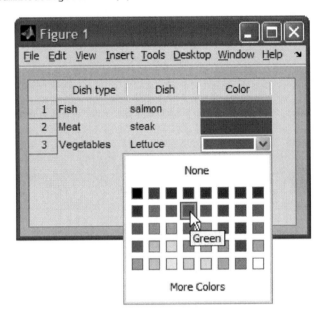

uitable with custom CellRenderer and CellEditor (**See color insert.**)

4.1.2 Table Callbacks

The old version of *uitable* (mtable) has a single callback, **DataChangedCallback**, which is called whenever a table value changes, the table structure changes (columns are added/deleted), or a table cell has been edited. As far as the table model is concerned, checkboxes are simply renditions of boolean (logical true/false) values and similarly for other `CellEditors`. Therefore, selecting a checkbox or drop-down value within a table is exactly the same as modifying a cell's content programmatically, causing firing (invocation) of the **DataChangedCallback**.[29]

The new version of *uitable* supports **CellEditCallback** (invoked when a table cell has changed), **CellSelectionCallback** (when a cell is selected), **KeyPressFcn** (when a keyboard key has been clicked), and **ButtonDownFcn** (when the mouse is clicked). In these callbacks, the EventData input parameter includes the relevant information in a MATLAB struct. This data includes cell indices [row,col], new/previous data, clicked key etc. (depending on the callback, some of the information may not be presented).

In our callback function, we need to prevent re-entry if the callback function is not re-entrant. For example, if the callback function checks the modified data and restores the old value in case of an invalid entry, the function needs to exit immediately when invoked due to the old value restoration. This can be done by setting some persistent value,[†] as shown below using the old ***uitable*** (adaptation to the new version is easy):

```
% Set a data-change callback function
try            % old uitable version
   set(mtable, 'DataChangedCallback',{@myDataChange_Callback,jtable});
catch          % new uitable version
   set(mtable, 'CellEditCallback', {@myDataChange_Callback,jtable});
end

% Sample myDataChange_Callback function
function myDataChange_Callback(mtable, eventdata, jtable)
   % Prevent re-entry if the callback function is not re-entrant
   % Note: reset DataChangedCallback within the CB func won't work
   persistent hash
   if isempty(hash), hash = java.util.Hashtable; end
   if ~isempty(hash.get(mtable)), return; end % exit upon reentry
   hash.put(mtable,1);

   % sanity check (in case the table got deleted by now)...
   if ~ishandle(mtable), return; end
   % Check modified value and restore old value if invalid
   try         % old uitable version
      modifiedRowIdx = eventdata.getEvent.getFirstRow;        % 0-based
      modifiedColIdx = eventdata.getEvent.getColumn;          % 0-based
   catch       % new uitable version
      modifiedRowIdx = eventdata.Indices(1) - 1;    % 1-based
      modifiedColIdx = eventdata.Indices(2) - 1;    % 1-based
   end
   if modifiedColIdx >=0 && modifiedRowIdx >=0
      data = get(mtable,'Data');
      newValue = data(modifiedRowIdx + 1,modifiedColIdx + 1);
      if (newValue < 0)
         jtable.setValueAt(-newValue,modifiedRowIdx,modifiedColIdx);
      end
   end

   % Release reentrancy flag
   hash.remove(mtable);
end % myDataChange_Callback
```

[†] We could use ***setappdata*** and ***getappdata*** instead, but these fail in some edge cases, whereas Hashtable always works.

Note that it is possible that multiple table cells have been modified at once (e.g., by pasting a block of values). In such a case, we will want to use methods such as eventDetails.*getFirstRow* and eventDetails.*getLastRow* and process the changed data as appropriate to our requirements.

Note that while this approach is simple and easily integrates within the MATLAB program, it cannot handle very rapid data-change events. At some point, when the event queue is overloaded, some callback events will get discarded. A simple solution is to temporarily disable the table object during callback processing, thus preventing concurrent updates. Alternately, temporarily disable the callback itself when updating rapidly changing table data. Both of these solutions can be done in pure MATLAB.[30] A more complex solution, suggested by a CSSM poster,[31] is to temporarily disable the JTable MATLAB listeners using Java or to set up a Java callback method rather than setting up a MATLAB one. These later approaches require Java knowledge and are advanced beyond the scope of this text.

The jtable object reference handle exposes all the standard Swing callbacks specified in Chapter 3. In addition, JTable exposes two additional callbacks: **CaretPositionChangedCallback** and **InputMethodTextChangedCallback**. These callbacks correspond to events that are supposed to be raised when the component's input caret position or content has changed, but apparently they are disconnected in MATLAB for some unknown reason.

4.1.3 *Customizing Scrollbars, Column Widths, and Selection Behavior*

A common action that many users need is controlling the scrollability of the table, vertically and/or horizontally. This can be done using code similar to the following:

```
jscroll = mtable.TableScrollPane;      % old uitable
jscroll = findjobj(mtable);            % new uitable
jscroll.setVerticalScrollBarPolicy(...
                    jscroll.VERTICAL_SCROLLBAR_AS_NEEDED);
jscroll.setHorizontalScrollBarPolicy(...
                    jscroll.HORIZONTAL_SCROLLBAR_AS_NEEDED);
```

The possible scrollbar policies are *_SCROLLBAR_AS_NEEDED, *_SCROLLBAR_ALWAYS, and *_SCROLLBAR_NEVER.[32] These policies are pretty much self-explanatory.

Horizontal scrolling policy is normally accompanied by setting the column auto-resize policy. This affects the interaction between the horizontal scrollbar and the column widths when the column widths are modified interactively (by dragging the column header boundary) or programmatically (see below). Possible auto-resize policy values are AUTO_RESIZE_ALL_COLUMNS, AUTO_RESIZE_LAST_COLUMN, AUTO_RESIZE_NEXT_COLUMN, AUTO_RESIZE_SUBSEQUENT_COLUMNS, and AUTO_RESIZE_OFF.[33] They are best understood by modifying the **AutoResizeMode** property then dragging a column header boundary sideways and checking the effect on the other columns and on the scrollbar:

```
jtable.setAutoResizeMode(jtable.AUTO_RESIZE_SUBSEQUENT_COLUMNS);
```

Setting of initial column sizes for all columns at once can be achieved easily:

```
mtable.setColumnWidth(100); % initial size for all columns
```

We often need to set initial and maximal column width. For example, a checkbox column should be non-resizable 10 pixels wide:

```
mtable.setCheckBoxEditor(1);                          % 1 = column A (Matlab)
jcol = jtable.getColumnModel.getColumn(0);  % 0 = column A (Java)
set(jcol,'Resizable','off','MaxWidth',10,'PreferredWidth',10);

% Alternative method: note that Matlab 'off' becomes a Java false
jcol.setResizable(false);
jcol.setMaxWidth(10);
jcol.setPreferredWidth(10);
```

Table columns may be resized to dynamically fit their data, using the following:

```
com.mathworks.mwswing.MJUtilities.initJIDE;
com.jidesoft.grid.TableUtils.autoResizeAllColumns(jtable);
```

Another interesting table policy is `ListSelectionModel`,[34] which affects the way in which multiple table cells may be selected together. Possible values are `SINGLE_SELECTION`, `SINGLE_INTERVAL_SELECTION`, and `MULTIPLE_INTERVAL_SELECTION`.

```
import javax.swing.ListSelectionModel;
selectPolicy = ListSelectionModel.MULTIPLE_INTERVAL_SELECTION;
jtable.setSelectionMode(selectPolicy);
```

The actual selected cells can be gotten via jtable.*getSelectedRows(), getSelectedColumns()*, and similar methods. The jtable.*getSelectionModel()* object contains additional information about the selected cells.

For programmatic cell selection, use jtable.*changeSelection(rowIndex, columnIndex, toggle-Flag, extendFlag)*.[†] For multi-cell selection (e.g., entire rows, columns, or cell blocks), set jtable.*setNonContiguousCellSelection(flag)* followed by *setRowSelectionAllowed(flag)* and *set ColumnSelectionAllowed(flag)*.[35] Cells blocks can then be selected using jtable.*setRowSelectionInterval(startCol,endCol)* and *setRowSelectionInterval(startRow,endRow)* for the first interval and *addRow/ColumnSelectionInterval(…)* for additional intervals. Intervals can be removed using *setColumnSelectionInterval(…)* or *clearSelection(); selectAll()* selects all table cells.

Selection colors can be set using *setSelectionBackground(color)* and *setSelectionForeground(color)*. For example,

```
jtable.setSelectionBackground(java.awt.Color(1.0,1.0,0));    %RGB values
jtable.setSelectionBackground(java.awt.Color.yellow);        %alternative
```

[†] http://download.oracle.com/javase/1.5.0/docs/api/javax/swing/JTable.html#changeSelection(int, int, boolean, boolean) (or http://bit.ly/g2i3f8); *toggleFlag* and *extendFlag* would normally be set to **false** or simply 0.

4.1.4 *Data Sorting and Filtering*

Users will normally try to sort columns by clicking the header. This has been a deficiency of JTable for ages. To solve this for the old (pre-R2008a) *uitable*, download one of several available JTable sorter classes or my TableSorter class.[36] The *TableSorter.jar* file must be located in the Java classpath as explained above. The result of the following code can be seen in the screenshot in Section 4.1.7.

```
% We want to use sorter, not data model...
% Unfortunately, UitablePeer expects DefaultTableModel not TableSorter
% so we need a modified UitablePeer class. But UitablePeer is a
% Matlab class, so use a modified TableSorter & attach it to the Model
if ~isempty(which('TableSorter'))

        % Add TableSorter as TableModel listener
        sorter = TableSorter(jtable.getModel());
        jtable.setModel(sorter);
        sorter.setTableHeader(jtable.getTableHeader());

        % Set the header tooltip (with sorting instructions)
        jtable.getTableHeader.setToolTipText('<html> <b>Click</b> to
        sort up; <b>Shift-click</b> to sort down<br> ...</html>');
else

        % Set the header tooltip (no sorting instructions...)
        jtable.getTableHeader.setToolTipText('<html> <b>Click</b>
        to select entire column <br> <b>Ctrl-click</b> (or
        <b>Shift-click</b>) to select multiple columns </html>');
end
```

Sorted columns can be set as follows (the primary sort column is specified first):

```
tableModel = jtable.getTableModel;
tableModel.setSortingStatus(3,tableModel.ASCENDING);  % 3 = Column D
tableModel.setSortingStatus(1,tableModel.DESCENDING); % 1 = Column B
```

The new (R2008a+) *uitable* is based on com.jidesoft.grid.SortableTable and so has built-in sorting support — all we need to do is to turn it on (note: the MATLAB handle has a hidden **Sortable** property, but it has no effect — use the Java property instead):

```
jtable.setSortable(true);              % or: set(jtable,'Sortable','on');
jtable.setAutoResort(true);
jtable.setMultiColumnSortable(true);
jtable.setPreserveSelectionsAfterSorting(true);
```

Sorting can be done programmatically as follows:

```
jtable.unsort();
jtable.resort();
jtable.unsortColumn(column);
jtable.sortColumn(column,newSortFlag,SortAscendingFlag);
```

where `column` indicates the column index (0-based) or column name; `newSortFlag` indicates whether this sorting is primary (true) or secondary (false); and `SortAscendingFlag` indicates whether to sort up (`true`) or down (`false`). Sorting can also be forced using *resort()* or canceled using *unsort()*.

The appearance of the sort icon can be controlled with several properties:

```
jtable.setSortArrowForeground(java.awt.Color.red);      % doesn't work...
jtable.setSortOrderForeground(java.awt.Color.blue);     % this works ok
jtable.setShowSortOrderNumber(true);
```

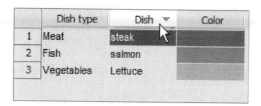

ShowSortOrderNumber = false

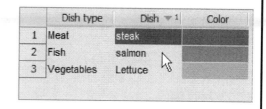

ShowSortOrderNumber = true

When more than one column is sorted, the sort-order number is always shown, regardless of the value of the **ShowSortOrderNumber** property:

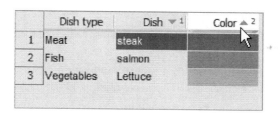

Multi-column sorting with blue sort-order numbers

Additional control over sorting can be achieved via the table's model, which is a `com.jidesoft.grid.SortableTableModel` object:

```
model = jtable.getModel;   % = com.jidesoft.grid.SortableTableModel obj
flag = model.isColumnSortable(columnIndex);  %columnIndex starts at 0
flag = model.isColumnSorted(columnIndex);
flag = model.isColumnAscending(columnIndex);
rank = model.getColumnSortRank(columnIndex);
sortCols = model.getSortingColumns; %returns [] or ArrayList of objs
realRowIndex = model.getSortedRowAt(displayedRowIdx);
model.reset();
model.resort();
set(model,'SortChangedCallback',@myFunc); % or: SortChangingCallback
```

The table's model also includes several useful sorting-related callbacks: **SortChanging-Callback**, **SortChangedCallback**, and **IndexChangedCallback**:

```
hmodel = handle(model, 'CallbackProperties');
set(hmodel,'SortChangedCallback',@myCallbackFunction);
```

Table data filtering can be done in several ways. MATLAB includes built-in filtering support using the `GlazedList` package (see Section 5.5.5 below). However, I believe that a much better table data filter package is the open-source `TableFilter` package by Luis Pena.[37] All we need to do to implement filtering is the following:

1. Download the latest `TableFilter` jar file (it will be named something such as *tablefilter-swing-4.1.4.jar*)[38]
2. Add this file to MATLAB's Java classpath, either statically (in *classpath.txt*) or dynamically (using the ***javaaddpath*** function), as described in Section 1.1.2
3. Call `net.coderazzi.filters.gui.TableFilterHeader(jtable)` in MATLAB:

```
javaaddpath('tablefilter-swing-4.1.4.jar');
filter = net.coderazzi.filters.gui.TableFilterHeader(jtable);
```

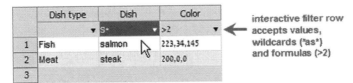

Automatic table filtering using `TableFilter`

`TableFilter` also provides the ability to automatically populate filter drop-down (combo-box, popup-menu) values from the list of distinct data values in each table column. Note that enabling this may take some time to run, depending on the amount of data. For this reason, this feature is turned off by default. To turn it on, run this:

```
filter.setAutoChoices(net.coderazzi.filters.gui.AutoChoices.ENABLED);
```

To retrieve the actual data in a filtered table, use the table's `RowSorter` object. For example, the following code returns the selected (filtered) row's data upon selection a specific table cell with the mouse:

```
set(jtable.getSelectionModel, 'ValueChangedCallback', ...
    {@selectionCallbackFcn,jtable});
function selectionCallbackFcn(jModel, jEventData, jtable)
    rowIdx = get(jModel,'LeadSelectionIndex');
    try
        rowIdx = jtable.getRowSorter.convertRowIndexToModel(rowIdx);
    catch
        % never mind: no filtering is used so stay with existing rowIdx
    end
```

```
    % Now do something useful with the selected row's data...
    data = jtable.getModel.getValueAt(rowIdx,1); % column #2

    % Alternate way to get the selected row's data...
    data = jtable.getActualRowAt(rowIdx);
    % ...
end % selectionCallbackFcn
```

4.1.5 JIDE Customizations

As noted above, table columns may be resized to dynamically fit their data:

```
com.jidesoft.grid.TableUtils.autoResizeAllColumns(jtable);
```

The `TableUtils` class provides many other useful methods for tables.† For example,

```
TableUtils.ensureRowSelectionVisible(jtable);
TableUtils.ensureRowVisible(jtable,rowIndex);
columnIndex = TableUtils.findColumnIndex(jtable.getModel,columnName);
```

The new *uitable*, as stated earlier, is based on JIDE's `SortableTable` class.[39] This class and its parent superclasses provide numerous possible customizations. Here, I will only mention a few of these possible customizations. For more details, the reader is referred to the online javadoc reference[40] and to JIDE's developer guide.[41]

One of the most annoying things with the standard `JTable` is that column widths can only be resized interactively by mouse-dragging column header boundaries (not grid boundaries), and row heights cannot be resized at all. JIDE easily fixes these issues:

```
jtable.setColumnResizable(true);
jtable.setRowResizable(true);
```

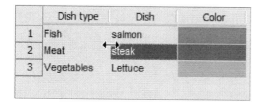

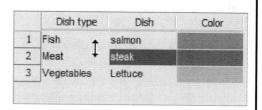

Resizing of column width   **Resizing of row height**

Also related to this is the fact that, while Swing's `JTable` enables setting row height (jtable. *setRowHeight(pixels)*), JIDE's table also enables setting different heights for different rows (jtable.*setRowHeights(com.jidesoft.grid.RowHeights)*).

---

† http://www.jidesoft.com/javadoc/com/jidesoft/grid/TableUtils.html (or http://bit.ly/ar9Wsy); do not confuse JIDE's useful `TableUtils` class with Mathworks' much less useful `com.mathworks.mwswing.TableUtils` class, which merely provides two methods: *adjustRowHeight(jtable)* and *getXForColumn(jtable,columnIndex)*.

JIDE's `SortableTable` class indirectly extends `CellSpanTable`,[42] and so if we use a `DefaultSpanTableModel`,[43] we can merge table cells by defining `CellSpan`[44] objects:

```
import com.jidesoft.grid.*;
data = {'Fish','salmon',12;'Meat','steak',23;'Vegetables','Lettuce',4};
cols = {'Dish type', 'Dish', 'Price'};
jtable.setModel(SortableTableModel(DefaultSpanTableModel(data,cols)))
jtable.getModel.getActualModel.addCellSpan(CellSpan(1,0,2,1));
jtable.getModel.getActualModel.addCellSpan(CellSpan(0,1,2,2));
```

Example of two table cell spans(1 × 2 and 2 × 2)

The cell spans were defined using `CellSpan`'s constructor: *startRow, startColumn, numRows, numColumns*. In this example, we defined 2×2 and 1×2 cell spans.

Note how we wrapped `DefaultSpanTableModel` within a `SortableTableModel` wrapper. This is standard practice in table models. In fact, we can chain several such model wrappers. Inner models are retrieved using *getActualModel()*. Even if we had not wrapped `DefaultSpanTableModel` within a `SortableTableModel`, such a wrap would automatically be applied by the JIDE code when we call `jtable.`*setSortable(true)*.

4.1.6 Controlling Table Structure (Adding/Deleting Rows)

In order to programmatically add or remove data rows, we can of course update the data and then set the MATLAB handle (`mtable`)'s **Data** property with the new data. Unfortunately, this redraws the entire data, causing a noticeable flicker and delay if the data set is large.

Unfortunately, we cannot (as far as I know) add and remove data rows from the model of the new *uitable*. We need to update the **Data** property, as noted above. And yes, this indeed causes flickering.

In the old uitable, we can programmatically add or remove table rows without redrawing the entire table. Simply call `jtable.`*getModel.addRow()* to append a new row at the table's bottom; or *insertRow()* to insert one row before another; or *removeRow()* to remove a specified row. Remember to stop editing the current cell:

```
% Stop editing the current cell
function stopEditing(jtable)
    component = jtable.getEditorComponent;
    if ~isempty(component)
        event = javax.swing.event.ChangeEvent(component);
        jtable.editingStopped(event);
    end
end  % stopEditing
```

```
% Insert a new row immediately above the currently-selected row
function tableInsertRow(mtable)
    % Stop any current editing
    jtable = mtable.getTable;
    stopEditing(jtable);

    % Insert the new row immediately before the current row
    newRowData = cell(1,mtable.getNumColumns); % empty data
    newRowIdx = max(0,jtable.getSelectedRow);
    jtable.getModel.insertRow(newRowIdx, newRowData);
end  % tableInsertRow

% Insert a new row as the last row in the table
function tableAppendRow(mtable)
    % Stop any current editing
    jtable = mtable.getTable;
    stopEditing(jtable);

    % Add a new row at the bottom of the data table
    newRowData = cell(1,mtable.getNumColumns); % empty data
    mtable.getTableModel.addRow(newRowData);

    % Move the selection to Column A of this new row
    jtable.changeSelection(jtable.getRowCount-1,0,false,false);
end  % tableAppendRow

% Delete the currently-selected row, if any rows are displayed
function tableDeleteRow(mtable)
    % Stop any current editing
    jtable = mtable.getTable;
    stopEditing(jtable);

    % If any rows are displayed
    rowCount = jtable.getRowCount;
    % rowCount might be 0 during slow processing & user double-click
    if (rowCount > 0)
        % Delete the currently-selected row
        row = max(0,jtable.getSelectedRow);
        col = max(0,jtable.getSelectedColumn);
        jtable.getModel.removeRow(row);
        if row >= rowCount-1
            jtable.changeSelection(row-1, col, false, false);
        elseif jtable.getSelectedRow < 0
            jtable.changeSelection(row, col, false, false);
        end
    end
end  % tableDeleteRow

% Delete all table rows
function tableDeleteAll(mtable)
    stopEditing(mtable.getTable);
    mtable.setNumRows(0);
end  % tableDeleteAll
```

4.1.7 Final Remarks

One bug final fix needs to be done to **uitable**s to solve a JTable bug,[45] which also affects **uitable**s. The following fix was suggested by Brad Phelan, following others:[46]

```
jtable.putClientProperty('terminateEditOnFocusLost', true);
```

I have written a wrapper[47] for **uitable** which facilitates its integration in MATLAB, including all the specific settings in the previous pages: feel free to download it from the File Exchange on MATLAB Central. Here is a sample usage script in MATLAB:

```
% Display the initial table
colHeaders = {'My','sortable','and selectable','column','names'};
data = {false, 1.3, 'abc', uint16(45), 'ert'; ...
        true, pi/2, 'def', uint16(13), 'test 123'; ...
        true, pi, 'ghi', uint16(0), '456...'};
mtable = createTable(hFig, colHeaders, data);

% Prepare an editable drop-down CellEditor
cb = javax.swing.JComboBox({'First','Last'});
cb.setEditable(true);
editor = javax.swing.DefaultCellEditor(cb);

% Attach the new CellEditor to table column E (=Java index 4)
mtable.getTable.getColumnModel.getColumn(4).setCellEditor(editor);
```

createTable **utility screenshot (note action buttons, sortable columns, and customized** CellEditor**) (See color insert.)**

Note how we have only modified column E's CellEditor and not its CellRenderer. This resulted in the drop-down appearing only for the currently selected cell, while all other cells in column E use the DefaultCellRenderer to display their data as a simple text field. Clicking outside the cell will revert its appearance to a standard text field.

4.2 Uitree

Since MATLAB 7 (R14), MathWorks have included the unsupported function **uitree** in the uitools folder. **uitree** uses the internal MATLAB Java widget com.mathworks.hg.peer. utils.UIMJTree, which derives from the standard Java Swing JTree[48] class (via com. mathworks.mwswing.MJTree — see Chapter 5).

Following *uitable*, which became documented and supported in the R2008a release, it seems that MathWorks plan to make *uitree* a documented function as well, at least judging from the following comment found in %matlabroot%/toolbox/local/hgrc.m:

> *Temporarily turn off old uitree and uitreenode deprecated function warning... When we introduce the new documented uitree to replace the old undocumented uitree, ...*

uitree appears to be far less popular than *uitable* on MATLAB's File Exchange and the CSSM forum–only two dozen answered threads over the years, mostly by the same people.[49] However, *uitree* is a top search term on the UndocumentedMatlab.com website. My explanation for this is that there is little available documentation and so few MATLAB developers are familiar with *uitree*s, but many would like to learn.

uitree sets up a scrollable Java Swing `JTree` onscreen without the hassle of setting up a scrollable viewport and other similar nuts and bolts. Also, *uitree* automatically knows how to read and display root objects of type Handle Graphics, Simulink model, or char string (interpreted as a file-system folder name).

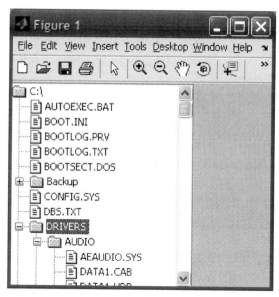

On the other hand, *uitree* is also a far less useful wrapper of the underlying Java object than *uitable*. If our data is one of the default types (HG object, Simulink model, or folder name), then *uitree* does little more than set up a tree within a `ScrollableViewport` with some drag-and-drop settings. As will be seen below, this is actually a very small part of the required setup.

uitree accepts an optional figure handle followed by P–V (property–value) pairs. Settable properties are **Root**, **ExpandFcn**, **SelectionChangeFcn**, **Parent**, and **Position**. As in *uitable*, a 'v0' input argument may be necessary to suppress a warning message.

uitree returns two arguments: a handle to the created tree (a `com.mathworks.hg.peer.-UITreePeer` Java object wrapped within a MATLAB handle) and an undocumented second optional argument holding a handle to the MATLAB GUI container of the created tree. These two arguments are exactly the two arguments returned from the *javacomponent* function (see Chapter 3). Use of the first return argument enables the user to specify the tree **Units** (e.g., 'Normalized'), **Position**, **Visible** (true/false), and a few other properties (see below). *uitree*'s help section implies that these properties may be passed directly as P–V pairs, just like **Root** or **ExpandFcn**, but this is, in fact, not so. A few callbacks may also be set, as described below.

The *uitree* **Units** and **Visibility** properties may also be set via the second (container) output argument. The container handle also enables changing other standard MATLAB handle properties, such as **Position**, **UserData**, **Tag**, **HandleVisibility, etc**. Note that the container handle is a simple MATLAB Handle-Graphics object, whereas the tree handle is a Java object wrapped in a MATLAB handle. This means that tree's **Visible** property accepts true/false (or 1/0) values, whereas the container expects 'on'/'off':

```
[mtree, container] = uitree('v0', 'Root','C:');
set(mtree, 'Visible', true);
set(container, 'Visible', 'on');
```

As in *uitable*, a *uitree*'s container handle can also be retrieved directly as follows:

```
mtree = uitree(...);
container = mtree.getUIContainer;
container = get(mtree,'UIContainer');         % equivalent method
```

and, conversely,

```
mtree = get(container,'JavaPeer'); % JavaPeer is a hidden property
```

Like all *javacomponent* containers, the container handle has a few hidden undocumented properties: **JavaPeer** (a handle to the mtree object), **FigureChild** (same), **PixelBounds**, **HelpTopicKey**, **Serializable,** and so on. Some properties, while settable, appear to have no effect (e.g., **Opaque** and **BackgroundColor**).

Note that the *uitree* is always created as a direct child of the containing figure and disregards creation-time **Parent** values. However, the **Parent** property can be modified following the tree's creation:

```
[mtree, container] = uitree(...,'Parent',hPanel); % Parent is ignored
set(container, 'Parent', hPanel);
```

As with *uitable*, it is useful to differentiate between the MATLAB handle wrapper returned by *uitree* (hereby called "mtree") and its underlying Java object ("jtree"):

```
mtree = uitree(...);
jtree = mtree.getTree;
jtree = get(mtree, 'tree'); % an alternative method
```

4.2.1 Customizing Uitree

The jtree object reference handle has a much richer variety of available methods and callbacks than mtree. This list is basically a superset of all Java Swing JTree functionalities, with a few MathWorks additions but also a few limitations.† The user is referred to the official documentation[50] or any good text about Java Swing. Use ***methodsview*** or ***uiinspect*** to see the full list of available methods.

If we need to create a custom tree hierarchy (i.e., our root node is not an HG object, Simulink model, or folder name), then we need to use the similarly semi-documented ***uitreenode*** function as follows:

```
>> root = uitreenode(handle(mtree),'my root','c:\root.gif',false);
>> mtree.setRoot(root);
>> set(mtree,'Root',root);  % alternative to mtree.setRoot()

>> mtree.Root = root; % lexically correct, but disallowed:
??? Changing the 'Root' property of javahandle_withcallbacks.
com.mathworks.hg.peer.UITreePeer is not allowed.
```

uitreenode accepts four arguments: a string or handle value (the node's "internal" value), a string description (shown next to the node's icon), an icon filename ([] will result in an icon assigned based on the node value), and a flag indicating whether the node is a leaf (no children) or not. ***uitreenode*** is little more than a MATLAB handle wrapper for `com.mathworks.hg.peer.UITreeNode` (which itself is a derived class of `javax.swing.tree.DefaultMutableTreeNode`[51]), with an added **UserData** property.

The root node can be hidden, making all its children nodes appear as top-level "roots". This can be done with the **RootVisible** property (default is 1 or true):

```
jtree.setRootVisible(false); % or: setRootVisible(0)
```

The jtree **ShowRootHandles** property, useful when **RootVisible** = false, controls the visibility of the "+" and "–" expansion signs next to root nodes (default is 0 or false):

```
jtree.setShowsRootHandles(true);   % or: setShowsRootHandles(1)
```

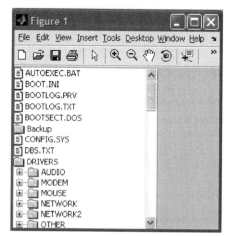 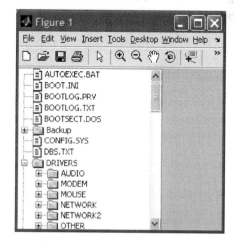

RootVisible = 0, ShowsRootHandles = 0   **RootVisible = 0, ShowsRootHandles = 1**

---

† For example, tooltips in ***uitree***s cannot be as easily set as in JTrees: http://www.mathworks.com/matlabcentral/newsreader/view_thread/153690 (or http://tinyurl.com/9amob6).

A tree is meaningless without children. Addition and removal of nodes can be done directly (see Section 4.2.3) or indirectly (by creating a custom callback function for node expansion, which returns a dynamic list of *uitreenode*s, that is then displayed). *uitree*'s help section shows an example for file-folder expansion (the indirect method). Here is a similar implementation for a GUI handle expansion:

```
set(mtree,'NodeExpandedCallback',@myExpandFcn);†
function nodes = myExpandFcn(tree, value)
    try
        iconpath = fullfile(matlabroot,'/toolbox/matlab/icons/');
        child = 0;
        ch = get(value,'children');
        for child = 1:length(ch)
            if ~isempty(get(ch(child),'children'))
                parent = 1;
                icon = [iconpath 'foldericon.gif'];
            else
                parent = 0;
                icon = [iconpath 'pageicon.gif'];
                % or set icon = [] to display the default icon
            end
            desc = get(ch(child),'type'); % textual description
            nodes(child) = uitreenode(ch(child),desc,icon,parent);
        end
        if ~child
            nodes = [];
        end
    catch
        error(['unknown uitree node type: ' value]);
    end
end % myExpandFcn
```

uitree, like *uitable*, is basically a Java object placed within a ScrollableViewport and wrapped within a MATLAB handle. As in *uitable*, *uitree*'s scrollbars can be customized. Refer to the *uitable* scrollbar customization section 4.1.3 for a full discussion:

```
scroll = mtree.ScrollPane; % note different syntax vs. uitable
vScrollPolicy = scroll.VERTICAL_SCROLLBAR_AS_NEEDED;
scroll.setVerticalScrollBarPolicy(vScrollPolicy);
hScrollPolicy = scroll.HORIZONTAL_SCROLLBAR_AS_NEEDED;
scroll.setHorizontalScrollBarPolicy(hScrollPolicy);
```

uitree (mtree) enables several callbacks:

- **NodeSelectedCallback** — invoked whenever any node is selected
- **NodeWillExpandCallback** — invoked just before any node is expanded
- **NodeExpandedCallback** — invoked right after a node has been expanded

---

† Note: *uitree*'s help section incorrectly calls this callback ExpandFcn. In fact, the property name is **NodeExpandedCallback**.

- **NodeWillCollapseCallback** — invoked just before any node is collapsed
- **NodeCollapsedCallback** — invoked right after a node has been collapsed
- **NodeDroppedCallback** — invoked after a node has been drag-and-dropped

The `jtree` object reference exposes some equivalent and many additional callbacks:

- **TreeWillExpandCallback** — invoked just before any node is expanded
- **TreeExpandedCallback** — invoked right after a node has been expanded
- **TreeWillCollapseCallback** — invoked just before any node is collapsed
- **TreeCollapsedCallback** — invoked right after a node has been collapsed
- **ValueChangedCallback** — invoked after a node value has changed
- as well as all the other regular GUI callbacks specified in Chapter 3:
  **AncestorAddedCallback, AncestorMovedCallback,
  AncestorRemovedCallback, AncestorResizedCallback,
  CaretPositionChangedCallback, ComponentAddedCallback,
  ComponentHiddenCallback, ComponentMovedCallback,
  ComponentRemovedCallback, ComponentResizedCallback,
  ComponentShownCallback, FocusGainedCallback, FocusLostCallback,
  HierarchyChangedCallback, InputMethodTextChangedCallback,
  KeyPressedCallback, KeyReleasedCallback, KeyTypedCallback,
  MouseDraggedCallback, MouseEnteredCallback, MouseExitedCallback,
  MouseMovedCallback, MousePressedCallback, MouseReleasedCallback,
  MouseClickedCallback, MouseWheelMovedCallback,
  PropertyChangeCallback,** and **VetoableChangeCallback**
- but note that `jtree` does not have equivalents to `mtree`'s useful **NodeSelectedCallback**
  and **NodeDroppedCallback**

4.2.2 Accessing Tree Nodes

Trees, unlike tables, cannot be read all at once. Instead, selected tree nodes can be accessed and read separately, often in a recursive manner:

```
function nodes = getNodeDescendants(root, nodes)
    if root.getChildCount > 0
        childrenVector = root.children;
        while childrenVector.hasMoreElements
            nodes(end + 1) = childrenVector.nextElement;
            nodes = getNodeDescendants(nodes(end), nodes);
        end
    end
end % getNodeDescendants
```

An important aspect of tree traversals is that only nodes which have previously been expanded can return their children. Nodes which have not been expanded yet return zero children (just like leaves), despite the fact that the tree knows that these nodes are not leaves.

There are many ways of traversing the tree: We can use the Vector-based approach shown above. An alternative is to use a loop-based approach using node.*getChildCount* (or node.*getSiblingCount*) together with node.*getChildAt*. Note the difference between node children (nodes whose direct Parent is this node) and siblings (nodes that share the same Parent node as this node). Alternatively, one can use node.*getNextNode*, *getNextLeaf*, or *getNextSibling*.

Alternatively again, we can use one of several predefined recursion strategies: `node. breadthFirstEnumeration`, `depthFirstEnumeration`, `postorderEnumeration`, `preorderEnumeration`, and `pathFromAncestorEnumeration`. For example,

```
root2 = root.breadthFirstEnumeration;
while root2.hasMoreElements
  node = root2.nextElement;
end
```

Note that in this and other similar cases involving collection enumeration, we have to store the collection in a temporary variable (here, root2) before traversing it. Otherwise, the collection enumeration will keep getting initialized in an endless loop:[†]

```
% The following code causes an endless loop - beware!
while root.breadthFirstEnumeration.hasMoreElements
    node = root.breadthFirstEnumeration.nextElement;
end
```

*UITreeNode*s have the following properties (read-only, except where noted):

- **AllowsChildren** — true if the node is not a leaf (read/write).[‡]
- **Leaf** — true if the node is a leaf.
- **LeafNode** — appears to be the same as Leaf, except for being read/write.
- **ChildCount** — the number of direct children (not their descendants).
- **SiblingCount** — the number of nodes sharing this node's Parent.
- **LeafCount** — the number of descendant leaves.
- **Depth** — depth beneath node (based on descendants expanded so far).
- **Level** — the node level beneath the root (root's level = 0).
- **FirstChild, LastChild** — node children (first/last) — used for traversal.
- **NextNode, PreviousNode** — next/prev child node — used for traversal.
- **NextSibling, PreviousSibling** — next/prev sibling node — used for traversal.
- **FirstLeaf, LastLeaf** — node children which are leaves — used for traversal.
- **NextLeaf, PreviousLeaf** — next/prev child leaf node — used for traversal.
- **Root** — true for the root node, false otherwise.
- **Parent** — parent node (empty for root node).
- **Path** — list of all node ancestors, up to and including the root node.
- **Value** — internal value of this node (read/write).

[†] See a discussion of this in Section 2.1.3.
[‡] Strictly speaking, this depends on the value of mtree.*getModel's* asksAllowsChildren property, but this property is normally unchanged from its default value of false.

- **Name** — node label (description), displayed next to icon (see below, R/W).
- **Icon** — node icon image (see below, read/write).
- **UserObject** — user-defined data storage (see below, read/write).
- **UserObjectPath** — list of **UserObject** values for all node ancestors, up to root.
- **UserData** — regular MATLAB user-defined data storage (read/write).

These properties are essentially inherited from Java Swing's `DefaultMutableTreeNode` class, from which `UITreeNode` derives. The user is referred to official Swing documentation[52] or any good Swing book for additional details.

4.2.3 Controlling Tree Nodes

In order to collapse and expand nodes programmatically, one could theoretically use *mtree. collapse(node)* and *mtree.expand(node)*. However, this often fails (I am unsure why), and we need to resort to using one of the following methods:

```
mtree.collapse(node);    % often fails
nodePath = javax.swing.tree.TreePath(node.getPath);
jtree.collapsePath(nodePath);  % this works
nodeRow = jtree.getRowForPath(nodePath);
jtree.collapseRow(nodeRow);    % an alternative method that works
```

JIDE's `com.jidesoft.tree.TreeUtils` class (see Section 5.7.2) provides static convenience methods that can be used instead: *expandAll(jtree,true/false)*, *expandAll(jtree, jTreePath,true/false)*. MATLAB's `com.mathworks.mwswing.TreeUtils` class provides similar convenience methods.

Nodes can be programmatically selected using *mtree.setSelectedNode(node)*. Multiple nodes may be selected using *mtree.setSelectedNodes*, if an earlier call to *mtree.- setMultipleSelectionEnabled(true)* was made (default is multiple-selection disabled):

```
mtree.setSelectedNode(root);  % root is a node
mtree.setSelectedNodes([root, node1, node2]);
```

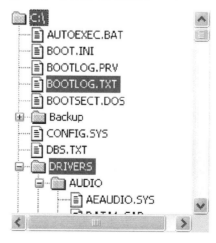

Programmatically selecting multiple tree nodes

The `jtree` object reference handle has several additional selection methods: *setSelection-Path*, *setSelectionPaths*, *setSelectionRow*, and *setSelectionRows*, as well as several other methods affecting selection. Use **methodsview** or **uiinspect** to see the full list of available methods.

Similar to **uitable**, tree node selection is controlled via a policy set in `TreeSelection Model`,[53] which affects the way in which multiple table cells may be selected together. Possible selection policy values are: `SINGLE_TREE_SELECTION` (the default), `CONTIGUOUS_ TREE_ SELECTION` and `DISCONTIGUOUS_TREE_SELECTION`:

```
import javax.swing.tree.TreeSelectionModel;
selectPolicy = TreeSelectionModel.CONTIGUOUS_TREE_SELECTION;
jtree.setSelectionMode(selectPolicy);
```

The currently selected node(s) can be accessed using `mtree`.*getSelectedNodes*, or one of several `jtree` methods: *getSelectionPath*, *getSelectionPaths*, or *getSelectionRows*. The first two methods return `javax.swing.tree.TreePath` Java object(s). `TreePath` is a vector of all nodes linking the selected node to the root node.

To extract the actual selected node (without its parents), use the following method:

```
node = jtree.getSelectedPath.getLastPathComponent;
```

Note that `jtree`.*getSelectionRows* returns the rows in 0-based indexing, so the top (root) node is row #0. Also note that row indexes are those appearing on screen, so any nonexpanded folder are only counted as a single row. Therefore, expanding/collapsing nodes will affect the row indexes of selected nodes that are below the expanded/collapsed node.

Node selection callbacks often require knowledge of the currently selected rows:

```
>> mtree = uitree (..., 'SelectionChangeFcn',@mySelectFcn);
>> set(mtree, 'SelectionChangeFcn',@mySelectFcn); % an alternative

function nodes = mySelectFcn(tree, value)
    selectedNodes = tree.getSelectedNodes;
    if ~isempty(selectedNodes)
        % ...
    end
end % mySelectFcn
```

Nodes can be added or removed programmatically using one of several related methods of the `TreeModel`.[†] **uitree** uses `javax.swing.tree.DefaultTreeModel`,[54] but we can use any hierarchical data class which implements the simple `javax.swing.tree.TreeModel` interface.[55] The **uitree** model is accessible via both `mtree`.*getModel* and `jtree`.*getModel*. The relevant `DefaultTreeModel` methods for node addition and removal are *insertNodeInto(childNode, parentNode, childIndex)* and *removeNodeFromParent(node)* respectively. A sample usage in MATLAB was posted a few years ago on CSSM.[56]

[†] The role of the Model in Swing components is explained at http://java.sun.com/products/jfc/tsc/articles/architecture/ (or http://tinyurl.com/atggc).

`DefaultTreeModel` has some useful self-explanatory callbacks that may be set by the user: **TreeNodesChangedCallback**, **TreeNodesInsertedCallback**, **TreeNodesRemovedCallback**, and **TreeStructureChangedCallback**. These can be set in the normal MATLAB manner:

```
set(jtree.getModel, 'TreeStructureChangedCallback', @myFunc);
```

Nodes can also be added, moved or removed by node methods: node.*add(anotherNode)* adds *anotherNode* to the end of this node's children list (possibly detaching it from its previous parent); node.*insert(anotherNode,index)* does the same but inserts *anotherNode* at a specific child index rather than at the end; node.*clone()* makes a duplicate of this node that can then be added to another node; node.*remove(index)* and node.*remove(node)* remove a specific node whereas node.*removeFromParent()* removes this node; node.*removeAllChildren()* removes all children, if any, of this node.

Finally, `mtree`.*add(parent,nodes)* allows adding a list of nodes to a parent node[57] and `mtree`.*remove(nodes)* removes the specified nodes.

4.2.4 Customizing Tree Nodes

Modification of a tree node's icon image can be done programmatically as follows:

```
myIcon = fullfile(matlabroot, '/toolbox/matlab/icons/foldericon.gif');
jImage = java.awt.Toolkit.getDefaultToolkit.createImage(myIcon);
node.setIcon(jImage);
node.setIcon(im2java(imread(myIcon))); % an alternative
```

Real-life programs should check and possibly update jImage's size to 16 pixels, before setting the node icon; otherwise, the icon might get badly cropped. This is how it can be done:

```
function iconImage = setIconSize(iconImage)
   try
      iconWidth  = iconImage.getWidth;
      iconHeight = iconImage.getHeight;
      if iconWidth > 16
         newHeight = fix(iconHeight * 16 / iconWidth);
         iconImage = iconImage.getScaledInstance(16,newHeight, ...
                                       iconImage.SCALE_SMOOTH);
      elseif iconHeight > 16
         newWidth = fix(iconWidth * 16 / iconHeight);
         iconImage = iconImage.getScaledInstance(newWidth,16, ...
                                       iconImage.SCALE_SMOOTH);
      end
   catch
       % never mind... - return original icon
   end
end  % setIconSize
```

Node icons can also be created programmatically (without requiring an existing icon image file), using MATLAB's ***im2java*** function.[58]

Nodes can be modified from leaf (nonexpandable) to parent behavior (=expandable) by setting their **LeafNode** property (a related property is **AllowsChildren**):

```
set(node,'LeafNode',false); % =expandable
node.setLeafNode(false); % an alternative
```

Nodes can contain user data in their **UserObject** property, which is similar to MATLAB's ubiquitous **UserData** property. Using the **UserObject** property is the preferred way to set applicative node data for *uitrees*. However, we need to be aware that it modifies the stored data in some rare cases. In such cases, use the node's **UserData** or **ApplicationData** property instead:

```
>> node.setUserObject(gcf)
>> isequal(node.getUserObject, gcf)   %ok
ans =
     1

>> node.setUserObject(handle(gcf))
>> isequal(node.getUserObject, handle(gcf))   % not ok!
ans =
     0

>> set(node,'UserData',handle(gcf))
>> isequal(get(node,'UserData'), handle(gcf)) % ok!: UserData is safe
ans =
     1
```

Node names (descriptions) can be set using node.*setName('...')*. Note that the horizontal space allotted for displaying the node name will not change until the node is collapsed or expanded. So, if the new name requires more than the existing space, it will be displayed as something like "abc...", until the node is expanded or collapsed.

Node names support HTML just like all Java Swing labels. Therefore, we can specify font size/face/color, bold, italic, underline, super-/sub-script, and so on:

```
txt1 = '<html><b><u><i>abra</i></u>';
txt2 = '<font color="red"><sup>kadabra</html>';
node.setName([txt1,txt2]);
```

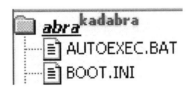

HTML-enriched tree nodes

If we include a <br> element in our label, the node name splits into two lines and does not have enough vertical space to display correctly. It does not help to expand or collapse the node in this case — we need to increase the vertical spacing from the default setting of 16 pixels. This can be done for all tree nodes at once using a positive pixel value, or dynamically for each cell by specifying a non-positive value:

```
jtree.setRowHeight(32);    % static height for all nodes
jtree.setRowHeight(0);     % dynamic height for each node
```

If we require fine-grained control over the tree node's appearance, we need to modify its `CellRenderer`. The concept of using a `CellRenderer` class to display cells and a `CellEditor` to edit cells is similar to the one described in *uitable*, and the reader is referred there (Section 4.1.1) for details. Swing texts and online documentation are also good references.[59] *uitree* uses `com.mathworks.hg.peer.UITreePeerRenderer` as a custom renderer class. User-defined renderer classes should normally derive from `javax.swing.tree.DefaultTreeCellRenderer`,[60] but we can always create our own implementation from scratch. After preparing the renderer, attach it to the tree:

```
renderer = javax.swing.tree.DefaultTreeCellRenderer;
renderer = jtree.getCellRenderer;    % an alternative based on existing
renderer.setOpenIcon(im2java(imread(myIcon)));
renderer.setIconTextGap(8);  % default = 4 [pixels]
jtree.setCellRenderer(renderer);
```

`CellEditors` derive from `javax.swing.tree.DefaultTreeCellEditor`.[61] *uitree* does not use any `CellEditor`, since its tree is read-only. To use a `CellEditor`, first enable cell editing and then specify an editor and attach it to the tree. Finally, assign a data-change callback to modify the displayed label and icon. Here is a simple example:

```
% Enable tree cell editing
jtree.setEditable(true);

% Define tree cell editor and attach it to the tree
cb = javax.swing.JComboBox({'First','Last'});
editor = javax.swing.DefaultCellEditor(cb);
%editor = javax.swing.DefaultCellEditor(javax.swing.JCheckBox);
jtree.setCellEditor(editor)

% Define a data-changed callback for the cell editor
callback = {@dataChangedCallback, jtree};
set(editor, 'EditingStoppedCallback',callback);

function dataChangedCallback(hEditor,eventData,jtree)
    currentNode = jtree.getSelectionPath.getLastPathComponent;
    currentNode.setName(get(hEditor, 'CellEditorValue'));
end    % dataChangedCallback
```

Unlike ***uitable***s, ***uitree***s do not have a default context menu, activated on mouse right-click. This is relatively easy to set up, using a Java Swing popup menu (Note the similarities and slight differences compared with the PopupMenu that was presented in Section 3.6):[62]

```matlab
% Prepare the context menu (note the use of HTML labels)
menuItem1 = javax.swing.JMenuItem('action #1');
menuItem2 = javax.swing.JMenuItem('<html><b>action #2');
menuItem3 = javax.swing.JMenuItem('<html><i>action #3');

% Set the menu items' callbacks
set(menuItem1,'ActionPerformedCallback',@myFunc1);
set(menuItem2,'ActionPerformedCallback',{@myfunc2,data1,data2});
set(menuItem3,'ActionPerformedCallback','disp ''action #3...'' ');

% Add all menu items to the context menu (with internal separator)
jmenu = javax.swing.JPopupMenu;
jmenu.add(menuItem1);
jmenu.add(menuItem2);
jmenu.addSeparator;
jmenu.add(menuItem3);

% Set the tree mouse-click callback
% Note: Default actions (expand/collapse) will still be performed
% Note: MousePressedCallback is better than MouseClickedCallback
%       since it fires immediately when mouse button is pressed,
%       without waiting for its release, as MouseClickedCallback does
set(jtree, 'MousePressedCallback', {@mousePressedCallback,jmenu});

% Set the mouse-press callback
function mousePressedCallback(hTree, eventData, jmenu)
    if eventData.isMetaDown % right-click is like a Meta-button
        % Get the clicked node
        clickX = eventData.getX;
        clickY = eventData.getY;
        jtree = eventData.getSource;
        treePath = jtree.getPathForLocation(clickX, clickY);
        try
            % Modify the context menu or some other element
            % based on the clicked node. Here is an example:
            node = treePath.getLastPathComponent;
            nodeName = ['Current node: ' char(node.getName)];
            item = jmenu.add(nodeName);

            % remember to call jmenu.remove(item) in item callback
            % or use the timer hack shown here to remove the item:
            timerFcn = {@removeItem,jmenu,item};
            start(timer('TimerFcn',timerFcn,'StartDelay',0.2));
        catch
            % clicked location is NOT on top of any node
            % Note: can also be tested by isempty(treePath)
        end
```

```
        % Display the (possibly-modified) context menu
        jmenu.show(jtree, clickX, clickY);
        jmenu.repaint;
    end
end   % mousePressedCallback


% Remove the extra context menu item after display
function removeItem(hObj,eventData,jmenu,item)
    jmenu.remove(item);
end   % removeItem


% Menu items callbacks must receive at least 2 args:
% hObject and eventData - user-defined args follow after these two
function myfunc1(hObject, eventData)
    % ...


function myFunc2(hObject, eventData, myData1, myData2)
    % ...
```

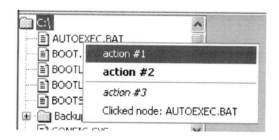

uitree node-specific context menu

4.2.5 FindJObj

One of my favorite File Exchange submissions, **FindJObj**,[63] uses **uitree** (well, actually its underlying UITreePeer) and **uitreenode** extensively. **FindJObj** explores the GUI hierarchy of a specified figure or container and displays it in a tree view (see Section 7.2.3).[†] Selection of a tree node displays the node's properties/callbacks. **FindJObj** uses custom node icons (taken from the components themselves, where available) and programmatically expands/collapses nodes. The reader is welcome to download **FindJObj** and read the sections dealing with the tree and its nodes.

Here is a screenshot of what **FindJObj** found for the sample **uitree** displayed in the screenshots above. Note that the **uitree** is apparently a UIMJTree object embedded in a JViewport

---

[†] **FindJObj** also has the ability to return the underlying Java component handle of the requested *MATLAB* handle (see Section 7.2.2).

with two `ScrollBars` (horizontal & vertical), all contained within an `MJScrollPane` container, which itself is contained in *javacomponent*'s `HGPanel`:

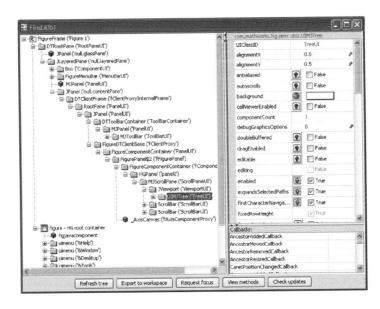

FindJObj presentation of the sample *uitree* from Section 4.2.4

4.3 Uitab

uitabgroup and the related *uitab* functions,[64] available since 2004 (R14 SP2, aka 7.0.4), use the internal MATLAB Java widget `com.mathworks.hg.peer.UITabGroupPeer`, which appears to use an internal `javax.swing.JTabbedPane`[65] which was extended with some extra methods (type 'methodsview com.mathworks.hg.peer.UITabGroupPeer' to see the list).

Unlike *uitable* and *uitree*, which use actual Java objects to both store and present the data, *uitabgroup* only sets up the Java object to display the tabs, whereas the tab contents themselves are placed in entirely unrelated MATLAB *uicontainer*s. MATLAB uses very clever double-booking to keep the Java and MATLAB objects synchronized. The ability to "switch" tabs is actually an optical illusion: in reality, a listener placed on the **SelectedIndex** property of the tab group causes the relevant MATLAB container to display and all the rest to become hidden. Other listeners control containers' position and size based on the tab group's. Addition and removal of tabs use similar methods to add/remove empty tabs to the `JTabbedPane`. Check the code in %matlabroot%/toolbox/matlab/@uitools/@uitabgroup/schema.m for details.

A drawback of this complex mechanism is the absence of a single customizable Java object as in *uitable* or *uitree*. The benefit is that it allows us to place any MATLAB content within the tabs, including plot axes that cannot be added to Java containers. Had *uitabgroup* been a Java container, we could not have added plot axes or images to its tabs.[66] The *UISplitPane* utility, described in Chapter 10, uses a similar solution.

Here is how a simple tab group is set up, adapted from ***uitabgroup***'s help section:[†]

```
hTabGroup = uitabgroup; drawnow;
tab1 = uitab(hTabGroup, 'title','Panel 1');
a = axes('parent', tab1); surf(peaks);
tab2 = uitab(hTabGroup, 'title','Panel 2');
uicontrol(tab2, 'String','Close', 'Callback','close(gcbf)');
```

Here, the returned ***uitabgroup*** object `hTabGroup` is actually a MATLAB container (extending ***uiflowcontainer***) that always displays two elements: the Java tab group and the active MATLAB ***uicontainer*** (the active tab's contents). The behavior of `hTabGroup`'s **FlowDirection** property is, therefore, clear: **FlowDirection** accepts one of the several enumerated types: the default is 'topdown', displaying the Java tab group on top of the MATLAB container. Other types are 'bottomup', 'lefttoright' (or 'auto'), and 'righttoleft' (or 'autoreverse'). Unfortunately, the Java tab orientation is not modified when changing **FlowDirection**, and so using this property is usually not a good idea.

A better way is to use the **TabLocation** property, which is a specific property for ***uitabgroup*** that accepts 'top', 'bottom', 'left', and 'right': the tab orientation is automatically arranged based on this location. So, for example, using a 'left' **TabLocation** causes the tabs to be arranged vertically, as expected.

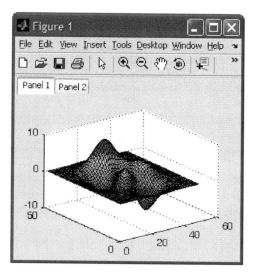

TabLocation = 'top'

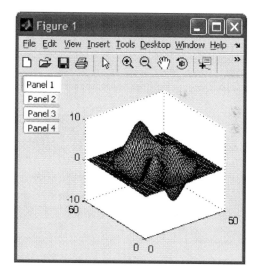

TabLocation = 'left'

Another `hTabGroup` property of interest is **Margin**, which sets the margin in pixels before each of the displayed elements — not just between them as might be expected: Increasing

---

[†] Recent MATLAB releases throw a warning when using this code: either add the 'v0' input arg to ***uitabgroup*** and ***uitab*** calls or suppress the MATLAB:uitabgroup:MigratingFunction warning in MATLAB versions up to 7.10 (R2010a); *remove* the 'v0' input arg and supress the MATLAB:uitabgroup:OldVersion waning in MATLAB 7.11 (R2010b) or newer.

Margin (default = 2 pixels) increases the gap between the tab group and the active tab's contents but also the gap between the tab group and the figure's edge:

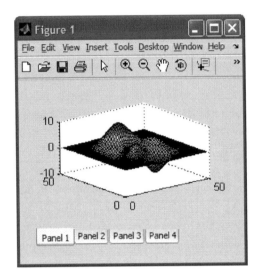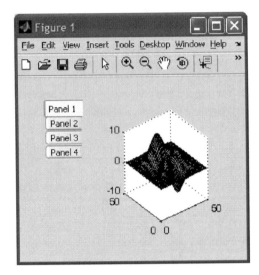

Margin = 20

Tab selection can be done programmatically, by setting hTabGroup's **SelectedIndex** property. This property is also readable, which is useful when setting callback actions based on the selected tab, using the **SelectionChangeFcn** callback:

```
set(hTabGroup,'SelectionChangeFcn',@myCallbackFcn);
set(hTabGroup,'SelectedIndex',2);      % activate second tab
```

In MATLAB 7.11 R2010b, a new hidden property called **SelectedTab** has been added: **SelectedTab** accepts a *uitab* handle rather than its index, and it can be used as an alternative to **SelectedIndex**. Also in R2010b, **SelectionChangeFcn** was renamed **SelectionChangeCallback**:

```
set(hTabGroup,'SelectionChangeCallback',@myCallbackFcn);   % R2010b +
set(hTabGroup,'SelectedIndex',2);        % activate second tab
set(hTabGroup,'SelectedTab',tab2);       % alternative on R2010b +
```

4.3.1 Customizing Tabs at the Java Level

Additional control over the tab group's behavior can be achieved by customizing the underlying UITabGroupPeer Java object. This object is not directly exposed by *uitabgroup*, but it can be found using the *FindJObj* utility (see above) or via the hidden **ApplicationData** or callbacks. Remember that Java uses 0-based indexing, so tab #1 is actually the second tab. HTML is accepted just as in other Swing components:

```
% Get the underlying Java reference using FindJObj
jTabGroup = findjobj('class','tabgroup');

% A direct alternative for getting jTabGroup
jTabGroup = getappdata(handle(hTabGroup),'JTabbedPane');

% Another direct alternative for getting jTabGroup
callback = get(getappdata(tab1,'TabGroupChildListener'),'Callback');
jTabGroup = callback{3}.getComponent;

% Now use the Java reference to set the title, tooltip etc.
jTabGroup.setTitleAt(1,'Tab #2');
jTabGroup.setTitleAt(1,'<html><b><i><font size=+2>Tab #2');
jTabGroup.setToolTipTextAt(1,'Tab #2');
```

Enabled is another useful property that is only settable using the Java object. **Enabled** can be set for the entire tab group using jTabGroup.*setEnabled(true/false)*, or for a specific tab using jTabGroup.*setEnabledAt(tabIndex,true/false)*. Note that setting the property value for a specific tab overrides the value set for ALL tabs, despite the fact that *setEnabled* is called after *setEnabledAt*:[67]

```
jTabGroup.setEnabledAt(1,false); % disable only tab #1 (=second tab)
jTabGroup.setEnabled(false); % disable all tabs
jTabGroup.setEnabled(true); % re-enable all tabs (except tab #1)
```

Similarly, we can set the tab font color using *setForeground* and *setForegroundAt* or via HTML (again, *setForegroundAt* overrides anything set by *setForeground*):

```
jTabGroup.setForegroundAt(1,java.awt.Color(1.0,0,0)); % tab #1
jTabGroup.setTitleAt(1,'<html><font color='red'><i>Panel 2');
jTabGroup.setForeground(java.awt.Color.red);
```

Unfortunately, the corresponding *setBackgroundAt(tabIndex,color)* method has no visible effect, and the MATLAB-extended tabs keep their white/gray backgrounds. A similar attempt to modify the tab's **BackgroundColor** property fails, since MATLAB made this property unmodifiable (='none').[†] The simple solution is to use CSS:[68]

```
jTabGroup.setTitleAt(1,'<html><div style="background:#ffff00;">Tab2');
jTabGroup.setTitleAt(1,'<html><div style="background:yellow;">Tab2');
```

[†] http://www.mathworks.com/matlabcentral/newsreader/view_thread/258700 (or http://bit.ly/dtisOw); The **BackgroundColor** property is actually modifiable, but a PropertyPostSet listener (see Appendix B) placed on the property reverts its value back to 'none'. At any rate, this property affects the MATLAB *uicontainer* (content)'s background color and not the tab's.

We can also set a background gradient image for the tabs, using the CSS *background-image* directive. Similarly, we can set foreground text color (*color* directive), font, and so on.

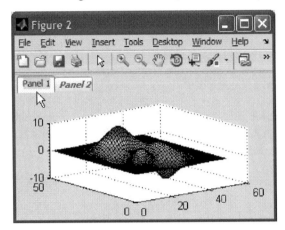

uitabgroup with non-default tab colors and fonts (**See color insert.**)

As explained in Section 3.3.2, jTabGroup's look-and-feel can be modified:

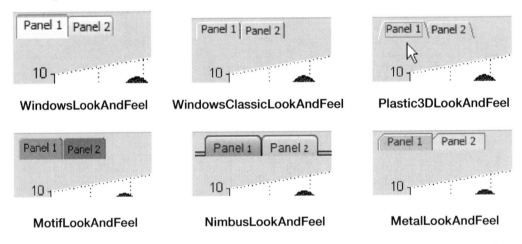

Icons and sub-components can be added to the tabs.[69] Unfortunately, for some reason that I do not fully understand, jTabGroup.*setIconAt* has no apparent effect. The solution is to set our own custom control as the requested tab, and then add our icon (or other customizations) to it. Section 4.6.4 has a detailed description of using icons. Here is a simple example:

```
% Add an icon to tab #1 (=second tab)
icon = javax.swing.ImageIcon('C:\Yair\save.gif');
jLabel = javax.swing.JLabel('Tab #2');
jLabel.setIcon(icon);
jTabGroup.setTabComponentAt(1,jLabel);    % Tab #1 = second tab

% Notice how the label and icon are automatically grayed when disabled
jTabGroup.setEnabledAt(1,false); % disable only tab #1
```

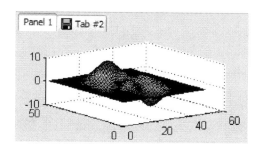 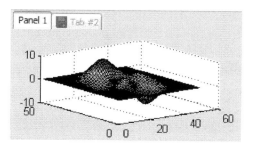

Tab with a custom icon (enabled and disabled) (See color insert.)

Now, let us try a more complex example, of adding a close ("x") button to one of the tabs. Generalization of this code snippet is left as an exercise to the reader:

```
% First let us load the close icon
jarFile = fullfile(matlabroot,'/java/jar/mwt.jar');
iconsFolder = '/com/mathworks/mwt/resources/';
iconURI = ['jar:file:/' jarFile '!' iconsFolder 'closebox.gif'];
icon = javax.swing.ImageIcon(java.net.URL(iconURI));

% Now let us prepare the close button: icon, size and callback
jCloseButton = handle(javax.swing.JButton,'CallbackProperties');
jCloseButton.setIcon(icon);
jCloseButton.setPreferredSize(java.awt.Dimension(15,15));
jCloseButton.setMaximumSize(java.awt.Dimension(15,15));
jCloseButton.setSize(java.awt.Dimension(15,15));
set(jCloseButton, 'ActionPerformedCallback', @(h,e)delete(tab2));

% Now let us prepare a tab panel with our label and close button
jPanel = javax.swing.JPanel; % default layout = FlowLayout
set(jPanel.getLayout, 'Hgap',0, 'Vgap',0); % default gap = 5 pixels
jLabel = javax.swing.JLabel('Tab #2');
jPanel.add(jLabel);
jPanel.add(jCloseButton);

% Now attach this tab panel as the tab-group's second tab component
jTabGroup.setTabComponentAt(1,jPanel);      % Tab #1 = second tab
```

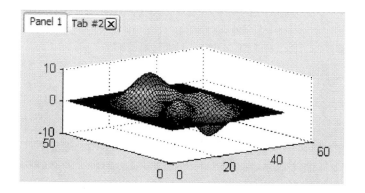

Tab with an attached close button (See color insert.)

There are other things that we can customize, such as setting mnemonics (keyboard shortcuts), etc. — refer to the official documentation[70] or any good text about Java Swing. The callbacks are the same as the standard Swing component callbacks, except for **StateChangedCallback**, which is automatically linked to the internal function that synchronizes between the Java tab group and the MATLAB ***uicontainer***s (in other words: it is not a good idea to mess with it).

Some jTabGroup functions that work well with standard JTabbedPane fail with ***uitab-group***, for example, jTabGroup.*setIconAt* or *setTabLayoutPolicy*.[71] I am unsure of the reason for this. Other limitations with ***uitabgroup*** are a reported problem when compiling any GUI that includes it;[72] a reported bug when reordering tabs;[73] a reported problem rendering some graphic object properties (e.g., clipping);[74] and a reported problem in displaying tabs containing ActiveX[75] or Java objects[76] (plus suggested solutions[77]). Interested readers can fix all these issues by modifying the m-files in the folders %matlabroot%/toolbox/matlab/@uitools/@uitabgroup and /@uitools/@uitab. At least some of these problems are fixed as of R2010a.

Readers might also be interested in the *Yet Another Layout Manager* utility.[78] This utility directly uses Swing's JTabbedPane object to implement tab panels, essentially mimicking the built-in ***uitab/uitabgroup*** functions.

4.3.2 Tabdlg and Other Alternatives

tabdlg is a related semidocumented and unsupported uitool that, unlike ***uitabgroup***, creates a tabbed user interface using plain MATLAB, without reliance on Java (all MATLAB GUI ultimately rely on Java, but ***tabdlg*** uses no Java beyond that). The end result looks less professional than ***uitabgroup***, but it works even when Java does not.

tabdlg has an extensive readable help section, so it will not be detailed here. In brief, the input parameters specify the tab labels, dimensions, offsets, callbacks, font, default tab, sheet dimensions, and parent figure. Whenever ***tabdlg*** is invoked without any input arguments, the following default sample tabs are shown:

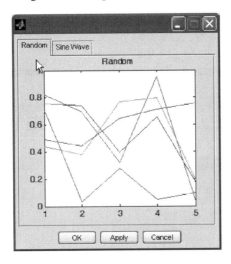

tabdlg left tab

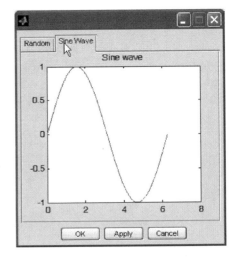

tabdlg right tab

There are many implementations of tab panels in MATLAB's File Exchange.[79] MATLAB's official Desktop Blog had an article about a specific example, which was that week's Pick of the Week,[80] relying on buttons that are easy to implement, but in my personal opinion, they are a far cry from our expectations of a tab panel. Better utilities are *Multiple Tab GUI*,[81] *Highlight Tab Objects easily*,[82] and best of all *uitabpanel*,[83] *TabPanel Constructor*,[84] or the excellent *GUI Layout Toolbox*.[85]

4.4 Uiundo

uiundo is a very useful tool for MATLAB GUI:[86] Whenever we have a MATLAB GUI containing user controls which the user may modify (edit boxes, checkboxes, sliders, toggle buttons, etc.), we may wish to include an undo/redo feature. This would normally be a painful programming task, but *uiundo* saves much of the setup work. Unlike uitools presented so far, *uiundo* is not Java based but rather uses MATLAB's classes and schema-based object-oriented approach. However, it is such a useful and undocumented MATLAB GUI concept, which is also used in the following section, that I thought readers of this book will benefit from a short discussion of this feature.

To use *uiundo*, invoke it in each *uicontrol* callback function (where we normally place GUI logic), with the undo/redo action name, the undo-ing action, and the redo-ing action (if it has been undone).[†] *uiundo* then takes care of adding this data to the figure's undo/redo options under Edit in the main figure menu.

Let us build a simple GUI consisting of a slider that controls the value of an edit box:

```
hEditbox = uicontrol('style','edit', 'position',[20,60,40,40]);
set(hEditbox, 'Enable','off', 'string','0');
hSlider = uicontrol('style','slider','userdata',hEditbox);
callback = @(h,e) set(hEditbox,'string',num2str(get(gcbo,'value')));
set(hSlider,'Callback',callback);
```

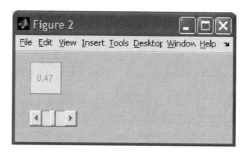

A simple GUI example

[†] Until R2010b, *uiundo* had a bug in a declared yet unset return value — this was fixed in R2011a by removing the return value.

Now, attach undo/redo actions to the slider's callback:

```
set(hSlider,'Callback',@test_uiundo);

% Main callback function for slider updates
function test_uiundo(varargin)
    % Update the edit box with the new value
    hEditbox = get(gcbo,'userdata');
    newValue = get(gcbo,'value');
    set(hEditbox,'string',num2str(newValue));

    % Retrieve and update the stored previous value
    oldValue = getappdata(gcbo,'oldValue');
    if isempty(oldValue), oldValue=0;  end
    setappdata(gcbo,'oldValue',newValue);

    % Prepare an undo/redo action
    cmd.Name = sprintf('slider update (%g to %g)',oldValue,newValue);

    % Note: the following is not enough since it only
    %       updates the slider and not the editbox...
    %cmd.Function       = @set;                    % Redo action
    %cmd.Varargin       = {gcbo,'value',newValue};
    %cmd.InverseFunction = @set;                   % Undo action
    %cmd.InverseVarargin = {gcbo,'value',oldValue};

    % This takes care of the update problem...
    cmd.Function       = @internal_update;         % Redo action
    cmd.Varargin       = {gcbo,newValue,hEditbox};
    cmd.InverseFunction = @internal_update;        % Undo action
    cmd.InverseVarargin = {gcbo,oldValue,hEditbox};

    % Register the undo/redo action with the figure
    uiundo(gcbf,'function',cmd);
end  % test_uiundo

% Internal update function to update slider & editbox
function internal_update(hSlider,newValue,hEditbox)
    set(hSlider, 'value',newValue);
    set(hEditbox,'string',num2str(newValue));
end  % internal_update
```

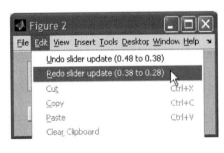
R2007a

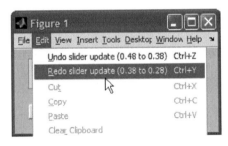
R2008a

Undo/redo functionality integrated in the figure's main menu-bar

Note that sometime in 2007, MATLAB added the standard Ctrl-Z/Ctrl-Y keyboard binding support to its undo/redo functionality, as seen in the above screenshots.

We can invoke the current Undo and Redo actions programmatically by calling ***uiundo*** with the 'execUndo' and 'execRedo' arguments:

```
uiundo(hFig,'execUndo');
uiundo(hFig,'execRedo');
```

When invoking the current Undo and Redo actions programmatically, we can ensure that this action is invoked only if it is a specific action that is intended:

```
uiundo(hFig,'execUndo','Save data');  % should equal cmd.Name
```

A little extra digging in undocumented territory enables additional customization of the undo/redo functionality: MATLAB stores all of a figure's undo/redo data in a hidden figure object, referenced by getappdata(hFig,'uitools_FigureToolManager'). This object, defined in %matlabroot%/toolbox/matlab/uitools/@uiundo/, uses a stack to store instances of the undo/redo cmd data structure introduced above

```
% Retrieve redo/undo object
undoObj = getappdata(hFig,'uitools_FigureToolManager');
if isempty(undoObj)
    undoObj = uitools.FigureToolManager(hFig);
    setappdata(hFig,'uitools_FigureToolManager',undoObj);
end
```

```
>> get(undoObj)
    CommandManager: [1x1 uiundo.CommandManager]
            Figure: [1x1 figure]
        UndoUIMenu: [1x1 uimenu]
        RedoUIMenu: [1x1 uimenu]
```

There are several interesting things that we can do with this undoObj. First, let us modify the main-menu items (see Section 4.6 for more details):

```
% Modify the main menu item (similarly for redo/undo)
if ~isempty(undoObj.RedoUIMenu)
    undoObj.RedoUIMenu.Position = 1; %default=2 (switch undo/redo)
    undoObj.RedoUIMenu.Enable = 'off';           % default='on'
    undoObj.RedoUIMenu.Checked = 'on';           % default='off'
    undoObj.RedoUIMenu.ForegroundColor = [1,0,0];   % =red
end
if ~isempty(undoObj.UndoUIMenu)
    undoObj.UndoUIMenu.Label = '<html><b><i>&Undo action';
    undoObj.UndoUIMenu.Separator = 'on';         % default='off'
    undoObj.UndoUIMenu.Checked = 'on';           % default='off'
    undoObj.UndoUIMenu.ForegroundColor = 'blue'; % default=black
end
```

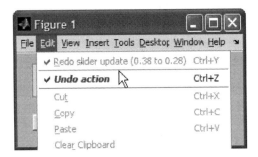

Note: &Undo underlines the 'U' and adds a keyboard accelerator, but unfortunately this only works if the label is non-HTML. In our case, we use HTML for the font effects (<b> <i>), so we lost the accelerator in the process.

Now, let us take a look at undoObj's CommandManager child (the Figure child is simply *handle(hFig)* and so is not very interesting):

```
>> undoObj.CommandManager.get
            UndoStack: [13x1 uiundo.FunctionCommand]
            RedoStack: [1x1 uiundo.FunctionCommand]
    MaxUndoStackLength: []
              Verbose: []

>> undoObj.CommandManager.UndoStack(end).get
              Parent: []
        MCodeComment: []
                Name: 'slider update (0.48 to 0.38)'
            Function: @internal_update
            Varargin: {[53.0037841796875]    [0.38]    [1x1 double]}
      InverseFunction: @internal_update
    InverseVarargin: {[53.0037841796875]    [0.48]    [1x1 double]}
```

This looks familiar: In fact, it is exactly the cmd data structure that is being passed to the *uiundo* function, with the additional (apparently unused) properties **Parent** and **MCodeComment**. CommandManager's **UndoStack** and **RedoStack** contain all stored undo/redo actions such that the latest action is at the end of these stack vectors.

We can inspect the latest undo/redo actions, without activating them, by using CommandManager's *peekundo()* and *peekredo()* methods, which return empty [] if no undo/redo action is available:

```
>> undoObj.CommandManager.peekredo.get      % first check if isempty...
              Parent: []
        MCodeComment: []
                Name: 'slider update (0.38 to 0.28)'
```

```
            Function: @internal_update
            Varargin: {[53.0037841796875]   [0.28]   [1x1 double]}
     InverseFunction: @internal_update
    InverseVarargin: {[53.0037841796875]   [0.38]   [1x1 double]}

>> undoObj.CommandManager.peekundo.get
              Parent: []
       MCodeComment: []
                Name: 'slider update (0.48 to 0.38)'
            Function: @internal_update
            Varargin: {[53.0037841796875]   [0.38]   [1x1 double]}
     InverseFunction: @internal_update
    InverseVarargin: {[53.0037841796875]   [0.48]   [1x1 double]}

>> undoObj.CommandManager.peekundo.Name
ans =
slider update (0.48 to 0.38)
```

We can undo/redo the latest action (last **UndoStack/RedoStack** element) by invoking CommandManager's *undo()/redo()* methods. This is actually what **uiundo** does behind the scenes when calling it with the 'execUndo' and 'execRedo' arguments:

```
undoObj.CommandManager.undo;
undoObj.CommandManager.redo;
```

We can clear the entire action stack by using CommandManager's *empty()* method:

```
undoObj.CommandManager.empty;
```

If we set CommandManager's **Verbose** property to a nonempty value, debug information is spilled onto the Command Window when new **uiundo** actions are added:

```
>> undoObj.CommandManager.Verbose = 1;
% now move the slider and see the debug info below:

internal_update(h_uicontrol, [0.48,], h_uicontrol); % Called by
slider update (0.28 to 0.48)
internal_update(h_uicontrol, [0.58,], h_uicontrol); % Called by
slider update (0.48 to 0.58)
```

Finally, CommandManager uses its **MaxUndoStackLength** property to limit the size of the undo/redo stacks. This property is defined as read-only in %matlabroot%\toolbox\matlab\uitools@ uiundo@CommandManager\schema.m line #12, and so if we wish to programmatically modify this property from its default value of empty (=unlimited), we will need to comment out that line.

Java Swing has a very similar UndoManager object,[87] which predates the MATLAB object by many years. It is obvious that the MATLAB's mechanism has its roots in this Java object. However, MATLAB's mechanism appears to be a separate implementation that does not directly extend Swing's UndoManager. MATLAB's undo manager has several advantages in

MATLAB compared with Java's `UndoManager` (automatic tie-in to the figure's Edit menu, acceptance of MATLAB actions, etc.). I, therefore, strongly suggest using MATLAB's undo manager in MATLAB applications rather than using Java's `UndoManager`.

uiundo is further revisited in Section 4.5.5, where it is used to illustrate adding dynamic undo/redo buttons to the figure toolbar.

Note: A utility called GUIHistory was submitted by Malcolm Lidierth to the MATLAB File Exchange.[88] This utility encapsulates MATLAB GUI undo/redo and can be used by readers who find MATLAB's *uiundo* functionality too complex.

4.5 Toolbars

MATLAB has elected to only partially document its toolbar-related functions: initially introduced in MATLAB 5.3, as undocumented internal functions,[89] *uitoolbar*, *uipushtool,* and *uitoggletool*, are now fully documented and supported functions that enable the user to specify a user-defined toolbar (and toolbar buttons). On the other hand, other related functions, *uigettool, uigettoolbar* and *uitoolfactory*, remain unsupported and only semidocumented (possessing a readable help section, but not online doc). These functions access the default MATLAB toolbar. MathWorks possibly feels more comfortable in letting users access and manipulate user-defined toolbars than the system toolbars. For this reason, the handles for the default toolbar and its buttons are hidden (found only by *findall* and not by *findobj*).

uigettool and *uigettoolbar* retrieve a specified toolbar button handle for a specified figure handle and action name (*uigettoolbar* was deprecated in MATLAB 7.4, but while the documentation says there is no replacement, apparently *uigettool* is the replacement). The actions are typically named <Group>.<Button>, for example, 'Annotation.InsertLegend' or 'Standard. EditPlot'. Beware: action names have changed between MATLAB versions: for example, the Zoom-in button ('Exploration.ZoomIn') was previously named 'figToolZoomIn'. Alternately, search all figure handles' **Tag** property:[90]

```
% Get the toolbar handle
hToolbar = findall(allchild(hFig),'flat','type','uitoolbar');
hToolbar = findall(hFig,'tag','FigureToolBar');        % equivalent

% Get the list (cell-array) of all toolbar actions
actionNames = get(findall(hToolbar),'tag');

% Access a specific toolbar button
hButton = uigettool(hFig, 'Exploration.ZoomIn');
hButton = uigettoolbar(hFig, 'Exploration.ZoomIn');    % equivalent
hButton = findall(hFig,'Tag', 'Exploration.ZoomIn');   % equivalent
```

Once a toolbar button handle is gotten, the button can be enabled/disabled (via its **Enabled** property), shown/hidden (**Visible** property), separated from or adjoined to its neighbors

(**Separated**), shown depressed (**State**), tooltip-ed (**Tooltip**), icon-customized (**CData**),[91] and deleted (by simply calling ***delete(hButton)***).

Several button callback properties are available: **ClickedCallback** is invoked upon button click; **OnCallback** and **OffCallback** are invoked when the button **State** changes on/off.[92] All the standard HG properties are also accessible: **UserData, Tag, Parent, ApplicationData**, and so on. Refer to the online documentation for details.[93]

Modification of the **ClickedCallback** is an easy way to change the default figure toolbar's behavior. For example, to modify the print button's default print action to a print-preview action:[94]

```
% Find the Print button's handle
hPrintButton = findall(hToolbar,'tag','Standard.PrintFigure');

% Modify the button's callback action & tooltip
set(hPrintButton, 'ClickedCallback','printpreview(gcbf)', ...
                  'TooltipString','Print Preview');
```

Modification of the Print button's default action

Similarly, the New Figure button ( ) can be customized to open a new data-entry dialog window; the Open button ( ) can be customized to only open files of specific formats; the Save button ( ) can be modified to Save As... (or to save using a non-*.*fig* format);[95] the Legend button ( ) can be used to customize the displayed legend's contents, appearance and location, and so on.

4.5.1 Uitoolfactory

uitoolfactory[96] creates a toolbar button (or button group) based on any predefined default figure button. This is useful when creating a user-defined toolbar that should contain only some of the default toolbar's buttons. Of course, an alternative could be to simply remove unneeded buttons from the default toolbar. Depending on the number of needed/unneeded default buttons, either approach could be used:

```
% Alternative 1: use uitoolfactory to add buttons to a new toolbar
hNewToolbar = uitoolbar ('parent',hFig);
hButton = uitoolfactory(hNewToolbar,'Standard.EditPlot');
hButton = uitoolfactory(hNewToolbar,'Exploration.ZoomIn');
set(hButton,'Separator','on');
hButton = uitoolfactory(hNewToolbar,'Exploration.ZoomOut');
hButton = uitoolfactory(hNewToolbar,'Exploration.Pan');
```

```
% Alternative 2: remove unneeded buttons from the default toolbar
% Note: in the findall() results below, toolbar is the 1st handle;
%       rightmost button is the 2nd; leftmost is the last
hDefaultButtons = findall(hToolbar); % leftmost button is last
delete(hDefaultButtons([2:7,end-3:end]));
set(hDefaultButtons(end-4),'separator','off');
```

A slimmed-down toolbar

uitoolfactory without args lists all registered toolbar buttons in the Command Window:

```
>> uitoolfactory
            TOOLBAR ITEMS
      Standard.NewFigure
      Standard.FileOpen
      Standard.SaveFigure
      Standard.PrintFigure
      Standard.EditPlot
      Exploration.ZoomIn
      Exploration.ZoomOut
      Exploration.Pan
      Exploration.Rotate
      Exploration.DataCursor
      Annotation.InsertColorbar
      Annotation.InsertLegend
      Annotation.InsertRectangle
      Annotation.InsertEllipse
      Annotation.InsertTextbox
      Annotation.InsertTextArrow
      Annotation.InsertDoubleArrow
      Annotation.InsertArrow
      Annotation.InsertLine
      Annotation.Pin
      Annotation.AlignDistribute
      Plottools.PlottoolsOff
      Plottools.PlottoolsOn
```

uitoolfactory's help section has a documentation flaw: it mentions that use of a second input parameter, 'getinfo', returns the list of all registered toolbar buttons whose handle is specified as the first parameter. In fact, it is only by NOT specifying a second parameter that this data can be found:[†]

```
>> hToolbar = findall(allchild(hFig),'flat','type','uitoolbar');
>> toolInfo = uitoolfactory(hToolbar)
toolInfo =
1x23 struct array with fields:
    name
    group
    constructor
    properties
    icon            <= files in [matlabroot '/toolbox/matlab/icons/']

>> toolInfo(1)
ans =
            name: 'NewFigure'
           group: 'Standard'
     constructor: 'uipushtool'
      properties: [1x1 struct]
            icon: 'newdoc'
>> toolInfo(1).properties
ans =
    ClickedCallback: 'filemenufcn(gcbf,'FileNew')'
            ToolTip: 'New Figure'

>> toolInfo(6)
ans =
            name: 'ZoomIn'
           group: 'Exploration'
     constructor: 'uitoggletool'
      properties: [1x1 struct]
            icon: 'view_zoom_in.gif'
>> toolInfo(6).properties
ans =
    ClickedCallback: 'putdowntext('zoomin',gcbo)'
            ToolTip: 'Zoom In'
```

 Note: *uitoolfactory* has a bug: using a second specified output parameter that is not mentioned in the help section, nor ever assigned within the code, causes a runtime error if used.[‡]

4.5.2 *Other Undocumented Toolbar Functions*

The uitools folder still contains an ancient (MATLAB 5, mid-1990s vintage) set of functions whose aim was to mimic toolbar buttons by patches drawn on invisible axes next to each other. These functions, probably left for backward compatibility, do not posses any noticeable

---

[†] This documentation flaw was fixed in R2011a.
[‡] This bug was also fixed in R2011a.

advantage over the newer toolbar functions, either documented (***uitoolbar, uipushtool,*** and ***uitoggletool***) or not (***uigettool*** and ***uitoolfactory***).

This set of obsolete functions includes ***btngroup, btnicon, btnresize, btnstate, btnpress, btndown***, and ***btnup***: ***btngroup*** sets the button group (toolbar); ***btnicon*** sets specific predefined icons for a button; ***icondisp*** displays a specific icon (or all icons if no icon name is specified); ***btnresize*** resizes a button; the other ***btn*** functions get or set the button's depressed state. Oddly, these ***btn*** functions are not indicated as obsolete or unsupported within their code or help section, although they have no online help. Interested readers can find a few ancient references in CSSM.[97]

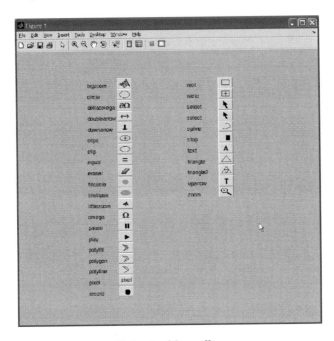

Output of *icondisp*

4.5.3 *Customizing Toolbars at the Java Level*

Usage of undocumented Java toolbar functionality enables additional customization of toolbars. MATLAB toolbars are `com.mathworks.mwswing.MJToolBar` objects that extend the standard `javax.swing.JToolBar`.[98] We can easily interlace the functionality of the Java reference handle with the functionality of the MATLAB handle.

For example, a CSSM user has once posted a question[99] asking how to add large buttons to the MATLAB toolbar. Apparently, MATLAB uses a 25-pixel-high toolbar and crops any button added to this toolbar in order to fit this height. This can be changed as follows:

First, let us enlarge the toolbar height from its default 25 pixels to 50:

```
hToolbar = findall(hFig,'tag','FigureToolBar');
jToolbar = get(get(hToolbar,'JavaContainer'),'ComponentPeer');
jToolbar.setPreferredSize(java.awt.Dimension(10,50));
jToolbar.revalidate;          % refresh/update the displayed toolbar
```

Now, let us add a 32×32 button:

```
icon = fullfile(matlabroot,'/toolbox/matlab/icons/warning.gif');
[cdata,map] = imread(icon);
cdata = ind2rgb(cdata,map);
hButton = uipushtool(hToolbar,'cdata',cdata,'tooltip','Warning');
```

Enlarged toolbar (50 pixels high) with a cropped warning button

Initially, MATLAB crops the button to 23×23, but we can modify its maximum size so that it looks perfect (increase the width if you wish wider margins between buttons):

```
% The requested button is the last component in the toolbar
numButtons = jToolbar.getComponentCount;
newSize = java.awt.Dimension(35,50);

% Remember that Java indexes start at 0, not 1...
jToolbar.getComponent(numButtons-1).setMaximumSize(newSize);
jToolbar.revalidate;
```

Enlarged toolbar with a non-cropped large warning button

Alternately, we can resize the button to the standard 16×16 toolbar icon size[†]

```
icon = jToolbar.getComponent(numButtons-1).getIcon;
iconImg = icon.getImage;       % a java.awt.image.BufferedImage object
newIconImg = iconImg.getScaledInstance(16,16,iconImg.SCALE_SMOOTH);
icon.setImage(newIconImg);
jToolbar.revalidate;           % refresh/update the displayed toolbar
```

Regular toolbar with a perfectly sized warning button

[†] If the white, instead of transparent, icon background disturbs your aesthetics, you can further customize this image as described in http://www.rgagnon.com/javadetails/java-0265.html (or http://tinyurl.com/dyspal).

Instead of using MATLAB's ***uipushtool/uitoggletool***, we can add JComponents directly into our jToolbar. This is particularly useful for drop-downs (javax.swing.JComboBox), checkboxes, and so on, which cannot be added to the toolbar using plain-vanilla MATLAB:

```
jToolbar = get(get(hToolbar,'JavaContainer'),'ComponentPeer');
jCheckBox = javax.swing.JCheckBox('Checkbox');
hCheckBox = handle(jCheckBox,'CallbackProperties');
set(hCheckBox, 'ActionPerformedCallback', @myCallbackFcn);
jToolbar.add(hCheckBox,5); % 5th position, after printer icon
jToolbar.revalidate;
```

Addition of a Java control to the standard MATLAB toolbar

A blog reader has asked[100] why buttons added to the toolbar in this manner appear "flat" instead of appearing three-dimensional like buttons created outside the toolbar. The answer is that the new buttons actually behave consistently with the default toolbar buttons (and the windowing system, for that matter): their border is painted only when we hover the mouse over the button. This is controlled by the JButton's **RolloverEnabled** property (which is true by default, but can be turned off at will):

```
jButton.setRolloverEnabled(false);
set(jButton,'RolloverEnabled','off');      % an alternative
```

To always display a border, override the button's default **Border** property:[101]

```
jBorder = javax.swing.BorderFactory.createRaisedBevelBorder;
jButton.setBorder(jBorder);
```

Default toolbar button

Toolbar button with non-default border

Remember that toolbars are simply containers for internal components, generally buttons and separators. These components can be accessed individually. An example of this can be found in my ***FindJObj*** utility that lists the individual figure components: whenever the user selects a toolbar button (or any other Java component), its **Border** is temporarily modified to a flashing red rectangle:

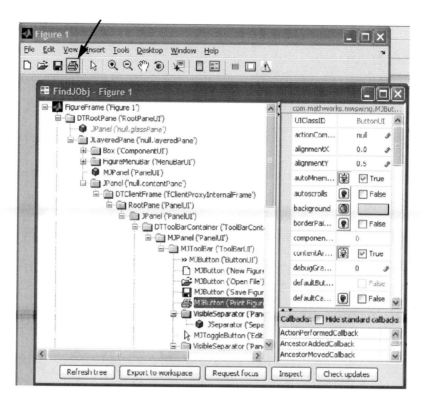

FindJObj **utility accessing a specific toolbar button (See color insert.)**

An additional interesting functionality of toolbars that can only be used in Java is enabling a floating toolbar, via jToolbar.*setFloatable(1)*.[102] The toolbar can then be dragged from its docked position at the top of the figure menu, becoming enclosed in an independent floating window (a non-modal javax.swing.JDialog child of the parent figure, to be exact). Since this toolbar window has a very small initial size and no name, a simple immediate fix is required as follows:

```
jToolbar.setFloatable(true);
hjToolbar = handle(jToolbar,'CallbackProperties'); % for callbacks
set(hjToolbar,'AncestorAddedCallback',@dockUndockCallbackFcn);

% Sample dockUndockCallbackFcn function
function dockUndockCallbackFcn(hjToolbar, eventData)
    if hjToolbar.isFloating
        jToolbarWin = hjToolbar.getTopLevelAncestor;
        jToolbarWin.setTitle('Toolbar');
        % jToolbarWin.setResizable(true); %to enable manual resize
        jToolbarWin.setPreferredSize(java.awt.Dimension(380,57));
        jToolbarWin.setSize(java.awt.Dimension(380,57));
        jToolbar.revalidate;          %repaint toolbar
        jToolbarWin.getParent.validate;    %repaint parent figure
```

```
        end
    end    % dockUndockCallbackFcn
```

Undocked toolbar   **...the same undocked toolbar after minor fixes**

Re-docking a floating toolbar is done by simply closing the floating window (clicking the "X" button at the window's top-right corner) — the toolbar then re-appears in its default (top) position within the parent figure window. The standard Java JToolBar, which MATLAB's MJToolBar extends, allows floating toolbars to be manually dragged and pinned to the window sides,[103] but apparently the MATLAB extension prevents it.

Here is how to programmatically place the toolbar on the window's bottom:

```
jPanel = jToolbar.getParent.getParent;
jPanel.add(jToolbar.getParent,java.awt.BorderLayout.SOUTH);
jPanel.revalidate;
```

Unfortunately, when we try to use the same method for placing the toolbar on the left/right sides of the window, the figure gets "frozen" and unresponsive. Perhaps, someone can enlighten me what needs to be changed in the following code:

```
jToolbar.setOrientation(jToolbar.VERTICAL);
jPanel.add(jToolbar.getParent,java.awt.BorderLayout.WEST);
jPanel.revalidate;
```

There are other interesting functions/properties available via the Java object. These can be explored using the ***methods***, ***methodsview***, ***inspect***, and ***uiinspect*** functions.

For example, *addGap()* can be used to add a transparent gap between the rightmost toolbar component and the window border: this gap is kept even if the window is shrunk — the rightmost components disappear, maintaining the requested gap.

setBackground() sets the background color that is seen beneath any transparent pixels of button images and gaps. Nontransparent (opaque or colored) pixels are not modified. If the button icons are improperly created, the result looks bad:

```
jToolbar.setBackground(java.awt.Color.cyan)        % or: Color(0,1.0,1.0)
```

Some of the default figure toolbar buttons having opaque backgrounds

We can modify the toolbar buttons to have a consistent background as follows:

```
color = java.awt.Color.cyan; %or: Color(0,1.0,1.0)
jToolbar.setBackground(color);
jToolbar.getParent.getParent.setBackground(color);
jtbc = jToolbar.getComponents;
for idx = 1 : length(jtbc)
     jtbc(idx).setOpaque(false);
     jtbc(idx).setBackground(color);
     for childIdx = 1 : length(jtbc(idx).getComponents)
             jtbc(idx).getComponent(childIdx-1).setBackground(color);
     end
end
```

Default figure toolbar buttons fixed in order to present a consistent background

setMorePopupEnabled() specifies the behavior when the window resizes to such a small width that one or more toolbar buttons disappear — by default (=1 or true), the chevron (>>) mark appears on the toolbar's right, enabling display of missing buttons, but this behavior can be overridden (0 or false) to simply crop the extra buttons.

setRollover() controls the behavior when the mouse passes ("rolls") over the toolbar buttons. The default value (1 or true) displays a three-dimensional button border, creating an embossing effect; this can be overridden (0 or false) for a different effect:

```
% Set non-default Rollover and MorePopupEnabled property values
jToolbar.setRollover(false);       % or: set(jToolbar,'Rollover','off')
jToolbar.setMorePopupEnabled(0);   % or: set(...,'MorePopupEnabled','off')
```

Default Rollover and MorePopupEnabled   **Nondefault Rollover and MorePopupEnabled**

4.5.4 Uisplittool and Uitogglesplittool

MATLAB 7.6 (R2008a) and onward contain a reference to *uisplittool* and *uitogglesplittool* in the javacomponent.m and %matlabroot%/bin/registry/hg.xml files. These are reported as built-in functions by the *which* function, although they have no corresponding m-file as other similar built-in functions:

```
>> which uisplittool
built-in (C:\Matlab\R2008a\toolbox\matlab\uitools\uisplittool)
```

These uitools are entirely undocumented, even as these lines are written in 2010, with MATLAB 7.12 (R2011a) around the corner. An acute reader (Jeremy Raymonds) suggested that they are related to toolbars, like other uitools such as *uipushtool* and *uitoggletool*. This turned out to be the missing clue that unveiled these useful tools:

Both *uisplittool* and *uitogglesplittool* are basic HG building blocks used in MATLAB tool-bars similarly to the well-documented *uipushtool* and *uitoggletool*. *Uisplittool* presents a simple drop-down, whereas *uitogglesplittool* presents a drop-down that is also selectable.

The plot-selection drop-down control on the Desktop's Workspace toolbar ( ) is an example of a *uisplittool*:

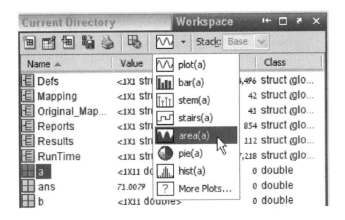

A uisplittool in action in the *MATLAB* Desktop

Other examples are the Publish and Run controls on the MATLAB Editor's toolbar (,). These controls, like other MATLAB toolbar controls, have a fly-over (hover) embossing effect on by default, as seen in this Publish control image . When we click the narrow arrow button, which is adjacent to the main control image, a drop-down appears with a control-specific functionality.

The Brush / Select-Data control on the figure's toolbar () is an example of a ***uitogglesplittool***. It behaves very similarly to a ***uisplittool*** in its drop-downs. In addition, it also supports a persistent selection:

	No mouse fly-over	Mouse fly-over
Unselected		
Selected		

Addition of ***uisplittool*** and ***uitogglesplittool*** to a toolbar is done in a similar manner to adding ***uipushtool***s and ***uitoggletool***s:

```
hToolbar = findall(hFig,'tag','FigureToolBar');
hUndo = uisplittool('parent',hToolbar);              % uisplittool
hRedo = uitogglesplittool('parent',hToolbar);        % uitogglesplittool
```

Default figure toolbar with additional user-created *uisplittool* and *uitogglesplittool* buttons

Just as with ***uipushtool***s and ***uitoggletool***s, the new buttons have an empty button-face appearance, until we fix their **CData, Tooltip,** and similar settable properties:

```
% Load the Redo icon
icon = fullfile(matlabroot,'toolbox/matlab/icons/greenarrowicon.gif');
[cdata,map] = imread(icon);

% Convert white pixels into a transparent background
map(find(map(:,1) + map(:,2) + map(:,3) ==3)) = NaN;

% Convert into 3D RGB-space
cdataRedo = ind2rgb(cdata,map);
cdataUndo = cdataRedo(:,[16:-1:1],:);

% Add the icon (and its mirror image = undo) to latest toolbar
set(hUndo, 'cdata',cdataUndo, 'tooltip','undo','Separator','on', ...
          'ClickedCallback','uiundo(gcbf,''execUndo'')');
set(hRedo, 'cdata',cdataRedo, 'tooltip','redo', ...
          'ClickedCallback','uiundo(gcbf,''execRedo'')');
```

uisplittool and ***uitogglesplittool*** buttons with nonempty icons

Note that the controls can be created with these properties in a single command:

```
hUndo = uisplittool('parent',hToolbar, 'cdata',cdataRedo, ...);
```

Let us now re-arrange our toolbar buttons. Unfortunately, at least in MATLAB versions 7.6–7.12 (R2008a–R2011a), an apparent bug causes ***uisplittool***s, ***uitogglesplittool***s, and any directly added Java component to always be placed flush-left when the toolbar's children are re-arranged. Therefore, we cannot re-arrange the buttons at the HG-children level (as shown in Section 4.5.5), but we can still re-arrange directly at the Java level:

```
jToolbar = get(get(hToolbar,'JavaContainer'),'ComponentPeer');
jButtons = jToolbar.getComponents;
for buttonIdx = length(jButtons)-3 : -1 : 7 % end-to-front
    jToolbar.setComponentZOrder(jButtons(buttonIdx), buttonIdx+1);
end
jToolbar.setComponentZOrder(jButtons(end-2), 5);    % Separator
jToolbar.setComponentZOrder(jButtons(end-1), 6);    % Undo
jToolbar.setComponentZOrder(jButtons(end), 7);         % Redo
jToolbar.revalidate;
```

uisplittool and ***uitogglesplittool*** button positions rearranged (not as simple as it may seem)

Now that we have added the controls, we need to specify their drop-down functionality: ***uisplittool*** and ***uitogglesplittool*** have a **Callback** property in addition to the standard **ClickedCallback** property that is available in ***uipushtool***s and ***uitoggletool***s. The standard **ClickedCallback** is invoked when the main button is clicked, while **Callback** is invoked when the narrow arrow button is clicked.[†] The accepted convention is that **ClickedCallback** should

---

[†] ***uitogglesplittool***, like ***uitoggletool***, also has settable **OnCallback** and **OffCallback** callback properties.

invoke the default control action (in our case, an Undo/Redo of the topmost undo action stack), while **Callback** should display a drop-down of selectable actions.

To set the drop-down functionality, we can use the **Callback** property to programmatically present a GUI of our choice to the user — we are definitely **NOT** confined to a simple drop-down. However, as noted above, the accepted convention is to present a selection drop-down. While this can be done programmatically using the **Callback** property, this functionality is already prebuilt into *uisplittool* and *uitogglesplittool* for our benefit. To access it, we need to get the control's underlying Java component. This is normally done using the *findjobj* utility (see Section 7.2.2), but in this case, we have a shortcut: the control handle's hidden **JavaContainer** property that holds the underlying com.mathworks.hg.peer.SplitButtonPeer (or .ToggleSplitButtonPeer) Java reference handle. This Java object's **MenuComponent** property returns a reference to the control's drop-down sub-component (a com.mathworks.mwswing.MJPopupMenu object):

```
>> jUndo = get(hUndo,'JavaContainer')
jUndo =
com.mathworks.hg.peer.SplitButtonPeer@f09ad5

>> jMenu = get(jUndo,'MenuComponent')   % or: =jUndo.getMenuComponent
jMenu =
com.mathworks.mwswing.MJPopupMenu[Dropdown Picker ButtonMenu,...]
```

Let us add a few simple textual options:

```
jOption1 = jMenu.add('Option #1');
jOption1 = jMenu.add('Option #2');

set(jOption1, 'ActionPerformedCallback', 'disp(''option #1'')');
set(jOption2, 'ActionPerformedCallback', {@myCallbackFcn, extraData});
```

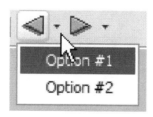

Setting up of *uisplittool* and *uitogglesplittool* popup menus

Popup menus are described in detail in Sections 4.2.4, 4.8.3, 6.6.2, and elsewhere.[104] Sections 4.6.3 and 4.6.4 explain how menu items can be set with icons and HTML markup. Sub-menus can also be added.

Like *uipushtool* and *uitoggletool*, *uisplittool* and *uitogglesplittool* also have unique **Type** property values, 'uisplittool' and 'uitogglesplittool', respectively. The handles can also be tested using the built-in *isa* function:

```
>> isa(handle(hUndo),'uisplittool')          % or: 'uitogglesplittool'
ans =
     1
>> class(handle(hUndo))
ans =
uisplittool
```

4.5.5 Adding Undo/Redo Toolbar Buttons

An important customization of the *uiundo* functionality (described in Section 4.4) is the addition of undo/redo buttons to the figure toolbar.[105] I am unclear why such an elementary feature was not included in the default figure toolbar, but this is a fact that can easily be remedied using the functionalities we explored earlier in this section. We start by adding simple undo/redo toolbar buttons:

```
% Load the Redo icon
icon = fullfile(matlabroot,'toolbox/matlab/icons/greenarrowicon.gif');
[cdata,map] = imread(icon);

% Convert white pixels into a transparent background
map(find(map(:,1) + map(:,2) + map(:,3) == 3)) = NaN;

% Convert into 3D RGB-space
cdataRedo = ind2rgb(cdata,map);
cdataUndo = cdataRedo(:,[16:-1:1],:);

% Add the icon (and its mirror image = undo) to latest toolbar
hUndo = uipushtool('cdata',cdataUndo, 'tooltip','undo', ...
                   'ClickedCallback','uiundo(gcbf,''execUndo'')');
hRedo = uipushtool('cdata',cdataRedo, 'tooltip','redo', ...
                   'ClickedCallback','uiundo(gcbf,''execRedo'')');
```

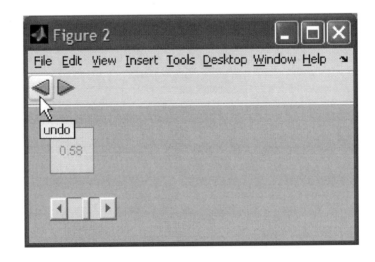

Setting up of simple undo/redo buttons

We would normally preserve hUndo and hRedo, and modify their **Tooltip** and **Enable** properties in runtime, based on availability and name of the latest undo/redo actions:

```
latestUndoAction = undoObj.CommandManager.peekundo;
if isempty(latestUndoAction)
    set(hUndo, 'Tooltip','', 'Enable','off');
else
    tooltipStr = ['undo' latestUndoAction.Name];
    set(hUndo, 'Tooltip',tooltipStr, 'Enable','on');
end
```

In the preceding screenshot, since no figure toolbar was previously shown, *uipushtool* added the undo and redo buttons to a new toolbar. Had the figure toolbar been visible, the buttons would have been added to its right end. Since undo/redo buttons are normally requested near the left end of toolbars, we need to re-arrange the toolbar buttons. This is done by re-arranging the buttons at the HG-children level:

```
hToolbar = findall(hFig,'tag','FigureToolBar');
%hToolbar = get(hUndo,'Parent'); % an alternative
hButtons = findall(hToolbar);

% all buttons need to be visible in order to be re-arrangeable
oldStatus = get(0,'showHiddenHandles');
set(0 ,'showHiddenHandles','on');
set(hToolbar,'children',hButtons([4:end-4,2,3,end-3:end])); %rearrange
set(0,'showHiddenHandles',oldStatus); % restore previous status
set(hUndo,'Separator','on');
```

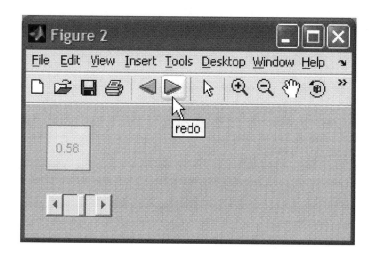

Simple undo/redo buttons added to the default figure toolbar

Note: Another re-arrangement method, which overcomes the ***uisplittool*** flush-left bug, is to use Java-level re-arrangement, as explained in Section 4.5.4.

More advanced customization is required to present the undo/redo actions in a drop-down (combo-box). Unfortunately, since MATLAB only enables adding ***uipushtool***s and ***uitoggletool***s to toolbars, we need to use a Java component. The drawback of using such a component is that it is inaccessible via the toolbar's **Children** property (implementation of the drop-down callback function is left as an exercise to the reader):

```
jToolbar = get(get(hToolbar,'JavaContainer'),'ComponentPeer');
if ~isempty(jToolbar)
    undoActions = get(undoObj.CommandManager.UndoStack,'Name');
    jCombo = javax.swing.JComboBox(undoActions(end:-1:1));
    set(jCombo, 'ActionPerformedCallback', @myUndoCallbackFcn);
    jToolbar(1).add(jCombo,5); % 5th position, after printer icon
    jToolbar(1).repaint;
    jToolbar(1).revalidate;
end

% Drop-down (combo-box) callback function
function myUndoCallbackFcn(hCombo,hEvent)
    itemIndex = get(hCombo,'SelectedIndex'); % 0 = topmost item
    itemName = get(hCombo,'SelectedItem');
    % user processing needs to be placed here
end   % myUndoCallbackFcn
```

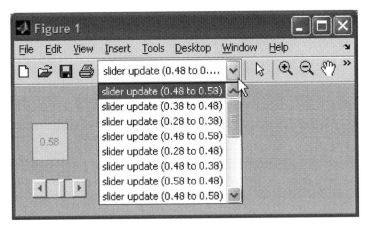

Undo/redo popup (actually, a drop-down) menu added to the default figure toolbar

 Note: Java components added directly to the Java toolbar in this fashion are not saved with the rest of the figure data/GUI when the figure is saved. A simple workaround for this limitation is to place the code that adds the Java components in the figure's **CreateFcn** callback. This callback value is stored with the rest of the figure's data/GUI and will be executed whenever the figure is reloaded from the disk.[106]

Let us now extend this example to use a small ***uisplittool*** rather than using a wide combo-box. A ***uisplittool*** will also enable us to easily undo/redo the latest action (like a simple toolbar button, by clicking the main ***uisplittool*** button) as well as select items from the actions drop-down (like a combo-box, by clicking the attached arrow button) — all this using a single Java component.

Finally, let us attach the undo actions to our Undo button's drop-down. This assumes that our GUI is already displayed — remember that the ***uicontrol*** function hides the figure toolbar and destroys all our previous customization work. Here is the complete code, which extends the ***uiundo*** code that was presented in Section 4.4:[†]

```
% Display our GUI
hEditbox = uicontrol('style','edit', 'position',[20,60,40,40], ...
                     'Enable','off', 'string','0');
hSlider  = uicontrol('style','slider', 'userdata',hEditbox, ...
                     'Callback',@sliderCallbackFcn);
[hUndo, hRedo] = prepareUndoRedoButtons();

% Callback function for slider movements
function sliderCallbackFcn(hSlider,hEventData)
    % Update the edit box with the new value
    newValue = get(hSlider,'value');
    set(hEditbox,'string',num2str(newValue));

    % Retrieve and update the stored previous value
    oldValue = getappdata(gcbo,'oldValue');
    if isempty(oldValue),  oldValue=0;  end
    setappdata(gcbo,'oldValue',newValue);

    % Prepare an undo/redo action
    cmd.Name = sprintf('slider update (%g to %g)', oldValue, newValue);
    cmd.Function = @internal_update;
    cmd.Varargin = {gcbo,newValue,hEditbox};
    cmd.InverseFunction = @internal_update;
    cmd.InverseVarargin = {gcbo,oldValue,hEditbox};

    % Register the undo/redo action with the figure
    uiundo(gcbf,'function',cmd);
    undoObj = getUndoObj(gcbf);
    setappdata(undoObj.CommandManager.UndoStack(end), 'oldValue',oldValue);
    setappdata(undoObj.CommandManager.UndoStack(end), 'newValue',newValue);
end  % sliderCallbackFcn

% Prepare the undo/redo toolbar buttons
function [hUndo, hRedo] = prepareUndoRedoButtons()
    % Display the figure toolbar, hidden by the uicontrol function
    set(gcf,'Toolbar','figure');

    % Add the Undo/Redo buttons
    hToolbar = findall(gcf,'tag','FigureToolBar');
    hUndo = uisplittool('parent',hToolbar);
```

[†] The reason for rearranging the buttons at the Java rather than the HG-children level was explained in Section 4.5.4.

```
    hRedo = uisplittool('parent',hToolbar);

    % Load the Redo icon
    icon = fullfile(matlabroot,'toolbox/matlab/icons/greenarrowicon.gif');
    [cdata,map] = imread(icon);

    % Convert white pixels into a transparent background
    map(map(:,1)+map(:,2)+map(:,3)==3) = NaN;

    % Convert into 3D RGB-space
    cdataRedo = ind2rgb(cdata,map);
    cdataUndo = cdataRedo(:, 16:-1:1, :);

    % Add the icon (and its mirror image = undo) to latest toolbar
    set(hUndo, 'ClickedCallback',{@undoRedoCallbackFcn,'Undo'}, ...
               'cdata',cdataUndo, 'enable','off', ...
               'tooltip','Nothing to undo', 'Separator','on');
    set(hRedo, 'ClickedCallback',{@undoRedoCallbackFcn,'Redo'}, ...
               'cdata',cdataRedo,'enable','off',...
               'tooltip','Nothing to redo');

    % Ensure everything is displayed onscreen otherwise Java won't work...
    drawnow;

    % Re-arrange the Undo/Redo buttons
    jToolbar = get(get(hToolbar,'JavaContainer'),'ComponentPeer');
    jButtons = jToolbar.getComponents;
    for buttonIdx = length(jButtons)-3 : -1 : 7  % end-to-front
        jToolbar.setComponentZOrder(jButtons(buttonIdx), buttonIdx+1);
    end
    jToolbar.setComponentZOrder(jButtons(end-2), 5);     % Separator
    jToolbar.setComponentZOrder(jButtons(end-1), 6);     % Undo
    jToolbar.setComponentZOrder(jButtons(end), 7);       % Redo
    jToolbar.revalidate;

    % Update the buttons whenever the undo/redo stack changes
    prepareUndoRedoListener(hUndo,'Undo');
    prepareUndoRedoListener(hRedo,'Redo');
end  % prepareUndoRedoButtons

% Prepare undo/redo listener
function prepareUndoRedoListener(hButton,actionType)
    undoObj = getUndoObj(gcf);
    hUndoCmd = handle(undoObj.CommandManager);
    hProp = findprop(hUndoCmd, [actionType 'Stack']);
    callback = {@updateUndoDropDown,hButton,actionType};
    hListener = handle.listener(hUndoCmd,hProp,'PropertyPostSet',callback);
    setappdata(hUndoCmd, [actionType 'Listener'],hListener);
end  % prepareUndoRedoListener

% Get the undo/redo object
function undoObj = getUndoObj(hFig)
    drawnow;
    undoObj = getappdata(hFig, 'uitools_FigureToolManager');
    if isempty(undoObj)
```

```matlab
            undoObj = uitools.FigureToolManager(hFig);
            setappdata(hFig,'uitools_FigureToolManager',undoObj);
        end
    end  % getUndoObj

    % Generic undo/redo for the requested action
    function undoRedoCallbackFcn(hButton,hEventData,actionType)
        hFig = gcbf;
        if isempty(hFig), hFig = gcf;  end
        uiundo(hFig,['exec' actionType]);
    end  % undoRedoCallbackFcn

    % Update a split-button's drop-down based on the undo/redo stack
    function updateUndoDropDown(ignore1,ignore2,hButton,actionType)
        hFig = ancestor(hButton,'figure');
        jButton = get(hButton,'JavaContainer');
        jMenu = get(jButton,'MenuComponent');
        undoObj = getUndoObj(hFig);
        actionsStack = undoObj.CommandManager.([actionType 'Stack']);
        jMenu.removeAll;
        if isempty(actionsStack)
            set(hButton, 'enable','off', 'Tooltip',['Nothing to ' actionType]);
        else
            actions = get(actionsStack,'Name');
            if ischar(actions)  % only a single undo/redo action
                actions = {actions};
            end
            for idx = length(actions) : -1 : 1     % end-to-front
                jAction = handle(jMenu.add(actions{idx}),'CallbackProperties');
                callback = {@undoMenuCallbackFcn, actionType};
                set(jAction, 'ActionPerformedCallback', callback);
            end
            set(hButton,'enable','on','Tooltip',[actionType ' '
              actions{end}]);
        end
        jButton.getComponentPeer.getParent.revalidate;
    end  % updateUndoDropDown


    % Drop-down callback function
    function undoMenuCallbackFcn(jActionItem,hEvent,actionType)
        jPopup = jActionItem.getParent;
        for idx = 0 : jPopup.getComponentIndex(jActionItem.java)
            undoRedoCallbackFcn([],[],actionType);
        end
    end  % undoMenuCallbackFcn

    % Internal update function to update slider & editbox
    function internal_update(hSlider,newValue,hEditbox)
        % Update slider & editbox
        set(hSlider, 'value',newValue);
        set(hEditbox,'string',num2str(newValue));
        setappdata(hSlider, 'oldValue', newValue);
    end  % internal_update
```

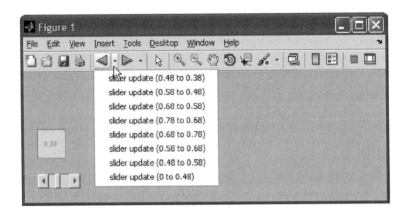

Undo/redo buttons implemented using *uisplittool* and *uiundo*

4.6 Menus

4.6.1 Accessing Menu Items

Figure menus may be accessed via hidden handles as shown below:[107]

```
% Add a context-sensitive Help menu option to the Help main menu
% Note: unlike other main menus, the Help menu tag is empty, so
% ^^^^ findall(hFig,'tag','figMenuHelp') is empty... Therefore,
%      we find this menu by accessing the Help/About menu item
helpAbout = findall(hFig,'tag','figMenuHelpAbout');
helpMenu = get(helpAbout,'parent');
csName = 'Context-sensitive help';
cbFcn = ['if strcmp(get(gcbo,''Checked''),''on''), ' ...
            'set(gcbo,''Checked'',''off''); ' ...
        'else, ' ...
            'set(gcbo,''Checked'',''on''); ' ...
        'end; ' ...
        'set(gcbf,''CSHelpMode'',get(gcbo,''checked''))'];
cshelp(hFig); % install the CSMode property and associated listeners
uimenu(helpMenu,'Label',csName,'Callback',cbFcn,'Separator','on');
```

Here is a snippet to find available menu items. Note that since some *uimenu*s have duplicate or empty tag names (probably a programming oversight), we may need to retrieve the handle using **Type**, **Label**, or another property rather than their **Tag**:[†]

```
% Find all available menu items
%hMenus = findall(hFig,'-regexp','tag','.*Menu.*'); % bad: empty tags!
>> hMenus = findall(hFig,'type','uimenu');
>> sort(get(hMenus,'tag'))
ans =
    'figMenuCameraToolbar'
```

---

† This may include some user-defined context menus as well, which should be separately filtered out.

```
    'figMenuDatatip'
    'figMenuDesktop'
    'figMenuEdit'
    'figMenuEditClear'
    'figMenuEditClearCmdHistory'
    'figMenuEditClearCmdWindow'
    'figMenuEditClearWorkspace'
    'figMenuEditColormap'
    'figMenuEditCopy'
    ...

% Find only the top-level menu items (except Help that's untagged)
% Note the absence of figMenuHelp (untagged) and odd presence of
%    figMenuDatatip and figMenuPan
>> hTopMenus = findall(hFig,'-regexp','tag','.*Menu[A-Z][a-z]*$');
>> sort(get(hTopMenus,'tag'))
ans =
    'figMenuDatatip'
    'figMenuDesktop'
    'figMenuEdit'
    'figMenuFile'
    'figMenuInsert'
    'figMenuOptions'
    'figMenuPan'
    'figMenuTools'
    'figMenuView'
    'figMenuWeb'
    'figMenuWindow'

% Find the Print menu item, which has an empty Tag:
hPrintMenuItem = get(findall(hFig,'Label','&Print...'));
```

Menu items' **Callback** property can be modified or invoked. Note that menu callbacks are kept in **Callback**, while toolbar callbacks are kept in **ClickedCallback**.

Menu callbacks generally use internal semi-documented functions (i.e., having a readable help section but no doc, online help, or official support), which are part of MATLAB's uitools folder. These functions are specific to each top-level menu tree: *filemenufcn, editmenufcn, viewmenufcn, insertmenufcn, toolsmenufcn, desktopmenufcn, winmenu,* and *helpmenufcn* implement the figure's eight respective top-level menu trees' callbacks.[108] These functions accept an optional figure handle (otherwise, *gcbf* is assumed), followed by a string specifying the specific menu item whose action needs to be run. *webmenufcn* implements the Help menu's Web Resources sub-menu callbacks in a similar manner.

Use of these **fcn* functions makes it easy to invoke a menu action directly from our MATLAB code: instead of accessing the relevant menu item and invoking its Callback, we simply find out the menu item string in advance and use it directly. For example,

```
filemenufcn FileClose;
editmenufcn(hFig,'EditPaste');
```

uimenufcn is a related fully-**un**documented (built-in) function, available since MATLAB R11 (late 1990s).[109] It accepts a figure handle (or the zero [0] handle to indicate the desktop) and action name. For example, the fully-documented *commandwindow* function uses the following code to bring the Command Window into focus:

```
uimenufcn(0, 'WindowCommandWindow');
```

A related now-useless *uitool* function, grandfathered in MATLAB 7.3, is *menubar*. Long ago, this function indicated the default figure menu bar. This is now done by simply setting the figure property **MenuBar** to 'none'.

4.6.2 Customizing Menus via Uitools

makemenu is another semi-documented *uitool* function that enables easy creation of hierarchical menu trees with separators and accelerators. It is a simple and effective wrapper for *uimenu*. *makemenu* is a useful function that has been made obsolete (grandfathered) without any known replacement.

makemenu accepts four parameters: a figure handle, a char matrix of labels ('>' indicating sub-item, '>>' indicating sub-sub-items, etc., '&' indicating keyboard shortcut, '^x' indicating an accelerator key, and '-' indicating a separator line), a char matrix of callbacks, and an optional char matrix of tags (empty by default). *makemenu* makes use of another semi-documented grandfathered function, *menulabel*, to parse the specified label components. *makemenu* returns an array of handles of the created *uimenu* items:

```
labels = str2mat('&File', ...       % File top menu
          '>&New^n', ...             % File=>New
          '>&Open', ...              % File=>Open
          '>>Open &document^d', ...    % File=>Open=>doc
          '>>Open &graph^g', ...      % File=>Open=>graph
          '>-------', ...            % File=>separator line
          '>&Save^s', ...            % File=>Save
          '&Edit', ...               % Edit top menu
          '&View', ...               % View top menu
          '>&Axis^a', ...            % View=>Axis
          '>&Selection region^r'); % View=>Selection
calls = str2mat('', ...              % no action: File top menu
          'disp(''New'')', ...
          '', ...                    % no action: Open sub-menu
          'disp(''Open doc'')', ...
          'disp(''Open graph'')', ...
          '', ...                    % no action: Separator
          'disp(''Save'')', ...
          '', ...                    % no action: Edit top menu
          '', ...                    % no action: View top menu
          'disp(''View axis'')', ...
          'disp(''View selection region'')');
handles = makemenu(hFig, labels, calls);
set(hFig,'menuBar','none');
```

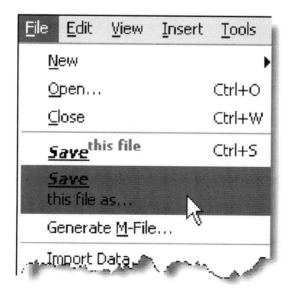

A simple figure menu

4.6.3 *Customizing Menus via HTML*

Since menu items share the same HTML/CSS support feature as all Java Swing labels, we can specify font size/face/color, bold, italic, underline, superscript/subscript, and practically any HTML formatting.

Note that some features, such as the font or foreground/background colors, have specific properties that we can set using the Java handle instead of using HTML. The benefit of using HTML is that it enables setting all the formatting in a single property. HTML does not require using Java — just pure MATLAB (see the following example).

Unlike TreeNodes (see Section 4.2.4), multi-line menu items can easily be done with HTML: simply include a <br> element in the label — the menu item splits into two lines and automatically resizes vertically when displayed (**See color insert.**):

```
txt1 = '<html><b><u><i>Save</i></u>';
txt2 = '<font color="red"><sup>this file</html>';
txt3 = '<br></b>this file as...</html>';
set(findall(hFig,'tag','figMenuFileSave'),   'Label',[txt1,txt2]);
set(findall(hFig,'tag','figMenuFileSaveAs'), 'Label',[txt1,txt3]);
```

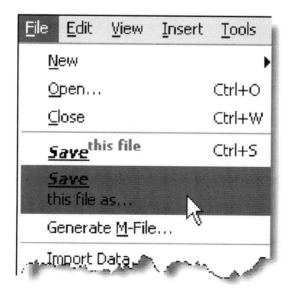

```
set(hMenuItem, 'Label',['<html>&2: C:\My Documents\doc.txt<br/>'
    '<font size="-1" face="Courier New" color="red">  '
    'Date: 15-Jun-2011 13:23:45<br/>   Size: 123 KB']);
```

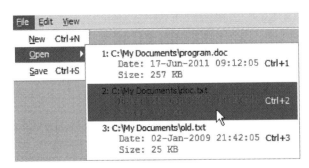

HTML-rendered menus

4.6.4 Customizing Menus via Java

MATLAB menus (*uimenu*) are basically simple wrappers for the much more powerful and flexible Java Swing JMenu and JMenuItem on which they are based.[110] Many important functionalities that are available in Java menus[111] are missing from the MATLAB *uimenu*s. For example, let us add DHTML-like behavior to the menu, such that the menu items will automatically be displayed when the mouse hovers over the item, without waiting for a user mouse click. First, let us get the Java reference for the menu:

```
jFrame = get(handle(hFig),'JavaFrame');
jMenuBar = jFrame.fHG1Client.getMenuBar;†
```

Now, set the **MouseEnteredCallback** to automatically simulate a user mouse click on each menu item using its *doClick()* method. Note that the callback is set on the reference's *handle()* wrapper, as explained in Section 3.4. Setting the callback should be done separately to each of the top-level menu components:

```
for menuIdx = 1 : jMenuBar.getComponentCount
    jMenu = jMenuBar.getComponent(menuIdx-1);
    hjMenu = handle(jMenu,'CallbackProperties');
    set(hjMenu,'MouseEnteredCallback','doClick(gcbo)');
end
```

As another example, MATLAB automatically assigns a nonmodifiable keyboard accelerator key modifier of <Ctrl>, while JMenus allow any combination of Alt/Ctrl/Shift/Meta (depending on the platform). Let us modify the default File/Save accelerator key from 'Ctrl-S' to 'Alt-Shift-S' as an example. We need a reference for the "Save" menu item. Note that unlike regular Java components, menu items are retrieved using the *getMenuComponent()* method and not *getComponent()*:

```
% File main menu is the first main menu item => index = 0
jFileMenu = jMenuBar.getComponent(0);
```

---

† In R2007b and earlier, use fFigureClient rather than fHG1Client.

```
% Save menu item is the 5th menu item (separators included)
jSave = jFileMenu.getMenuComponent(4); %Java indexes start with 0!
inspect(jSave)      => just to be sure: label = 'Save' => good!
```

Finally, set a new accelerator key for this menu item:

```
% Set a new accelerator key
jAccelerator = javax.swing.KeyStroke.getKeyStroke('alt shift S');
jSave.setAccelerator(jAccelerator);
```

That is all there is to it — the label is modified automatically to reflect the new keyboard accelerator key. More info on setting different combinations of accelerator keys and modifiers can be found on the official Java documentation for KeyStroke.[112]

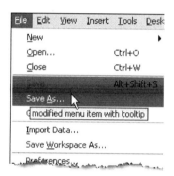

Modification of menu-item accelerators and tooltips (See color insert.)

Note that the Save menu-item reference can only be retrieved after opening the File menu at least once earlier; otherwise, an exception will be thrown when trying to access the menu item. The File menu does NOT need to remain open — it only needs to have been opened sometime earlier, for its menu items to be rendered. This can be done either interactively (by selecting the File menu) or programmatically:

```
% Simulate mouse clicks to force the File main-menu to open & close
jFileMenu.doClick;  % open the File menu
jFileMenu.doClick;  % close the menu

% Now the Save menu is accessible:
jSave = jFileMenu.getMenuComponent(4);
```

There are many customizations that can only be done using the Java handle: setting icons, several dozen callback types, tooltips, background color, font, text alignment, and so on. Interested readers may wish to ***get/set/inspect/methodsview/uiinspect*** the jSave reference handle and/or to read the documentation for JMenuItem.[113]

Let us now set an icon for some menu items. Many of MATLAB's icons reside in either the [matlabroot '/toolbox/matlab/icons/'] folder or the [matlabroot '/java/jar/*mwt.jar*'] file (a JAR file is simply a zip file that includes Java classes and resources such as icon images). Let us create icons from the latter to keep a consistent look-and-feel with the rest of MATLAB (we could just as easily use our own external icon files):

```
% External icon file example
jSave.setIcon(javax.swing.ImageIcon('C:\Yair\save.gif'));

% JAR resource example
jarFile = fullfile(matlabroot,'/java/jar/mwt.jar');
iconsFolder = '/com/mathworks/mwt/resources/';
iconURI = ['jar:file:/' jarFile '!' iconsFolder 'save.gif'];
iconURI = java.net.URL(iconURI); % not necessary for external files
jSave.setIcon(javax.swing.ImageIcon(iconURI));
```

Note that setting a menu item's icon automatically re-aligns all other items in the menu, including those that do not have an icon (an internal bug that was introduced in R2010a causes a misalignment, as shown below).

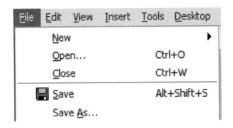

Menu item with a custom Icon (R2009b)   **... and the same in R2010a onward**

The empty space on the left of the menu is reserved for the checkmark. Each MATLAB menu item is check-able, since it is an object that extends the `com.mathworks.mwswing.MJCheckBoxMenuItem` class. I have not found a way to eliminate this empty space, which is really unnecessary in the File-menu case (it is only actually necessary in the View and Tools menus).

Icons can be customized: modify the gap between the icon and the label with the **IconTextGap** property (default = 4 [pixels]); place icons to the right of the label by setting **HorizontalTextPosition** to `jSave.LEFT` (=2), or centered using `jSave.CENTER` (=0). Note that the above-mentioned misalignment bug does not appear in these cases:

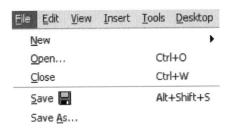
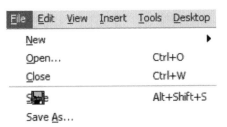

`jSave.`*setHorizontalTextPosition*`(jSave.LEFT)`   `jSave.`*setHorizontalTextPosition*`(jSave.`
CENTER)

Note how the label text can be seen through (or on top of) the icon when it is centered. This feature can be used to create stunning menu effects as shown below. Note how the width and height of the menu item automatically increased to accommodate my new 77×31 icon size (icons are normally sized 16×16 pixels):

Overlaid icon (HorizontalTextPosition = CENTER)

To resize an icon programmatically before setting it in a Java component, we can use the following example:

```
myIcon = fullfile(matlabroot,'/toolbox/matlab/icons/matlabicon.gif');
imageToolkit = java.awt.Toolkit.getDefaultToolkit;
iconImage = imageToolkit.createImage(myIcon);
iconImage = iconImage.getScaledInstance(32,32,iconImage.SCALE_SMOOTH);
jSave.setIcon(javax.swing.ImageIcon(iconImage));
```

Remember when rescaling images, particularly small ones with few pixels, that it is always better to shrink than to enlarge images: enlarging a small icon image might introduce a significant pixelization effect:

16×16 icon image resized to 32×32

Separate icons can be specified for a different appearance during mouse hover (**RolloverIcon**; requires **RolloverEnabled** = 1), item click/press (*PressedIcon*), item selection (**SelectedIcon, RolloverSelectedIcon, DisabledSelectedIcon**), and disabled menu item (**DisabledIcon**). All these properties are empty ([]) by default, which applies a predefined default variation (image color filter) to the main item's Icon. For example, let us modify **DisabledIcon**:

```
myIcon = 'C:\Yair\Undocumented Matlab\Images\save_disabled.gif';
jSaveAs.setDisabledIcon(javax.swing.ImageIcon(myIcon));
jSaveAs.setEnabled(false);
```

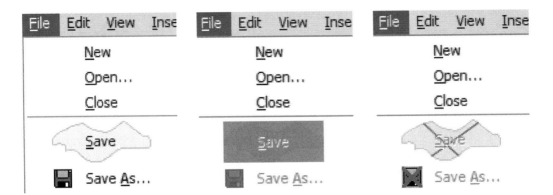

Enabled, main Icon   Disabled, default Icon variation Disabled, custom DisabledIcon

Note the automatic graying of disabled menu items, including their icon.† When we use a non-default custom **DisabledIcon**, it is used instead of the gray icon variant.

Several additional `JMenu` and `JMenuItem` properties could be useful in applications:

- **Armed** — a flag (on/off, default = *off*); when it is *on*, the menu item is highlighted just as when it is selected (see the save as... item in the screenshot above). On a Windows system, this means a blue background. When the item is actually selected and then de-selected, **Armed** reverts to *off* value.

- **State** — a flag (on/off, default = *off*); when it is *on*, the menu item is indicated with an attached checkmark. This property corresponds to the MATLAB handle's **Checked** property. Note that if an icon is set for the item, both the icon and the checkmark will be displayed, side by side:

```
set(findall(hFig,'tag','figMenuFileSave'), 'Checked','on');
jSave.setState(true);              % this is equivalent
```

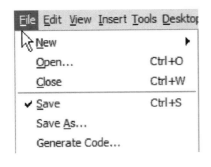

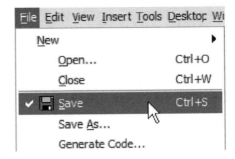

State = true, Icon = []   **State = true, Icon = custom**

---

† This effect can also be achieved programmatically using the static methods in `com.mathworks.mwswing.` `IconUtils`: changeIconColor(), createBadgedIcon(), createGhostedIcon(), and createSelectedIcon().

- **ToolTipText** — for some unknown reason, MathWorks failed to include this useful property in its MATLAB menu handle, so we must use the Java handle:

```
jSave.setToolTipText('Save this figure...');
```

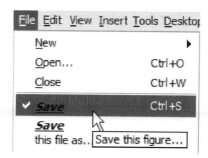

Modified menu-item tooltip

In addition to the standard Swing control callbacks discussed in Chapter 3, menu items possess several additional callbacks, including

- **ActionPerformedCallback** — fired when the menu item is invoked
- **StateChangedCallback** — fired when the menu item is selected or deselected
- **MenuDragMouseXXXCallback**   (XXX = Dragged/Entered/Exited/Released)   — fired when the menu item is dragged for the corresponding event
- **MenuKeyXXXCallback** (XXX = Pressed/Released/Typed) — fired when a keyboard click event occurs (the menu item's accelerator was typed)

4.7 Status Bar

getstatus and *setstatus* appear to be early attempts made by MATLAB to enable users an access to a figure's status bar (the text bar at the bottom of the figure). In these early attempts, MATLAB assumes that the user prepares a text label having a tag of 'status'. *getstatus* then returns this label's string, while *setstatus* modifies it:

```
uicontrol('Parent',hFig,'Style','text','Tag','Status');
setstatus(hFig, 'Goodbye');
string = getstatus(hFig);
```

Nothing prevents the user from placing the label anywhere in the figure, and also from having multiple such labels at once, adding to the confusion. The result is inconsistent with normal windowing practices, and this is probably the reason that MathWorks have grandfathered these functions in MATLAB 7.4 (R2007a). It would be much more logical for MATLAB to have the status bar accessible via a figure property, and perhaps this will happen in some future version.

A better, consistent, and more flexible access to the figure (and desktop) status bar can be achieved by using some undocumented Java functions:[114]

```
jFrame = get(handle(hFig),'JavaFrame');
jRootPane = jFrame.fHG1Client.getWindow;'
statusbarObj = com.mathworks.mwswing.MJStatusBar;
jRootPane.setStatusBar(statusbarObj);
statusbarObj.setText('please wait - processing...');
```

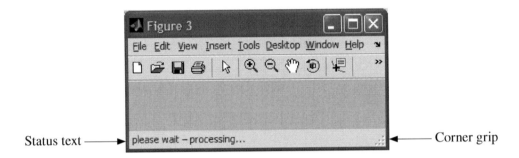

A simple status bar with corner grip

Here, `statusbarObj` is actually a component contained within a parent container (a `com.mathworks.mwswing.MJPanel` object). When `statusbarObj` is created and attached to the figure via the frame's root pane's *setStatusBar()* method, it is this parent container which is actually attached, and the `statusbarObj` is then added to it. In addition to `statusbarObj` (which includes an internal `JPanel`‡ to display the text), the parent also contains a corner grip (`com.mathworks.mwswing.MJCornerGrip`) object:

```
>> statusbarObj.getParent.list

com.mathworks.mwswing.MJPanel[,0,81,291x20,layout=java.awt.Border...]
  com.mathworks.mwswing.MJCornerGrip[,279,0,12x20,alignment...]
  com.mathworks.mwswing.MJStatusBar[,0,0,279x20,layout=...]
    com.mathworks.mwswing.MJStatusBar$1[,0,2,277x16,...,text=please wait -
processing...,verticalAlignment=CENTER,verticalTextPosition=CENTER]
```

Each of these status bar components can be accessed and customized separately by traversing the component hierarchy tree. For example, let us hide the corner grip and set a colored text background (note the alternative ways of specifying colors):

```
% Hide the corner grip
cornerGrip = statusbarObj.getParent.getComponent(0);
cornerGrip.setVisible(false);       % or: (cornerGrip,'Visible','off')

% Set a red foreground & yellow background to status bar text
statusbarTxt = statusbarObj.getComponent(0);
```

---

† In R2007b and earlier, use `fFigureClient` rather than `fHG1Client`.

‡ Actually, an inner class of type `com.mathworks.mwswing.MJStatusBar$1`.

```
statusbarTxt.setForeground(java.awt.Color.red);
set(statusbarTxt,'Background','yellow');
set(statusbarTxt,'Background',[1,1,0]); % an alternative...
```

A colored status bar with no corner grip

Note that the status bar is not an HG object, and cannot be accessed via *findobj*, *findall*, or the figure's HG hierarchy. Similarly, *clf* does not delete the status bar. In short, it is not a regular MATLAB handle.

Also, note that the status bar is 20 pixels high across the entire bottom of the figure. It hides everything between pixel heights 0–20, even parts of uicontrols, regardless of which was created first or the relative **ComponentZOrder** in the frame's **ContentPane**:

```
uicontrol('string','click me!', 'position',[10,10,70,25]);
```

Status bars overlap all other figure components in pixel heights 0–20

The `statusbarObj` object is simply a container. As such, we can add any other Swing component or container to it. Let us decorate our status bar with a progress bar:[115]

```
jProgressBar = javax.swing.JProgressBar;
set(jProgressBar, 'Value',73);        % value in % [0-100]
jProgressBar.setValue(73); % an alternative...
statusbarObj.add(jProgressBar,'West');
statusbarObj.revalidate;
```

A status bar with progress bar (West)   **...and East**

Note that if the window resizes to a width smaller than that required to display the status text, the text is automatically cropped:

Effects of resizing

As with other objects in MATLAB and Java, the progress bar object can be customized with min/max values and other internal properties. A very useful property is **StringPainted** — when this property is set to 'on' (default = 'off'), the progress bar appears continuous with the internal value displayed:[†]

```
jProgressBar.setStringPainted(true);
```

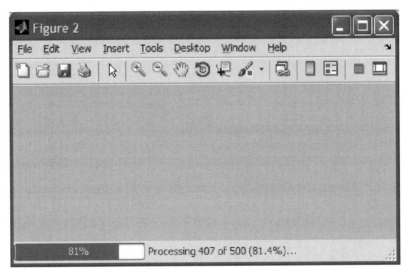

Status bar with a continuous progress bar using the StringPainted property

---

[†] Additional customizations of progress bars are discussed in detail in Section 3.3.1; additional progress bars and busy indications are presented in Sections 5.4 and 5.5.1.

In order to update the desktop's status bar (as opposed to a figure's status bar), we need to set up a delayed-activation timer. The reason for this is that during regular MATLAB processing, the desktop's status bar text is modified to display "Busy" and this will immediately override any status bar update that we code. The delayed-activation timer solves this by updating the desktop's status bar after the MATLAB command returns to the Command Prompt to await further commands:

```
%% Set the status bar text of the Matlab desktop
function setDesktopStatus(statusText)
    % Set the desktop status (differently for Matlab 6 & Matlab 7)'
    try
        % Matlab 7
        desktop = com.mathworks.mde.desk.MLDesktop.getInstance;
    catch
        % Matlab 6
        desktop = com.mathworks.ide.desktop.MLDesktop.getMLDesktop;
    end
    % Schedule a timer to update the status text, because an
    % immediate update will be overridden by Matlab's 'busy' message
    try
        start(timer('Name','statusbarTimer', ...
                    'TimerFcn',{@setText,desktop,statusText}, ...
                    'StartDelay',0.05, ...
                    'ExecutionMode','singleShot'));
    catch
        % Probably an old Matlab version: still doesn't have timer
        desktop.setStatusText(statusText);
    end
end   % setDesktopStatus

%% Utility function used as internal timer's callback
function setText(hSrc,hEvent,targetObj,statusText)
    targetObj.setStatusText(statusText);
end   % setText
```

I have created a wrapper function, aptly called *statusbar*, which encapsulates all the above with some additional error checking, MATLAB version compatibility checks, and so on. The *statusbar* utility file is posted on the MathWorks File Exchange and is available for free download at http://www.mathworks.com/matlabcentral/fileexchange/14773 (or http://tinyurl.com/akcwf9). Readers are encouraged to look at *statusbar*'s source code for examples of statusbar manipulation.

---

† See a discussion of desktop references in Chapter 8 (The MATLAB Desktop).

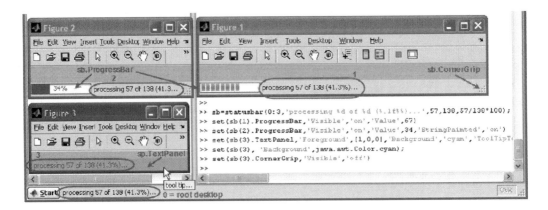

statusbar usage examples (**See color insert.**)

References

1. http://www.mathworks.com/matlabcentral/fileexchange/10045 (or http://tinyurl.com/2agvb3), http://www.mathworks.com/matlabcentral/fileexchange/11201 (or http://tinyurl.com/ywnlm6).
2. http://www.mathworks.com/matlabcentral/fileexchange/6734 (or http://tinyurl.com/2y9d97).
3. http://www.mathworks.com/matlabcentral/fileexchange/10782 (or http://tinyurl.com/2ar4vv).
4. http://www.mathworks.com/matlabcentral/fileexchange/6889 (or http://tinyurl.com/368mw2), http://www.mathworks.com/matlabcentral/fileexchange/3477 (or http://tinyurl.com/35nf5m).
5. http://www.mathworks.com/matlabcentral/fileexchange/5752 (or http://tinyurl.com/2xvx53).
6. http://www.mathworks.com/matlabcentral/fileexchange/15372 (or http://tinyurl.com/2nzzby).
7. http://www.mathworks.com/matlabcentral/fileexchange/1113 (or http://tinyurl.com/2omrak), http://www.mathworks.com/matlabcentral/fileexchange/7026 (or http://tinyurl.com/3yl2nn).
8. http://java.sun.com/docs/books/tutorial/uiswing/components/table.html (or http://tinyurl.com/29n77).
9. http://blogs.mathworks.com/desktop/2008/06/02/tables-in-matlab-with-uitable/ (or http://bit.ly/ac0EW6).
10. http://java.sun.com/j2se/1.4.2/docs/api/javax/swing/table/DefaultTableModel.html (or http://tinyurl.com/ypql9n).
11. The role of the Model in Swing components is explained at http://java.sun.com/products/jfc/tsc/articles/architecture/ (or http://tinyurl.com/atggc).
12. http://java.sun.com/docs/books/tutorial/uiswing/components/html.html (or http://tinyurl.com/5v38m).
13. http://www.mathworks.com/matlabcentral/answers/1132-display-an-icon-in-png-format-in-a-cell-of-a-uitable (or http://bit.ly/pECRz4).
14. http://www.mathworks.com/matlabcentral/newsreader/view_thread/239788 (or http://bit.ly/dFRQ7b).
15. http://www.mathworks.com/matlabcentral/newsreader/search_results?search_string=uitable&dur=all (http://bit.ly/bBYljW).
16. http://www.mathworks.com/matlabcentral/newsreader/view_thread/147554 (or http://tinyurl.com/ywgndk).
17. http://java.sun.com/docs/books/tutorial/uiswing/components/table.html#renderer (or http://tinyurl.com/3x24op).
18. http://java.sun.com/products/jfc/tsc/articles/architecture/ (or http://bit.ly/9G7AMZ).
19. http://en.wikipedia.org/wiki/Model%E2%80%93view%E2%80%93controller (or http://bit.ly/d7QY5o).
20. http://en.wikipedia.org/wiki/Delegation_pattern (or http://bit.ly/da7Oa4).
21. http://java.sun.com/javase/6/docs/api/javax/swing/table/DefaultTableCellRenderer.html (or http://tinyurl.com/2aue8y).
22. http://www.mathworks.com/matlabcentral/newsreader/view_thread/153934 (or http://tinyurl.com/2t7ljr).
23. http://www.mathworks.com/matlabcentral/newsreader/view_thread/150507 (or http://bit.ly/brXusn); download a zip file with source and class files from: http://UndocumentedMatlab.com/files/ColoredFieldCellRenderer.zip (or http://bit.ly/dpMVhI).

24. http://java.sun.com/docs/books/tutorial/uiswing/components/table.html#editrender (or http://tinyurl.com/4kvy8).
25. http://java.sun.com/javase/6/docs/api/javax/swing/DefaultCellEditor.html (or http://tinyurl.com/2ry2xa).
26. http://java.sun.com/docs/books/tutorial/uiswing/components/table.html#combobox (or http://tinyurl.com/2j8h5m).
27. http://UndocumentedMatlab.com/files/LookupFieldCellEditor.zip (or http://bit.ly/aiHumG).
28. http://UndocumentedMatlab.com/files/ColorCell.zip (or http://bit.ly/ddmqsa), which contains both the source and the class files for the `ColorCellRenderer` and the `ColorCellEditor` classes.
29. http://www.mathworks.com/matlabcentral/newsreader/view_thread/150532 (or http://tinyurl.com/a8kh2p).
30. http://www.mathworks.com/matlabcentral/newsreader/view_thread/164149 (or http://tinyurl.com/darxm2).
31. http://www.mathworks.com/matlabcentral/newsreader/view_thread/235046 (or http://tinyurl.com/4k7qcc).
32. http://java.sun.com/docs/books/tutorial/uiswing/components/scrollpane.html#scrollbars (or http://tinyurl.com/22wvy5).
33. http://java.sun.com/javase/6/docs/api/javax/swing/JTable.html#field_summary (or http://tinyurl.com/24oroo).
34. http://java.sun.com/javase/6/docs/api/javax/swing/ListSelectionModel.html (or http://tinyurl.com/38b2tk).
35. http://www.mathworks.com/matlabcentral/newsreader/view_thread/268448#702699 (or http://bit.ly/fecP2c).
36. http://www.mathworks.com/matlabcentral/fileexchange/14225 (or http://tinyurl.com/24kyjr).
37. http://coderazzi.net/tablefilter (or: http://bit.ly/lRqhgL); http://code.google.com/p/tablefilter-swing/ (or: http://bit.ly/iUrzKW).
38. http://www.coderazzi.net/tablefilter/download.html (or: http://bit.ly/kyh4ZD); http://repo2.maven.org/maven2/net/coderazzi/tablefilter-swing/ (or: http://bit.ly/jZUC2C).
39. http://www.jidesoft.com/javadoc/com/jidesoft/grid/SortableTable.html (or http://bit.ly/aOyv6U).
40. http://www.jidesoft.com/javadoc/ (or http://bit.ly/bcevRu).
41. http://www.jidesoft.com/products/JIDE_Grids_Developer_Guide.pdf (or http://bit.ly/a88Xzt).
42. http://www.jidesoft.com/javadoc/com/jidesoft/grid/CellSpanTable.html (or http://bit.ly/bo3sGT).
43. http://www.jidesoft.com/javadoc/com/jidesoft/grid/DefaultSpanTableModel.html (or http://bit.ly/dwQSIV).
44. http://www.jidesoft.com/javadoc/com/jidesoft/grid/CellSpan.html (http://bit.ly/9u0MdX).
45. http://bugs.sun.com/bugdatabase/view_bug.do;:WuuT?bug_id=4709394 (or http://tinyurl.com/2z79od).
46. http://xtargets.com/snippets/posts/show/37 (currently offline — use this cached version: http://bit.ly/amtwcL); also see http://www.mathworks.com/matlabcentral/newsreader/view_thread/284958#755942 (or http://bit.ly/dBvayJ).
47. http://www.mathworks.com/matlabcentral/fileexchange/14225 (or http://tinyurl.com/24kyjr).
48. http://java.sun.com/docs/books/tutorial/uiswing/components/tree.html (or http://tinyurl.com/937xd); http://www.java2s.com/Code/Java/Swing-JFC/Tree.htm (or http://tinyurl.com/clr7mh).
49. http://www.mathworks.com/matlabcentral/newsreader/search_results?search_string=uitree&dur=all (or http://bit.ly/dv3XzJ).
50. http://java.sun.com/javase/6/docs/api/javax/swing/JTree.html (or http://tinyurl.com/2nv3nx).
51. http://java.sun.com/javase/6/docs/api/javax/swing/tree/DefaultMutableTreeNode.html (or http://tinyurl.com/2fpd9c).
52. http://java.sun.com/javase/6/docs/api/javax/swing/tree/DefaultMutableTreeNode.html (or http://tinyurl.com/2fpd9c).
53. http://java.sun.com/docs/books/tutorial/uiswing/events/treeselectionlistener.html (or http://tinyurl.com/yqpvt9).
54. http://java.sun.com/javase/6/docs/api/javax/swing/tree/DefaultTreeModel.html (or http://tinyurl.com/2x9y3a).
55. http://java.sun.com/docs/books/tutorial/uiswing/components/tree.html#data (or http://tinyurl.com/2cruyh).
56. http://www.mathworks.com/matlabcentral/newsreader/view_thread/104957#269485 (or http://tinyurl.com/yt4y2k).
57. http://www.mathworks.com/matlabcentral/newsreader/view_thread/138971 (or http://tinyurl.com/2zjq9f).
58. http://www.mathworks.com/matlabcentral/newsreader/view_thread/164189#417341 (orhttp://bit.ly/4LhgrP) for an example of programmatically creating checked and unchecked node icon images.
59. http://java.sun.com/docs/books/tutorial/uiswing/components/tree.html#display (or http://tinyurl.com/2e766m).

60. http://java.sun.com/javase/6/docs/api/javax/swing/tree/DefaultTreeCellRenderer.html (or http://tinyurl.com/yttzjp).
61. http://java.sun.com/javase/6/docs/api/javax/swing/tree/DefaultTreeCellEditor.html (or http://tinyurl.com/2ar3vg).
62. Brad Phelan suggested a similar approach at http://xtargets.com/cms/Tutorials/Matlab-Programming/Adding-Popups-To-The-Tree-Control.html (currently offline — cached version: http://bit.ly/9pQ3Av). Also see: http://UndocumentedMatlab.com/blog/adding-context-menu-to-uitree/ (or http://bit.ly/b3vRjI).
63. http://www.mathworks.com/matlabcentral/fileexchange/14317 (or http://tinyurl.com/bnprwc).
64. http://UndocumentedMatlab.com/blog/tabbed-panes-uitab-and-relatives/ (or http://bit.ly/9Mfu3x).
65. http://java.sun.com/docs/books/tutorial/uiswing/components/tabbedpane.html (or http://tinyurl.com/ad2lv).
66. There are several non-Java tab-panel implementations on the File Exchange, for example, http://www.mathworks.com/matlabcentral/fileexchange/1741, http://www.mathworks.com/matlabcentral/fileexchange/11546, http://www.mathworks.com/matlabcentral/fileexchange/2852, http://www.mathworks.com/matlabcentral/fileexchange/4780, and http://www.mathworks.com/matlabcentral/fileexchange/6996. Some of these are highly elaborate and excellent submissions, but unfortunately they still suffer from the same limitation regarding incorporation of axes/plots/images.
67. http://UndocumentedMatlab.com/blog/uitab-customizations/ (or http://bit.ly/bEfGgJ).
68. http://www.w3schools.com/css/css_background.asp (or http://bit.ly/cYByfm).
69. http://UndocumentedMatlab.com/blog/uitab-colors-icons-images/ (or http://bit.ly/bi5qoj).
70. http://java.sun.com/javase/6/docs/api/javax/swing/JTabbedPane.html (or http://tinyurl.com/2662ue).
71. http://java.sun.com/j2se/1.4.2/docs/guide/swing/1.4/tabbedPane.html (or http://bit.ly/dcKKyp).
72. http://www.mathworks.com/matlabcentral/newsreader/view_thread/158711 (or http://tinyurl.com/37qw4t).
73. http://www.mathworks.com/matlabcentral/newsreader/view_thread/156065#392032 (or http://tinyurl.com/3b6746).
74. http://www.mathworks.com/matlabcentral/newsreader/view_thread/235274#597793 (or http://tinyurl.com/48o4lk).
75. http://www.mathworks.com/matlabcentral/newsreader/view_thread/130100 (or http://tinyurl.com/3dgwar).
76. http://www.mathworks.com/matlabcentral/newsreader/view_thread/110949 (or http://tinyurl.com/2o2j8e); http://www.mathworks.com/matlabcentral/newsreader/view_thread/268477 (or http://bit.ly/7q0M1D).
77. http://UndocumentedMatlab.com/blog/common-javacomponent-problems (or: http://bit.ly/q8o2vJ); http://www.mathworks.com/matlabcentral/newsreader/view_thread/162430 (or http://tinyurl.com/22sstq), http://www.mathworks.com/matlabcentral/newsreader/view_thread/235274#597813 (or http://tinyurl.com/4hmzqq).
78. http://www.mathworks.com/matlabcentral/fileexchange/20218-yet-another-layout-manager (or http://bit.ly/cMHrP3).
79. http://www.mathworks.com/matlabcentral/fileexchange/?term=tab+gui (or http://bit.ly/cXPvWo).
80. http://blogs.mathworks.com/desktop/2010/02/15/putting-the-tab-into-a-gui/ (or http://bit.ly/bUCK5c); an older (2002) tab-panel implementation was also Pick-of-the-Week: *Tab panel example* (http://www.mathworks.com/matlabcentral/fileexchange/1741-tab-panel-example or http://bit.ly/dbS3qQ).
81. http://www.mathworks.com/matlabcentral/fileexchange/25938-multiple-tab-gui (or http://bit.ly/8ZrMMU).
82. http://www.mathworks.com/matlabcentral/fileexchange/22488-highlight-tab-objects-easily (or http://bit.ly/arlFKD).
83. http://www.mathworks.com/matlabcentral/fileexchange/11546-uitabpanel (or http://bit.ly/dm8zFL).
84. http://www.mathworks.com/matlabcentral/fileexchange/6996-tabpanel-constructor (or http://bit.ly/a4p5k6).
85. http://www.mathworks.com/matlabcentral/fileexchange/27758-gui-layout-toolbox (or http://bit.ly/cpoTAI).
86. http://UndocumentedMatlab.com/blog/uiundo-matlab-undocumented-undo-redo-manager/ (or http://tinyurl.com/yhzm9xy).
87. http://java.sun.com/javase/6/docs/api/javax/swing/undo/UndoManager.html (or http://bit.ly/8X1yXA); http://www.javaworld.com/javaworld/jw-06-1998/jw-06-undoredo.html (or http://bit.ly/b5ltso); http://www.java2s.com/Code/Java/Swing-JFC/Undomanager.htm (or http://bit.ly/aLkJK5).
88. http://www.mathworks.com/matlabcentral/fileexchange/28322-GUIHistory (or http://bit.ly/9FuQtE).
89. http://www.mathworks.com/matlabcentral/newsreader/view_thread/21676 (or http://tinyurl.com/2g75vw).

90. http://UndocumentedMatlab.com/blog/modifying-default-toolbar-menubar-actions/#Item-handles (or http://bit.ly/9KBREq).
91. http://www.mathworks.com/matlabcentral/newsreader/view_thread/33887 (or http://tinyurl.com/yoyogr).
92. http://www.mathworks.com/matlabcentral/newsreader/view_thread/35735 (or http://tinyurl.com/2477ct).
93. http://www.mathworks.com/help/techdoc/ref/uitoggletool_props.html (or http://tinyurl.com/2bygbk).
94. http://UndocumentedMatlab.com/blog/modifying-default-toolbar-menubar-actions/#Print (or http://bit.ly/bsgjSi).
95. http://www.mathworks.com/matlabcentral/newsreader/view_thread/286235 (or http://bit.ly/aEVP63).
96. First reported online in http://www.mathworks.com/matlabcentral/newsreader/view_thread/81390 (or http://bit.ly/cT1fpD).
97. http://www.mathworks.com/matlabcentral/newsreader/search_results?search_string=btngroup&page=1&dur=all (or http://bit.ly/94nWV5); http://groups.google.com.by/group/comp.soft-sys.matlab/search?q=btngroup (or http://bit.ly/azQ8G4).
98. http://java.sun.com/javase/6/docs/api/javax/swing/JToolBar.html (or http://bit.ly/d9Lm2d).
99. http://www.mathworks.com/matlabcentral/newsreader/view_thread/163829 (or http://tinyurl.com/2enfpo).
100. http://UndocumentedMatlab.com/blog/figure-toolbar-components/#comment-4342 (or http://tinyurl.com/yb25qhm).
101. http://java.sun.com/docs/books/tutorial/uiswing/components/border.html (or http://tinyurl.com/3a2n6e).
102. http://UndocumentedMatlab.com/blog/figure-toolbar-customizations/ (or http://tinyurl.com/yfnxbxz).
103. http://java.sun.com/docs/books/tutorial/uiswing/components/toolbar.html (or http://bit.ly/bSJTSY).
104. http://java.sun.com/docs/books/tutorial/uiswing/components/menu.html (http://bit.ly/cMm0Ke) and http://java.sun.com/javase/6/docs/api/javax/swing/JPopupMenu.html (http://bit.ly/bCiQCC).
105. http://UndocumentedMatlab.com/blog/figure-toolbar-components/ (or http://tinyurl.com/yfxkp43).
106. http://UndocumentedMatlab.com/blog/figure-toolbar-components/#comment-13897 (or http://bit.ly/dcShDr).
107. http://UndocumentedMatlab.com/blog/modifying-default-toolbar-menubar-actions/#Item-handles (or http://bit.ly/9KBREq).
108. http://www.mathworks.com/matlabcentral/newsreader/view_thread/27281#67933 (or http://tinyurl.com/n7at7p).
109. http://www.mathworks.com/matlabcentral/newsreader/view_thread/12736#27281 (or http://tinyurl.com/np6rf5).
110. http://www.mathworks.com/matlabcentral/newsreader/view_thread/282197 (or http://bit.ly/b73IFp).
111. http://java.sun.com/docs/books/tutorial/uiswing/components/menu.html (or http://tinyurl.com/dsxgl); or in www.java2s.com: http://www.java2s.com/Tutorial/Java/0240_Swing/0400_JMenuBar.htm (or http://tinyurl.com/c4k43v), http://www.java2s.com/Tutorial/Java/0240_Swing/0380_JMenu.htm (or http://tinyurl.com/cjk7wg) and http://www.java2s.com/Tutorial/Java/0240_Swing/0420_JMenuItem.htm (or http://tinyurl.com/dacdce).
112. http://java.sun.com/javase/6/docs/api/javax/swing/KeyStroke.html#getKeyStroke(java.lang.String) (or http://bit.ly/9Iw0Ti).
113. http://java.sun.com/javase/6/docs/api/javax/swing/JMenuItem.html (or http://tinyurl.com/2xe3lj).
114. http://UndocumentedMatlab.com/blog/setting-status-bar-text/ (or http://tinyurl.com/yjqgv52).
115. http://UndocumentedMatlab.com/blog/setting-status-bar-components/ (or http://tinyurl.com/ygmrkom).

Chapter 5

Built-In MATLAB® Widgets and Java Classes

5.1 Internal MATLAB Java Packages

MATLAB has several Java packages (com.mathworks.*) which contain Swing-extended components and resources (icon images), which are used internally by the MATLAB application and can also be used in user applications. All these components reside in packages that have the com.mathworks domain prefix. The packages are grouped based on utility and topic and are all located in the %matlabroot%/java/jar/ folder in separate JAR (Java archive zip) files.† These Java archive files are all included in the static Java classpath, within the *classpath.txt* file. This means that all internal MATLAB classes are immediately accessible from within our MATLAB m-code.

 Important warning: All classes included in MATLAB packages, their internal functionalities, and even the packages themselves and the location of their containing JAR files, are entirely unsupported and are prone to change without any warning between MATLAB releases. Extensive use of exception handling must, therefore, be done in any code which uses these classes/packages. The description in this chapter focuses on release R2008a (7.6); other releases may well vary.

Note that in all the ensuing discussions, MATLAB's internal classes are not hacked or decompiled — an act which could violate MATLAB's license agreement. Instead, we shall only use commands available from within MATLAB or snippets copied from openly accessible MATLAB m-code files.

When asked about these packages, Mike Katz said the following on the official MATLAB Desktop blog (http://blogs.mathworks.com/desktop):[1]

"com.mathworks. represent Java classes that we've built here. These mostly represent the libraries and widgets used by the Desktop and toolboxes, and as such are not considered part of the MATLAB language and are largely [un]documented. Occasionally we mention some of these in technical support solutions as workarounds to bug fixes. Some of the classes are documented for use for xml, web, and database I/O, and a whole bunch are meant for use in with the MATLAB Builder JA product for deploying MATLAB code to work with Java programs."*

5.1.1 Inspecting Package Contents

Since jar files are simple zip files, they can be inspected using any zip client application, including MATLAB's standard ***unzip*** command.‡ For example,

```
zipFilename = fullfile(matlabroot, '/java/jar/mwswing.jar');
classFilenames = unzip(zipFilename, outputFolder);
```

This will spill all of *mwswing.jar*'s files (which are the requested classes and some static resources such as icons) into a tree structure in outputFolder (i.e., outputFolder/com/mathworks/mwswing/...) and return a cell-array of the file paths.

---

† For example, C:\Program Files\MATLAB\R2008b\java\jar\mwswing.jar.

‡ Matlab's built-in ***unzip.m*** itself uses standard Java zip-processing classes and methods.

However, although it may help our understanding, we do not really need the physical class files. We only need to know <u>which</u> files are available in the mwswing package. So, let us use only a small segment of ***unzip.m***:

```
>> zipJavaFile = java.io.File(zipFilename);
>> zipFile = org.apache.tools.zip.ZipFile(zipJavaFile);

>> files = zipFile.getEntries
files =
java.util.Hashtable$Enumerator@e4bf0e

>> while files.hasMoreElements, disp(files.nextElement.getName); end
com/mathworks/mwswing/MJMenu$1.class
com/mathworks/mwswing/MJOptionPane$ButtonAction.class
com/mathworks/mwswing/MJMultilineRadioButton.class
com/mathworks/mwswing/MJToolBar$Gap.class
com/mathworks/mwswing/MJMenu$MWindowsMenuUI.class
com/mathworks/mwswing/ScrollablePopupList$PopupDismisser.class
com/mathworks/mwswing/SelectAllOnFocusListener$1$1.class
com/mathworks/mwswing/MJScrollStrip.class
com/mathworks/mwswing/dialog/MJGotoDialog.class
. . .
com/mathworks/mwswing/MJToggleButton.class
com/mathworks/mwswing/MJSlider.class
com/mathworks/mwswing/MJSpinner.class
. . .
```

Remember to close the zipFile after reading is done, in order to free resources (this is done by ***unzip.m*** in its cleanup phase):

```
>> zipFile.close(); % close file even if exception happens!
```

Note that the reverse operation, namely finding the source JAR of a specific Java object, can be obtained using standard Java reflection. For example, let us find the JAR source of the Java object that underlies a simple pushbutton ***uicontrol***:

```
>> hButton = uicontrol('String','Yair');
>> jButton = java(findjobj(hButton))
jButton =
com.mathworks.hg.peer.PushButtonPeer$1[...]
>> jButton.getClass.getProtectionDomain.getCodeSource
ans =
(file:/C:/Program%20Files/MATLAB/R2010b/java/jar/hg.jar ...)
```

5.1.2 *Inspecting an Internal MATLAB Class*

We are now faced with two important questions: Which of the internal MATLAB packages and classes are useful, and how can we use them?

The first question can be answered to a small degree by simple inspection of the file names. It is pretty obvious that `com.mathworks.mwswing.MJToggleButton` implements a

MathWorks toggle-button that extends the standard `javax.swing.JToggleButton`.[2] All MathWorks classes and resources are contained in a `com.mathworks.*` package. The "M" prefix in the class name is MATLAB's standard prefix for all its internal classes; "Mclassname" would then be MATLAB's extension of "classname" and in this case, MATLAB's extension of `javax.swing.JToggleButton`. Similarly, `com.mathworks.mwswing.MJSlider` implements a `JSlider`-extended slider.

The easiest way to visualize these components is to display them onscreen, using ***javacomponent***:

```
javacomponent('com.mathworks.mwswing.MJSlider',[10,10,100,20]);
```

This looks exactly like a `javax.swing.JSlider`. In fact, this is not surprising, considering that `MJSlider` extends `JSlider`. While the MathWorks components may look similar to their Swing ancestors, some have important functionality extensions that merit their use, rather than using their Swing ancestors.[†]

I have created a utility called ***checkClass*** that looks at the MathWorks class and reports its modifications versus the ancestor class. ***checkClass*** is available for download on the UndocumentedMatlab.com website[3] and the MATLAB File Exchange.[4]

Here is a sample output for the `com.mathworks.mwswing.MJToggleButton` class:

```
>> checkClass(com.mathworks.mwswing.MJToggleButton)

Class: com.mathworks.mwswing.MJToggleButton

Superclass: javax.swing.JToggleButton

Methods in MJToggleButton missing in JToggleButton:
    hasFlyOverAppearance() : boolean
    hideText()
    isAutoMnemonicEnabled() : boolean
    setAutoMnemonicEnabled(boolean)
    setFlyOverAppearance(boolean)
    setFocusTraversable(boolean)

Methods inherited & modified by MJToggleButton:
    getForeground() : java.awt.Color
    getParent() : java.awt.Container
    isFocusTraversable() : boolean
    paint(java.awt.Graphics)
    setAction(javax.swing.Action)
    setModel(javax.swing.ButtonModel)
    setText(java.lang.String)
```

[†] For those interested, Matlab makes extensive use of the *Decorator* and *Façade* design patterns in these classes.

```
Interfaces in JToggleButton missing in MJToggleButton:
    javax.accessibility.Accessible
```

From this output, we learn that `MJToggleButton` added the **FlyOverAppearance** and **AutoMnemonicEnabled** properties, together with their associated getter/setter methods. We also learn that `MJToggleButton` added a non-property method: *hideText()*.

Finally, *setFocusTraversable()* was added as a setter method to the **FocusTraversable** property: the getter method *isFocusTraversable()* is already defined in the ancestor `JToggleButton` class — `MJToggleButton` simply added a setter method. These properties can also be accessed using the familiar MATLAB notation:

```
% Matlab notation:
>> get(jButton,'FlyOverAppearance')
ans =
off                                      % <= a
string
>> set(jButton,'FlyOverAppearance','on');); % causes a warning on R2010b+

% Java notation:
>> jButton.hasFlyOverAppearance
ans =
     0                                   % = false
>> jButton.setFlyOverAppearance(true);
>> jButton.setFlyOverAppearance(1);      % an equivalent alternative
```

Some classes (e.g., `MJSlider`) are just simple wrappers for their Swing ancestor:

```
>> checkClass('com.mathworks.mwswing.MJSlider')

Class: com.mathworks.mwswing.MJSlider

Superclass: javax.swing.JSlider

Methods inherited & modified by MJSlider:
    setModel(javax.swing.BoundedRangeModel)

Interfaces in JSlider missing in MJSlider:
    javax.accessibility.Accessible
    javax.swing.SwingConstants
```

In other cases, MathWorks has extensively modified the base class's functionality:

```
>> checkClass com.mathworks.mwswing.MJToolBar

Class: com.mathworks.mwswing.MJToolBar

Superclass: javax.swing.JToolBar

Methods in MJToolBar missing in JToolBar:
    addGap()
    addGap(int)
    addToggle(javax.swing.Action)) : javax.swing.JToggleButton
    configureButton(com.mathworks.mwswing.MJButton) (static)
    configureButton(com.mathworks.mwswing.MJToggleButton) (static)
    createMacPressedIcon(javax.swing.Icon): javax.swing.Icon (static)
```

```
    dispose()
    dispose(javax.swing.JToolBar) (static)
    getFlyOverBorder() : javax.swing.border.Border (static)
    getToggleFlyOverBorder() : javax.swing.border.Border (static)
    isFloating() : boolean
    isMarkedNonEssential(javax.swing.JComponent) : boolean (static)
    isMorePopupEnabled() : boolean
    markAsNonEssential(javax.swing.JComponent) (static)
    setArmed(boolean)
    setInsideToolbarBorder()
    setMorePopupEnabled(boolean)

Methods inherited & modified by MJToolBar:
    add(javax.swing.Action) : javax.swing.JButton
    addSeparator()
    addSeparator(java.awt.Dimension)
    doLayout()
    getMinimumSize() : java.awt.Dimension
    getPreferredSize() : java.awt.Dimension
    removeAll()

Interfaces in JToolBar missing in MJToolBar:
    javax.accessibility.Accessible
    javax.swing.SwingConstants

Static fields in MJToolBar missing in JToolBar:
    MORE_BUTTON_NAME          = 'MoreButton'
    NON_ESSENTIAL_PROPERTY_KEY = 'NonEssentialComponent'

Sub-classes in MJToolBar missing in JToolBar:
    com.mathworks.mwswing.MJToolBar$VisibleSeparator
```

Class methods can also be inspected using ***methodsview*** or preferably ***uiinspect*** (see Section 1.3); properties and callbacks can be inspected by using ***get***, ***inspect***, or again ***uiinspect***. Users may also try to search the MATLAB-supplied m-files for sample usage of these classes.[†] For example, MATLAB 7.2's ***uisetfont.m***[‡] uses com.mathworks.mwswing.MJButton, and ***datacursormode.m*** uses MJOptionPane.

Note that different MATLAB releases use internal classes differently. For example, ***datacursormode.m*** on newer MATLAB releases no longer uses MJOptionPane. Therefore, searching the m-file code base of separate MATLAB releases may yield additional clues.

5.1.3 Standard MATLAB Packages

Here is the full list of standard MATLAB packages on the R2011a release, excluding packages dedicated to installation, activation, and licensing. As noted above, a package called XYZ,

---

[†] This can easily be done using MATLAB's editor Find Files tool — search *.m for "com.mathworks." in the "Entire MATLAB path".

[‡] Or %matlabroot%/toolbox/matlab/uitools/private/uisetfont_deprecated.m in relatively new MATLAB releases.

including all its classes and sub-packages, will be included in a file called *XYZ.jar* located under the %matlabroot%/java/jar/ folder.[†]

- **bde**.jar — block-diagram editor (see Section 5.8.3).[‡]
- **beans**.jar — creates automatic JavaBeans for external objects and includes specialized editors to handle some common built-in classes.
- **common**.jar — repository for common icons.
- **comparisons**.jar — file and data comparison tools.
- **desktop**.jar — MATLAB Desktop support classes (new in R2011a).
- **fatalexit**.jar — displays an error message following a MATLAB crash.
- **foundation_libraries**.jar — tiny package with some i18n-support classes.
- **hg.jar** — Handle-Graphics-related Java classes. Contains Java peers for all UI components, as well as class interfaces for plotting functions that are done via compiled native libraries.
- **ide**.jar — Integrated Development Environment classes, leftover from old MATLAB releases (v5 & 6). **ide**.jar was replaced in MATLAB 7 with **mde**.jar (see below), but some **ide**.jar classes still remain to this day.
- **jmi**.jar — Java-to-MATLAB Integration classes (see Chapter 9).
- **mde**.jar — MATLAB 7 Desktop (see Chapter 8).
- **mlservices**.jar — MATLAB services (see Section 5.6).
- **mlwebservices**.jar — services that communicate with MathWorks.com.
- **mlwidgets**.jar — MATLAB widgets (= GUI controls; see Section 5.4).
- **mwswing**.jar — MathWorks Swing extensions (see Section 5.2).
- **mwt**.jar — MathWorks AWT extensions (see Section 5.3).
- **net**.jar — network proxy support classes.
- **page**.jar — functionality that wraps MATLAB plots, for example: plot editor, statistical data fitting, and page/print setup.
- **services**.jar — contains several utility classes.
- **timer**.jar — implements MATLAB timers as Java threads.
- **toolstrip**.jar — supports a tool-strip (ribbon) container that has modern stylish behavior and appearance (new in R2011a).[5]
- **util**.jar — contains a list of support classes.
- **webintegration**.jar — services to check for product updates and display an HTML start page.
- **widgets**.jar — another set of MATLAB widgets (see Section 5.5).
- **wizard**.jar — framework that enables the creation of interactive wizard screens.
- **xml**.jar — utility classes to parse MATLAB-related XML files.

[†] For example, C:\Program Files\MATLAB\R2011a\java\jar\mwswing.jar.

[‡] The bde.jar package was removed in R2011b.

MATLAB toolboxes normally use separate JAR files, located under %matlabroot%/java/jar/ toolbox/. For example,

- toolbox/compiler.jar — the MATLAB Compiler Toolbox
- toolbox/instrument.jar — the MATLAB Instrument Control Toolbox

A couple of standard MATLAB packages are located under %matlabroot%/java/jar/toolbox/ matlab/ and .../java/jar/toolbox/shared/:[†]

- toolbox/shared/controllib — plot property-editing control components
- toolbox/matlab/guide.jar — the MATLAB GUIDE (GUI Design Editor)
- toolbox/matlab/audiovideo.jar — audio/video recorder/player control[‡]

MATLAB also uses external Java package archives. These are also located in %matlabroot%/java/, under the separate subfolders org/ and jarext/. Here is the list for R2008a. Most jars are open-source packages that are easily found documented online:

- jar/org/netbeans.jar[§] — Java development platform and desktop framework[6]
- jar/org/openide.jar — portion of NetBeans dealing with dev environment (IDE)
- jarext/abbot.jar — Java-automated GUI testing framework[7]
- jarext/access-bridge.jar — Windows support for accessibility functionality[8]
- jarext/activation.jar — Java Activation Framework (JAF)[9]
- jarext/ant.jar — Java project build automation[10]
- jarext/avalon.jar — component framework for container (server) applications[11]
- jarext/axis.jar — Java SOAP-based webservice framework (aka JWS)[12]
- jarext/batik-svggen.jar — Java toolkit for SVG images[13]
- jarext/collections.jar — the standard `java.util.collections` package[14]
- jarext/commapi/comm.jar — communication toolkit (serial/parallel ports)[15]
- jarext/commons-codec.jar — Java encoders/decoders (codecs) package[16]
- jarext/commons-discovery.jar — Java service/interface discovery package[17]
- jarext/commons-el.jar — JSP Expression-Language interpreter package[18]
- jarext/commons-httpclient.jar — Java-based rich HTTP client infrastructure[19]
- jarext/commons-io.jar — Java library of I/O functionality[20]
- jarext/commons-logging.jar — Java event logging facilities[21]
- jarext/commons-net-1.4.1.jar — Java internet protocols toolkit[22]
- jarext/dtdparser121.jar — Java DTD parser toolkit[23]
- jarext/glazedlists_java15.jar — Java lists/table text-filtering package[24]
- jarext/jaccess-1_4.jar — Java accessibility utilities[25]
- jarext/jakarta-oro-2.0.8.jar — regular expression support utilities[26]
- jarext/jakarta-regexp-1.2.jar — another package for regular expressions[27]

[†] I do not know why these were not placed in the root %matlabroot%/java/jar/ folder together with the other standard packages.

[‡] The audiovideo.jar package was removed in R2011a for some unknown reason.

[§] I do not know why MathWorks chose to place org.netbeans and org.openide in jar/ rather than jarext/ as would be expected.

- jarext/jaxrpc.jar — Java-based webservice support utilities[28]
- jarext/jdom.jar — Java XML parsing utilities[29]
- jarext/jemmy.jar — Java GUI testing framework[30]
- jarext/jfcunit.jar — an extension to JUnit (see below) for testing Java GUI[31]
- jarext/jgoodies-forms.jar — elegant Java form panels (preferences, etc.)[32]
- jarext/jgoodies-looks.jar — high-fidelity GUI look-and-feel appearance[33]
- jarext/jox116.jar — JavaBeans-to-XML serialization/deserialization support[34]
- jarext/junit.jar — Java unit-testing framework package[35]
- jarext/lucene-*.jar — Java-based text-search engine[36]
- jarext/mail.jar — Java mail support (Sun's standard JavaMail API)[37]
- jarext/mwucarunits.jar — support for conversion between different units[38]
- jarext/nekohtml.jar — simple HTML parser[39]
- jarext/saaj.jar — Java SOAP support utilities[40]
- jarext/saxon*.jar — XSLT/XQuery/XPath XML-processing support utilities[41]
- jarext/spring-*.jar — Java application framework[42]
- jarext/wsdl4j.jar — Java support utilities for WSDL[43]
- jarext/xalan.jar — XSLT/XPath XML-processing support utilities[44]
- jarext/xercesImpl.jar — Java XML parser[45]
- jarext/xml-apis.jar — Java XML-processing support utilities (aka JAXP)[46]

In a few cases (JIDE being perhaps the most notable), the packages are commercially licensed products:

- jarext/ice/*.jar — packages that implement the commercial ICEsoft browser[47]
- jarext/J2PrinterWorks.jar — printing Java documents (commercial)[48]
- JIDE packages:†
  - jarext/jide/jide-action.jar — dockable command bars.
  - jarext/jide/jide-common.jar — tabbed/option panes; calendar, gripper, and other UI controls; PLAFs; multipage dialogs; and so on. This was the largest JIDE package bundled in MATLAB in past releases, but as of R2011a, jide-grids.jar (see below) is now the largest package.
  - jarext/jide/jide-components.jar — tabbed-panes; document panes; status bars; animations, and so on.
  - jarext/jide/jide-dialogs.jar — wizard; multipage; and other dialog windows.
  - jarext/jide/jide-dock.jar — support for dockable frame windows.
  - jarext/jide/jide-grids.jar — specialized `JTree` and `JTable` classes.
  - jarext/jide/jide-shortcuts — support for keyboard shortcuts (R2010a+).
- jarext/vb20.jar — Java online presentation solutions[49]
- jarext/webrenderer.jar — Java web browser[50]

---

† The JIDE packages are described in detail in Section 5.7.

In comparison, let us compare this list of packages with the corresponding list for R2011a. We find that the R2011a list is similar, with many additions and removals (all under jarext/), and some other packages being updated to a newer version:

- <u>Added:</u> AnimatedTransitions, annotations, commons-collections, commons-compress, commons-lang, commons-math, commons-net, dws_client, felix, foxtrot, freehep-*, google-collect, jxlayer, mwaws_client, RXTXcomm, saxon9*, scr, tablelayout, TimingFramework, xstream, jide/jide-shortcut, guice/*, axis2/*, wn32/jogl, win32/gluegen-rt, webservices/service_request_client, and webservices/loginws_client
- <u>Moved</u> (to the jarext/axis2/ subfolder):[†] activation, commons-httpclient, commons-codec, mail, wsdl4j, and xml-apis
- <u>Removed:</u> abbot, avalon, axis, batik-svggen, collections, commons-discovery, commons-el, commons-net-1.4.1, dtdparser121, J2PrinterWorks, jakarta-*, jaxrpc, jemmy, jfcunit, jox116, junit, mwucarunits, saaj, saxon8*, spring-*, vb20, wembeddedframe, and xalan

A major lesson that can be learned from this comparison is that the list of internal JAR files, not to mention their contents, can change dramatically from one Matlab release to another. For this reason, we must exercise caution when relying on any of these internal components within our code.

The following sections in this chapter will describe interesting classes that I found, and which have remained relatively stable across releases, grouped by their containing package. The list is by no means complete: I will be happy to hear from readers who find other interesting or useful classes and features.

5.2 MWSwing Package

5.2.1 Enhancements of Standard Java Swing Controls

 Note: For space considerations, only a handful of MWSwing classes can be described here. Readers are encouraged to look at the MWSwing package themselves and search for usable classes.

Most of the MWSwing classes are simple extensions of the corresponding Java Swing class. For example,

```
>> checkClass(com.mathworks.mwswing.MJSpinner)

Class: com.mathworks.mwswing.MJSpinner

Superclass: javax.swing.JSpinner

Methods in MJSpinner missing in JSpinner:
    setDefaultEditorAccessibleName(java.lang.String)
```

[†] Along with some additional packages that were added to this new subfolder.

```
Methods inherited & modified by MJSpinner:
   setEditor(javax.swing.JComponent)
   updateUI()

Interfaces in JSpinner missing in MJSpinner:
   javax.accessibility.Accessible
```

This specific control, `MJSpinner`, deserves a short description, since it does not have any built-in MATLAB *uicontrol* counterpart, unlike most other Swing controls. Like its `JSpinner` superclass,[51] `MJSpinner` is basically an editbox with two tiny adjacent up/down buttons. Spinners are similar in functionality to a combo-box (aka drop-down or popup menu), where a user can switch between several preselected values. Spinners are often used when the list of possible values is too large to display in a combo-box menu. Like combo-boxes, spinners too can be editable (meaning that the user can type a value in the editbox) or not (the user can only "spin" the value using the up/down buttons).

Spinners use an inner model, similarly to `JTree` and `JTable` and other complex controls. The default model is `javax.swing.SpinnerNumberModel`, which defines a min/max value (unlimited = [] by default) and step-size (1 by default). Additional predefined models are `javax.swing.SpinnerListModel` (which accepts a cell array of possible string values) and `javax.swing.SpinnerDateModel` (which defines a date range and step unit). The spinner **Value** can be set using the editbox or by clicking on one of the tiny arrow buttons. To attach a data-change callback, set the spinner's **StateChangedCallback** property.

I have created a small MATLAB demo, *SpinnerDemo*,[52] which demonstrates usage of `JSpinner` in MATLAB figures. Each of the three predefined models (number, list, and date) is presented, and the spinner values are interconnected via their callbacks. Readers are welcome to download this demo and reuse its source code.

Sample usage of three spinner models in a MATLAB figure

`MJButton` and `MJToggleButton` added the **AutoMnemonicEnabled, FlyOverAppearance**, and **FocusTraversable** properties (see Sections 6.1 and 6.2) to `JButton` and `JToggleButton`, respectively.[53]

MJRadioButton added the **AutoMnemonicEnabled** property (see Section 6.3) to the standard JRadioButton.[54]

MJCheckBox also added the **AutoMnemonicEnabled** property (see Section 6.4) to the standard JCheckBox.[55] A specific trick enables setting a tri-state (mixed) checkbox mode to either MJCheckBox or JCheckBox as described in detail in Section 6.4:

```
% Display the checkbox (UNSELECTED state at first)
import com.mathworks.mwswing.checkboxtree.*
jCB = com.mathworks.mwswing.MJCheckBox('MJCheckbox - mixed',0);
javacomponent(jCB, [10,70,150,20], hFig);

% Update the checkbox state to MIXED
jCB.setUI(TriStateButtonUI(jCB.getUI));
jCB.putClientProperty('selectionState', SelectionState.MIXED);
jCB.repaint;
```

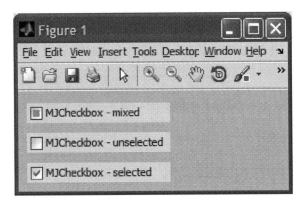

Tri-state checkbox (yes/no/maybe)

MJTextField added the ContextMenu and SelectAllOnFocus functionality methods, as well as the **TipWhenTruncatedEnabled** property (see Sections 6.5.1 and 6.5.2) to the standard JTextField.[56]

MJFormattedTextField added the ContextMenu, SelectAllOnFocus (see Sections 6.5.1 and 6.5.2), and EnterTriggersDefaultButton functionality methods to the standard JFormattedTextField.[57] This enables setting up text-entry (edit-box) fields that display and accept values in a formatted manner, such as: "3.45%", "123,456", "12:34:45", "May 30, 2011", "(212) 123.456.789" or "($123.45)". The control stores the actual value separate from the presented string. It automatically converts the base value to text string for presentation, and from string to base value upon user input.

MJList added the CellPainter and SelectionAppearanceReflectsFocus functionalities, as well as the **CellViewerEnabled, DragSelectionEnabled**, and **RightSelectionEnabled** properties (see Section 6.6) to JList.[58]

MJComboBox added the PopupWidthConstrained functionality and the **TipWhenTruncated-Enabled** property (see Section 6.7) to JComboBox.[59]

MJLabel added the **TipWhenTruncatedEnabled** property and mnemonic functionality (see Section 6.9) to the standard JLabel.[60]

MJMenuItem added the mnemonic functionality to the standard JMenuItem, by adding a second optional flag to the MJMenuItem constructors, specifying whether or not to treat the first "&" found within the menu item's text as the mnemonic indicator. As in MJLabels, the default flag value is ***true*** (more details in Section 6.9).

MJMenuBar added the **MoreMenuEnabled** property (see Section 6.9) to the standard JMenuBar.[61] This property affects the behavior of the menu bar when the width of its containing window is shrunk and so not all menu items can be fully seen:

```
jFrame = get(handle(hFig),'JavaFrame');
jMenuBar = jFrame.fHG1Client.getMenuBar;'
jMenuBar.setMoreMenuEnabled(true);
```

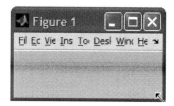

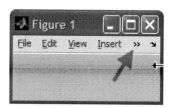

**Standard menu
(MoreMenuEnabled = false)**   **MoreMenu enabled
(MoreMenuEnabled = true)**

MJScrollPane added the *anchorToBottom()* method, which is very useful for logger-type scroll-boxes, where the data is constantly updated at the bottom of the box and the scroll-pane needs to keep showing the bottom item (also see Section 6.5.2).

MJTabbedPane added the *indexAtLocation()* and *indexFromMouse()* methods, and the ability to add/remove tab mouse listeners, to the standard JTabbedPane.[62]

MJTable added the **FillEmptyColumnHeader, CellViewerEnabled, HorizontalAuto-ScrollEnabled, MiddleSelectionEnabled, RightSelectionEnabled, and SelectionAppearanceReflectsFocus** flag properties, and a few methods, to the standard JTable.[63] Here is a sample use of the **FillEmptyColumnHeader** property:‡

```
cols = {'a1','b2','c3'};
data = mat2cell(magic(3),[1,1,1],[1,1,1]);
jTable = com.mathworks.mwswing.MJTable(data,cols);
jTable.setAutoResizeMode(jTable.AUTO_RESIZE_OFF);
jTable.setFillEmptyColumnHeader(true);
jScrollPane = com.mathworks.mwswing.MJScrollPane(jTable);
[jhScroll,hContainer] = javacomponent(jScrollPane,[10,10,300,50],gcf);
```

---

† In R2007b and earlier, use fFigureClient rather than fHG1Client.

‡ Ken Orr, former official MATLAB Desktop blog co-owner (http://blogs.mathworks.com/desktop/) and a Java enthusiast, mentions this specific feature in his personal blog: http://explodingpixels.wordpress.com/2009/05/18/creating-a-better-jtable/.

a1	b2	c3
8.0	1.0	6.0
3.0	5.0	7.0

a1	b2	c3
8.0	1.0	6.0
3.0	5.0	7.0

FillEmptyColumnHeader = true   **FillEmptyColumnHeader = false (default)**

`MJToolBar` added many methods and properties to the standard `JToolBar`.[64] These include the *addGap()*, *markAsNonEssential(component)*, *setArmed(flag)*, and *setInsideToolbarBorder()* methods as well as the **MorePopupEnabled** property.

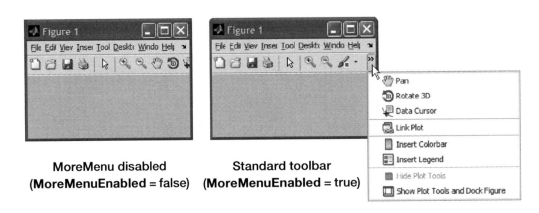

MoreMenu disabled   **Standard toolbar**
(MoreMenuEnabled = false)   (MoreMenuEnabled = true)

The *markAsNonEssential(component)* method can be used to specify which toolbar component will be the first to "disappear" from view when the window is shrunk. By default, no toolbar component is marked as nonessential, and so the buttons disappear according to their position, from right to left. But if we set any toolbar buttons as non-essential, they will disappear first (and will NOT appear in the More menu).

The *setArmed(flag)* method sets whether or not hot keys (mnemonics) should be enabled for toolbar labels (true by default). Compare these main Desktop toolbars:

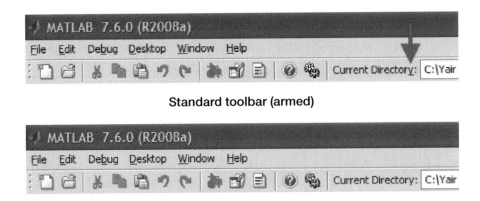

Standard toolbar (armed)

Dis-armed toolbar

`MJSplitPane` added the **ProportionalResizeEnabled, IntersectionDragEnabled**, and **SuppressingBorderIfNested** flag properties (which shall not be described here) to the standard `JSplitPane`.[65] Java split-panes (and `JPanels` in general) are problematic in MATLAB, since they cannot contain MATLAB plots, but only other Java components. Readers can use the *UISplitPane* utility (see Section 10.1) as an alternative.

`MJTree` is an extension of the standard `JTree`.[66] If we use the ***checkClass*** utility, then we will notice many new and modified methods compared with the standard `JTree`. Unfortunately, I could not find any which would be particularly useful for use applications, which would merit a description here. Trees, in general, are very important GUI components. It is regrettable that MathWorks did not see fit to include trees in its standard set of uicontrols. However, MATLAB user can use the built-in ***uitree*** function (described in Section 4.2) or any Java tree component (`JTree`, `MJTree`, `JideTree`, etc.). Both the ***FindJObj*** and the ***UIInspect*** utilities, which are used extensively in this book, use a Java tree component that is placed onscreen using the built-in ***javacomponent*** function. In this case, I have chosen to use `com.mathworks.hg.peer.UITreePeer`,[†] but I could have used any other tree class:

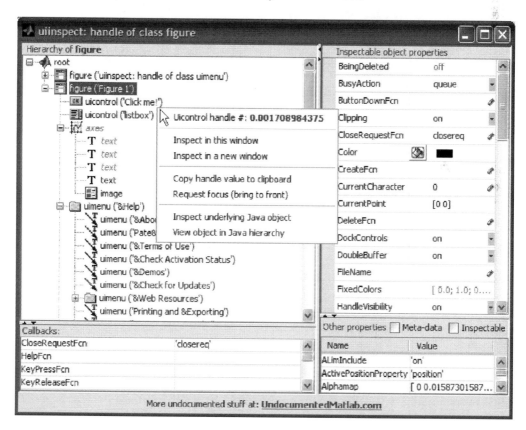

A Java tree component incorporated within a MATLAB figure (See color insert.)

[†] This is the class that underlies the built-in ***uitree*** function.

MJFileChooser extends JFileChooser[67] with the **IncludeFilterExtension**, **Show OverwriteDialog**, and **UseAWTFileDialog** write-only boolean properties. MJFileChooser can be used both as an embedded non-modal component within an existing figure window, or as a standalone modal dialog window (where the **UseAWTFileDialog** flag property comes into play):†

```
% Place a file-selection component as part of a GUI
jc = javaObjectEDT('com.mathworks.mwswing.MJFileChooser');
jc.setMultiSelectionEnabled(true); % = false by default
[jhc,hContainer] = javacomponent(jc,[10,10,500,400],gcf); % embedded

% The following section is optional: set user-defined filter(s)
filter = com.mathworks.mwswing.FileExtensionFilter( ...
            'Image File', {'png','gif','jpg'}, true);
jc.setFileFilter(filter); % more filters can be added to jc if needed
```

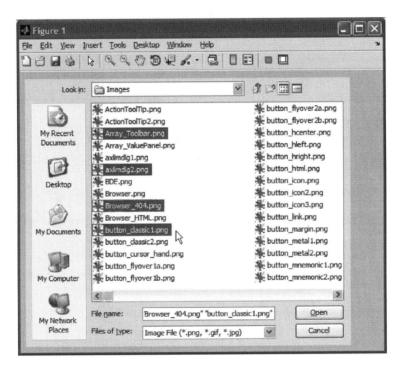

MJFileChooser **component embedded in a figure window**

```
jc.setUseAWTFileDialog(true);        % old-style AWT (not newer Swing)
jc.showDialog([],'');                % stand-alone modal dialog window
```

---

† A very detailed sample usage can be found in %matlabroot%\toolbox\matlab\timeseries@tsguis@allExportdlg\ exportsinglefile.m. Note that an internal Java bug (http://bit.ly/oRJRQu) causes the top-right icons to disappear in Windows 7. This was fixed only in JVM 1.6_18, so MATLAB users who wish to use the fix need to retrofit this JVM version, as explained in Section 1.8.2.

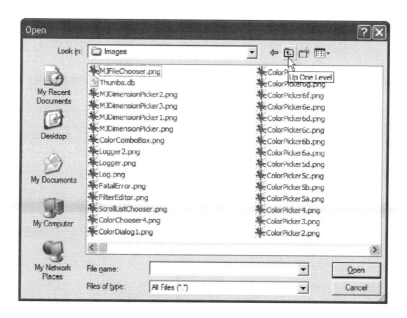

`MJFileChooser` **as a standalone modal dialog window (old-style AWT, rather than newer Swing)**

Similarly, there is a `com.mathworks.mwswing.dialog.MJFolderChooser`. I do not know why `MJFolderChooser` was placed in the dialog sub-package rather than in the root package next to `MJFileChooser`. The dialog sub-package also contains `MJGotoDialog`.

`MJDialog` added the **CloseOnEscapeEnabled** property and some useful methods to the standard `JDialog`:[68] *getHWnd(), getSpecifiedTitle(), setFocusTarget(component),* and *setNonBlockingVisible(flag)*; `MJDialogParent` and `MJFrame` both added many properties and methods to the standard `JFrame` that they extend. The related `MJOptionPane` added hot keys and some static button strings to the standard `JOptionPane`.[69]

`MJTextArea` added the ability to add/remove a context (right-click) menu to the standard `JTextArea`.[70] Similarly, `MJTextPane` added the context-menu functionality and the *setWrapping(flag)* method to the standard `JTextPane`.[71]

`MJEditorPane` added the Wrapping and ContextMenu functionality methods to the standard `JEditorPane`.[72] The reader is referred to the documentation[73] for an explanation of the differences between `JEditorPane` and its `JTextPane` subclass. Yet another built-in MATLAB class that extends `JEditorPane` is `com.mathworks.webintegration. startpage.framework.view.ResourcePane`, which has useful methods for setting HTML title, subtitles, text, hyperlinks, and image resources.

5.2.2 Entirely New Java Controls

The MWSwing package also contains several entirely new controls:

`MJGrip` is used to display a gripper for toolbars and the Editor's document bar:

Vertical and horizontal `MJGrip`

`MJCornerGrip` is used to display a gripper for resizing documents and figure windows. The gripper is often placed in the status bar, at the figure's or document's bottom-right corner (see Section 4.7), although this is not strictly necessary:

```
jc = javaObjectEDT('com.mathworks.mwswing.MJCornerGrip');
pos = getpixelposition(gcf);
gripPos = [pos(3)-20,1,20,20];
[jhc,hContainer] = javacomponent(jc,gripPos,gcf);
```

`MJCornerGrip`

Both `MJGrip` and `MJCornerGrip` have methods that enable their functional correlation with their tied component: `MJGrip.`*setComponentToMove(component)* and `MJCornerGrip.`*setComponentToResize(component)*.

`MJStatusBar` is a `JPanel` that is normally placed at the bottom of windows to display non-intrusive content-sensitive information. The use of status bars in MATLAB figures is described in detail in Section 4.7. Here is a simple example:

```
jFrame = get(handle(gcf),'JavaFrame');
jRootPane = jFrame.fHG1Client.getWindow;†
statusbarObj = com.mathworks.mwswing.MJStatusBar;
jRootPane.setStatusBar(statusbarObj);
statusbarObj.setText('please wait − processing...');
```

Status text →   ← Corner grip

A simple status bar with corner grip

Here, statusbarObj is actually a component contained within a parent container (a com.mathworks.mwswing.MJPanel object). When statusbarObj is created and attached to the

---

† In R2007b and earlier, use fFigureClient rather than fHG1Client.

figure via the frame's root pane's *setStatusBar()* method, it is this parent container which is actually attached, and the statusbarObj is then added to it. In addition to statusbarObj (which includes an internal `JPanel`[†] to display the text), the parent also contains a corner grip (`com.mathworks.mwswing.MJCornerGrip`) object.

`DefaultSortableTable` is a significantly extended `javax.swing.JTable` that enables lexical (alphanumeric) data column sorting. While a significant improvement over the standard `JTable`, this component has many limitations: only one column is sortable and cannot be unsorted; sorting is lexical and not numeric, and so on:

```
data = mat2cell(int16(magic(4)),[1,1,1,1],[1,1,1,1]);
headers = {'1','b','c23','#4'};
jTable = com.mathworks.mwswing.DefaultSortableTable(data,headers);
jScrollPane = com.mathworks.mwswing.MJScrollPane(jTable);
[jComp,hc] = javacomponent(jScrollPane,[10,10,150,110],gcf);
```

A `DefaultSortableTable`

For anything but very simple tasks, I suggest using a different sortable table than `DefaultSortableTable`. For example, JIDE tables (we can use `com.jidesoft.grid.SortableTable` or any of its subclasses as direct replacement for `DefaultSortable-Table` — see Section 5.7 for more details), or one of the solutions presented in Section 4.1:

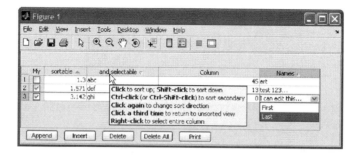

An entirely different approach to table sorting (see Section 4.1 for details)

[†] Actually, an inner class of type `com.mathworks.mwswing.MJStatusBar$1`.

`MJTreeTable` is an `MJTable` (see above) with additional functionality to support basic tree-like hierarchical cells, as shown in the following screenshots:

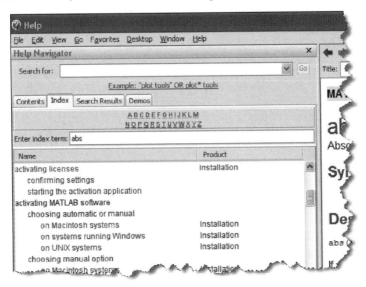

An example of an `MJTreeTable`†

A much more sophisticated tree-table implementation is provided by JIDE's `TreeTable` class, which is described in Section 5.7.2:

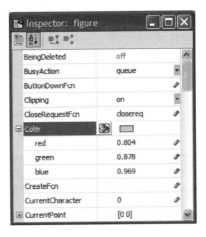

Another tree-table example (this time a JIDE `TreeTable`)

`MJColorComboBox` is a `JColorComboBox` that is suited for color selection (see Section 5.4.1). Unfortunately, this class was removed in MATLAB 7.11 (R2010b).

---

† Note that the Help Index tab was removed in MATLAB R2009b. Vociferous user protests have demanded its return, so perhaps the tab will indeed return in some future release.

`MJDimensionPicker` is a `JPanel` dedicated to interactive selection of a table grid size:

```
jc = com.mathworks.mwswing.MJDimensionPicker(java.awt.Dimension(3,4),1);
jc.setAutoGrowEnabled(true);
jc.setSizeLimit(java.awt.Dimension(5,5));
[jhc,hContainer] = javacomponent(jc,[100,100,150,200],gcf);
>> jc.getSelectedSize
ans =
java.awt.Dimension[width = 2,height = 3]
```

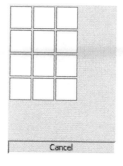  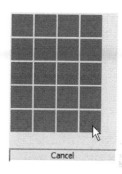

Initial 4x3   **Selecting 3x2 cells**   **Table auto-growth (up to 5x5)**
`MJDimensionPicker`

`MJMultilineLabel`, `MJMultilineToggleButton`, `MJMultilineRadioButton`, and `MJMultilineCheckBox` enable setting a text label spanning multiple lines if necessary. This contrasts with the default `JLable` behavior of truncating excess characters. However, it should be remembered that HTML labels are automatically line-wrapped as well. Therefore, all we need to do to achieve label text-wrapping is to enclose the label text with a "<html>" tag, without needing these four new `MJMultiline*` classes:

```
% Non-HTML (single-line, truncated) label
str = 'This is an MJLabel string';
jLabel = javaObjectEDT('com.mathworks.mwswing.MJLabel',str);
jLabel.setTipWhenTruncatedEnabled(true);
[hcomponent,hcontainer] = javacomponent(jLabel,[10,50,50,30],gcf);

% HTML (multi-line) label — see discussion in Section 6.9 below
str = '<html>&&#8704;&beta; <b>bold</b><i><font color="red">label';
jLabel.setText(str);
```

 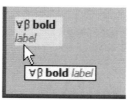 

Truncated single-line   **Multiline (HTML) label**   Multiline (HTML) radio
(non-HTML) label   button & checkbox

MJScrollStrip is used to display components in a panel that has arrow buttons at each end to enable scrolling when the panel's size is not enough to display all components. MJScrollStrip has both Horizontal and Vertical display modes (set via its **Orientation** property). This is used, for example, in the MATLAB Editor:

Horizontal MJScrollStrip **in the MATLAB Editor**

By default, the Editor has its MJScrollStrip setup to automatically scroll the components in the arrow's direction upon mouse hover on one of the arrows. This behavior can be switched off via the **ScrollOnMouseOver** property.

MJTiledPane is a container that enables to define a nonuniform matrix of subpanels, each containing its own Java components. This is used extensively by the MATLAB Editor in "Tiled" mode (see Section 8.1.1), but it can also be used in use applications:

```
jc = com.mathworks.mwswing.MJTiledPane(java.awt.Dimension(3,4));
[jhTiledPane,hContainer] = javacomponent(jc,[10,10,300,200],gcf);
```

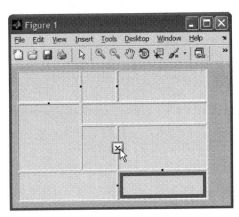

MJTiledPane **(4x3) with some cells merged and some inner borders moved**

MJUtilities is not a displayable component but rather contains a set of ancillary methods that are used throughout the system. These include self-explanatory methods such as *beep()*, *getCaretBlinkRate()*, *getCheckBoxIndent()*, *getDoubleClickInterval()*, *getScreenBounds()*, *getVirtualScreenBounds()*, and *hasMultipleMonitors()*.

Several MJUtilities methods handle mnemonics (hot keys): *setMnemonicFrom-Text(component,labelString)*, *exciseMnemonic(labelString)*, *findMnemonic(labelString)*, *indexOfMnemonic(labelString)*, and a few others.

There are also some EDT-related methods (see Section 3.2): *invokeLater(...)*, *runOnEvent-DispatchThread(...)*, *assertIsEventThread()*, *getThreadWarningCount()*, *setThreadingChecks-Enabled(flag)*, and *threadSafetyWarningStackTraceSuppressed()*.

Finally, the *initJIDE()* method is necessary for using JIDE functionality (see below).

Until R2010b, `MJUtilities` also included several font-related methods. In R2011a, these methods were transferred to `com.mathworks.mwswing.FontUtils`.

`MouseUtils` contains the single self-explanatory method *isDoubleClickEvent(mouseEvent)*, which is useful in mouse callback functions.

`ColorUtils` contains color management methods: *darker(), brighter(), intensify()*, and so on.

`TableUtils` provides convenience methods for controlling a `JTable`: *adjustRowHeight-(jtable)* and *getXForColumn(jtable, columnIndex)*. The corresponding JIDE class, `com.jidesoft.grid.TableUtils`, provides many other complementary methods: *autoResizeAllColumns, autoResizeAllRows, autoResizeColumn, autoResizeRow, ensureRowSelectionVisible, findColumnIndex, getViewPositionForRow, loadRowHeights, loadSelection(...)*, and many others.[†]

Similarly, `TreeUtils` provides convenience methods for controlling a `JTree`: *childExists(...), collapseAllNodes(jtree), expandAllNodes(jtree), nodeToPath(...)*, and *pathExists(...)*. There is a corresponding JIDE class, `com.jidesoft.tree.TreeUtils`, which provides complementary methods: *expandAll(jtree,flag), findTreeNode(...), getLeafCount(...), isDescendant(...), loadSelection(...)*, and a few others.[‡]

`SimpleDOMParser` provides, as its name suggests, simple DOM-based XML parsing support. Its main method is *parse(java.io.Reader)*, which returns a `SimpleElement` object, which includes all the important parsing methods: DOM traversal is done via *getChildNodes(), getChildrenByTagName(name), getElementsByTagName(name), getFirstChild(), getParentNode()*, and *hasChildNodes()*; XML node inspection is done via *getAttribute(name), getAttributes(), getNodeName(), getNodeValue(), getTagName(), getText()*, and *hasAttribute(name)*.

The MWSwing package contains several subpackages with specialized controls:

`com.mathworks.mwswing.checkboxlist.CheckBoxList` is an `MJList` extension that displays a list of labels in a list with a checkbox next to each label.[§] The labels' checkboxes can be set, unset, and queried using methods supplied by the `CheckBoxList` class or its `com.mathworks.mwswing.checkboxlist.DefaultListCheckModel` model:

```
jList = java.util.ArrayList;   % any java.util.List will be ok
jList.add(0,'First');
jList.add(1,'Second');
jList.add(2,'Third');
jList.add(3,'and last');
jCBList = com.mathworks.mwswing.checkboxlist.CheckBoxList(jList);
jScrollPane = com.mathworks.mwswing.MJScrollPane(jCBList);
[jhCBList,hContainer] = javacomponent(jScrollPane, [10,10,80,65],gcf);
set(jCBList, 'ValueChangedCallback', @myMatlabCallbackFcn);
jCBModel = jCBList.getCheckModel;
jCBModel.checkAll;
```

[†] See Section 4.1.5 for sample usage.

[‡] See Section 4.2.3 for sample usage.

[§] There is also an unrelated JIDE equivalent: `com.jidesoft.swing.CheckBoxList` (http://www.jidesoft.com/java-doc/com/jidesoft/swing/CheckBoxList.html or http://bit.ly/9d2GSJ).

```
jCBModel.uncheckIndex(1);
jCBModel.uncheckIndex(3);

>> jCBList.getCheckedValues
ans =
[First, Third]
>> jCBList.getCheckedIndicies'
ans =
                   0        2
>> jCBModel.isIndexChecked(0)
ans =
       1
```

CheckBoxList **example**

Similarly, com.mathworks.mwswing.checkboxtree.CheckBoxTree is an MJTree extension that displays tree nodes with a checkbox next to each label.[†] In the following example, a regular MJTree is presented next to a CheckBoxTree:

```
import com.mathworks.mwswing.checkboxtree.*
jRoot = DefaultCheckBoxNode('Root');
l1a = DefaultCheckBoxNode('Letters'); jRoot.add(l1a);
l1b = DefaultCheckBoxNode('Numbers'); jRoot.add(l1b);
l2a = DefaultCheckBoxNode('A'); l1a.add(l2a);
l2b = DefaultCheckBoxNode('b'); l1a.add(l2b);
l2c = DefaultCheckBoxNode('<html><b>&alpha;'); l1a.add(l2c);
l2d = DefaultCheckBoxNode('<html><i>&beta;'); l1a.add(l2d);
l2e = DefaultCheckBoxNode('3.1415'); l1b.add(l2e);

% Present the standard MJTree:
jTree = com.mathworks.mwswing.MJTree(jRoot);
jScrollPane = com.mathworks.mwswing.MJScrollPane(jTree);
[jComp,hc] = javacomponent(jScrollPane, [10,10,120,110],gcf);

% Now present the CheckBoxTree:
jCheckBoxTree = CheckBoxTree(jTree.getModel);
jScrollPane = com.mathworks.mwswing.MJScrollPane(jCheckBoxTree);
[jComp,hc] = javacomponent(jScrollPane, [150,10,120,110],gcf);
```

[†] There is also an unrelated JIDE equivalent: com.jidesoft.swing.CheckBoxTree (http://www.jidesoft.com/ javadoc/com/jidesoft/swing/CheckBoxTree.html or http://bit.ly/93K6up); http://www.mathworks.com/matlabcentral/ newsreader/view_thread/300225 (or http://bit.ly/hG79AE).

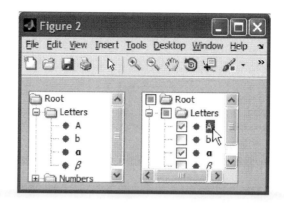

A regular `MJTree` **(left) and a** `CheckBoxTree` **(right)**

Unlike `MJTree` or `CheckBoxList`, `CheckBoxTree` does not have a separate data model. Instead, it relies on the base `MJTree`'s model, which is a `javax.swing.tree.DefaultTreeModel`[74] by default (JIDE's `CheckBoxTree` does have its own model).

Node checkboxes can be set, unset, and queried using the nodes' **SelectionState** property (or the corresponding *get/set* accessor methods). The three possible values are specified by the `com.mathworks.mwswing.checkboxtree.SelectionState` class: `SelectionState.SELECTED`, `SelectionState.NOT_SELECTED` and `SelectionState.MIXED`:

```
set(l2a,'SelectionState',SelectionState.SELECTED);        % select 'A'
jCheckBoxTree.repaint;

>> isequal(jRoot.getSelectionState, SelectionState.MIXED)
ans =
     1
```

The `SelectionState` value is used by the checkboxes' UI. To see this UI, run the following <u>before</u> rendering the tree onscreen:

```
>> jCheckBoxTree.list
com.mathworks.mwswing.checkboxtree.CheckBoxTree[...]
 javax.swing.CellRendererPane[,0,0,0x0,invalid,hidden]
  javax.swing.JPanel[...]
    com.mathworks.mwswing.MJCheckBox[...]
    ...

>> jCheckBoxTree.getComponent(0).getComponent(0).getComponent(0).getUI
ans =
com.mathworks.mwswing.checkboxtree.TriStateButtonUI@c98a94
```

`com.mathworks.mwswing.checkboxtree.TriStateButtonUI` displays a distinctive pattern, which is PLAF (Look-&-Feel) dependent. In Section 6.4, we shall use `TriStateButtonUI` to set a similar tri-state for regular checkboxes.

Lastly, the `com.mathworks.mwswing.desk` sub-package, which shall not be detailed here, includes numerous classes, many of which extend the basic MWSwing components described above, and which are used as MATLAB Desktop components. In R2011a, this sub-package was renamed `com.mathworks.widgets.desk`.

5.2.3 *Other MWSwing Controls*

Finally, there are a few custom MATLAB controls that, as far as I could see, do not add much important functionality to their Swing superclass. These include `MJComponent`, `MJButtonGroup`, `MJSlider`, `MJSpinner`, `MJScrollBar`, `MJProgressBar`, `MJColorChooser`, `MJMenu`, `MJPopupMenu`, `MJRadioButtonMenuItem`, `MJPanel`, `MJLayeredPane`, and `MJWindow`.

I believe that it is safe to say that these control classes detract nothing from their superclass parents. I, therefore, recommend using the MathWorks classes instead of their corresponding Swing counterparts, even in cases which have no apparent benefit.

5.3 MWT Package

The MWT package contains components that directly access the base GUI (AWT), bypassing the standard Swing family of components. Many MWT components have corresponding MWSwing components. In practice, the MWSwing/Swing components usually look more "polished" and professional than their MWT counterparts.

Without additional information, it appears to me that MWT was an early attempt made by MathWorks to implement a Java-based GUI, which was later replaced with the MWSwing package. Perhaps, the reason for this package being kept in new MATLAB releases is the support that it provides to backward-compatibility with code written for earlier releases.

However, the MWT components sometimes have specific features that are missing in MWSwing. Therefore, in specific circumstances, users may actually prefer using MWT. For example, consider the `com.mathworks.mwt.MWCheckbox` component, which is `com.mathworks.mwswing.MJCheckBox`'s counterpart (note the different capitalization):

```
jMWCheckbox = com.mathworks.mwt.MWCheckbox('MWCheckbox',1);
javacomponent(jMWCheckbox,[10,10,100,20],gcf);

jMJCheckBox = com.mathworks.mwswing.MJCheckBox('MJCheckBox',1);
javacomponent(jMJCheckBox,[10,40,100,20],gcf);

jMWCheckbox = com.mathworks.mwt.MWCheckbox('MWCheckbox - mixed',0);
jMWCheckbox.setMixedState(true);'
javacomponent(jMWCheckbox,[120,10,140,20],gcf);

jMWCheckbox = com.mathworks.mwt.MWCheckbox('MWCheckbox - radio',1);
jMWCheckbox.setAppearance(jMWCheckbox.RADIO_BUTTON);
javacomponent(jMWCheckbox,[120,40,140,20],gcf);
```

---

† The checkbox **State** must be false for the **MixedState** to have a visible effect. Note that the mixed state is only available in checkbox appearance and not in radio appearance.

MWT's `MWCheckbox` **vs. MWSwing's** `MJCheckBox`

In this example, we see that although the `MJCheckBox` appears more "polished" (e.g., anti-aliased font), the `MWCheckbox` enables setting a mixed-state checkbox, as well as a radio-button appearance. Specific use-cases that require a mixed-state or radio appearance may, therefore, favor using `MWCheckbox` over `MJCheckBox` (Section 6.4 shows how to set a tri-state mode also for regular `MJCheckBox`).

In some cases, the **MWT** components have different names than their standard counterparts. For example, `MWSplitter` is similar to a `JSplitPane`; `MWListbox` is really just a simple table;[75] `MWCardPanel` is similar to a **MATLAB** Frame *uicontrol*; and so on.

An **MWT** component that may be of interest, and which does not have an **MWSwing** counterpart, is `MWRuler`, which was removed in R2011a. `MWRuler` creates a pixel ruler that can be useful for graphical-editing applications. In fact, it is used by **MATLAB**'s own GUIDE (GUI Design Editor), as detailed in Section 8.7.3. Both horizontal and vertical rulers can be specified, and the tick & label intervals can be customized:

```
javacomponent(com.mathworks.mwt.MWRuler,[30,150,250,20],gcf);
javacomponent(com.mathworks.mwt.MWRuler(0,5,25),[10,10,20,142],gcf);
```

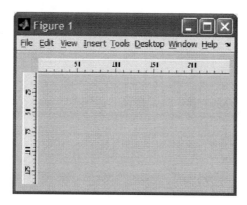

Horizontal and vertical ruler controls

Note: MWT is an undocumented "feature" of the built-in *usejava*, *javachk* functions:

```
if ~usejava('mwt'), ...                    % alternative #1
    error(javachk('mwt','title msg'));     % alternative #2
```

5.4 MLWidgets Package

The MLWidgets package contains miscellaneous components (widgets) used throughout MATLAB. The components are divided into sub-packages based on usage. So, for example, the array sub-package (com.mathworks.mlwidgets.array.*) includes components and support classes used by the MATLAB Array Editor.

Here are some interesting MLWidgets sub-packages:

- **actionbrowser** — appears to relate to function-hints tooltip, introduced in R2008b (MATLAB 7.7):

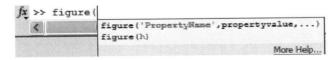

- **array** — relates to the Array Editor. For example, the following snippet displays variable data in a figure panel:

```
data = magic(5);
jValuePanel = com.mathworks.mlwidgets.array.ValuePanel('data');
javacomponent(jValuePanel,[10,10,400,150],gcf);
```

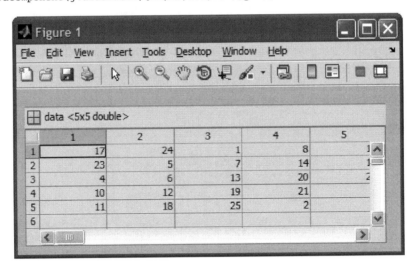

The VariableToolBar component implements the array-editor toolbar:

- **configeditor** — relates to the run/publish configuration-editor dialog window.
- **cwd** — CwdDisplayPanel is the component that displays the current folder in the main desktop toolbar. It can also be displayed as a standalone component:

■ **debug** — Editor's Debug menu's "Stop if Errors/Warnings" dialog window.

■ **dialog**—Editor'sinteractivefile-runningsupportdialogwindows:`PathUpdateDialog` presents the following dialog window:

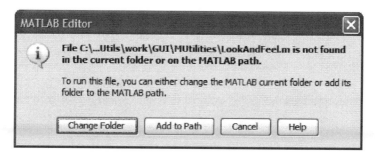

`SpecifyNewFilenameDialog` presents the following dialog window:

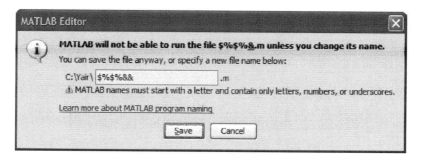

`ProgressBarDialog` presents an animated progress-bar dialog window, similar to MATLAB's built-in ***waitbar*** function but with an animated circular busy icon (see `BusyAffordance` in Section 5.5.1):

```
d = com.mathworks.mlwidgets.dialog.ProgressBarDialog.createProgressBar
    ('test...',[]);
d.setValue(0.75);                        % default = 0
d.setProgressStatusLabel('testing...');  % default = 'Please Wait'
d.setSpinnerVisible(true);               % default = true
d.setCircularProgressBar(false);         % default = false
d.setCancelButtonVisible(true);          % default = true
d.setVisible(true);                      % default = false
```

`ProgressBarDialog` — **separate dialog window**

The progress window can also appear as an internal window pane within a MATLAB figure, as follows:[†]

```
jFrame = get(gcf,'JavaFrame'); surf(peaks); drawnow;
jWindow = jFrame.fHG1Client.getWindow;'
d = com.mathworks.mlwidgets.dialog.ProgressBarDialog.createProgressBar.
       createHeavyweightInternalProgressBar(jWindow,'test...',[]);
d.setValue(0.37);
d.setVisible(true);
```

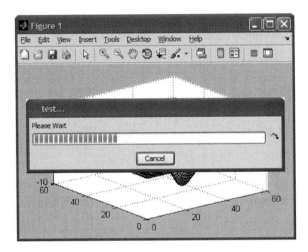

ProgressBarDialog — **internal window pane**

- **explorer** — Current-Folder explorer panel.
- **graphics** — contains color and plot-type selection panels and controls, described in the following sections.
- **help** — classes that deal with the different aspects of MATLAB help, parsing, and displaying internal or user-defined[76] help docs, and so on. One of the controls, HelpPopup, is presented in detail in Section 8.3.2. Another control, HelpPanel, can be used to embed a panel that displays doc pages in user GUI.

  The search subpackage handles the help-contents search functionality, which is apparently based on the well-known open-source Lucene engine (com.mathworks. mlwidgets.help.search.lucene).[77]
- **html** — contains the classes for presenting and displaying HTML webpages in MATLAB-branded containers via HTMLBrowserPanel, for example, *doc* pages or the MATLAB webbrowser (invoked with the built-in *web* function), as detailed in Section 8.3.2.

[†] Note: More advanced animated busy indications, including automated percentage and time-remaining labels, can be specified using JBusyComponent (http://code.google.com/p/jbusycomponent/ or http://bit.ly/gvPyAv), which is a JXLayers (https://jxlayer.dev.java.net/ or http://bit.ly/edIeym) decorator that can be applied to any displayable component. Also, see status-bar progress-bars (Section 4.7), Swing's JProgressBar (Section 3.3.1), and the built-in *waitbar* function.

[‡] In R2007b and earlier, use fFigureClient rather than fHG1Client.

Browser preferences are stored in the global preferences file (see Section 8.2) and usually have an "HTML" prefix (e.g., HTMLUseProxy, HTMLProxyHost, HTMLMaxFileSize), with a few preferences that do not follow this convention (e.g., SystemBrowser, SystemBrowserOptions). Section 8.2 shows how these preferences can be set programmatically, but in this case there is another alternative, using the built-in com.mathworks.mlwidgets.html.HTMLPrefs class: HTMLPrefs has static *get/set* accessor methods for all the relevant preferences, for example, HTMLPrefs.*setUseProxy(flag),setProxyHost(hostName),   setMaxFileSize(value), setSystemBrowser(browserName)*, and *setSystemBrowserOptions(optionsString)* — see urlread.m/urlwrite.m for sample usage. After setting proxy preferences, we must call HTMLPrefs.*setProxySettings()*, or restart MATLAB.

Another possibly useful class is com.mathworks.mlwidgets.html. HTMLUtils, which contains static methods to encode/decode URLs, and so on.

- **inspector** — contains classes used to implement the built-in property inspector. MATLAB's inspector is based on JIDE grid tables, which are presented in Section 5.7.2. The property inspector can be embedded as a standalone control in user GUIs. I have done so in my *UIInspect* (see Section 1.3) and *FindJObj* (see Sections 4.2.5 and 7.2.3) utilities. Interested users are encouraged to download these utilities and see how the inspector panel was configured, added to the GUI, and then assigned various object handles for inspection.

- **interactivecallbacks** — contains the InteractiveCallbackEditor class that presents an interactive callback-function editor panel. This could be used to input code segments from the user that are automatically syntax-highlighted and mlinted (also see the related SyntaxTextPane component in Section 5.5.1):

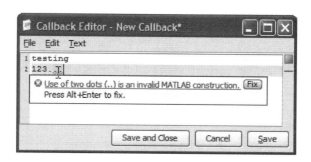

InteractiveCallbackEditor

- **io** — contains com.mathworks.mlwidgets.io.InterruptibleStreamCopier. Used by all the built-in MATLAB I/O functions, this is a faster and more powerful I/O class than the standard java.net.url.[78]

- **mlservices** — this com.mathworks.mlwidgets.mlservices sub-package should not be confused with the main com.mathworks.mlservices package (described in Section 5.6). The `mlwidgets.mlservices` sub-package is mainly responsible for the source-control integration, which is located in its scc sub-package (`com.mathworks.mlwidgets.mlservices.scc`).

- **path** — the `PathUtils` class has a few static methods that display path-related error messages:

```
jFrame = java.awt.Frame;
jFrame.setLocation(java.awt.Point(400,300));
import com.mathworks.mlwidgets.path.PathUtils
PathUtils.showChangeNotificationDialog(jFrame, PathUtils.PATH_CHANGE)[79]
```

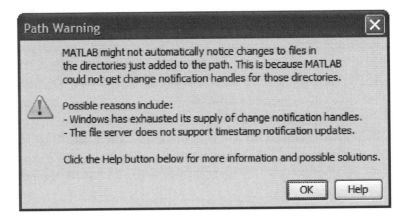

Path change notification warning

```
PathUtils.showInvalidPathEntryDialog(javax.swing.JButton,'Error msg')
```

Path change error message

- **prefs** — contains components that help manage the MATLAB system preferences in a dedicated window. These components are simply GUI representations of the preferences, which can be modified programmatically as explained in detail in Section 8.2. Among these components, we can find `PrefsDialog`, which is the main dialog window. It can be used to present a specified preferences panel (e.g., in response to user action):

```
jPrefsDialog = com.mathworks.mlwidgets.prefs.PrefsDialog;
jPrefsDialog.showPrefsDialog('Command Window');
```

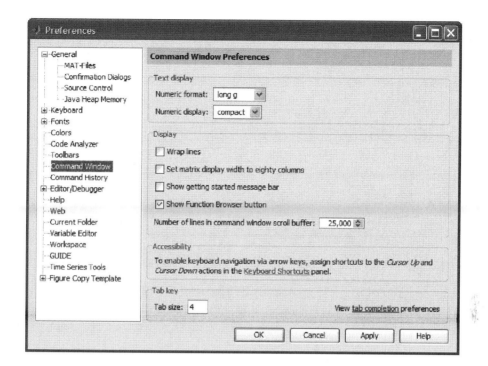

MATLAB main preference window

As an alternative to `PrefsDialog.`*showprefsDialog()*, we can use the built-in ***preferences*** function with the undocumented input argument of the preference name, for example, ***preferences('Command Window')*** or ***preferences('General.Source Control')***.

`PrefsDialog`'s *registerPanel(name1,name2)* method apparently enables adding a new preference panel to the Preferences dialog window. This method accepts two names that are presumably the panel title ('Command Window') and its class name (or possibly vice versa). Posted stack-traces[80] seem to indicate that preference panel classes are expected to have a *createPrefsPanel()* method that returns the displayable `JPanel` component.

Many preference panel classes are located in this package, under self-explanatory names such as `GeneralMatFilePrefsPanel` or `FontPrefsPanel`. The keyboard shortcut preferences panel, available since R2009b, received a dedicated sub-package (`com.mathworks.mlwidgets.prefs.binding`) due to its complex implementation. Other panels are located in other packages: `com.mathworks.mlwidgets.mlservices.scc.SccPrefsPanel` ("General/Source Control"), `com.mathworks.mlwidgets.text.mcode.MLintPrefsPanel` ("Code Analyzer"), `com.mathworks.mlwidgets.explorer.ExplorerPrefsPanel` ("Current Folder"), several panels under `com.mathworks.mde.*` and elsewhere.

- **shortcuts** — this sub-package includes classes and components used to handle Desktop Toolbar shortcuts. The pivot class appears to be `com.mathworks.mlwidgets.shortcuts.ShortcutUtils`, which is described in Section 8.1.4, together with the `ShortcutTreePanel` component.
- **stack** — the stack functions drop-down combo-box and surrounding panel.

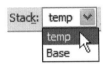

- **tabcompletion** — handles the MATLAB Desktop and Editor tab completions (the popup that displays when we type <TAB> in the Command Prompt or Editor, with a list of possible completions). The customization of tab completion is explored in Section 8.3.4.
- **tex** — the `TestTexDraw` class displays a Java window frame with a canvas that displays TₑX-formatted strings,[81] apparently as a test for valid strings (if the string is invalid, the frame contents will appear empty):

```
f = com.mathworks.mlwidgets.tex.TestTexDraw;
f.setString('$T_ex$ \bf example: $\sqrt{x^2+2x+1}=x+1$');
f.repaint;
```

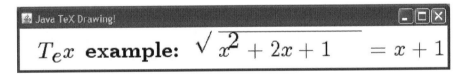

`TestTexDraw` **window**

Note that there are plenty of TeX-testing alternatives online, including some on the MATLAB File Exchange.[82]

- **text** — the `com.mathworks.mlwidgets.text.mcode.*` sub-package contains the preference panel and support classes for the MATLAB Mlint ("Code Analyzer").
- **util** — product information utilities.
- **workspace** — contains the components and support classes used by the Workspace panel in the main MATLAB Desktop. Customization of the Workspace is described in Section 8.6. One of the support classes, `com.mathworks.mlwidgets.workspace.MatlabCustomClassRegistry`, is used by the toolboxes' prefspanel.m function to register class callbacks. Another support class, `WhosInformation`, is used by ***arrayviewfunc***.m and by ***workspacefunc***.m. Lastly, `ImportFileChooser` is used by the built-in ***uiimport***.

5.4.1 Color-Selection Components[83]

MATLAB's fully-documented *uisetcolor* function uses a modal dialog window, whereas we often need to integrate color-selection components as a sub-component of an existing GUI. This is not supported by *uisetcolor*.

Luckily, MATLAB contains several internal color-selection components that can be integrated in our GUI. These include the beans package's `ColorPicker`, and the MLWidgets package's `ColorPicker` and `ColorDialog`. In addition, external Java components, such as Swing's standard `JColorChooser`,[84] can also be used (note that `JColorChooser` should have a minimum size of about 425×325 pixels to appear uncropped). They can be added to our GUI using the built-in *javacomponent* function. For example,

```
>> cc=javax.swing.JColorChooser;

>> [jColorChooser,container]=javacomponent(cc,[1,1,450,325],gcf);

>> jColorChooser.getColor
ans=
java.awt.Color[r=102,g=153,b=255]
```

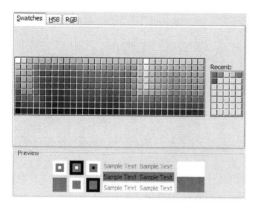
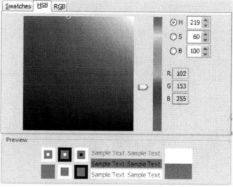

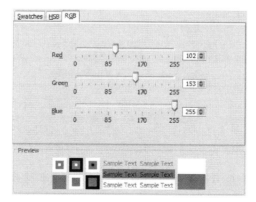

Swing's standard `JColorChooser` **panel**

The beans package's `ColorPicker` is an alternative component provided by MATLAB:

```
>> cp = com.mathworks.beans.editors.ColorPicker;

>> [jColorPicker,container]=javacomponent(cp,[1,1,400,200],gcf);

>> jColorPicker.getSelectedColor
ans=
java.awt.Color[r=255,g=86,b=158]
```

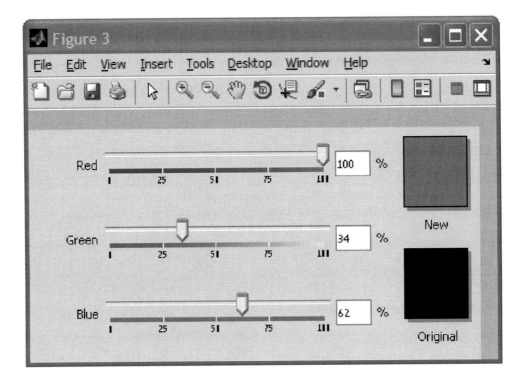

MATLAB's `ColorPicker`

The selection color can be specified using the `ColorPicker`'s *setSelectedColor()* method:

```
cp.setSelectedColor(0,0,255);            % = blue
cp.repaint; % necessary if the component is already displayed
```

Similarly, the original color can be specified using the *setInitialColor()* method:

```
cp.setInitialColor(255,0,0);             % = red
cp.repaint; % necessary if the component is already displayed
```

`ColorPicker` is used by the property inspector as the editor for color properties of Java objects (MATLAB color properties use another editor — see below):

```
inspect(cp)
```

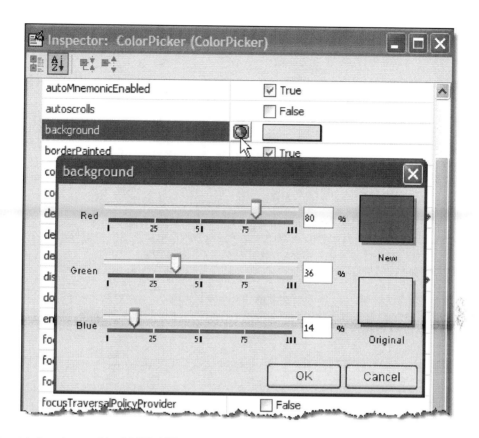

ColorPicker is used by MATLAB's property inspector as the editor for color properties of Java objects

Note how MATLAB's implementation of the color-selection window blended the window and *javacomponent* background colors, and how simple <OK> and <Cancel> buttons were added to the window's bottom. This is a good example of blending Java components/controls into a MATLAB GUI.[†]

The com.mathworks.beans.editors.ColorPicker object should not be confused with the com.mathworks.mlwidgets.graphics.ColorPicker object. The Beans editor ColorPicker is a simple standalone Java component that can be embedded in GUI figures, whereas the MLWidgets ColorPicker is a Java button control that is used to present a popup selection similar to a ColorDialog:

```
options = 0; icon = 0;
cp = com.mathworks.mlwidgets.graphics.ColorPicker(options,icon,'');
[jColorPicker,hContainer] = javacomponent(cp,[10,220,30,20],gcf);
```

[†] Actually, this is an entirely Java-based window, but a Matlab figure would look exactly the same except for the window *decoration* (window icon and minimization/maximization buttons).

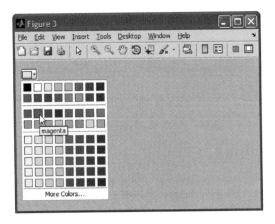

A different MATLAB ColorPicker

The `com.mathworks.mlwidgets.graphics.ColorPicker` component enables specifying different button icons and popup options. The list of supported icons is as follows:

Enumeration name	Value	Icon
ColorPicker.NO_ICON	0	
ColorPicker.FILL_ICON	1	
ColorPicker.LINE_ICON	2	
ColorPicker.TEXT_ICON	3	

The icons need to be supplied to the `ColorPicker` constructor, as shown above, as either a numeric value or as a more readable static enumeration name. For example,

```
icon = com.mathworks.mlwidgets.graphics.ColorPicker.LINE_ICON; % = 2
cp = com.mathworks.mlwidgets.graphics.ColorPicker(options,icon,'');
```

Similarly, for the popup options, the list is as follows:

Enum Name	ColorPicker.NO_OPTIONS	ColorPicker.AUTO	ColorPicker.MARKER
Value	0	1	2
Displayed popup			

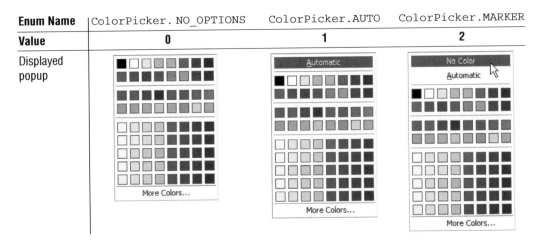

ColorPicker. SURFACE_AND_PATCH	ColorPicker. SURFACE_FACECOLOR	ColorPicker. SURFACE_MARKER	ColorPicker.NONE
3	**4**	**5**	**6**

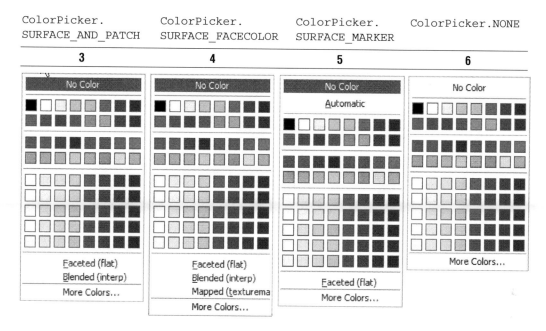

Depending on the selected popup-menu element, the `ColorPicker`'s **Value** property will return a `java.awt.Color` object or a string representation of the selected option:

```
>> get(cp,'Value')
ans =
java.awt.Color[r=0,g=0,b=255]    <= a java.awt.Color object
>> jColorPicker.getValue         % equivalent to: get(cp,'Value')
ans =
interp                           <= a string value
```

And similarly, we can also set the `ColorPicker`'s **Value** to a `java.awt.Color` or string:

```
set(cp,'Value',java.awt.Color.red)          % or: java.awt.color(1,0,0)
jColorPicker.setValue('flat');
```

We can programmatically display the popup menu using `ColorPicker`'s *showMenu()* method. For example, let us set the button so that the popup menu is displayed whenever the mouse hovers over the button. Recall from Section 3.4 that we should use the ***handle*** wrapper for the Java object, to prevent memory leaks:

```
>> jhColorPicker = handle(jColorPicker,'CallbackProperties')
jhColorPicker =
   javahandle_withcallbacks.com.mathworks.mlwidgets.graphics.ColorPicker
>> set(jhColorPicker,'MouseEnteredCallback', @(obj,evd) obj.showMenu);
```

Advanced programmers can customize the color-selection popup menu using its object handle returned from the *getPopupMenu()* method. Although the popup menu looks like a complex container, it is really a very simple menu list that includes com.mathworks.mwswing. MJMenuItem items and a few javax.swing.JSeparator objects:

```
>> jColorPicker.getPopupMenu.list
com.mathworks.mlwidgets.graphics.ColorPicker$ColorPickerMenu[ColorPickerMenu,0,
0,150x284,...]
 com.mathworks.mwswing.MJMenuItem[No Color,5,5,140x16,...]
 com.mathworks.mwswing.MJMenuItem[Automatic,5,25,140x16,...]
 javax.swing.JSeparator[,5,25,140x2,...,orientation=HORIZONTAL]
 com.mathworks.mwswing.MJMenuItem[black,4,30,16x16,...]
 com.mathworks.mwswing.MJMenuItem[white,22,30,16x16,...]
 com.mathworks.mwswing.MJMenuItem[0.94,0.94,0.94,40,30,16x16,...]
 com.mathworks.mwswing.MJMenuItem[0.83,0.82,0.78,58,30,16x16,...]
 ...
 javax.swing.JSeparator[,3,68,144x2,...,orientation=HORIZONTAL]
 com.mathworks.mwswing.MJMenuItem[0.85,0.16,0,4,74,16x16,...]
 com.mathworks.mwswing.MJMenuItem[0.93,0.93,0.93,22,138,16x16,...]
 ...
```

To modify this popup menu, follow the steps outlined in Sections 4.6.3 and 4.6.4.

JIDE's com.jidesoft.combobox.ColorComboBox (see Section 5.7.2) is very similar to ColorPicker. Despite its name and appearance as a combo-box, it actually extends the basic JComponent and not JComboBox. It includes three separately customizable sub-components: a color label, the color values, and the drop-down arrow button. All are shown by default (the color values may be hidden if the control is set too narrow), and each of the sub-components can easily be hidden:

Default   **ColorValueVisible = 0**   **ColorIconVisible = 0**   **ButtonVisible = 0**
ColorComboBox

ColorComboBox has a very nice feature, enabling manual modification of the color values (RGB) — the label's color automatically changes once a new value has been entered (the <Enter> key is pressed).

Another, much simpler, color-selection drop-down control (which was most regrettably removed in MATLAB 7.11 R2010b), is com.mathworks.mwswing.MJColorComboBox, which is a simple extension of the standard Swing javax.swing.JComboBox:

```
jc = com.mathworks.mwswing.MJColorComboBox;
jc.addColor(java.awt.Color.red,'red');
jc.addColor(java.awt.Color.blue,'blue');
jc.addColor(java.awt.Color.green.darker,'green');
jc.addColor(java.awt.Color(.8,.5,.3),'???');
jc.setSelectedColor(java.awt.Color.green.darker);
[jhc,hContainer] = javacomponent(jc,[50,50,60,20],gcf);

selectedIndex = jc.getSelectedIndex;
```
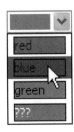

```
selectedColor = jc.getSelectedColor;
selectedColorName = jc.getSelectedName;
```

The property inspector uses a different control to edit color properties of MATLAB HG objects. As can be seen below, editing a figure handle's **Color** property displays a modal com. mathworks.mlwidgets.graphics.ColorDialog window; when clicking its <More Colors...> button, a new window containing a Swing JColorChooser is presented:

```
>> inspect(gcf)
```

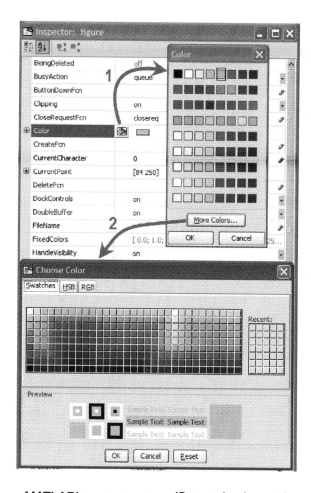

MATLAB's ColorDialog **(See color insert.)**

A ColorDialog can also be presented programmatically from within our GUI, regardless of any property inspector. To present a ColorDialog, pass a parent Java container to its *showDialog()* method. Any container will do, even a null container:

```
>> cd=com.mathworks.mlwidgets.graphics.ColorDialog('Yair''s Colors');
>> color=cd.showDialog([]) % pass a null value as the container
```

```
(the entire MATLAB application waits until the dialog window closes)
color=
java.awt.Color[r=255,g=0,b=153]
```

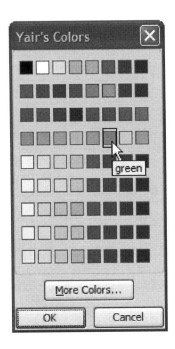

MATLAB'S `ColorDialog`

The `ColorDialog` window is modal, meaning that the entire MATLAB application is inaccessible until the dialog window is closed. When it closes, the *showDialog()* method returns the selected color value or the initial color if no other value was selected.

Despite its name, `ColorDialog` can also be presented in a GUI panel — not just in a stand-alone dialog window. Instead of using *showDialog()*, use the *getPickerPanel()* method to get the contents panel (a `com.mathworks.widgets.color.ColorPickerPanel` object), and then display this panel using *javacomponent*.

Like `com.mathworks.beans.editors.ColorPicker`, `ColorDialog` also has a *setInitialColor()* method. This *setInitialColor()* method can be used to specify the initially displayed color. Note that unlike `ColorPicker`'s method that accepts three integer values, `ColorDialog`'s variant expects a `java.awt.Color` object. Also, note that if the specified color does not match any displayed color, the user will NOT be informed — the `ColorDialog` will simply display the previously stored color.

`ColorDialog` also has a *setTitle()* method that can be used to update the title from the one supplied during `ColorDialog`'s creation. `ColorDialog` has a default constructor that sets a 'Choose Color' title, and so we can use *setTitle()* to override this default title.

Finally, there is a color-chooser panel class in the `com.mathworks.hg.util` package:

```
initialColor = java.awt.Color.cyan;
dc = com.mathworks.hg.util.dColorChooser(initialColor,{},'','');
dcPanel = dc.getContentPanel;
[jColorPicker,hContainer] = javacomponent(dcPanel,[1,1,400,200],gcf);
```

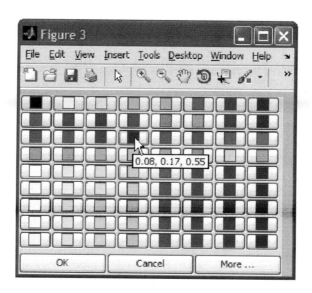

```
com.mathworks.hg.util.dColorChooser
```

To access the currently selected color, read the control's **Color** property:

```
>> get(dc,'Color')
color =
      0.992156862745098      0.917647058823529      0.796078431372549
>> dc.getColor
color=
java.awt.Color[r=253,g=234,b=203]
```

In summary, MATLAB programmers have several distinct ways of presenting a visually appealing color-selection dialog:

1. In a standalone window (`com.mathworks.mlwidgets.graphics.ColorDialog`)
2. In a drop-down button (`com.mathworks.mlwidgets.graphics.Color-Picker`, `com.mathworks.mwswing.MJColorComboBox` (pre-R2010b), JIDE `ColorComboBox`)
3. As an embedded component that can also be shown in a separate window: (`com.math-works.beans.editors.ColorPicker` or `javax.swing.JColorChooser`)
4. As an embedded cell component within data tables (see Section 4.1.1)

5.4.2 *Plot-Type Selection Components*

MATLAB contains several graphical plot-type selection components that can be used in our GUI. PlotPicker is a combo-box (drop-down/popup) control, which displays a list of possible plot functions 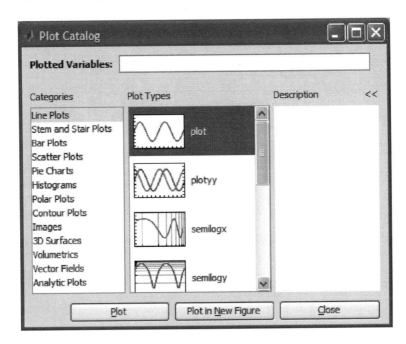; PlotCatalog is a dialog window that presents the plot functions catalog:

```
>> com.mathworks.mlwidgets.graphics.PlotCatalog.getInstance.show
```

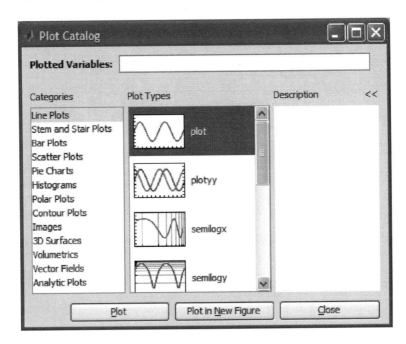

Another plot-function selector is PlotTypeCombo, which is my personal favorite control for embedding in a MATLAB application:

```
% Present the PlotCombo control
import com.mathworks.mlwidgets.graphics.*
jPlotCombo = PlotTypeCombo;
pos = [100,100,170,50];
[jhPlotCombo,hPanel] = javacomponent(jPlotCombo,pos,gcf);
set(jhPlotCombo,'ActionPerformedCallback',@myCBFcn);

% Callback function to process PlotCombo selections
function myCBFcn(jObject,jEventData) newPlotFunc =
jObject.getSelectedItem.getName.char;
    %Now do something useful with the selected function
end % myCBFcn
```

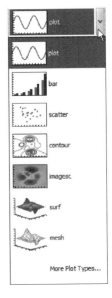

For the IDS application (see Section 10.2 for details), I implemented a dynamic report page that enables users to select the plotting function using a `PlotTypeCombo`. When users select a nondefault function (e.g., ***stairs*** or ***loglog***), that function is automatically added to the combo-box for possible future reuse.[85] I implemented this as follows:

```
% Add the specified plotFunc to a PlotTypeCombo control
function updatePlotCombo(plotCombo, plotFunc)

   % Convert plotFunc (a Matlab string) into a Java PlotSignature
   import com.mathworks.mlwidgets.graphics.*
   plotSig = PlotMetadata.getPlotSignature(plotFunc);

   % Get the list of all existing plot types in the combo box
   existingPlotTypes = {};
   for plotIdx = 0 : plotCombo.ItemCount-1
      nextItem = plotCombo.getItemAt(plotIdx);
      if isjava(nextItem)
         nextItem = char(nextItem.getName);
      end
      existingPlotTypes = [existingPlotTypes, nextItem];
   end

   % If the new plotType is NOT already in the list
   if isempty(strmatch(plotType,existingPlotTypes,'exact'))
      % Add the new plotType to the list just prior to the end,
      % so that "More plots..." will always be last
      plotCombo.insertItemAt(plotSig,plotCombo.ItemCount-1);
   end

   % Set the currently-selected item to be the requested plotType
   % Note: temporarily disable callbacks to prevent involuntary action
   plotCombo.ActionPerformedCallback = [];
   plotCombo.setSelectedItem(plotSig);
   plotCombo.ActionPerformedCallback = @myCBFcn; % selection callback

end % updatePlotCombo
```

Usage of this function would then be as simple as:

```
updatePlotCombo(jhPlotCombo,'stairs');
```

The astute reader would have noticed from the *updatePlotCombo()* function above that the `PlotTypeCombo` items are plot-signature objects. Different plot functions expect a different set of input arguments (signature). This information is kept and can be queried from the `PlotMetadata` class:[86]

```
>> com.mathworks.mlwidgets.graphics.PlotMetadata.listAllSignatures
ans =
[area, bar, bar (stacked), barh (stacked), barh, comet, compass, errorbar,
feather, loglog, plot, plot N series, plot N series against T, plot3,
plotmatrix, plotyy, polar, quiver, quiver3, ribbon, scatter, scatter3, semilogx,
semilogy, stairs, stem, stem3, null, contour, contour3, contourf, image,
```

```
imagesc, mesh, meshc, meshz, pcolor, plot against first column, surf, surfc,
surfl, waterfall, null, contour, contour3, contourf, image, imagesc, mesh,
meshc, meshz, pcolor, plot against first column, surf, surfc, surfl, waterfall]

>> plotFunc = 'surf'; % for example
>> plotSig = PlotMetadata.getPlotSignature(plotFunc);
>> args = cell(plotSig.getArgs)'
args =
    [1x1 com.mathworks.mlwidgets.graphics.PlotArgDescriptor]
    [1x1 com.mathworks.mlwidgets.graphics.PlotArgDescriptor]
    [1x1 com.mathworks.mlwidgets.graphics.PlotArgDescriptor]
    [1x1 com.mathworks.mlwidgets.graphics.PlotArgDescriptor]

>> argsStruct = []; % initialize
>> for argsIdx = 1 : length(args)
       argsStruct(argsIdx).axis = char(args{argsIdx}.getAxis);
       argsStruct(argsIdx).name = char(args{argsIdx}.getName);
       argsStruct(argsIdx).label = char(args{argsIdx}.getLabel);
       argsStruct(argsIdx).dims = args{argsIdx}.getNumDimensions;
       argsStruct(argsIdx).reqObj = args{argsIdx}.getRequired;
       req = argsStruct(argsIdx).reqObj;
       argsStruct(argsIdx).requiredFlag = req.equals(req.REQUIRED);
   end

>> argsStruct(1)                    % Info about the first input argument
ans =
            axis: 'X'
            name: 'X'
           label: 'X Data Source'
            dims: [2x1 int32]
          reqObj: [1x1
com.mathworks.mlwidgets.graphics.PlotArgDescriptor$RequiredType]
    requiredFlag: 0

>> {argsStruct.axis}                % axis info of all input arguments
ans =
     'X'     'Y'     'Z'     ''

>> {argsStruct.name}                % name of all input arguments
ans =
     'X'     'Y'     'Z'     'C'

>> {argsStruct.label}               % data label of all input arguments
ans =
     'X Data Source'    'Y Data Source'    'Z Data Source'    'Color'

>> {argsStruct.dims}                % dimensionality of all input args
ans =
     [2x1 int32]    [2x1 int32]    [2]    [2]

>> {argsStruct.requiredFlag}        % is any input arg mandatory?
ans =
     [0]     [0]     [1]     [0]
```

5.5 Widgets Package

The `com.mathworks.widgets` package contains component classes similar to those found in the MLWidgets package (see Section 5.4). The widgets are normally composed of several basic Swing or MWSwing (see Section 5.3) components.

5.5.1 *Widget Components*

`ClosableToolTip` is a simple tooltip window that can be displayed anywhere (regardless of "parent" component). It was first introduced for Editor mlint messages in R2009a[87] and then added to the Desktop's Current-Folder pane in R2010a:[88]

```
cttd = com.mathworks.widgets.ClosableToolTipData('name','Tooltip msg');
dirsEnums = javaMethod('values', ...
     'com.mathworks.widgets.tooltip.BalloonToolTip$ArrowDirection');
dirs = java.util.ArrayList;
dirs.add(dirsEnums(1)); % NORTH
dirs.add(dirsEnums(2)); % EAST
pos = java.awt.Rectangle(300,600,500,500);
com.mathworks.widgets.ClosableToolTip.show(cttd,pos,dirs);
```

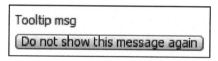

ClosableToolTip

The displayed tooltip contains the requested message text as well as a button that (when clicked) not only closes this tooltip message but also stores a system preference to never again show similar tooltips (which have the same 'name'). This preference can always be modified programmatically:

```
cttd.setPreferenceToClose(false); % true will prevent display
currentState = cttd.isPreferenceSetToClose(); % true/false
```

For the inquisitive reader, the preference is stored in the matlab.prf file in the user's ***prefdir*** folder under the key "ClosedToolTip<tooltip-name>":

```
ClosedToolTipname=Bfalse†
```

Note the narrow rounded-edge gradient-shade button used in the `ClosableToolTip` above. This is another component in the Widgets package, called `LightButton`, an extension of `com.mathworks.mwswing.MJButton`. It has the appealing look-and-feel that reverses the shade gradient when the button is clicked:

```
jButton = com.mathworks.widgets.LightButton('My LightButton');
[hjButton,hContainer] = javacomponent(jButton,[20,20,83,13],gcf);
```

---

† The preference file and its accessor methods are described in detail in Section 8.2. The `Dialogs` class, described below, uses a similar way to prevent message from reappearing following a user indication.

My LightButton

A regular shaded `LightButton`...

My LightButton

...and the same button depressed

Informational messages can also be displayed using another lightweight widget, `Light weightWindow`. `LightweightWindow` is not a displayable component but rather a class that creates such a component. I do not know the reason for this somewhat-cumbersome usage:

```
% Create a light-weight dialog window
jf = java.awt.Frame;
jsb = com.mathworks.mwswing.MJStatusBar;
jb = javax.swing.JButton('click me!');
jlww = com.mathworks.widgets.LightweightWindow(jf,'Yair #2',jb,jsb);
jDialogWindow = jlww.getWindow;

% Customize the dialog window's size, status-bar text etc., then show
jsb.setText('This is the status-bar');
jDialogWindow.setSize(200,100)
jDialogWindow.show;
```

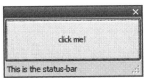

LightweightWindow

In the lightweight widget family, there is also a `LightScrollPane` component that extends com.mathworks.mwswing.MJScrollPane to achieve a different look-and-feel:

```
% Regular scroll-pane:
cols = {'a1','b2','c3'};
data = mat2cell(magic(3),[1,1,1],[1,1,1]);
data = [data;data;data;data];
jTable = com.mathworks.mwswing.MJTable(data,cols);
jScrollPane = com.mathworks.mwswing.MJScrollPane(jTable);
[jhScroll,hContainer] = javacomponent(jScrollPane,[10,10,200,150],gcf);

% Light scroll-pane:
jScrollPane = com.mathworks.widgets.LightScrollPane(jTable);
[jhScroll,hContainer] = javacomponent(jScrollPane,[10,10,200,150],gcf);
```

com.mathworks.mwswing.

MJScrollPane

com.mathworks.widgets.

LightScrollPane

DropdownButton is a button that masquerades as a drop-down/popup menu. Menu items can be attached to the DropdownButton's internal MJPopupMenu, just as for any Java JMenu[89] (see Section 4.6.4 for additional details and examples):

```
% Create the drop-down button
jDDButton = com.mathworks.widgets.DropdownButton('Select here');
[jhButton,hContainer] = javacomponent(jDDButton,[10,100,100,20],gcf);

% Customize the popup menu
import javax.swing.* java.awt.event.*
jDDMenu = jDDButton.getPopupMenu; % com.mathworks.mwswing.MJPopupMenu
menuItem = JMenuItem('Text-only menu item');
menuItem.setAccelerator(KeyStroke.getKeyStroke(KeyEvent.VK_1, ...
                        ActionEvent.ALT_MASK));
menuItem.getAccessibleContext().setAccessibleDescription('do nothing')
jDDMenu.add(menuItem);
myIcon = fullfile(matlabroot,'/toolbox/matlab/icons/warning.gif');
menuItem = JMenuItem('Text and icon', javax.swing.ImageIcon(myIcon));
menuItem.setAccelerator(KeyStroke.getKeyStroke(KeyEvent.VK_2, ...
                   bitor(ActionEvent.CTRL_MASK,ActionEvent.ALT_MASK)));
jDDMenu.add(menuItem);
jDDMenu.addSeparator;
cbMenuItem = JCheckBoxMenuItem('A check box menu item');
jDDMenu.add(cbMenuItem);
```

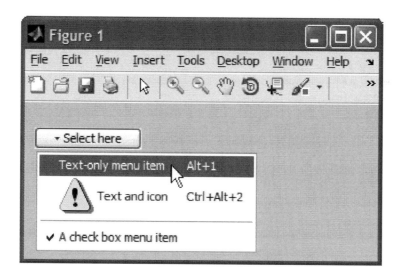

DropdownButton **in action**

The popup menu can also be aligned to the right rather than to the default left:

```
set(hContainer,'pos',[200,100,100,20]);
aligns = javaMethod('values', ...
         'com.mathworks.widgets.DropdownButton$PopupAlignment');
jDDButton.setPopupMenuAlignment(aligns(2)); % RIGHT
```

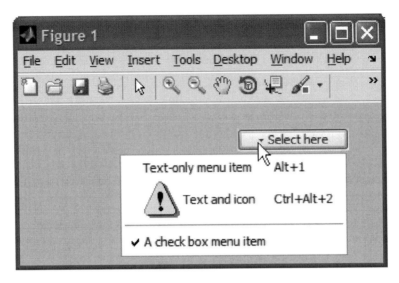

A right-aligned `DropdownButton`

`FormPanel` is another `MJPanel` extension. `FormPanel` automatically lays out specified subcomponents in a table manner, labels to the left and controls/subcomponents to the right. Such a panel is often called a "Form", hence its class name. For example,

```
jPanel = com.mathworks.widgets.FormPanel;
jPanel.addRow('Row #1:', javax.swing.JCheckBox('test1'))
jPanel.addRow('A very long label:', javax.swing.JCheckBox)
jPanel.addRow('#3:',javax.swing.JButton('Click me!'))
jPanel.addRow('Row #4:',javax.swing.JComboBox({'red','green','blue'}))
[jhPanel,hContainer] = javacomponent(jPanel,[10,10,200,110],gcf);
```

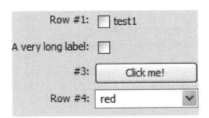

`FormPanel`

 Note: MATLAB comes bundled with the JGoodies Forms package, which is even more powerful and flexible than `FormPanel` (see Section 5.8.2).

`FormPanel` automatically sets **HorizontalSpace** and **VerticalSpace** of 5 pixels. These properties can be modified to achieve a tighter- or more expanded-looking layout.

If we look closely at the screenshot above, then we will notice that the button and drop-down subcomponents are stretched in the horizontal direction but not in the vertical direction. This is the basic behavior, which can be modified separately for any of the subcomponents by specifying a third (optional) numeric argument, which is one of FormPanel.STRETCH_NONE (=0), FormPanel.STRETCH_HORIZONTAL (=1, the default value), FormPanel. STRETCH_VERTICAL (=2), and FormPanel.STRETCH_BOTH (=3):

```
jPanel = com.mathworks.widgets.FormPanel;
jPanel.addRow('Row #1:', javax.swing.JCheckBox('test1'))
jPanel.addRow('A very long label:', javax.swing.JCheckBox)

% Unstretched button
jButton = javax.swing.JButton('Click me!');
jPanel.addRow('#3:', jButton, jPanel.STRETCH_NONE)

% Vertically-streched combo-box
jPanel.addRow('Row #4:',javax.swing.JComboBox({'red','green','blue'}),2)
[jhPanel,hContainer] = javacomponent(jPanel,[10,10,200,110],gcf);
```

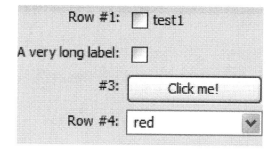 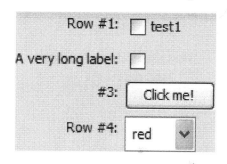

Original FormPanel   FormPanel **with modified**
(default stretching)   **stretching behavior**

FormPanel's labels are right-aligned by default, meaning that they are flushed rightward to the invisible vertical divider between the label column and the subcomponent column. This behavior can be modified by modifying the **RightAligned** property from its default value of *true*:

```
jPanel = com.mathworks.widgets.FormPanel;
jPanel.addRow('Row #1:', javax.swing.JCheckBox('test1'))
jPanel.addRow('A very long label:', javax.swing.JCheckBox)
drawnow; % allow time for Java EDT to process...

% Labels #3 & #4 are left-aligned:
jPanel.setRightAligned(false);
jPanel.addRow('#3:',javax.swing.JButton('Click me!'))
jPanel.addRow('Row #4:',javax.swing.JComboBox({'red','green','blue'}))
[jhPanel,hContainer] = javacomponent(jPanel,[10,10,200,110],gcf);
```

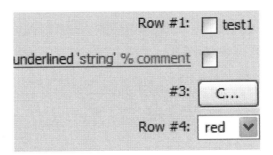

Left-aligned `FormPanel` **labels**

As a final customization, we note that the labels themselves can be customized. For example, let us use the `SyntaxTextLabel` class that will be introduced immediately below (actually, any component that extends `javax.swing.JLabel` will be accepted):

```
str = 'underlined ''string'' % comment';
codeType = com.mathworks.widgets.SyntaxTextLabel.M_STYLE;
jCodeLabel = com.mathworks.widgets.SyntaxTextLabel(str,codeType);
jCodeLabel.setUnderlined(true);

jPanel = com.mathworks.widgets.FormPanel;
jPanel.addRow('Row #1:', javax.swing.JCheckBox('test1'))
jPanel.addRow(jCodeLabel, javax.swing.JCheckBox)
jPanel.addRow('#3:',javax.swing.JButton('Click me!'))
jPanel.addRow('Row #4:',javax.swing.JComboBox({'red','green','blue'}));
[jhPanel,hContainer] = javacomponent(jPanel,[10,10,200,110],gcf);
```

Styled `FormPanel` **labels (See color insert.)**

`SyntaxTextLabel` is used to display a single-line text label that is syntax-highlighted according to the specified programming language: C_STYLE, HTML_STYLE, JAVA_STYLE, PLAIN_STYLE and, of course, M_STYLE for MATLAB code:

```
s = 'for id = 1:3, set(h(id),''string'',num2str(id)); end % Matlab code';
codeType = com.mathworks.widgets.SyntaxTextLabel.M_STYLE;
jCodeLabel = com.mathworks.widgets.SyntaxTextLabel(s,codeType)
[jhLabel,hContainer] = javacomponent(jCodeLabel,[10,10,300,20],gcf);
```

SyntaxTextLabels **(different code styles) (See color insert.)**

More flexibility in the displayed label styles can be achieved with HTML/CSS, and the bundled com.jidesoft.swing.StyledLabel[90] provides even more flexibility (JIDE is described in Section 5.7):

```
import java.awt.*
import com.jidesoft.swing.*
str = 'Mixed Underlined Strikethrough Super and Subscript combo Styles';
jStyledLabel = StyledLabel(str);
styles = [StyleRange(0,5,  Font.BOLD, Color.BLUE), ...
          StyleRange(6,10, Font.PLAIN,StyleRange.STYLE_UNDERLINED),...
          StyleRange(17,13,Font.PLAIN, Color.RED, ...
                          StyleRange.STYLE_STRIKE_THROUGH), ...
          StyleRange(31,5, Font.PLAIN, Color.BLUE, ...
                          StyleRange.STYLE_SUPERSCRIPT), ...
          StyleRange(37,3, Font.ITALIC, Color.BLACK), ...
          StyleRange(41,9, Font.PLAIN, Color.BLUE, ...
                          StyleRange.STYLE_SUBSCRIPT), ...
          StyleRange(51,5, Font.PLAIN, StyleRange.STYLE_WAVED + ...
                          StyleRange.STYLE_STRIKE_THROUGH)];
jStyledLabel.setStyleRanges(styles);
[jhLabel,hContainer] = javacomponent(jStyledLabel,[10,10,300,20],gcf);
```

JIDE StyledLabel **(different font styles) (See color insert.)**

JIDE also provides the convenient StyledLabelBuilder,[91] which enables easy multi-style text construction. Our StyledLabel example could thus be coded as follows (note the different equivalent ways of specifying the blue font color):

```
str = ['{Mixed:b,f:blue} {Underlined:u} {Strikethrough:s,f:red} ' ...
       '{Super:sp,f:(0,0,255)} {and:i} {Subscript:sb,f:#00f} ' ...
       '{combo:s,w} Styles'];
jStyledLabel = StyledLabelBuilder.createStyledLabel(str);
[jhLabel,hContainer] = javacomponent(jStyledLabel,[10,10,300,20],gcf);
```

Finally, JIDE also provides the `ClickThroughStyledLabel`,[92] a `StyledLabel` extension that allows setting a target component, so that mouse clicks on the label will actually trigger the target component. This can be useful in forms such as the `FormPanel` component described above, where components have adjacent descriptive labels.

Multi-line syntax-highlighted code can be displayed with the `SyntaxTextPane` component. `SyntaxTextPane` uses MIME types[93] rather than styles for syntax-highlighting, but the end-result is essentially the same:

```
jCodePane = com.mathworks.widgets.SyntaxTextPane;
codeType = jCodePane.M_MIME_TYPE;    % = 'text/m-MATLAB'
jCodePane.setContentType(codeType)
str = ['% create a file for output\n' ...
       '!touch testFile.txt\n' ...
       'fid = fopen(''testFile.txt'', ''w'');\n' ...
       'for i = 1:10\n' ...
       '    % Unterminated string:\n' ...
       '    fprintf(fid,''%6.2f \\n, i);\n' ...
       'end'];
str = sprintf(strrep(str,'%','%%'));
jCodePane.setText(str)
jScrollPane = com.mathworks.mwswing.MJScrollPane(jCodePane);
[jhPanel,hContainer] = javacomponent(jScrollPane,[10,10,300,100],gcf);
```

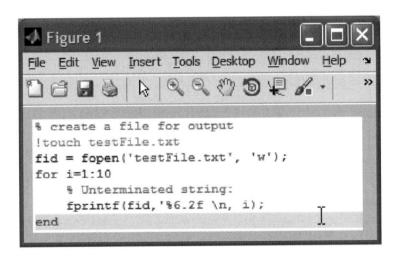

`SyntaxTextPane` **panel (MATLAB MIME type) (See color insert.)**

The nice thing about `SyntaxTextPane` is that it syntax-highlights on-the-fly as we type or edit in the `SyntaxTextPane` (assuming that we have not disabled editing with the *setEditable(flag)* method). This is exactly the behavior that we have come to expect in the full-blown MATLAB editor, and it can now be embedded as a simple panel in our GUI.

Despite its misleadingly simple look, `SyntaxTextPane` actually has most capabilities of the full-blown editor and not just syntax-highlighting. This includes multiple undo/redo actions;

smart indentation and commenting; automatic indication of corresponding block elements (if-end, for-end, etc. — also known as delimiter matching); drag-and-drop and cut-copy-paste support; and many others.

Line numbers can be added as a separate *glyph* gutter-panel.[94] This mimics the MATLAB Editor, which also uses a separate panel for line numbers and code folding.

Interested readers should use the ***uiinspect*** and ***checkClass*** utilities to explore the full capabilities offered by `SyntaxTextPane`. In this respect, it would be helpful to also look at its superclass (`SyntaxTextPaneBase`) and the related `SyntaxTextPaneUtilities` class.

Additional classes appear to provide wrappers for `SyntaxTextPane`, to enable the MATLAB Editor functionality. These include `MCommentWrapper`, `StateMRUFiles`, `STPPrefsManager`, `STPStateManagerFactory`, `STPStateManagerImpl`, `Syntax-TextPaneMultiView`, and a few others.

`SyntaxDelimiterPanel` is a related class, which displays the Delimiter Matching panel in the Keyboards tab of the Preferences window. Its constructor accepts six input args, which correspond to the logical or numeric values seen in the panel:

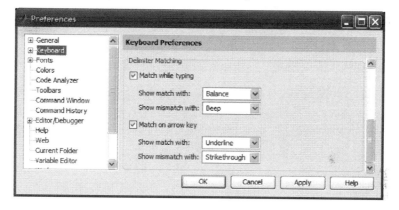

SyntaxDelimiterPanel

`FileExtensionsPanel` is a class that creates a panel with the specified list of strings to be used as file extensions. This is used in several places in the MATLAB Preferences window. Note that we need the *getComponent()* method to get the displayable panel:

```
jExtObj = com.mathworks.widgets.FileExtensionsPanel('',{'htm','HTML'});
jExtPanel = jExtObj.getComponent;
[jhPanel,hContainer] = javacomponent(jExtPanel,[10,10,200,100],gcf);
```

FileExtensionsPanel

`AutoCompletionList` also creates a component in a similar way. This component has a list of string items where the user can type in the header row and the cursor automatically selects the corresponding list item and autocompletes the rest of the entry. Invalid user entries are greeted with a beep and are not allowed by default (unless *setStrict(false)* is called). Items can be selected either by entering text in the header row, or by selecting a list item:

```
strs = {'This','is','test1','test2'};
strList = java.util.ArrayList;
for idx = 1 : length(strs), strList.add(strs{idx}); end
jPanelObj = com.mathworks.widgets.AutoCompletionList(strList,'');
javacomponent(jPanelObj.getComponent, [10,10,200,100],gcf);
```

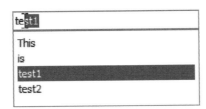

```
AutoCompletionList
```

`BusyAffordance` is another similar class that creates a visible panel internally. In this case, the panel presents an animated (spinning) icon and optional text label as long as the panel's object is in the "started" mode (the mode can be started/stopped numerous times). When the object is *stop()*ed, the icon and label are removed by default, but can be displayed un-animated (non-spinning) via the **PaintsWhenStopped** property:[†]

```
iconsClassName = 'com.mathworks.widgets.BusyAffordance$AffordanceSize';

iconsSizeEnums = javaMethod('values',iconsClassName);[‡]
SIZE_32x32 = iconsSizeEnums(2);   % (1) = 16x16
jObj = com.mathworks.widgets.BusyAffordance(SIZE_32x32,'testing...');
jObj.setPaintsWhenStopped(true);
javacomponent(jObj.getComponent, [10,10,80,80],gcf);
jObj.start;
     % do some long operation...
jObj.stop;
```

[†] Note: More advanced animated busy indications, including automated percentage and time-remaining labels, can be specified using `JBusyComponent` (http://code.google.com/p/jbusycomponent/ or http://bit.ly/gvPyAv), which is a JXLayers (https://jxlayer.dev.java.net/ or http://bit.ly/edIeym) decorator that can be applied to any displayable component. Also, see `ProgressBarDialog` control (Section 5.4), Swing's `JProgressBar` (Section 3.3.1), and the built-in ***waitbar*** function.

[‡] This causes an error on MATLAB R2009b and earlier versions. On these MATLAB releases, try the following instead: c1 = java.awt.Color(1,0,0); c2 = java.awt.Color(0,0,0); jObj = com.mathworks.widgets.BusyAffordance(c1,c2);

`BusyAffordance` **started...**   **...stopped**   **...stopped**
(animated spinning icon)   **(PaintsWhenStopped =** *false*)   **(PaintsWhenStopped =** *true*)

`SearchTextField` is yet another component that uses this internal panel creation behavior. The created component provides a text-entry box that is normally used to enter search terms in the Help Browser, but it can actually be used for entirely other purposes as an embedded component in our GUI. An optional light-gray prompt automatically disappears when the user clicks in the entry box and reappears when the focus leaves an empty entry box; a search icon appears on the right for as long as the entry box is empty and turns into a deletion icon when something is entered:

```
jObj = com.mathworks.widgets.SearchTextField('Enter search term:');
jPanel = jObj.getComponent;
[jhPanel,hContainer] = javacomponent(jPanel,[10,10,150,25],gcf);
```

Enter search term: 🔍	❘ I 🔍	help me!❘ I ✕

`SearchTextField`   **...user clicks in entry box —**   **...user types something — the**
initial view   **gray prompt disappears**   **search icon changes**
                                                    (clicking it will clear the text)

`SearchTextField`'s viewable panel is composed of a `PromptingTextField` editbox and the right-side search/clear icon. If the icon is not needed, use the editbox directly:

```
jEditBox = com.mathworks.widgets.PromptingTextField('Enter search:');
[jhEditBox,hContainer] = javacomponent(jEditBox,[10,10,150,25],gcf);
```

In order to process user entries, we need to trap the editbox's <Enter> event. This can be done on the internal editbox's **KeyTypedCallback** property, by specifying a callback function that will check whether the typed key was <Enter> or not:[†]

```
set(jPanel.getComponent.getComponent(0),'KeyTypedCallback',@myCbFunc)
```

The search box is accompanied in the MATLAB Help Browser with a popup that shows recent and suggested searches. This is apparently supported by the `SearchTextFieldHint` and `SearchTextFieldIntelliHints` widget classes.

[†] There must be a better way to do this, but this works for me in all practical use-cases.

`HyperlinkTextLabel` is a class that creates a `com.mathworks.mwswing.` `MJEditorPane` containing the requested HTML label. Remember to modify the background color from its default white to the actual background color at the label's position:

```
htmlStr = '<a href="">click this link!</a>'
jObj = com.mathworks.widgets.HyperlinkTextLabel(htmlStr);
jPanel = jObj.getHTMLPane;
color = get(gcf,'color');
jObj.setBackgroundColor(java.awt.Color(color(1),color(2),color(3)));
[jhPanel,hContainer] = javacomponent(jPanel,[10,10,150,25],gcf);

jObj.setEnabled(false);    % to disable the hyperlink
```

`HyperlinkTextLabel` **in regular, hover, and disabled modes**

There are other ways to present HTML-aware labels (see Section 6.9). However, `HyperlinkTextLabel` has an advantage for displaying hyperlinks: it automatically modifies the link color and cursor shape during mouse hover over the link, as shown in the screenshots above.

Hyperlinks are useless if they do nothing when clicked. In Java code, we can attach a hyperlink handler to the label, either as an optional second argument to the object constructor, or later by modifying its **HyperlinkHandler** property. In practice, it is much easier in MATLAB to simply set the object's **MouseClickedCallback** property:

```
hjObj = handle(jObj,'CallbackProperties');
set(hjObj,'MouseClickedCallback',@myHyperlinkCallbackFcn);
% an alternative:
set(hjObj,'MouseClickedCallback','web(''www.google.com'')');
```

This label can contain hyperlinks, hence the class's name, but the label can actually contain any HTML segment, and even multiple lines, as shown below:[†]

```
htmlStr = ['What <b>a bloody <i>mess!</i></b><br/>' ...
           '<a href = "http://www.google.com">click this link!'];
```

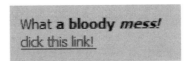

Multi-line `HyperlinkTextLabel`

[†] Additional methods of displaying hyperlinks are discussed in Sections 3.3.1, 6.5.2, 6.9, 8.3.1, and 8.3.2.

Note that **MouseClickedCallback** applies to the entire label area, including the invisible empty margins around the actual hyperlink and the non-hyperlink HTML segments. This is often the expected behavior, but for tight-fitting click sensitivity, use a tight-fitting label size or set the **HyperlinkHandler** property.

`TextPrintPanel` extends `MJPanel` to present the familiar MATLAB Editor page-setup panel. This can be integrated in our GUI:

```
options = com.mathworks.widgets.TextPrintPanel.SYNTAX_STYLE;
jPanel = com.mathworks.widgets.TextPrintPanel(options,true, ...
             'My selection...',[],'');
[jhPanel,hContainer] = javacomponent(jPanel,[10,10,250,250],gcf);
```

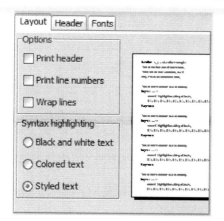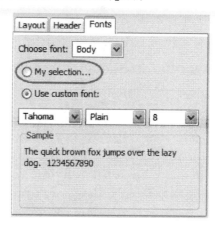

TextPrintPanel

5.5.2 Font-Selection Components

`DesktopFontPicker` is a component that extends `MJPanel` to present a font selection panel:

```
font = java.awt.Font('Tahoma',java.awt.Font.PLAIN,11);
jFontPanel = com.mathworks.widgets.DesktopFontPicker(true,font);
[jhPanel,hContainer] = javacomponent(jFontPanel,[10,10,250,170],gcf);
```

DesktopFontPicker **panel**

Instead of the "Use desktop font" label, we can use our own label:

```
jFontPanel.setUseDesktopFontLabel('Use Yair''s standard font...')
```

Non-standard `DesktopFontPicker` **panel**

There are several alternatives to `DesktopFontPicker`: we can use `FontPrefsPanel` (see Section 5.4), `com.mathworks.widgets.fonts.FontDialog`, or the fully-documented *uisetfont* function (which is basically a simple wrapper for `FontDialog`):

***uisetfont* dialog window**

When the window closes, we can retrieve the user's selected font using the *getSelectedFont()* and *getUseDesktopFont()* methods.

Font selection can also be shown with drop-downs (combo-boxes), rather than with lists as in `DesktopFontPicker`, `FontPrefsPanel`, or ***uisetfont***. Use of drop-downs significantly reduces the display "real-estate" required by the control. This is extremely useful in forms where the font selection is only one of several user-configurable options and where enough space must be reserved for other configuration controls. We can do this using the com.math-works.widgets.fonts.FontPicker class, which is an extension of the standard MATLAB MJPanel. FontPicker's constructor accepts optional parameters of a pre-selected font (a java.awt.Font object), boolean flag indicating whether to display sample text using the selected font, layout indicator, and list of selectable font names. Several screenshots of different parameter combinations are shown below:

```
jFontPicker = com.mathworks.widgets.fonts.FontPicker(font, sampleFlag, layout);
[hjFontPicker, hContainer] = javacomponent(jFontPicker, position, gcf);
```

font =	[]	java.awt. Font('Tahoma', java. awt.Font.PLAIN, 8)	[]
sampleFlag =	false	false	true
layout =	FontPicker. GRID_LAYOUT (=1)	FontPicker.LONG_ LAYOUT (=2)	FontPicker.LONG_ LAYOUT (=2)
position =	[10,200,140,40]	[10,200,225,20]	[10,200,225,80]

As before, the selected font can be retrieved using FontPicker.*getSelectedFont()*. These component variants are used by the MATLAB Fonts Preferences dialogs:

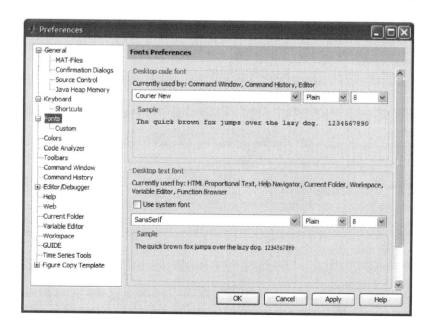

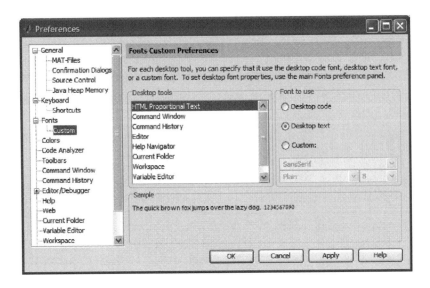

As a final alternative for font selection, we can use the JIDE font-selection component bundled in the jide-grids.jar file (see Section 5.7.2). This component has two variants: as a drop-down/combo-box (com.jidesoft.combobox.FontComboBox) and as a standard JPanel (com.jidesoft.combobox.FontChooserPanel):

```
jFont = java.awt.Font('arial black',java.awt.Font.PLAIN,8);
jFontPicker = com.jidesoft.combobox.FontComboBox(jFont);
[hjFontPicker, hContainer] = javacomponent(jFontPicker,position,gcf);
set(jFontPicker, 'ItemStateChangedCallback', @myCallbackFunction);
```

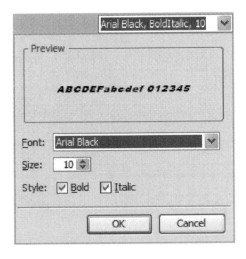

JIDE's FontComboBox

Within the callback function, use *getSelectedFont()* to retrieve the updated font. There is also a corresponding *setSelectedFont(font)* to programmatically update the control.

The combo-box presents a `FontChooserPanel`, which can be accessed (via the **PopupPanel** property or the corresponding *getPopupPanel()* method) after it has been initially created. Thereafter, the panel can be customized. For example, the preview text can be modified via the panel's **PreviewText** property (or the *setPreviewText* method).

The same `FontChooserPanel` can also be displayed as a standalone font-selection panel, unrelated to any combo-box. Different GUI requirements might prefer using a compact combo-box approach or the larger standalone panel.

5.5.3 Dialogs

`om.mathworks.widgets.Dialogs` is a class that provides seven static methods for displaying preset modal dialog windows (message boxes). Five of these methods display dialogs whose text labels are provided as input arguments, but whose button texts cannot be modified. The order of the input arguments is inconsistent, as can be seen in the example below. The different methods are also inconsistent in their return values: some return a `com.mathworks.widgets.Dialogs$Option` enumeration objects, while others return a numeric value. For all these reasons, it is advisable to use the documented *questdlg* function instead, since it is much more customizable:

```
>> import com.mathworks.widgets.Dialogs
>> choice = Dialogs.showEditAnyway([],'My title','My prompt')
choice =
EDIT_ANYWAY        <= a com.mathworks.widgets.Dialogs$Option enum
>> choice = Dialogs.showSaveDirtyFile([],'My prompt','My title')
choice =
2                  <= a numeric value: <Yes> = 0, <No> = 1, <Cancel> = 2
```

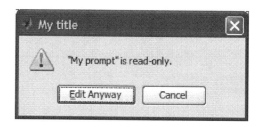 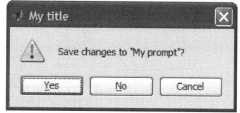

Dialogs-**presented message box** (Dialogs.*showEditAnyway* **and** Dialogs.*showSaveDirtyFile*)

 Note: The `Dialogs` class's interface has changed in R2010b: the *showEditAnyway()* method has been removed along with *showOverwriteCancel()*, and an additional input parameter was added to several methods — use *methodsview, uiinspect,* or *checkClass* to determine the correct interface on your specific target Matlab release.

Two `Dialogs` methods may still be useful, as they provide functionality unavailable in *questdlg*: the ability to skip message presentation based on a globally-stored preference flag, and the ability for the user, from within the message box, to indicate that the global preference should be modified to never again show this message:

showOptionalMessageDialog([], prompt, title, icon, prefName, defaultFlag) presents an optional message box with the specified *prompt* (a string or any Java component), title (a string), and icon (see below), depending on the logical value of the specified *prefName*, which is stored in the global preferences file (see Section 8.2). In this case, the preferences file will include an entry such as **myPrefName = Bfalse** following user selection of the checkbox:

```
import com.mathworks.widgets.Dialogs
icon = javax.swing.JOptionPane.ERROR_MESSAGE; % = 0
Dialogs.showOptionalMessageDialog([],'My prompt','My title',icon, ...
                                  'myPrefName',true);'
```

An **optional message box**

The possible icons are `javax.swing.JOptionPane.ERROR_MESSAGE` (=0), `INFORMATION_MESSAGE` (=1), `WARNING_MESSAGE` (=2), `QUESTION_MESSAGE` (=3), and `PLAIN_MESSAGE` (=1). Either the enumerated or the numeric values may be specified. If the chosen icon is `PLAIN_MESSAGE` (=1), then no icon will be displayed:[95]

```
Dialogs.showOptionalMessageDialog([],'My prompt','My title',-1, ...
                                  'myPrefName',true);
```

An icon-less optional message box

---

† The `ClosableToolTips` class, described above, uses a similar way to prevent a message from reappearing following a user indication (in that case, use a `LightButton` rather than a checkbox).

Note: Unlike the standard dialog windows above (which are best replaced with **questdlg**), this optional message box does NOT return any value. The window can only be dismissed, with no optional buttons for user selection, and so there are no data to convey in a return value. Still, the inconsistency is disturbing.

Instead of a prompt string, we can display any Java component. This would normally be a `JPanel` component that contains several subcomponents in some layout. To illustrate the point, let us display a simple `JButton` instead:

```
promptObject = javax.swing.JButton('click me!');
Dialogs.showOptionalMessageDialog([],promptObject,'My title',icon, ...
                           'myPrefName',true);
```

An optional message box with a custom prompt component

Similarly, *showOptionalConfirmDialog([], prompt, title, options, icon, prefName, default-Value, defaultFlag)* presents an optional dialog with the specified prompt, title, option buttons, and icon, depending on the logical value of the specified *prefName*. The possible button *options* are `javax.swing.JOptionPane.DEFAULT_OPTION` (=–1), `YES_NO_OPTION` (=0), `YES_NO_CANCEL_OPTION` (=1), and `OK_CANCEL_OPTION` (=2). If *prefName* is already set (i.e., user chose to never show the dialog again), then the method returns *defaultValue*; otherwise, it returns the selected button's numeric value:

```
options = javax.swing.JOptionPane.YES_NO_CANCEL_OPTION;      % = 1
icon = javax.swing.JOptionPane.QUESTION_MESSAGE;             % = 3
defRetVal = -1;
Dialogs.showOptionalConfirmDialog([], 'My prompt', 'My title', ...
                  options, icon, 'myPrefName', defRetVal, true);
```

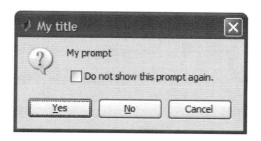

An optional confirmation dialog

showOptionalMessageDialog and *showOptionalConfirmDialog* correspond to standard Swing methods having similar names. Refer to the documentation for additional information about these and related methods.[96] If the required dialog customization is greater than what is possible to achieve with the `Dialogs` class, the user should try to use `JOptionPane`[97] (or its MATLAB extension `com.mathworks.mwswing.MJOptionPane`) or `JDialog`[98] (or its MATLAB extension `com.mathworks.mwswing.MJDialog`). In these cases, place the prompt and checkbox within a panel, and present this panel to the user using these generic Swing classes.

5.5.4 Closable (Collapsible) Panels

`com.mathworks.widgets.ClosablePanel` extends `com.mathworks.mwswing.MJPanel`, so that the panel can be collapsed/opened programmatically, or by clicking its title.[†] The following example contains a simple button component, but any Java object (including complex options/data panels) can be added to the `ClosablePanel`'s content:

```
jButton = com.mathworks.mwswing.MJButton('MJButton');
jPanel = com.mathworks.widgets.ClosablePanel('name','my Title',jButton);
[jhPanel,hContainer] = javacomponent(jPanel,[20,20,100,100],gcf);
jhPanel.setOpen(false);        % programmatically close the panel
jhPanel.setEnabled(false);     % disable the panel
```

Open, closed, and disabled `ClosablePanel`

The title component (a `ClosablePanel$ClosablePanelButton`) is actually implemented as a checkbox (!), with the arrow icons replacing the standard checkbox square icons. This is a good example of disguising a component to look like another, in order to make use of the former's functionality (in this case, checkbox's click handling).

`ClosablePanels` are usually grouped and contained within a `ClosablePanelContainer`, which is an `MJPanel` extension (like `ClosablePanel`), providing wrapper functionality:

```
jp = com.mathworks.widgets.ClosablePanelContainer;
```

[†] JIDE's `com.jidesoft.pane.CollapsiblePane`, `CollapsiblePanes` (Section 5.7.1) have a similar functionality.

```
jp.addPanel('panel1','my Title #1',javax.swing.JButton('click me!'));
jp.addPanel('panel2','my Title #2',javax.swing.JCheckBox('select?'));
options = {'Red','Green','Blue'};
jp.addPanel('panel3','my Title #3',javax.swing.JComboBox(options));
[jhScroll,hContainer] = javacomponent(jp,[10,10,200,130],gcf);
```

Note how scrollbars appear/disappear automatically in a `ClosablePanelContainer`, whenever it is resized or subpanels are opened/closed (panel 2 in these screenshots):

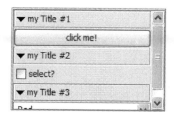 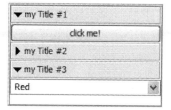

`ClosablePanelContainer` **with auto-scrollbars**

`ClosablePanelContainer` can add subpanels in two manners: *addPanel()* adds a `ClosablePanel` at the bottom; *insertPanel(name,title,contents,index)* adds it at the specified position index (0 = top). There is also a *removePanel(name)* method.

Individual subpanels can be opened and closed via the *setPanelOpen(name,flag)* method, and can be hidden/shown using the *setPanelVisible(name,flag)* method. There are also corresponding *isPanelOpen(name)* and *isPanelVisible(name)* methods.

The `ClosablePanelContainer`'s *setEnabled(flag)* method was overridden to provide the functionality of disabling all the contained subpanels individually. Subpanels can also be disabled/enabled individually, but we will need to travel down the panel hierarchy (via the *getComponent(index)* method) to access the individual subpanels.

`ClosablePanel` and `ClosablePanelContainer` objects can be stacked in a multi-level (hierarchical) manner, as shown below. In some cases, we may need to programmatically call the top panel's *revalidate()* method after modifying the layout of internal subpanels to force the subpanels, to "stick" together without gaps:

```
% Top ClosablePanelContainer
jp1 = com.mathworks.widgets.ClosablePanelContainer;
jp1.addPanel('panel1','my Title #1',javax.swing.JButton('click me!'));
jp1.addPanel('panel2','my Title #2',javax.swing.JCheckBox('select?'));
options = {'Red','Green','Blue'};
jp1.addPanel('panel3','my Title #3',javax.swing.JComboBox(options));

% Bottom ClosablePanelContainer
jp2 = com.mathworks.widgets.ClosablePanelContainer;
jp2.addPanel('panel1','my Title #1',javax.swing.JButton('click me!'));
```

```
jp2.addPanel('panel2','my Title #2',javax.swing.JCheckBox('select?'));
jp2.addPanel('panel3','my Title #3',javax.swing.JComboBox(options));

% Add the two ClosablePanelContainers in a larger container
jp3 = com.mathworks.widgets.ClosablePanelContainer;
jp3.addPanel('cp1','ClosablePanelContainer #1',jp1);
jp3.addPanel('cp2','ClosablePanelContainer #2',jp2);
[jhScroll,hContainer] = javacomponent(jp3,[10,10,200,250],gcf);
jp3.revalidate;

% Set individual panels properties
jp1.setPanelOpen('panel2',false);
jp2.setPanelOpen('panel1',false);
jp2.setEnabled(false);
```

Multi-level (hierarchical) `ClosablePanelContainer` **with auto-scrollbars**

5.5.5 Specialized Widgets

The widgets class has several specialized sub-packages, each of which deals with a particular aspect. Some of the interesting ones are

- **color** — the com.mathworks.widgets.color sub-package contains color-selection components that mirror the equivalent components in the com.mathworks.mlwidgets package, explained in Section 5.4.1. The relevant classes are ColorPicker, ColorPickerPanel, ColorPickerUtils, and ColorDialog. Since the results look similar, I suggest using the mlwidgets version rather than widgets.color, since mlwidgets appears to be newer and, therefore, possibly more powerful and/or less buggy.

- **datatransfer** — classes supporting MATLAB's Drag-and-Drop and Cut-Copy-Paste operations (which use similar mechanisms). See Section 3.7 for details.

- **find** — classes that implement MATLAB's Find/Replace functionality. This includes the standard popup dialog (also see related Section 8.7.2):

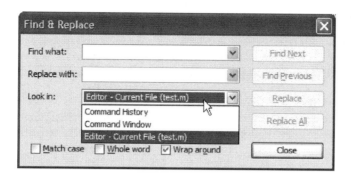

Standard Find/Replace dialog window

The com.mathworks.widgets.find.FindClientRegistry class apparently enables registering a specific Find/Replace functionality (defined in a class that implements the com.mathworks.widgets.find.FindClientInterface interface), but I have not tested this myself.

- **fonts** — see the description in Section 5.5.2.
- **glazedlists** — provides the com.mathworks.widgets.glazedlists. GlazedTableSupport support class, which helps set up Glazed-List functionality[99] on JTables. This is the open-source functionality that enables real-time (interactive) dynamic sorting and filtering on table data.[100] The GlazedLists JAR file is prebundled in the MATLAB installation, and so it can be used directly.

A sample usage of glazed lists can be found in the MLint (Code Analyzer) preferences panel, which uses a com.mathworks.mlwidgets.text.mcode.MLintTable that relies on glazed list:

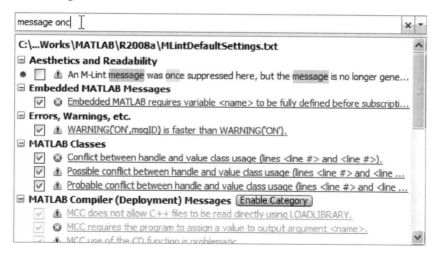

Glazed lists in action (note the interactive filtering and highlighting)

Note: a different (I think better) table filtering is described in Section 4.1.4.

- **grouptable** — implements the GroupTable extension of JIDE's TreeTable. See Section 5.7.2 for a description of JIDE tables.
- **incSearch** — the com.mathworks.widgets.incSearch.IncSearch class implements incremental search, used in the Command Window and the Editor.
- **jidesoft.grid** — this package holds only two classes: com.mathworks.widgets. jidesoft.grid.MWSortableTable and MWCellSpanTable, which are, respectively, wrappers for JIDE's corresponding com.jidesoft.grid. SortableTable and com.jidesoft.grid.CellSpanTable. See Section 5.7.2 for a description of JIDE tables.
- **text** — this package and its subpackages contain numerous classes that support the Editor, including syntax-highlighting, support for different languages, code folding, printing, formatting, etc.
- **wizard** — com.mathworks.widgets.wizard.WizardFrame appears to create a dedicated window Frame for implementation of multipanel wizard dialogs. The code requires a Java class that implements the com.mathworks.widgets.wizard. IWizardContents interface that defines methods for navigating back and forth in the wizard. Java-savvy programmers will be able to relatively easily create Java classes that implement this interface and then use the infrastructure provided by WizardFrame; MATLAB programmers will need to rely on other means (such as the undocumented built-in *wizard* function).

5.6 MLServices Package

The com.mathworks.mlservices package contains a variety of support services to major MATLAB components: the Desktop, Command Window, Editor, Variable Editor, and so on. All these classes have static methods that supply their functionalities. The classes apparently use an internal registry (MLServicesRegistry) to register themselves, but, in practice, we do not need to use this registry — we can access their static support methods directly.

The available support classes in MATLAB 7.12 (R2011a) are

- MatlabDebugServices
- MatlabDesktopServices
- MLArrayEditorServices
- MLCommandHistoryServices
- MLCommandWindowServices
- MLEditorServices
- MLExecuteServices
- MLHelpServices
- MLInspectorServices
- MLLicenseChecker [†]

[†] Note the missing "Services" suffix for the MLLicenseChecker class.

- `MLPathBrowserServices`
- `MLPrefsDialogServices`
- `MLWorkspaceServices`
- `FileExchangeServices`†

`MatlabDebugServices` provides support functionality for the MATLAB debugger. All of the debugger functionalities can be accessed by regular built-in MATLAB functions (***dbstep, dbcont, dbstatus, mdbstatus***, etc.) or by the `MLExecuteServices` class (see below). I could not find any actual use of `MatlabDebugServices` for MATLAB users as opposed to using the fully-supported built-in MATLAB functions.

`MatlabDesktopServices` provides access to the main MATLAB Desktop components: *getDesktop()* returns a `com.mathworks.mde.desk.MLDesktop` object (which can also be obtained via `com.mathworks.mde.desk.MLDesktop.getInstance`), which is explored in detail in Chapter 8; *showCommandHistory(), showCommandWindow(), showFileBrowser(), showHelpBrowser(), showProfiler()*, and *showWorkspaceBrowser()* bring the requested panel/window into focus, displaying it if it was previously closed or hidden; *closeCommandHistory(), closeCommandWindow(), closeFileBrowser(), closeHelpBrowser(), closeProfiler()*, and *close-WorkspaceBrowser()* close the requested panel/window; finally, *setCommandAndHistoryLayout(), setCommandOnlyLayout()*, and *setDefaultLayout()* organize the desktop component in a predetermined manner (see Section 8.1.3).

`MLArrayEditorServices` provides support services for the Variable Editor (called the Array Editor prior to R2008a (MATLAB 7.6)). See Section 8.7.5 for more details.

`MLCommandHistoryServices` supports the Desktop's Command History panel. The support methods are *getAllHistory(), getSessionHistory(), removeAll()*[101] and *save()*. Note that *getAllHistory()* and *getSessionHistory()* return a Java `String` array that should be converted to a MATLAB string cell array using the built-in ***cell*** function:

```
% The following are equivalent:
com.mathworks.mlservices.MLCommandHistoryServices.getAllHistory.cell
cell(com.mathworks.mlservices.MLCommandHistoryServices.getAllHistory)
```

The following returns the last entered Command Window expression (command):[102]

```
import com.mathworks.mlservices.MLCommandHistoryServices;
history = MLCommandHistoryServices.getSessionHistory;
lastCommand = char(history(end));
```

Any previous command can similarly be retrieved and executed programatically:[103]

```
eval(char(history(end-10))); % execute the 10th previous command
```

---

† `FileExchangeServices` was added in MATLAB 7.11 (R2010b) and is unavailable in earlier MATLAB releases.

MLCommandHistoryServices is closely related to the com.mathworks.mde. cmdhist.* classes (particularly CmdHistoryWindow and CmdHistory), which you may find useful. Just remember that they may change in future MATLAB releases without prior notice.

MLCommandWindowServices provides the *hasFocus()* and *isJavaCWInitialized()* methods: *hasFocus()* determines whether the MATLAB Command Window is currently in focus (e.g., as opposed to some GUI window); *isJavaCWInitialized()* determines whether the Command Window (CW) is Java-based, as opposed to a native CW when MATLAB is started with the "-nojvm" startup switch. Some CW functionality (e.g., tab completion and line-wrapping) is only available in Java.[104]

MLCommandWindowServices is closely related to the com.mathworks.mde.cmdwin.* classes, similarly to MLCommandHistoryServices described above. As above, interested readers may find these additional classes useful. See Section 8.3 for additional details.

MLEditorServices provides access to the MATLAB Editor. *getEditorApplication()* returns a com.mathworks.mde.editor.MatlabEditorApplication object (which can also be obtained via com.mathworks.mde.editor.MatlabEditorApplication. *getInstance*). Several other methods are available to open/access/save/reload/close documents. The full list of these functions is detailed in Section 8.4.1.

MLExecuteServices provides Command-Window evaluation functions. This is useful in GUI callbacks that wish to run MATLAB commands exactly as if they were manually entered at the Command Window. For example, I often use a generic error-handling routine (**handleError**) that presents an informative **msgbox** about the details of trapped errors (exceptions). Let us take a simple example (with annotated line-numbers):

```
test.m file:
1:    function test
2:        ...
3:        data = 1 : 5;
4:        test2(data)
5:        ...
10:       function test2(data)
11:          try
12:             b = data(6);    < = Error! Index exceeds matrix dimensions
13:          catch
14:             handleError;
15:          end
16:       end % test2
17:    end % test


handleError.m file:
1:    function handleError
2:        err = lasterror;
3:        msgbox(...);
4:        ...          < = place breakpoint here, following msgbox display
9:    end % handleError
```

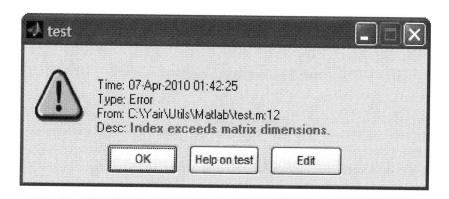

In this generic ***msgbox***, the <Edit> button has a callback that opens the Editor at the offending error location (in this case, test.m line #12). It would be beneficial to place a breakpoint following the ***msgbox***'s presentation, so that the user could debug the cause of the error (in this case, check why the index exceeded the matrix dimensions). For this, we need to issue the ***dbup*** command, in order to move from the generic error-handler (handleError.m) workspace context to the offending function's (test.m):

```
K>> dbstack
> In handleError at 5        <= breakpoint is placed here
  In test > test2 at 14      <= try-catch error handler calls handleError
  In test at 4
```

Unfortunately, adding '***dbup***' to the callback string has no effect — the Editor/Debugger immediately returns the stack focus to the breakpoint location (handleError.m) rather than to the error location (test.m). We need to actively enter "dbup" at the Command Window. MLExecuteServices comes to our rescue, by simulating a Command-Window entry without need for any actual user interaction:

```
% Prevent a dbup error if not in debug mode by using try-catch:
dbupStr = 'try dbup catch, end';

% Prepare an MLExecuteServices command
dbupStr = ['com.mathworks.mlservices.MLExecuteServices.consoleEval('''
           dbupStr '''); ']; % note double quotes on internal command

% Move the editor cursor to the offending error line (#12)
% which is *NOT* the line that called handleError (#14):
setLineStr = sprintf('opentoline(''%s'',%d);', fileName, lineNumber);

% Assemble everything in the <Edit> button callback string
set(hEditButton, 'Callback', [dbupStr,dbupStr,setLineStr]);
```

Interested readers are referred to Chapter 9 for other mechanisms of accessing the MATLAB Workspace and Command Prompt from within Java. I suspect that `MLExecuteServices` itself uses JMI, one of these mechanisms.

`MLHelpServices` provides support functions for the Help Browser: *invoke()* brings the Help Browser into focus, displaying it if it was not already shown (like the built-in ***helpbrowser*** and ***doc*** functions); *hideNavigator(), showNavigator(),* and *toggleNavigator()* open/close the navigation panel (which contains the Contents and Search Results); *isNavigatorShowing()* returns the panel's state. Note that unlike `MatlabDesktopServices` and `MLEditorServices`, `MLHelpServices` does not provide a getter method for the help browser object. However, it can be gotten directly via `com.mathworks.mde.help.HelpBrowser`.*getInstance()*.

getDocRoot() returns the root folder containing the help files. *getDocRoot* is equivalent to the semidocumented built-in ***docroot*** function, except that forward slashes (/) are replaced in ***docroot*** by backslashes (\):

```
>> com.mathworks.mlservices.MLHelpServices.getDocRoot
ans =
C:/Program Files/Matlab/R2010a/help

>> docroot
ans =
C:\Program Files\Matlab\R2010a\help
```

There is also a corresponding *setDocRoot(folderName)* (or: ***docroot***(*foldername*)), but using it will disable MATLAB's online help, so please use it with caution.

Note that the Help Browser component displays the help pages (which are really HTML webpages) using an integrated browser. It is, therefore, not surprising that many of the functions have web-browser relationships:

getCurrentLocation() returns the webpage's URL, while *setCurrentLocation(url)* sets it; *setCurrentLocationAndHighlightKeywords(url,{keywords})* sets the URL and also highlights the requested keywords/phrases:

```
>> url = com.mathworks.mlservices.MLHelpServices.getCurrentLocation
url =
jar:file:///C:/Program Files/Matlab/R2010a/help/techdoc/help.jar! /matlab_prog/
bq9l448-1.html#bq9tdlq-1
```

We can display help URLs in any MATLAB-based browser, as explained in Section 8.3.2. An alternative is to display it in a standalone window (a Java `JFrame`):

```
docRoot = char(com.mathworks.mlservices.MLHelpServices.getDocRoot);
url = ['jar:file:///' docRoot '/techdoc/help.jar!/ref/lasterror.html'];
com.mathworks.mlservices.MLHelpServices.cshDisplayFile(url);
```

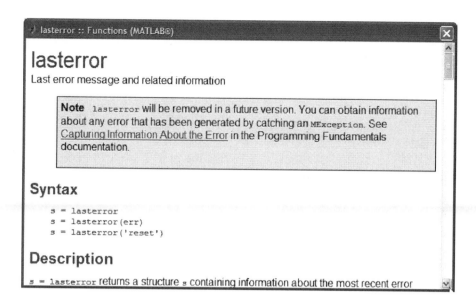

cshSetSize(width,height) and *cshSetLocation(xFromLeft,yFromTop)* set the displayed window frame's size and location, respectively.

getHtmlText() returns the HTML contents currently being displayed in the HTML Browser; the corresponding method *setHtmlText(text)* updates the displayed contents, and *setHtmlText-AndHighlightKeywords(text,{keywordsCellArray})* does the same while also highlighting the requested keywords/phrases, as discussed in Section 8.3.2.

`MLHelpServices` provides many other methods, but all the useful ones already have the corresponding built-in MATLAB functions (***demo, docsearch***, and ***doc***). There are also some related semidocumented pure-MATLAB functions that users may be interested to explore: ***demowin, helpview, help2html, helpdesk***, and ***helpinfo***.

`MLInspectorServices` provides support methods for the built-in property inspector. *inspectObject(handle)* presents a property inspection table for the specified handle, which is a Java object reference or a MATLAB ***handle*** (i.e., inspect ***handle(gcf)*** rather than ***gcf***); *inspectObject(handle,flag)* accepts an optional second *flag* argument (default = true), which indicates whether or not to bring the inspector window forward (into focus); *inspectIfOpen(handle)* is similar, but it only inspects the object if the inspector window is currently being displayed; *inspectObjectArray(handles,flag)* inspects an array of objects and displays the values of their common properties (properties which are unavailable in some of the objects are not displayed). The *handles* can be specified using two MATLAB formats: [handle1, handle2, …] for similarly typed handles, and {handle1, handle2, …} for handles of unrelated classes.

isUDDObjectInJava(object) and *isUDDObjectArrayInJava(handles)* indicate whether the supplied objects are UDD (MATLAB HG ***handle***) objects or not; *getRegistry()* provides access to the inspector's properties registry (see Sections 5.7.3 through 5.7.6 and the ***uiinspect*** utility); *selectProperty(propName)* selects a specific property.

activateInspector() and *invoke()* have a similar functions — to display and focus on the inspector window; *isInspectorOpen()* indicates whether the inspector is displayed; *refreshIfOpen()* refreshes the object's property values; *toFront()* brings a visible inspector window forward (into focus); *closeWindow()* closes the inspector window.

setUseNewInspector(flag) indicates whether to use the new MDE inspector version (com. mathworks.mde.inspector.Inspector, used since MATLAB R2006a+ or 7.2) or the old IDE inspector version (com.mathworks.ide.inspector.Inspector, used up to MATLAB R14SP3 or 7.1).[105] By default, *flag* = true (new MDE). *isNewInspector()* returns the current inspector's type (true=MDE; false=IDE). The IDE inspector was removed in R2011a along with these two methods:

```
jObject = javaObjectEDT('com.mathworks.mwswing.MJFileChooser');
com.mathworks.mlservices.MLInspectorServices.inspectObject(jObject);
com.mathworks.mlservices.MLInspectorServices.setUseNewInspector(false)
com.mathworks.mlservices.MLInspectorServices.inspectObject(jObject);
```

We can also pass these settings and queries directly to MATLAB's **inspect** function (again, up to R2010b):

```
isOldFlag = inspect('-isnewinspector');
inspect('newinspector','off'); % default = 'on'
inspect(jObject);
```

Alternatively, we can always invoke a specific inspector version directly:

```
com.mathworks.ide.inspector.Inspector.activateInspector; % old
com.mathworks.mde.inspector.Inspector.activateInspector; % new
```

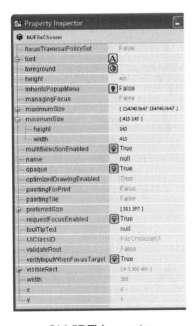

Old (IDE) inspector   **New (MDE) inspector**

Section 5.7.3 provides additional details about MATLAB's property inspector. The inspector object happens to be a very interesting MATLAB component from a technical standpoint. It is also a prime example of using JIDE tables (property grid to be exact) within MATLAB.

`MLLicenseChecker` checks whether or not the specified toolbox(es) is/are included in the MATLAB license, using *hasLicense(toolboxName)* and *hasLicenses(toolboxNames).*

`MLPathBrowserServices` provides MATLAB path browser functionality. In reality, this class only has a single static *invoke()* method that presents the path browser window:

```
com.mathworks.mlservices.MLPathBrowserServices.invoke;
```

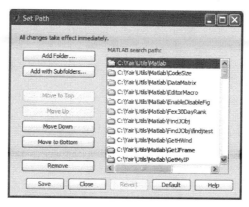

MATLAB's path browser

`MLPrefsDialogServices` provides functionality relating to the Preferences dialog window. *showPrefsDialog()* simply displays this dialog (or brings it into focus); *showPrefsDialog(prefs-PanelName)* also sets the shown preferences panel to the one requested. Some of the preference panels are children of other panels — their names are specified as parentPanelName.childPanelName. For example,

```
panel = 'Editor/Debugger.Code Folding';
com.mathworks.mlservices.MLPrefsDialogServices.showPrefsDialog(panel)
```

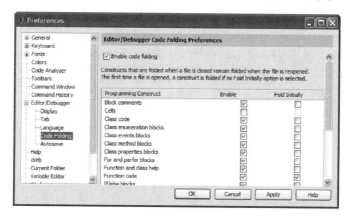

Displaying a specific preference panel

`MLPrefsDialogServices` also provides the *registerPanel(...)* and *unregisterPanel(...)* methods, which are apparently accessor methods for adding and removing preference panels as explained in Section 5.4.

`MLWorkspaceServices` provides functionality relating to the Desktop Workspace Browser. *invoke()* brings the Workspace Browser panel into focus (this can also be done via `com.mathworks.mlservices.MatlabDesktopServices.showWorkspace-Browser`); *getSelectedClasses()*, *getSelectedNames()*, and *getSelectedSizes()* return information about the selected variables in the Worspace Browser panel. The returned data is a Java `String` array that should be converted to a MATLAB string cell array using the built-in ***cell*** function, as discussed above. See Section 8.6 for additional details about MATLAB's Workspace Browser.

Finally, `FileExchangeServices`, newly added in MATLAB 7.11 (R2010b), provides functionality to search the MATLAB File Exchange (known in the MATLAB community as FEX),[106] functionality that was integrated into the MATLAB Desktop with much fanfare in R2009b.[107] The Desktop's FEX integration existed since R2009b, but `com.mathworks.mlservices.FileExchangeServices` was only added in R2010b. Its single static method *search(string)* searches FEX for utilities that contain the specified search string and presents the results in a dockable Desktop panel/window:

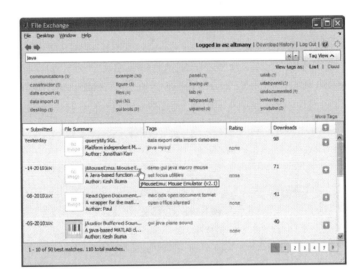

MATLAB Desktop's File Exchange search-results window

5.7 JIDE

JIDE is a commercial set of packages from www.jidesoft.com that is bundled in the MATLAB installation. JIDE classes significantly extend the basic Swing functionality. They are widely used by MATLAB and can also be used in any user MATLAB application.

The MATLAB-bundled JIDE packages include (see Section 5.1):

- jarext/jide/jide-action.jar — command bars (JIDE Action Framework).[108]
- jarext/jide/jide-common.jar — tabbed/option panes; calendar, gripper, and other UI controls; PLAFs; multipage dialogs; and so on.[109] This package is open-source.[110]
- jarext/jide/jide-components.jar — tabbed-panes, document panes, status bars, animations, balloon tips, and so on.[111]
- jarext/jide/jide-dialogs.jar — wizard, multipage, and other dialog windows.[112]
- jarext/jide/jide-dock.jar — support for dockable frame windows.[113]
- jarext/jide/jide-grids.jar — specialized `JTree`, `JList`, and `JTable` classes, with support for sorting, filtering, grouping, nesting, cell-spanning, and so on.[114]
- jarext/jide/jide-shortcuts — keyboard shortcuts support.

JIDE has extensive documentation, both in book (PDF) form† and in online reference (javadoc) form.[115] There is also a very active user forum,[116] with 50K posts.

 Note 1: JIDE docs refer to the latest JIDE version, but MATLAB uses an older version, (`com.jidesoft.utils.Lm.`*getProductVersion()*) for code-freeze configuration reasons. For example, R2011b uses JIDE version 2.8.7 from June 2010, not the newest 3.2.2 as of September 2011.[117] This may cause inconsistencies between the documentation and code.

Note 2: JIDE is a commercial product. We may not use it (except for the open-source jide-common package) without JIDESoft's permission outside the MATLAB environment. For any licensing questions, contact sales@JIDESoft.com.

5.7.1 Important JIDE Classes

There is not enough space in this book for a full description of the numerous features available in the bundled JIDE packages (this would require an entire book). Instead, I will just list the classes that I believe are more important and then describe JIDE tables in more detail. Many JIDE classes have specialized derived classes (e.g., renderers for Tree/List/Table) or related classes (builders, factories, or interfaces). Interested readers are encouraged to use the documentation resources that were given above for additional details.

Usage notes: all JIDE components, even those that have internal *show()* methods, require *java-component* to be displayed in a MATLAB figure window. Some JIDE components (e.g., `com.jidesoft.tooltip.BalloonTip`) have mechanisms of automatic component removal (upon mouse click, etc.), events that are not detected by MATLAB. So, unless we programmatically check for it, odd artifacts will appear in the GUI when it repaints. I tried using property listeners and callbacks to detect this event, but this did not work, so we are left with a timer that periodically checks the component's **Visible** property value and updates the component's container handle

---

† The online links to the PDFs were specified in the references above, separately for each of the JIDE packages.

appropriately. Also, note that although some components have a transparent background, they display with opaque backgrounds in MATLAB (see Section 7.3.3 for an explanation).

- com.jidesoft.action (part of jide-action.jar)
  - CommandBar — dockable/floatable/shrinkable/closable toolbar
  - CommandMenuBar — JMenuBar extension supporting floating/shrinking
  - JideMenu — JMenu extension supporting floating/shrinking
- com.jidesoft.swing (part of jide-common.jar)
  - CheckBoxList/CheckBoxTree — a JList/JTree that supports a checkbox in the row/tree node (see Section 5.2.2).
  - Calculator — component used by CalculatorComboBox (part of JIDE grids).
  - JideSwingUtilities — general functionality common to all swing components. This is a very important class, well worth exploring.[118]
  - JideButton, JideToggleButton, and so on: common button components.
  - JideSplitButton — button/popup-menu combination (see Section 4.5.4).
  - JidePopupMenu — extends JPopupMenu, ensuring that the popup-menu contents are within the screen boundary. If the popup menu is too long, it will automatically display scroll buttons at the menu's top and bottom.
  - JideMenu — extends JMenu to allow lazy creation of menu items and to allow specification of the popup menu's alignment.
  - JideSplitPane — a split pane that supports multiple splits.
  - JideScrollPane — extends JScrollPane to support RowFooter, ColumnFooter, and corner components on either side of the scrollbars.
  - SimpleScrollPane — extends JScrollPane to use four scroll buttons to do the scrolling; tt has no scrollbar.
  - JideTabbedPane — extends JTabbedPane to support many different tab styles, tab resize modes, tab leading/trailing component, and so on.
  - JideBoxLayout — similar to BoxLayout but can support different constraints to give child components of different resize weights.
  - JideBorderLayout — extends BorderLayout just to make the north and south components of the same width as the center component.
  - PartialLineBorder,PartialEtchedBorder,PartialGradientLine Border — extend LineBorder/EtchedBorder to only paint lines on certain sides.
  - AutoResizingTextArea — a JTextArea that automatically resizes vertically as text is added into it.
  - LabeledTextField — a JTextField that supports a JLabel in front of it. See related com.mathworks.widgets.FormPanel in Section 5.5.1.
  - MultilineLabel — a JTextArea that looks like a JLabel but supports multiple lines; also see Sections 5.5.1 and 6.9.
  - RangeSlider — a JSlider with two thumbs to specify a min/max range.

- `StyledLabel` — extends `JLabel` to support different text styles (see details and sample usage in Section 5.5.1).
- `TristateCheckBox` — a checkbox that has three states (on/off/maybe; see Sections 5.2 and 6.4).
- `FolderChooser` — a folder-selection component that extends `JFileChooser` with additional functionality.
- `MarqueePane` — continuously scrolling data pane (e.g., stock tickers and news).
- `TitledSeparator` — a `JSeparator`-like component with a title.
- `AutoCompletion` — auto text-completion for `JComboBox`, `JText-Component`.
- `ResizablePanel`, `ResizableWindow`, `ResizableFrame`, `Resizable-Dialog`.
- `Searchable` — search support for `ComboBox/List/Table/Tree/Text-Component`.
- `SearchableBar` — a searching component similar to the Firefox browser's search bar, using the `Searchable` functionality.
- `Overlayable` — enables placing one component on top of another to display icons/text such as validation error and process indicator.
- `com.jidesoft.dialog` (part of jide-common.jar)
  - `StandardDialog` — a `JDialog` supporting commonly used dialog standards
  - `ButtonPanel` — supports a panel for buttons in an OS-aware way
- `com.jidesoft.hints` (part of jide-common.jar)
  - `IntelliHints` — dynamic hints to help user typing (see `AutoCompletion`), used for IDE IntelliSense (use `ListDataIntelliHints`, `FileIntelliHints`)
- `com.jidesoft.icons` (part of jide-common.jar)
  - `ColorFilter` — creates a color filter that can dim/brighten images
- `com.jidesoft.document` (part of jide-components.jar)
  - `DocumentPane` — a tabbed-document implementation
- `com.jidesoft.pane` (part of jide-components.jar)
  - `CollapsiblePane` — also see Section 5.5.4
  - `FloorTabbedPane` — extends `JTabbedPane` for an Outlook 2000 navigation-pane appearance
  - `OutlookTabbedPane` — extends `JTabbedPane` (Outlook 2003 appearance)
- `com.jidesoft.status` (part of jide-components.jar)
  - `StatusBar` — a tabbed-document implementation, similar (but entirely unrelated) to the MATLAB Editor
  - `StatusBarSeparator` — a separator component within a `StatusBar`
  - `StatusBarItem` — superclass for `StatusBar` components that are all called `xxxStatusBarItem`, where `xxx` = `Button`, `ComboBox`, `Empty`, `Label`, `Memory`, `OvrIns`, `Progress`, `Time`, and `Resizable` (=window gripper)

- `com.jidesoft.alert` (part of jide-components.jar)
  - `Alert` — a popup alert message, similar to Outlook's new-email alert
- `com.jidesoft.animation` (part of jide-components.jar)
  - `CustomAnimation` — supports component entry/exit animation using a variety of customizable fly/zoom/fade effects
- `com.jidesoft.tooltip` (part of jide-components.jar)
  - `BalloonTip` — a cartoon-baloon-like tooltip/alert message, using a variety of shadows (`com.jidesoft.tooltip.shadows.*`) and shapes (`.shapes.*`)
- `com.jidesoft.tipoftheday` (part of jide-dialogs.jar)
  - `TipOfTheDayDialog` — creates a standard Tip-of-the-Day dialog window, similar to MATLAB's unrelated internal ***tipoftheday*** function
- `com.jidesoft.wizard` (part of jide-dialogs.jar)
  - Several classes that facilitate creating standard-looking multipage dialog windows (aka "*wizard*"), with banner, navigation panel, steps, and so on
- `com.jidesoft.docking` (part of jide-dock.jar)
  - Several classes that support dockable windows. Interested developers can try using these classes to achieve the so-far-impossible task of docking Matlab figures in other MATLAB figures.[119] Note that I have never tried this and I do not know if this is even possible (I would be happy to learn ...)
- `com.jidesoft.plaf` (part of jide-common.jar)
  - Several Look-and-Feel framework classes (see Section 3.3.2 for details).

5.7.2 *JIDE Grids*

JIDE grids is a class library that contains JIDE extensions for the standard Swing `JTable` and `JTree`. The library resides in the jide-grids.jar file, which is the largest JIDE file bundled with MATLAB as of R2011a.

Since Tables and Trees are such an important and useful GUI concept, this JIDE library deserves more detailing. This subsection provides an overview of JIDE grids, and subsequent subsections provide sample usage and detailed examples.

jide-grids.jar includes the following classes:

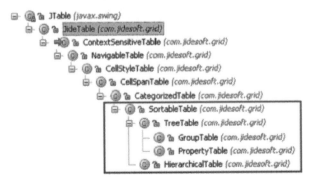

JIDE Grids class hierarchy. We would normally use only one of the marked classes

- com.jidesoft.grid
  - JideTable extends JTable for functionality that is important in any table and should have been part of the basic JTable: nested table column header; date validation support; customizable cell selection/editing behavior; scrolling support; automatic row/column resizing; and so on.
  - ContextSensitiveTable extends JideTable to enable different cell renderers and cell editors for each cell.
  - NavigableTable extends ContextSensitiveTable to enable definition of table navigation keys.
  - CellStyleTable extends NavigableTable to enable cell styles.
  - CellSpanTable extends CellStyleTable to enable cell merging.
  - CategorizedTable extends CellSpanTable to enable data rows grouping.
  - SortableTable extends CategorizedTable to enable column sorting. This class is used by MATLAB for the new (R2008a+) *uitable* version.
  - TreeTable extends SortableTable to enable collapsible/expandable row groups (as expected from trees) while showing multiple columns for the displayed rows (as expected from tables).
  - PropertyTable extends TreeTable to display a two-column property-value TreeTable. This is used by MATLAB's built-in *inspect* function and is described in detail in the following sections.
  - HierarchicalTable extends SortableTable to enable display of any component, including other tables, in child rows. Unlike TreeTable, the child components do not need to have the same class or format.
  - TableUtils provides several utility methods for controlling table appearance and behavior (see Section 4.1.5 for some examples; also see com.mathworks. mwswing.TableUtils).
- com.jidesoft.tree
  - SortableTreeModel — a tree model wrapper that supports sorting.
  - FilterableTreeModel — a tree model wrapper that supports node filters.
  - QuickTreeFilterField — adds filtering support to any tree model.
  - StyledTreeCellRenderer—a tree cell renderer based on JIDE StyledLabel (see Section 5.5.1); note: included in jide-common.jar, not jide-grids.jar.
  - TreeModelWrapperUtils — utility class with useful tree model methods.
  - TreeSelectionModelGroup — enables creating rules for single/multiple node selections, similarly to checkbox/radio-button groups.
  - TreeUtils — a utility class that provides several useful JTree methods (see Section 4.2.3; also see com.mathworks.mwswing.TreeUtils).
- com.jidesoft.list
  - DualList — a component that contains two lists: a left list of selectable items and a right list of selected items

- `GroupList` — a list with groupable items (similar to a tree with collapsible/ expandable nodes)
- `ImagePreviewList` — a list whose selectable items are image previews (`DefaultPreviewImageIcon` objects) with a description and title
- `StyledListCellRenderer` — a list cell renderer based on JIDE `StyledLabel` (see Section 5.5.1); note: included in jide-common.jar not in jide-grids.jar
- `ListUtils` — a utility class that provides several useful `JList` methods

- `com.jidesoft.combobox`

Note: Each of the following classes has both a combo-box control and a corresponding panel that is presented when the combo-box arrow button is clicked (activated). The panel is normally called `xxxChooserPanel` and its corresponding combo-box is called `xxx-ComboBox`. After the combo-box is created, the panel can be accessed and customized via the combo-box's **PopupPanel** property or the corresponding *getPopupPanel()* method. The panel can also be displayed as a standalone control. GUI designers can choose whether to use a compact combo-box or the full-size panel control.

- `CalculatorComboBox` — a combo-box control that displays the result of simple arithmetic calculations.

- `ColorComboBox` — see the discussion in Section 5.4.1.
- `DateComboBox` — presents a date-selection combo-box, whose attributes (day/ month names, display format, etc.) are taken from the system definitions. JIDE's date-selection components are discussed in Section 5.7.8.

- `DateSpinnerComboBox` — presents a date-selection combo-box that includes both the `DateComboBox` and a spinner control.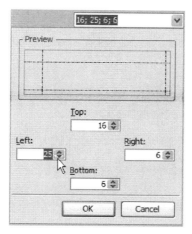
- `MonthComboBox` — a month-selection combo-box, similar to `DateComboBox` but without the ability to select individual days.
- `FileChooserComboBox` — a combo-box that displays a file-selection window when activated.
- `FolderChooserComboBox` — a combo-box that displays a folder-selection window when activated.
- `FontComboBox` — see Section 5.5.2.
- `InsetsComboBox` — enables interactive definition of insets (margins).

- `ListComboBox` — a simple combo-box that accepts a list (cell array) of selectable values (e.g., string values).
- `TableComboBox` — a combo-box that displays selectable data in table format (i.e., multiple columns).
- `TreeComboBox` — a combo-box that displays selectable data in tree format (i.e., a hierarchy of items).
- `MultilineStringComboBox` — a combo-box that enables entering multiple lines of text (string) items; its corresponding panel is `MultilineStringPopupPanel`.

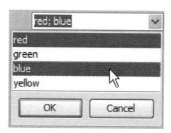

- **MultiSelectListComboBox** — a combo-box that enables selecting multiple items using a combination of <Shift> and <Ctrl> keys.
- **com.jidesoft.field**
  - **IPTextField** — a text-field that presents, validates, and accepts IP data (#.#.#.# format): `123.255.55.3`
  - **creditcard.CreditCardTextField** — a text-field that presents, validates, and accepts credit card numbers. The card issuer is automatically inferred from the card number; by default, Visa, MasterCard, DinersClub, JCB, Discover, and American Express are supported (this can be changed):
- accepted value:[†] `VISA ******9659`
- invalid value: `123456`
  - **creditcard.VISA** (and similarly **MasterCard**, **DinersClub**, **JCB**, **Discover**, and **AmericanExpress**) contains methods specific to that credit-card type: *getName(), getIcon()*, and *isCardNumberValid(string)*

 Note: You should <u>not</u> rely on JIDE's control for credit-card validation. The control provides only rudimentary checks of the entered number validity. It does not check whether the card itself is valid, stolen etc.

- **com.jidesoft.hssf**
  - **HssfTableUtils** — a utility class that enables exporting table data to Excel (XLS) format using the open-source POI-HSSF library.[120]

For a full description of the classes, please refer to the documentation resources listed at the beginning of the JIDE section. The following subsections will detail the usage of one particularly useful JIDE class, the `PropertyTable`.

5.7.3 MATLAB's PropertyInspector[121]

We often wish to edit the properties of heterogeneous objects using a common interface. MATLAB's property inspector, invoked with the ***inspect***[122] function, answers this need.

The inspector is based on a two-column table of property names and values. Properties and their values are populated automatically, and the user can edit values in-place. The inspector enables property categorization, sub-categorization, and sorting, which help users find and modify properties easily. For each property, the inspector displays a matching edit control: editbox/combobox/checkbox, and so on. This simplifies property value editing and prevents illegal value entry.

MATLAB's GUI builder, GUIDE,[123] uses the inspector to let users edit GUI properties such as position and color. It is also used by other tools such as the Plot Editor.[124]

[†] The Visa trademark is being used with permission. Visa is a registered trademark of Visa International Service Association. For the avoidance of doubt, Visa's permission does not constitute Visa's endorsement of this book nor of the JIDE control.

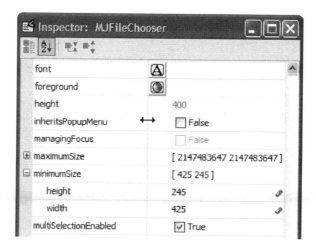

MATLAB's built-in property inspector

The MATLAB inspector can be embedded, with not-too-much effort, within MATLAB GUI applications. Examples of this can be found in the ***FindJObj*** and ***UIInspect*** utilities, which are described elsewhere in this book.

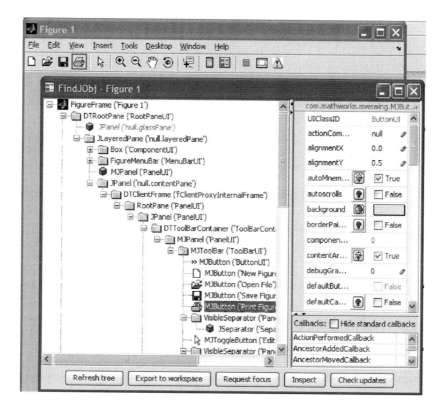

FindJObj — embedded property inspector

To embed the inspector within the *findjobj* and *uiinspect* utilities, for example, the following (simplified) code was used:

```
% prepare and populate the properties table
inspectorPane = com.mathworks.mlwidgets.inspector.PropertyView;
inspectorPane.setObject(hInspectedObject);
inspectorPane.setAutoUpdate(true);

inspectorTable = inspectorPane;
try
    while ~isa(inspectorTable,'javax.swing.JTable')
        inspectorTable = inspectorTable.getView;
    end
catch      % R2010a
    scrollPane = inspectorPane.getComponent(0).getScrollPane;
    inspectorTable = scrollPane.getViewport.getComponent(0);
end

% prevent JIDE alert by run-time (not load-time) evaluation
com.mathworks.mwswing.MJUtilities.initJIDE;
jideTableUtils = eval('com.jidesoft.grid.TableUtils;');
jideTableUtils.autoResizeAllColumns(inspectorTable);
inspectorTable.setRowAutoResizes(true);
inspectorTable.getModel.setShowExpert(true);

% Display onscreen
javacomponent(inspectorPane,position,hFig);

% Update the inspected data in the inspector table
if (numHandles == 1) % only one selected object
    inspectorPane.setObject(thisHandle)
else   % multiple objects — only inspect the common properties
    % (jArray is an array of inspected handles)
    jArray = javaArray('java.lang.Object', numSelections);
    ... % populate jArray with the list of selected handles)
    inspectorPane.getRegistry.setSelected(jArray, true);
end
```

Unfortunately, MATLAB's property inspector is limited to Handle Graphics, Java, and COM objects. It cannot be used for structures or user-defined MATLAB classes. We shall see below how to set up our own property grid, populate it with data, and subscribe to property change events.

This is a rather procedural approach. It is usually more convenient to use a declarative approach in which a structure or MATLAB class is passed to a function that automatically discovers its properties and their meta-information. The Property Grid[125] utility at MATLAB File Exchange provides these services.

5.7.4 JIDE's PropertyTable

 Note: The following discussion only works on MATLAB 7.6 (R2008a) onward. Read the comment mentioned in the reference[126] for additional details and workarounds.

MATLAB's property inspector is based on a JIDE property grid control. Recall that JIDE Grids is a components library bundled with MATLAB. JIDE Grids includes the `Property-Table` class, which is a fully customizable property grid component. Details on JIDE Grids can be found in the Developer Guide[127] and Javadoc documentation.[128]

Several related classes are associated with `PropertyTable`:[129] `PropertyTableModel`[130] encapsulates all the properties that are visualized in the property grid. Each property derives from the `Property`[131] abstract class, which features some common actions to properties, most notably to get and set the property value. `DefaultProperty`[132] is a default concrete subclass of `Property`. Finally, `PropertyPane`[133] decorates a property grid with icons for alphabetical sorting (rather than grouped), as well as for expanding and collapsing categories; a description text box at the bottom can be shown or hidden.

Here are the `DefaultProperty` fields and their respective roles:

Field	Role
Name	Internal property name, not necessarily displayed, used as a key to identify the property
DisplayName	A short property name shown in the property grid's left column
Description	A concise description of the property, shown at the bottom of the property pane, below the grid
Type	The Java type associated with the property, used to invoke the appropriate cell renderer or editor (see Section 4.1.1 for details)
EditorContext	An editor context object. If set, both type and context are used to look up the renderer or editor to use. This lets, for instance, one flag value to display as a true/false label, while another as a checkbox
Category	A string specifying the property's category, for grouping purposes
Editable	Specifies whether the property value is modifiable or read-only
Value	The current property value, as a Java object

Like any Java object, these fields may either be accessed with Java get/set semantics (e.g., *getName()*, *setName(name)*), or MATLAB ***get/set*** semantics (e.g., ***get**(prop,'Name')* and *set(prop,'Name',name)*). When using the MATLAB syntax, remember to wrap the Java object in a ***handle()*** call, in order to prevent a memory leak, as explained in Section 3.4.

To use a property grid in MATLAB, first construct a set of `DefaultProperty` objects for each of the grid properties. For each object, set at least the name, type, and initial value. Next, add the properties to a table model. Finally, construct a property grid with the given table model and encapsulate in a property pane:

```
% Initialize JIDE's usage within Matlab
com.mathworks.mwswing.MJUtilities.initJIDE;

% Prepare the properties list
list = java.util.ArrayList();
prop1 = com.jidesoft.grid.DefaultProperty();
```

```
prop1.setName('stringProp');
prop1.setType(javaclass('char',1));
prop1.setValue('initial value');
prop1.setCategory('My Category');
prop1.setDisplayName('Editable string property');
prop1.setDescription('A concise description for my property.');
prop1.setEditable(true);
list.add(prop1);

prop2 = com.jidesoft.grid.DefaultProperty();
prop2.setName('flagProp');
prop2.setType(javaclass('logical'));
prop2.setValue(true);
prop2.setCategory('My Category');
prop2.setDisplayName('Read-only flag property');
prop2.setEditable(false);
list.add(prop2);

% Prepare a properties table containing the list
model = com.jidesoft.grid.PropertyTableModel(list);
model.expandAll();
grid = com.jidesoft.grid.PropertyTable(model);
pane = com.jidesoft.grid.PropertyPane(grid);

% Display the properties pane onscreen
hFig = figure;
panel = uipanel(hFig);
javacomponent(pane, [0 0 200 200], panel);

% Wait for figure window to close & display the prop value
uiwait(hFig);
disp(prop1.getValue());
```

Here, com.mathworks.mwswing.MJUtilities.*initJIDE* is called to initialize JIDE's usage within MATLAB. Without this call, we may see a JIDE warning message. We only need to *initJIDE* once per MATLAB session, but there is no harm in repeated calls.

javaclass is a function (included in the Property Grid utility or directly downloadable[134]) that returns a Java class for the corresponding MATLAB type with the given dimension: *javaclass('logical')* or *javaclass('logical',0)* (a single logical flag value) returns a java.lang.Boolean class; *javaclass('char',1)* (a character array) returns a java.lang.String class; *javaclass('double',2)* (a matrix of double-precision floating point values) returns double[][].

javacomponent is the undocumented built-in MATLAB function that adds Java Swing components to a MATLAB figure (see Section 3.1.1). When the user closes the figure, prop. *getValue()* fetches and displays the new property value.

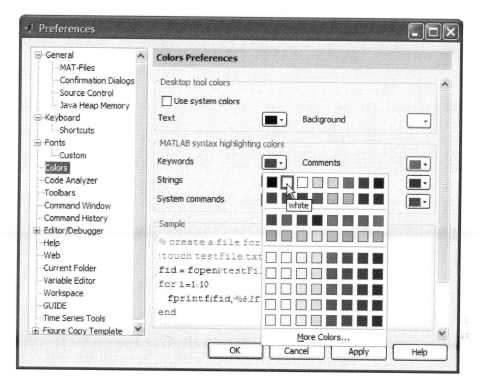

Stylish window using Swing controls (see Section 3.0)

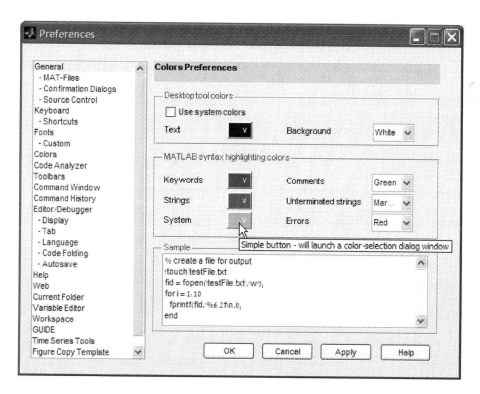

Simulation of the same dialog window, using standard MATLAB uicontrols
(see Section 3.0)

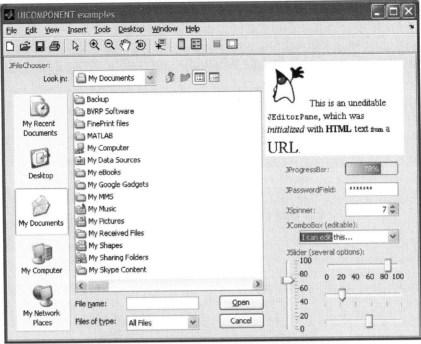

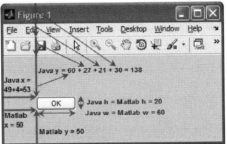

Sample Swing components integrated in a MATLAB figure window (see Section 3.1)

A Swing JScrollBar component (top) and a MATLAB slider uicontrol (bottom) — (see Section 3.1.2)

The effect of Look-and-Feel (L&F) on some standard GUI controls (see Section 3.3.2)

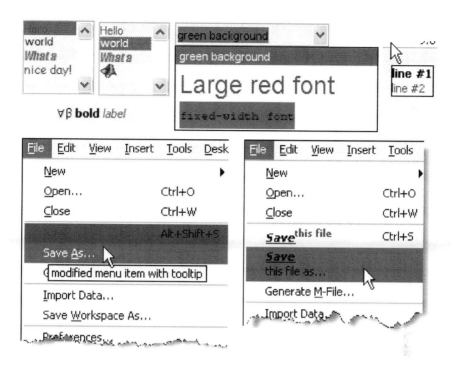

HTML in standard MATLAB listboxes, labels, popup-menus, tooltips, and menus
(see Section 3.3.3)

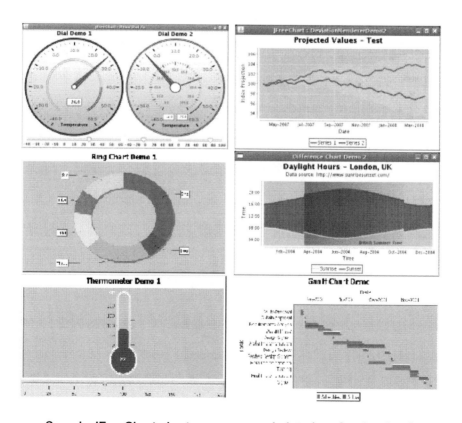

Sample JFreeChart charts, gauges, and plots (see Section 3.5.1)

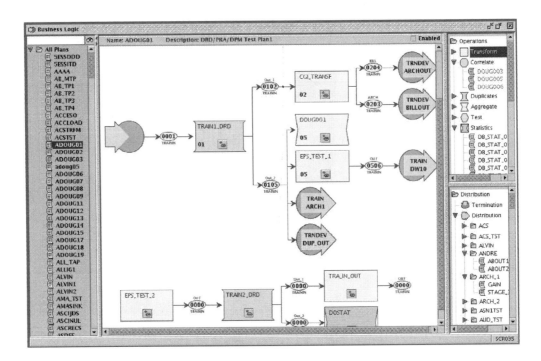

JGraph sample report (see Section 3.5.3)

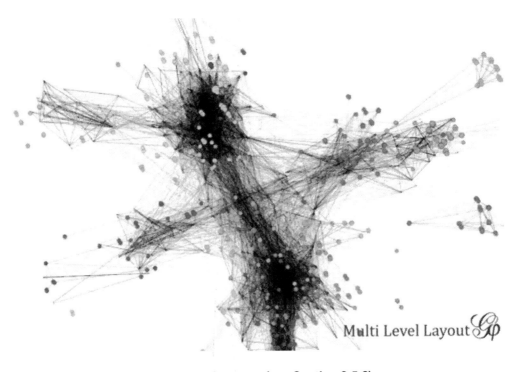

Gephi sample report (see Section 3.5.3)

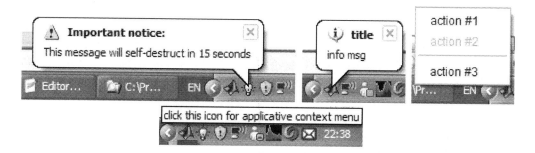

System tray-icon messages (top-center), context-menu (top-right), and tooltip (bottom) — (see Section 3.6)

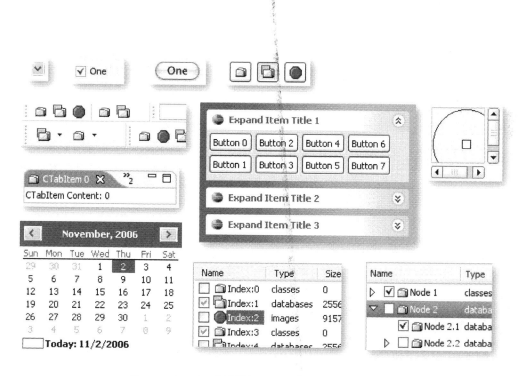

Some standard SWT controls (see Section 3.9)

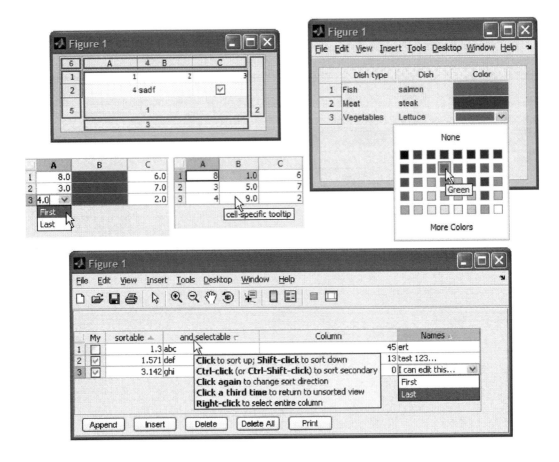

uitable customizations (see Section 4.1)

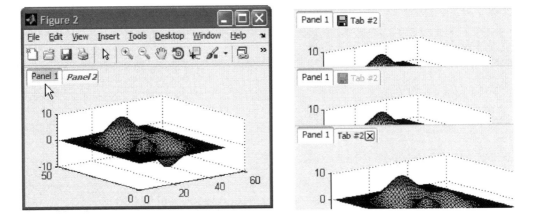

uitab customizations (see Section 4.3)

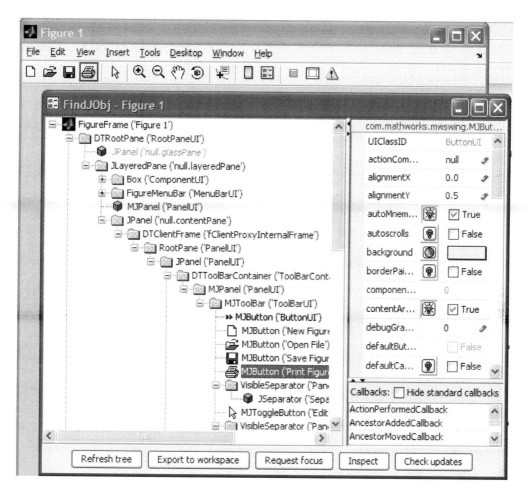

findjobj utility accessing a specific toolbar button (see Section 4.5)

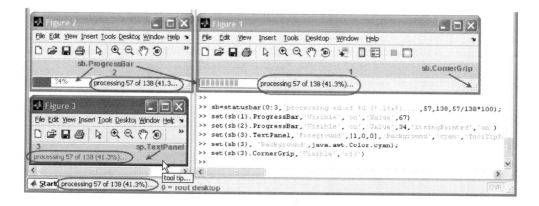

statusbar usage examples (see Section 4.7)

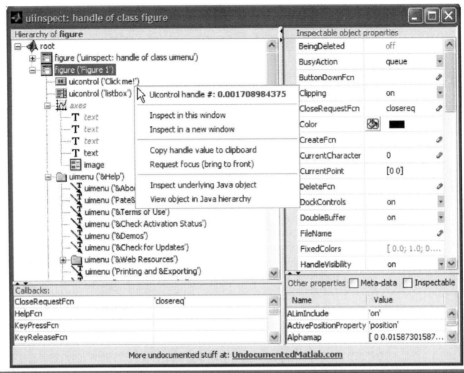

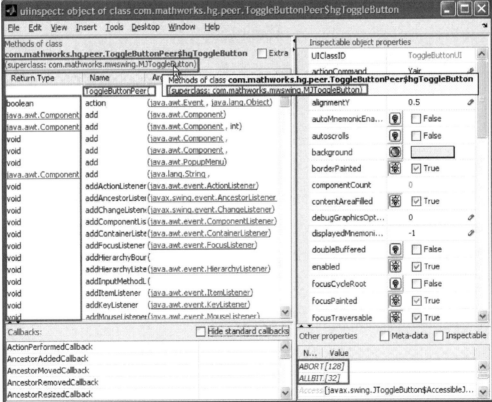

uiinspect usage examples (see Sections 5.2 and 6.0)

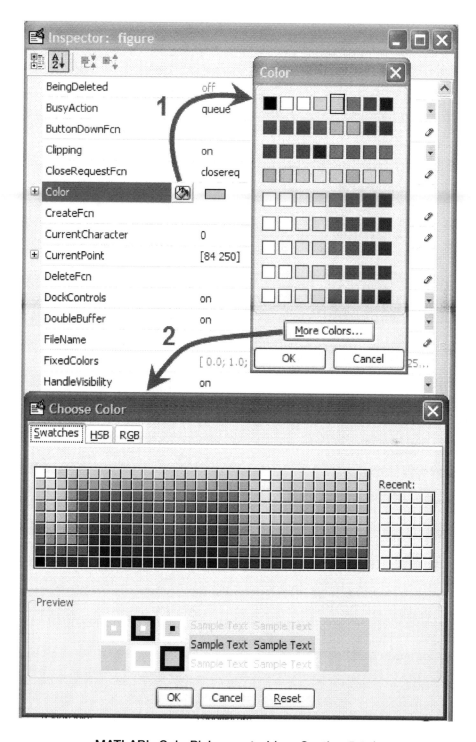

MATLAB's ColorPicker control (see Section 5.4.1)

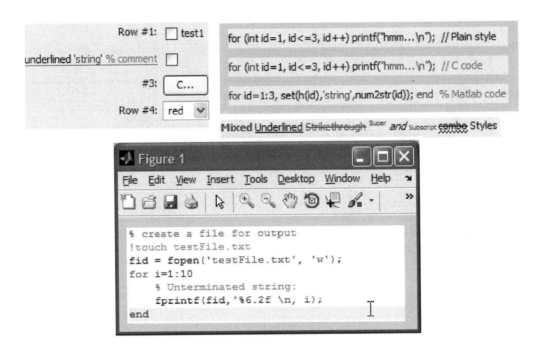

Different GUI controls displaying syntax-highlighted text (see Section 5.5.1)

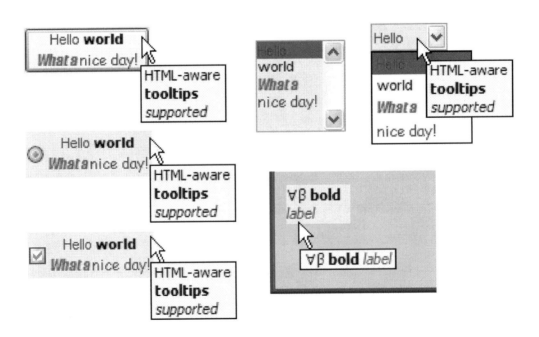

HTML-enriched MATLAB uicontrols (see Chapter 6)

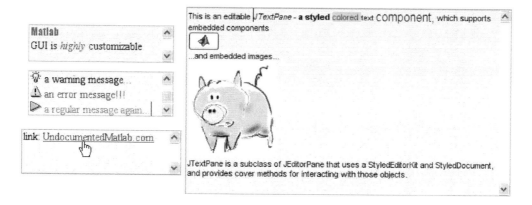

Editbox customizations (see Section 6.5)

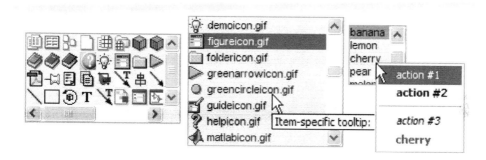

Listbox customizations (see Section 6.6)

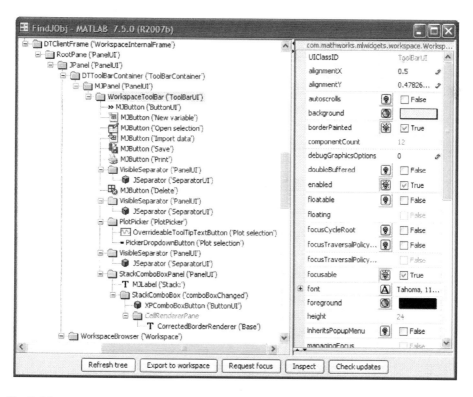

findjobj presentation of a toolbar with non-standard controls (see Section 7.3.5)

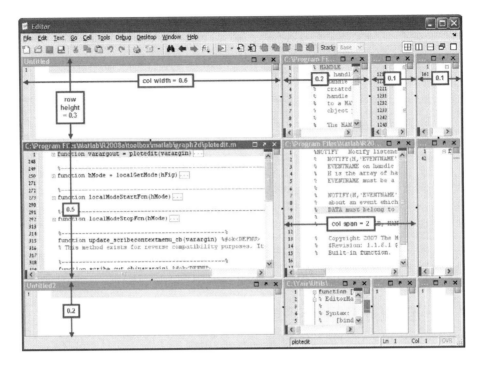

Document cell width, height, and span example (see Section 8.1.1)

```
>>
>>
>>    cprintf('text',    'regular black text');
       cprintf('hyper',   'followed %s ','by');
       cprintf('k',       '%d colored', 4);
       cprintf('-comment','& underlined');
       cprintf('err',     'elements\n');
       cprintf('cyan',    'cyan');
       cprintf('-green',  'underlined green');
       cprintf(-[1,0,1],  'underlined magenta');
       cprintf([1,0.5,0],'and multi-\nline orange\n');

regular black text followed by 4 colored & underlined elements
cyan underlined green underlined magenta and multi-
line orange
>> |
```

cprintf — display styled formatted text in the Command Window (see Section 8.3.1)

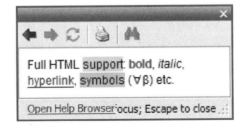

Help popup with custom HTML content and highlighting (see Section 8.3.2)

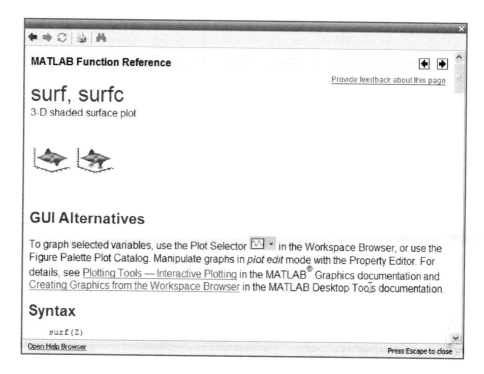

Help popup with a standard MATLAB doc page (see Section 8.3.2)

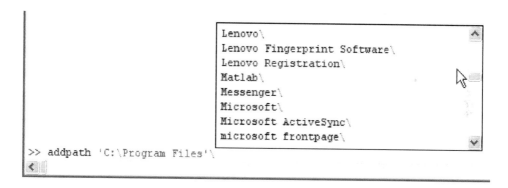

Tab-completion (see Section 8.3.4)

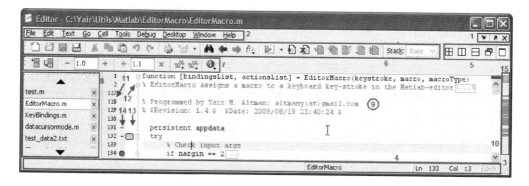

MATLAB Editor components (see Section 8.4.2)

Function listing
Color highlight code according to [time ▼]

time	calls	mem	unjitted	line	
				275	function hF
				276	
				277	% Pre
0.21	1	432k/376k/14.4k	X	278	metho
0.06	1	400k/345k/20.5k	X	279	[call
0.34	1	689k/615k/17.6k	X	280	[prop
0.44	1	4.11m/3.98m/18.4k	X	281	child
				282	
				283	% Pre
				284	impor
< 0.01	1	2.8k/2.26k/556b	X	285	right

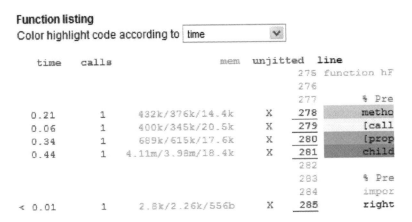

Memory and JIT information in MATLAB's Profiler (see Section 8.7.1)

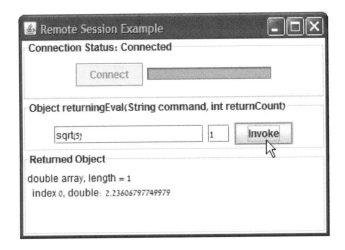

Invoking MATLAB commands from Java using RMI (see Section 9.4)

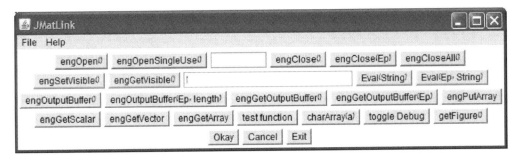

JMatLink's built-in testing GUI (see Section 9.5)

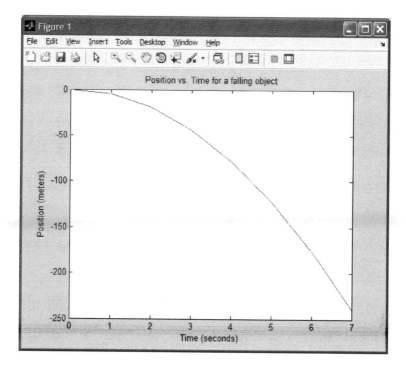

MATLAB plot generated from Java using JNI/JNA (see Section 9.5)

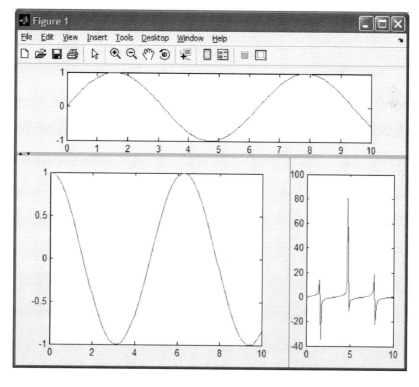

A horizontal *UISplitPane* contained within a vertical *UISplitPane* (see Section 10.1)

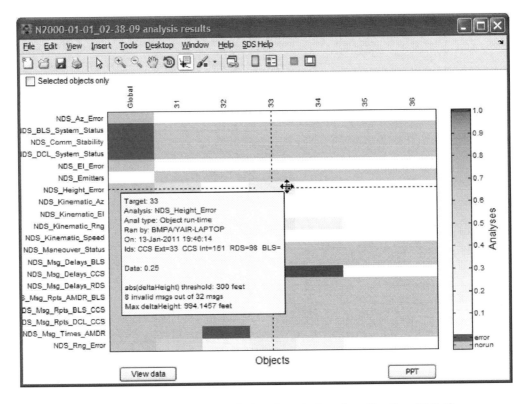

Dynamic data-tips in the IDS Results window (see Section 10.2.5)

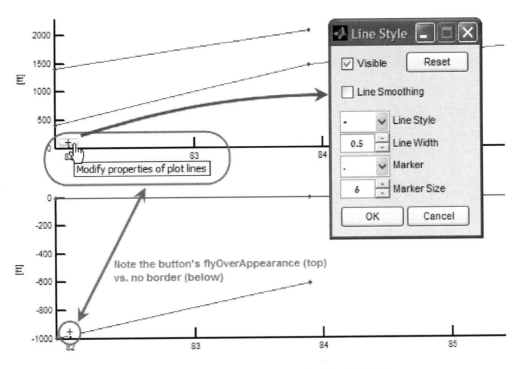

Customizing result plots (see Section 10.2.5)

A simple user-defined property grid

5.7.5 Nonstandard Property Renderers and Editors[135]

Cell renderers and cell editors[136] (see Section 4.1.1) are the backbone of JTable implementations, and this includes JIDE's property grid. Each property is associated with a type, and a renderer and an editor may be registered for a type. The cell renderer controls how the property value is displayed, while the editor determines how it is edited. For example, flags (Java Booleans) are often both rendered and edited using a checkbox, but they can also use a text renderer (displaying the string "true" or "false") with a combo-box editor (for selecting a true or false value). PropertyTable automatically assigns a default renderer and editor to each property, based on its type: Flags are assigned a combo-box editor (true/false), and similarly for other types.

Let us modify the preassigned editor: First, assign spinners[137] (SpinnerCellEditor[138]) for numbers and checkboxes[139] (BooleanCheckBoxCellEditor[140]) for logical flags:

```
% Prepare the properties list:
% First two logical values (flags)
list = java.util.ArrayList();
prop1 = com.jidesoft.grid.DefaultProperty();
prop1.setName('mylogical');
prop1.setType(javaclass('logical'));
prop1.setValue(true);
list.add(prop1);

prop2 = com.jidesoft.grid.DefaultProperty();
prop2.setName('mycheckbox');
prop2.setType(javaclass('logical'));
```

```
prop2.setValue(true);
cbContext = com.jidesoft.grid.BooleanCheckBoxCellEditor.CONTEXT;
prop2.setEditorContext(cbContext);
list.add(prop2);

% Now integers (note the different way to set property values):
prop3 = com.jidesoft.grid.DefaultProperty();
javatype = javaclass('int32');
set(prop3,'Name','myinteger','Type',javatype,'Value',int32(1));
list.add(prop3);

prop4 = com.jidesoft.grid.DefaultProperty();
set(prop4,'Name','myspinner','Type',javatype,'Value',int32(1));
set(prop4,'EditorContext',com.jidesoft.grid.SpinnerCellEditor.CONTEXT);
list.add(prop4);

% Prepare a properties table containing the list
model = com.jidesoft.grid.PropertyTableModel(list);
model.expandAll();
grid = com.jidesoft.grid.PropertyTable(model);
pane = com.jidesoft.grid.PropertyPane(grid);

% Display the properties pane onscreen
panel = uipanel(gcf);
javacomponent(pane, [0 0 200 200], panel);
```

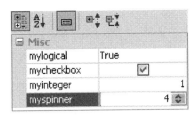

A property grid with checkbox and spinner controls

Notice how the **EditorContext** is used to specify a nonstandard renderer/editor for myspin-ner and mycheckbox: the mylogical flag displays as a string label, while mycheckbox displays as a checkbox; myinteger uses a regular integer editor that accepts whole numbers, while myspinner uses a spinner control to modify the value.

In addition to BooleanCheckBoxCellEditor and SpinnerCellEditor, JIDE pre-defines many other cell editors. All these classes implement the EditorContextSupport[141] interface and have a class name of <type>CellEditor.[142] Here is the full list of supported cell editor <type>s: Boolean, Byte, Calculator, CheckBoxListComboBox, Color, ContextSensitive, Date, Dimension, Double, Enum, File, FileName, Float, Folder, Font, FontName, FormattedTextField, Insets, Integer, IPAddress, ListComboBox, Long, Month, MultilineString, MultilineTable, MultilineEnum, Number, Password, Point, Rectangle, Short, Slider, Spinner, String, StringArray, TableComboBox, TextField, and TreeComboBox.

Note that instead of creating an entirely new properties list and table, we could have run the previ-ous section's example, modified *list*, and then simply called *model.refresh()* to update the display.

MATLAB types are automatically converted to Java types, but we must ensure that the conversion matches our *setType* declaration: the logical value ***true*** correctly converts to a `java.lang.Boolean`; however, the value 1 would be a double, MATLAB's standard numeric type, and so an ***int32(1)*** cast is required to force a `java.lang.Integer` conversion.

Spinners with indefinite value bounds are seldom useful. The following shows how to register a new editor to restrict values to a fixed range:

```
import com.jidesoft.grid.*;
javatype = javaclass('int32');
value = int32(0);
minVal = int32(-2);
maxVal = int32(5);
step = int32(1);
spinner = javax.swing.SpinnerNumberModel(value, minVal, maxVal, step);
editor = SpinnerCellEditor(spinner);
context = EditorContext('spinnereditor');
CellEditorManager.registerEditor(javatype, editor, context);

prop = DefaultProperty();
set(prop, 'Name','myspinner', 'Type',javatype, ...
        'Value',int32(1), 'EditorContext',context);
```

Note how we registered a specific cell editor for a specific property context, using the `CellEditorManager.`*registerEditor()* method. Similarly, we can register specific cell renderer using `CellRendererManager.`*registerRenderer()*, which accepts similar input parameters (javatype, renderer, context), where renderer is a `javax.swing.table.TableCellRenderer` (such as `DefaultTableCellRenderer` — see Section 4.1.1). The principle is the same for combo-boxes:

```
import com.jidesoft.grid.*;
javatype = javaclass('char', 1);
options = {'spring', 'summer', 'fall', 'winter'};
editor = ListComboBoxCellEditor(options);
context = EditorContext('comboboxeditor');
CellEditorManager.registerEditor(javatype, editor, context);

prop = com.jidesoft.grid.DefaultProperty();
set(prop, 'Name','season', 'Type',javatype, ...
        'Value','spring', 'EditorContext',context);
```

A property grid with a combobox control

Both `CellEditorManager` and `CellRendererManager` are global objects that live as long as our JVM session (i.e., the entire MATLAB session). We can unregister an editor or renderer when it is no longer used, thereby freeing memory, as follows:

```
com.jidesoft.grid.CellEditorManager.unregisterEditor(javatype,context);
```

5.7.6 Nested Properties

Properties can act as a parent node for other properties. A typical example is an object's dimensions: a parent node value may be edited as a 2×1 matrix, but width and height may also be exposed individually. Nested properties are created as regular properties. However, rather than adding them directly to a `PropertyTableModel`, they are added under a `Property` instance using its *addChild* method:

```
propdimensions = com.jidesoft.grid.DefaultProperty();
propdimensions.setName('dimensions');
propdimensions.setEditable(false);

propwidth = com.jidesoft.grid.DefaultProperty();
propwidth.setName('width');
propwidth.setType(javaclass('int32'));
propwidth.setValue(int32(100));
propdimensions.addChild(propwidth);

propheight = com.jidesoft.grid.DefaultProperty();
propheight.setName('height');
propheight.setType(javaclass('int32'));
propheight.setValue(int32(100));
propdimensions.addChild(propheight);
```

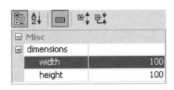

A property grid with nested properties

`PropertyTableModel` accesses properties in a hierarchical naming scheme. This means that the parts of nested properties are separated with a dot (.). In the above example, these two fully-qualified names are *dimensions.width* and *dimensions.height*.

5.7.7 Trapping Property Change Events

Sometimes, it is desirable to subscribe to the **PropertyChange** event. This event is fired by `PropertyTableModel` whenever any property value is updated. To expose Java events to MATLAB, we use the two-parameter form of the ***handle*** function with the optional CallbackProperties parameter.

```
hModel = handle(model, 'CallbackProperties');
set(hModel, 'PropertyChangeCallback', @callback_onPropertyChange);
```

The callback function receives two input arguments: the first is the `PropertyTableModel` object that fired the event, and the second is a `PropertyChangeEvent` object with properties **PropertyName**, **OldValue**, and **NewValue**. The `PropertyTableModel`'s *getProperty (PropertyName)* method may be used to fetch the `Property` instance that has changed.

Callbacks enable property value validation: **OldValue** can be used to restore the original property value, if **NewValue** fails to meet some criteria that cannot be programmed into the cell editor. We may, for instance, set the property type to a string and then, in our callback function, use *str2num* as a validator to try to convert **NewValue** to a numeric matrix. If the conversion fails, we restore the **OldValue**:

```
function callback_onPropertyChange(model, event)
    string = event.getNewValue();
    [value, isvalid] = str2num(string); %#ok
    prop = model.getProperty(event.getPropertyName());
    if isvalid % standardize value entered
        string = mat2str(value);
    else % restore previous value
        string = event.getOldValue();
    end
    prop.setValue(string);
    model.refresh(); % refresh value onscreen
end  % callback_onPropertyChange
```

Now combine the nested properties and callbacks to update the parent property according to the changes made to child property. First, set the parent property's initial value:

```
propdimensions.setValue('[100,100]')
```

Next, set a callback on the model's PropertyChange event:

```
hModel = handle(model, 'CallbackProperties');
set(hModel, 'PropertyChangeCallback', @onPropertyChangeFcn);
```

And in this callback function, update the parent's value whenever one of its child values changes:

```
function onPropertyChangeFcn(model, event)
    propName = event.getPropertyName;
    switch propName
        case {'dimensions.width', 'dimensions.height'}
            newWidth = model.getProperty('dimensions.width').getValue;
            newHeight = model.getProperty('dimensions.height').getValue;
            parentProp = model.getProperty(propName).getParent;
            parentProp.setValue(sprintf('[%d,%d]',newWidth,newHeight));
        otherwise
            % Some other property has changed...
        end
    model.refresh(); % refresh value onscreen
end % onPropertyChangeFcn
```

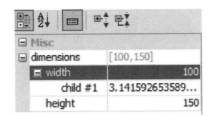

Deeply nested properties with updated parent value

 Note: Levente Hunyadi, who wrote the preceding subsections about JIDE's property table, also created two utilities that facilitate property visualization in MATLAB: ***PropertySheet***[143] (based on the open-source Fraeser project[144]) and ***PropertyGrid*** (based on JIDE's `PropertyTable`).[145]

5.7.8 Date-Selection Components[146]

MATLAB has many built-in date-handling functions (***calendar, date, datestr, datenum, datetick, datevec,*** etc.). Unfortunately, this built-in support does not extend to MATLAB GUI. If we need a date-selection drop-down or calendar panel, we have to design it ourselves, or use a third-party Java component or ActiveX control.[147]

Luckily, we have a much better alternative, right within MATLAB. This relies on JIDE Grids, which includes the following date-selection controls:

- `DateChooserPanel`:[148] an extension of Swing's `JPanel` that displays a single month and enables selecting one or more days
- `CalendarViewer`:[149] a similar panel, which displays several months in a table format (e.g., 4 × 3 months)
- `DateComboBox`:[150] a combo-box (drop-down/popup menu) that presents a `DateChooserPanel` for selecting a date
- `DateSpinnerComboBox`:[151] presents a date-selection combo-box that includes both the `DateComboBox` and a spinner control
- `MonthChooserPanel`:[152] a panel that enables selection of entire months (not specific dates)
- `MonthComboBox`:[153] a month-selection combo-box, similar to `DateComboBox` but without the ability to select individual days

Usage of these controls is very similar, so I will just show the basics here. First, to present any control, we need to use the built-in ***javacomponent*** function or the ***uicomponent*** utility:[154]

```
% Initialize JIDE's usage within Matlab
com.mathworks.mwswing.MJUtilities.initJIDE;
```

```
% Display a DateChooserPanel
jPanel = com.jidesoft.combobox.DateChooserPanel;
[hPanel,hContainer] = javacomponent(jPanel,[10,10,200,200],gcf)
```

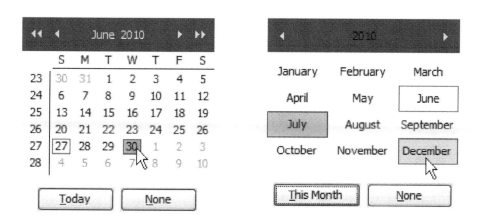

DateChooserPanel **and** MonthChooserPanel **components**

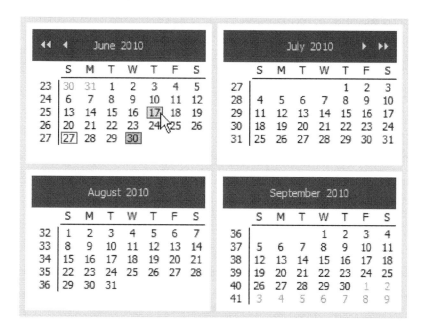

2 × 2 CalendarViewer **component**

Just as with any Java object, properties may be accessed with either the Java get/set semantics (e.g., *getName()* or *setName(name)*) or the MATLAB get/set semantics (e.g., **get**(prop,'Name') or *set*(prop,'Name',value)). When using the MATLAB syntax, remember to wrap the Java object in a *handle()* call, in order to prevent a memory leak (i.e., use hPanel, not jPanel — see Section 3.4):

```
jPanel.setShowWeekNumbers(false); % Java syntax
set(hPanel,'ShowTodayButton',true); % Matlab syntax
```

Retrieving the selected date is easy:

```
>> selectedDate = jPanel.getSelectedDate;
selectedDate =
Sun Jun 27 00:00:00 IDT 2010

% Note: selectedDate is a java.util.Date object:
>> selectedDate.get
        Class = [ (1 by 1) java.lang.Class array]
        Date = [27]
        Day = [0]
        Hours = [0]
        Minutes = [0]
        Month = [5]
        Seconds = [0]
        Time = [1.27759e + 012]
        TimezoneOffset = [-180]
        Year = [110]
```

We can enable selection of multiple dates (MULTIPLE _ INTERVAL_SELECTION = 2, SINGLE_INTERVAL_SELECTION = 1, SINGLE_SELECTION = 0):

```
jModel = hPanel.getSelectionModel; % => a DefaultDateSelectionModel
jModel.setSelectionMode(jModel.MULTIPLE_INTERVAL_SELECTION);

>> jModel.getSelectedDates
ans =
java.util.Date[]:
    [java.util.Date]
    [java.util.Date]
    [java.util.Date]
```

We can, of course, set a callback for user modification of the selected date(s):

```
hModel = handle(hPanel.getSelectionModel, 'CallbackProperties');
set(hModel, 'ValueChangedCallback', @myCallbackFunction);
```

For the combo-box (drop-down/popup menus) controls, we obviously need to modify the displayed size (in the *javacomponent* call) to something much more compact, such as [10,10,100,20]. These components display one of the above panels as their popup selection panels. Users can access these panels using the combo-box control's *getPopupPanel()* function (or **PopupPanel** property).

DateComboBox **and** DateSpinnerComboBox **components**

5.8 Miscellaneous Other Internal Classes

This section details, in no particular order, several interesting classes that can be found in the standard MATLAB packages and possibly used in our user applications.

Several possibly useful components can be found in the com.mathworks.hg.util subpackage. One such component, dColorChooser, was described in Section 5.4.1. com.mathworks.hg.util.FontChooser is another similar self-explanatory component.

Another interesting component is com.mathworks.hg.util.StringScrollList-Chooser, which presents a component that is composed of an automatic-scrolling listbox with an attached editbox. The editbox contents are updated whenever any item is selected in the listbox, but the user can also type any nonlisted text value in the editbox.

The control's **SelectedValue** property returns the current text in the Edit box. The editbox contents can be set directly using the *setSelectedItem(string)* method or via the listbox via *setSelectedValue(string, shouldScrollToSelectionFlag)*. Note that if the value string specified to *setSelectedValue()* is not found in the listbox, neither the editbox nor the listbox will be updated. So, to specify non-listed values, use the *setSelectedItem* method:

```
strList = {'Oh','what','a','wonderful','world!'};
slc = com.mathworks.hg.util.StringScrollListChooser(strList);
[jComponent,hContainer] = javacomponent(slc,[1,1,300,85],gcf);
```

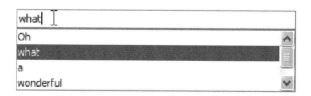

StringScrollListChooser

Note: While this may seem like a regular property, be careful to always use the *setSelectedItem(string)* method and <u>never use the alternative *set(slc, 'SelectedItem', string)* since it **crashes MATLAB**,</u>† whereas using *setSelectedItem(string) or setSelectedValue(string,flag)* is entirely safe.

`com.mathworks.webintegration.startpage.framework.view.ResourcePane` can be used to extend `JEditorPane` with some very useful methods for setting HTML title, subtitles, text, hyperlinks, and image resources.

`com.mathworks.page.utils.ClipboardHandler` is an easy-to-use wrapper for system-clipboard copy/paste operations, used by the built-in MATLAB function *clipboard.m*.

The `com.mathworks.services` package contains several useful classes that are described in more detail elsewhere in this book: `com.mathworks.services.Prefs` handles system preferences (Section 8.2); `com.mathworks.services.AntialiasedFontPrefs` handles font anti-aliasing (Section 6.5.3); `com.mathworks.services.binding.MatlabKeyBindings` can be used to load the default Editor and Command Window key bindings (Section 8.5).

The `com.mathworks.util` package also contains several useful classes:

`com.mathworks.util.StringUtils` contains a set of useful string-processing functions that complement those available in the standard `java.lang.String`.

`com.mathworks.util.Range` enables setting/getting start and end integer duo values that define a numeric range. `Range` contains relevant methods such as min/max of the duo values and union/intersection/adjacency with another `Range` object.

`com.mathworks.util.PlatformInfo` retrieves information about the current platform. Methods such as *isWindowsXP*, *isIntelMac64*, *isMacOSXTiger*, and *isUnix* describe the platform and operating system; methods such as *isWindowsVistaAppearance* and *getWindowsColorScheme* determine (or modify) the operating system's appearance; and methods such as *isVersion118* return information about the Java version.

 Note: Do not confuse `com.mathworks.util.PlatformInfo` with the unrelated `com.mathworks.mlwidgets.util.productinfo.ProductInfoUtils` class.[155]

Finally, there is a `com.mathworks.util.Timer` class that should not be confused with the very similar yet unrelated `com.mathworks.timer.Timer` class. I believe that the latter `Timer` class is used to implement MATLAB's built-in *timer* function. I do not know why another implementation of a Timer class was needed. Note that we can also use the standard Java `javax.swing.Timer` class.[156]

---

† At least on MATLAB 7.5 (R2007b) through 7.12 (R2011a) running JVM 1.6 on a Windows XP PC. I suspect this may be due to a missing *getSelectedItem* method in MATLAB's implementation.

5.8.1 *Logging Utilities*

MATLAB contains several utilities/components that can be used for logging events.

The hg package contains a logger utility, which enables displaying log messages in a separate Java window. The logger window is always available, yet hidden:

```
frames = java.awt.Frame.getFrames;
for idx = 1 : length(frames)
    if strcmp(frames(idx).getTitle,'Logger')
        f = frames(idx); f.setVisible(true); break;
    end
end
```

The logger logs HG events based on the debugging options specified via com.mathworks. hg.peer.DebugUtilities.*setDebugOptions(dbgOptions)*. The possible options are enumerated static fields in the DebugUtilities class. They are powers of 2, so apparently are bitwise flag representations that can be combined — a well-known programming practice. By increasing value, the enumerated options are

```
DEBUG_CMD_WINDOW      =       1 = 2⁰
DEBUG_LOG_FILE        =       2 = 2¹
DEBUG_LOG_WINDOW      =       4 = 2²
(missing)             =       8 = 2³
DEBUG_SURPRISES       =      16 = 2⁴
DEBUG_TRACE_METHODS   =      32 = 2⁵
DEBUG_TRACE_INVOKES   =      64 = 7
DEBUG_SIZE            =     128 = 2⁷
DEBUG_LISTENERS       =     256 = 2⁸
DEBUG_EVENTS          =     512 = 2⁹
DEBUG_DESKTOP         =    1024 = 2¹⁰
DEBUG_MODAL           =    2048 = 2¹¹
DEBUG_TREELOCK        =    4096 = 2¹²
DEBUG_TEMP            =    8192 = 2¹³
DEBUG_ACTIVATE        =   16384 = 2¹⁴
DEBUG_JOGL            =   32768 = 2¹⁵
```

It appears that the first (lowest value) options determine the logging destination (the MATLAB Command Window and/or a *figure_log.txt* file in the current folder (***pwd***) and/or the

Logger window). The rest of the options determine which events should be logged. For example, to log all events to the Logger window only:

```
com.mathworks.hg.peer.DebugUtilities.setDebugOptions(2^16-1-(1 + 2));
```

or equivalently (a lot more coding, but much more readable/maintainable):

```
dbg = com.mathworks.hg.peer.DebugUtilities;
options = dbg.DEBUG_LOG_WINDOW      + dbg.DEBUG_SURPRISES + · · ·
        dbg.DEBUG_TRACE_METHODS + dbg.DEBUG_TRACE_INVOKES + · · ·
        dbg.DEBUG_SIZE          + dbg.DEBUG_LISTENERS + · · ·
        dbg.DEBUG_EVENTS        + dbg.DEBUG_DESKTOP + · · ·
        dbg.DEBUG_MODAL         + dbg.DEBUG_TREELOCK + · · ·
        dbg.DEBUG_TEMP + dbg.DEBUG_ACTIVATE + dbg.DEBUG_JOGL;
dbg.setDebugOptions(options);
```

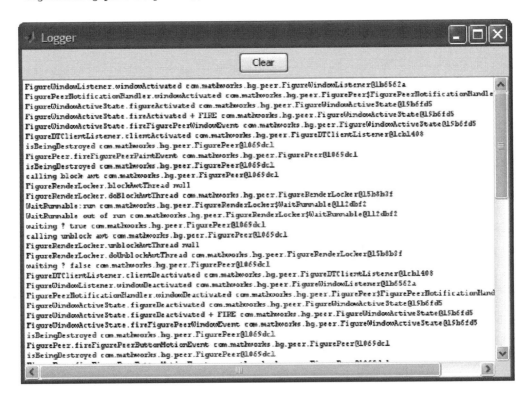

We can also log our custom events using DebugUtilities. This is done with the *logMessage(destination, description, object)* method, where *destination* is a combination of DEBUG_CMD_WINDOW, DEBUG_LOG_FILE, and DEBUG_ LOG_WINDOW; *description* is a descriptive string, and *object* is the logged object reference or MATLAB value. Note that only *destinations* that were previously enabled with *setDebugOption()* will display the new message — non-enabled destinations will simply silently ignore it.

For example, the following will output a log message only to the Command Window but not to the Logger window (since it was not logger-enabled):

```
>> dbg.setDebugOptions(dbg.DEBUG_CMD_WINDOW + dbg.DEBUG_LOG_FILE);
>> dbg.logMessage(7,'Yair: ', jFrame)
Yair: com.mathworks.hg.peer.FigureFrameProxy$FigureFrame[...]

>> dbg.logMessage(7,'Yair: ', hFig)
Yair: 171.0030517578125

>> dbg.logMessage(7,'Yair: ', magic(3))
Yair: [[D@1f5e413

>> dbg.logMessage(7,'Yair: ',{1,2,3,'abc'})
Yair: [Ljava.lang.Object;@9cde9a

>> dbg.logMessage(7,'Yair: ', 2+3 == 5)
Yair: true
```

This logger utility should not be confused with another logger utility, provided by the com.mathworks.util.Log class: Log provides a very simple logging class, which enables displaying log messages in a separate Java window. For example,

```
com.mathworks.util.Log.setLogging(true);
com.mathworks.util.Log.printLn('testing 12345...');
```

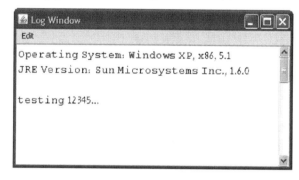

A simple log window

In addition to the hg.peer.DebugUtilities and util.Log classes, we can implement our own logging utility. An example of this is shown in Section 6.5.2:

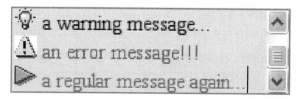

A custom icon and color-enriched log panel

Finally, we can use external logger packages, such as the pre-bundled simple loggers in the *jarext/commons-logging.jar* or the well-established open-source Log4J.[157]

5.8.2 JGoodies

Section 5.5.1 introduced `FormPanel`, a convenient form layout preparation class by MathWorks. MATLAB installation prebundled the open-source JGoodies Forms package by www.jgoodies.com (*%matlabroot%/java/jarext/jgoodies-forms.jar*).[158] This package, well known in the Java community, enables easy and consistent form creation. Using JGoodies Forms, we can create forms that have aligned components, shrinking appropriately in any or all dimensions as their container shrinks or grows.

Let us recreate the `FormPanel` example of Section 5.5.1 using JGoodies Forms:

```
jLayout = com.jgoodies.forms.layout.FormLayout('r:p, 4dlu, p:g');
jFormBuilder = com.jgoodies.forms.builder.DefaultFormBuilder(jLayout);
jFormBuilder.append('Row #1:', javax.swing.JCheckBox('test1'))
jFormBuilder.append('A very long label:', javax.swing.JCheckBox)
jFormBuilder.append('#3:', javax.swing.JButton('Click me!'))
jComboBox = javax.swing.JComboBox({'red','green','blue'});
jFormBuilder.append('Row #4:',jComboBox);
jPanel = jFormBuilder.getPanel;
[jhPanel,hContainer] = javacomponent(jPanel,[10,10,200,110],gcf);
```

A simple form created using JGoodies `FormLayout`

This may appear to be a very simple case where JGoodies Forms are not really needed, but when alignment and resizing behavior requirements kick in, these are easy to implement in `FormLayout`, but are much more difficult using other Swing layouts. It is possible that JGoodies Forms were also used for Mac MATLAB's unified toolbar.[159]

As an alternative to Java-based forms, consider using Kesh's excellent *Enhanced Input Dialog Box* utility on the MATLAB File Exchange.[160]

In addition to JGoodies Forms, MATLAB also comes bundled with the JGoodies Looks library (*%matlabroot%/java/jarext/jgoodies-looks.jar*),[161] which contains a set of L&F classes, including the Plastic Look-and-Feel that was introduced in Section 3.3.2.

5.8.3 Additional Others

The `com.mathworks.mde` and `com.mathworks.desktop` packages and subpackages contain classes that are used to implement the MATLAB Desktop and its associated tools (Editor, Command History, etc.). These are described in detail in Chapter 8.

The `com.mathworks.fatalexit` package contains a single class, `FatalExitFrame`, which displays the dreaded MATLAB crash message:

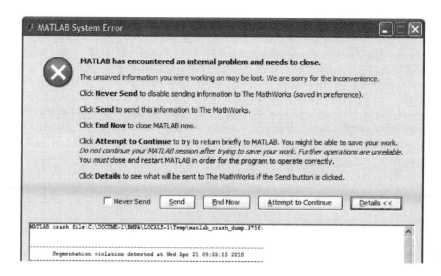

In some cases, MATLAB decides that an attempt to continue is impossible:

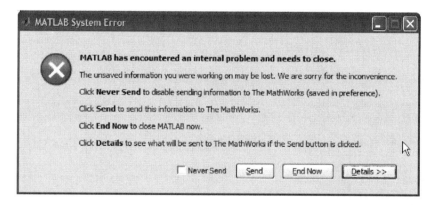

We can programmatically invoke this crash display using com.mathworks.fatalexit. FatalExitFrame.main('xxx'), where 'xxx' is the filename of the crash details file, which is then presented in the window's text box. However, **closing the message window will have the side effect of closing MATLAB** ....

The com.mathworks.ide.filtermgr.FilterEditor and FilterManager classes (removed in R2011a) enable managing object properties. I have not encountered this in normal MATLAB usage and do not know its intended usage. It can be regenerated as follows:

```
fm = com.mathworks.ide.filtermgr.FilterManager('Yair');
fm.editFilter('javax.swing.JButton', 'MyFilter');'
```

[†] or javacomponent(com.mathworks.ide.filtermgr.FilterEditor(fm,[]),[1,1,300,285],gcf);

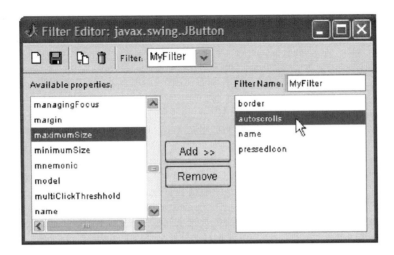

The BDE (Block Diagram Editor?) package is a graphical block-diagram editor available up to R2011a. I do not know the diagrams' purpose (note that BDE was redesigned in R2009b, breaking the code below):[162]

```
bde = com.mathworks.bde.clients.BDEDesktop; % this fails in R2009b +
diagram = com.mathworks.bde.diagram.Diagram;
client = com.mathworks.bde.clients.DiagramViewDTClient('YMA',diagram);
bde.addClient(client,'YMA');
```

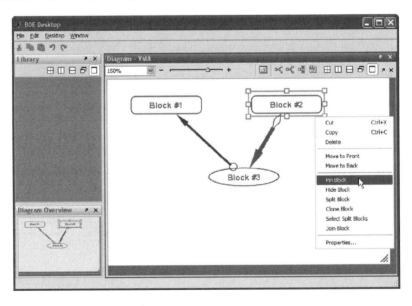

References

1. http://blogs.mathworks.com/desktop/2009/07/06/calling-java-from-matlab/#comment-6836 (or http://bit.ly/cUbHnQ).

2. http://java.sun.com/javase/6/docs/api/javax/swing/JToggleButton.html (or http://bit.ly/davduF).
3. http://UndocumentedMatlab.com/files/checkClass.m (or http://bit.ly/cLwCHs).
4. http://www.mathworks.com/matlabcentral/fileexchange/26947-checkclass (or http://bit.ly/98151H).
5. http://UndocumentedMatlab.com/blog/toolstrip (or http://bit.ly/nlU9PX).
6. http://platform.netbeans.org/ (http://bit.ly/dqIav2); http://en.wikipedia.org/wiki/NetBeans (or http://bit.ly/aDBvU0).
7. http://abbot.sourceforge.net/ (or http://bit.ly/15j85L); also see jarext/jimmy.jar below.
8. http://java.sun.com/javase/technologies/accessibility/accessbridge/index.jsp (or http://bit.ly/a41Qxg); also see jaccess below.
9. http://www.oracle.com/technetwork/java/javase/jaf-136260.html (or http://bit.ly/azwbIS).
10. http://ant.apache.org/ (http://bit.ly/coyRzw); http://en.wikipedia.org/wiki/Apache_Ant (or http://bit.ly/a3BGiM).
11. http://avalon.apache.org/ (http://bit.ly/dhHxpN); the Avalon project has been superseded by http://excalibur.apache.org/ (http://bit.ly/955LYU) — I do not know why MATLAB keeps the older Avalon. Avalon was apparently removed in R2011b.
12. http://ws.apache.org/axis/ (or http://bit.ly/9RGXoy); http://en.wikipedia.org/wiki/Apache_Axis (or http://bit.ly/dyEDz7); also see jarext/jaxpc.jar below.
13. http://xmlgraphics.apache.org/batik/ (or http://bit.ly/aOl2en); http://www.w3.org/Graphics/SVG/ (or http://bit.ly/bO0UJK); http://en.wikipedia.org/wiki/Batik_(softyes ware) (or http://bit.ly/9WcILg).
14. http://java.sun.com/javase/6/docs/technotes/guides/collections/index.html (or http://bit.ly/9Hwq6O); see the discussion in Section 2.1.
15. http://en.wikibooks.org/wiki/Serial_Programming/Serial_Java (or http://bit.ly/bdYZl3).
16. http://commons.apache.org/codec (or http://bit.ly/bfxtK0).
17. http://commons.apache.org/discovery (or http://bit.ly/a4PHgB).
18. http://commons.apache.org/el/ (or http://bit.ly/ds7R6S).
19. http://hc.apache.org/httpclient-3.x/ (or http://bit.ly/aS6hBN).
20. http://commons.apache.org/io/ (or http://bit.ly/bMmcHx).
21. http://commons.apache.org/logging/ (or http://bit.ly/bLoxXv); http://en.wikipedia.org/wiki/Log4j (or http://bit.ly/ainRgK).
22. http://commons.apache.org/net/ (or http://bit.ly/9mT1a9).
23. http://www.wutka.com/dtdparser.html (or http://bit.ly/ailDg9). This Web page is currently offline, so use: http://bit.ly/adj8bz.
24. http://publicobject.com/glazedlists/ (or http://bit.ly/bqkgQM); http://www.javaworld.com/javaworld/jw-10-2004/jw-1025-glazed.html (or http://bit.ly/cIY8HA); see the discussion in Section 5.5.5.
25. http://java.sun.com/javase/technologies/accessibility/index.jsp (or http://bit.ly/ahfFFM); http://java.sun.com/javase/technologies/accessibility/docs/jaccess-1.3/doc/ (or http://bit.ly/c0gAAf) — note that this is a documentation of version 1.3 not of 1.4 (there is no official documentation for the 1.4 version); also see access-bridge above.
26. http://jakarta.apache.org/oro/ (or http://bit.ly/dd4WKy).
27. http://jakarta.apache.org/regexp/ (or http://bit.ly/cWNDIS).
28. https://jax-rpc.dev.java.net/ (or http://bit.ly/ahGigm); http://en.wikipedia.org/wiki/JAX-RPC (or http://bit.ly/9mF1FR); also see jarext/axis.jar above.
29. http://www.jdom.org/; http://en.wikipedia.org/wiki/JDOM (or http://bit.ly/ar871s).
30. https://jemmy.dev.java.net/ (or http://bit.ly/9gGenk); also see jarext/jabot.jar above.
31. http://jfcunit.sourceforge.net/ (or http://bit.ly/dcTafp).
32. http://www.jgoodies.com/freeware/forms/index.html (http://bit.ly/byJQjj).
33. http://www.jgoodies.com/freeware/looks/index.html (http://bit.ly/9x0xvz).
34. http://www.wutka.com/jox.html (http://bit.ly/bt7Sd9). This webpage is currently offline, so use http://bit.ly/98T1IR instead.
35. http://junit.org/; http://en.wikipedia.org/wiki/JUnit (or http://bit.ly/dlkh5Y).
36. http://lucene.apache.org/java/docs/ (or http://bit.ly/9ut3Zf); http://en.wikipedia.org/wiki/Lucene (or http://bit.ly/aseA2K).
37. http://java.sun.com/products/javamail/ (http://bit.ly/axPlpR); http://en.wikipedia.org/wiki/Javamail (or http://bit.ly/9HtmOg).

38. http://www.unidata.ucar.edu/software/netcdf-java/v2.2.22/javadocAll/ucar/units/package-summary.html (or http://bit.ly/bM0TSN); the "mw" prefix indicates a MathWorks-modified package to support MATLAB functionality.
39. http://nekohtml.sourceforge.net/ (or http://bit.ly/a0RxtA).
40. https://saaj.dev.java.net/ (or http://bit.ly/aKTYju); http://en.wikipedia.org/wiki/SAAJ (or http://bit.ly/dsl884).
41. http://saxon.sourceforge.net/ (or http://bit.ly/ccuzZx); http://en.wikipedia.org/wiki/Saxon_XSLT (or http://bit.ly/aFoTNo); also see jarext/xalan.jar below.
42. http://www.springsource.org/ (or http://bit.ly/aoVAn9); http://en.wikipedia.org/wiki/Java_Spring (or http://bit.ly/9WoUq9).
43. http://sourceforge.net/projects/wsdl4j/ (or http://bit.ly/9OmEot).
44. http://xalan.apache.org/ (or http://bit.ly/aBPAJE); also see jarext/saxon.jar above.
45. http://xerces.apache.org/ (or http://bit.ly/bhR3lT); http://en.wikipedia.org/wiki/Xerces (or http://bit.ly/aNmqeq).
46. https://jaxp.dev.java.net/ (or http://bit.ly/9mo9Gi); http://java.sun.com/developer/codesamples/xml.html (or http://bit.ly/9ZljMp); http://java.sun.com/j2ee/1.4/docs/tutorial/doc/JAXPIntro.html (or http://bit.ly/bMadFt).
47. http://www.icesoft.com/products/icebrowser.html (or http://bit.ly/ai0dMs). ICE browser solves problems with the standard (built-in) MATLAB browser: http://www.mathworks.com/support/solutions/en/data/1-91A14Y/ (or http://bit.ly/a7C9xS); but it does have limitations: http://www.mathworks.com/support/solutions/en/data/1-5AJ3DV (or http://bit.ly/gIGaAm).
48. http://www.wildcrest.com/Software/J2PrinterWorks/J2PrinterWorksREADME.html (or http://bit.ly/a2j620).
49. http://www.qarbon.com/presentation-software/solution/ (or http://bit.ly/9z7eCS).
50. http://www.webrenderer.com/ (or http://bit.ly/a3AFiR).
51. http://java.sun.com/docs/books/tutorial/uiswing/components/spinner.html (or http://bit.ly/b1gg56).
52. http://www.mathworks.com/matlabcentral/fileexchange/26970-spinnerdemo (or http://bit.ly/bAhlpw).
53. http://java.sun.com/docs/books/tutorial/uiswing/components/button.html (or http://bit.ly/ahRKa7).
54. http://java.sun.com/docs/books/tutorial/uiswing/components/button.html#radiobutton (or http://bit.ly/bbYbtz).
55. http://java.sun.com/docs/books/tutorial/uiswing/components/button.html#checkbox (or http://bit.ly/9oxWwX).
56. http://java.sun.com/docs/books/tutorial/uiswing/components/textfield.html (or http://bit.ly/9Vl3Qh).
57. http://java.sun.com/docs/books/tutorial/uiswing/components/formattedtextfield.html (or http://bit.ly/axEi5c).
58. http://java.sun.com/docs/books/tutorial/uiswing/components/list.html (or http://bit.ly/9qFQU0).
59. http://java.sun.com/docs/books/tutorial/uiswing/components/combobox.html (or http://bit.ly/b0e4ta).
60. http://java.sun.com/docs/books/tutorial/uiswing/components/label.html (or http://bit.ly/dCx3uq).
61. http://java.sun.com/docs/books/tutorial/uiswing/components/menu.html (or http://bit.ly/cMm0Ke).
62. http://java.sun.com/docs/books/tutorial/uiswing/components/tabbedpane.html (or http://bit.ly/aAFMyM).
63. http://java.sun.com/docs/books/tutorial/uiswing/components/tree.html (or http://bit.ly/dBH5Mt).
64. http://java.sun.com/docs/books/tutorial/uiswing/components/toolbar.html (or http://bit.ly/9zkpxO).
65. http://java.sun.com/docs/books/tutorial/uiswing/components/splitpane.html (or http://bit.ly/b3LKpb).
66. http://java.sun.com/docs/books/tutorial/uiswing/components/table.html (or http://bit.ly/aysVb3).
67. http://java.sun.com/docs/books/tutorial/uiswing/components/filechooser.html (or http://bit.ly/d3YLAr).
68. http://java.sun.com/docs/books/tutorial/uiswing/components/dialog.html (or http://bit.ly/cNtSGs).
69. http://download.oracle.com/javase/6/docs/api/javax/swing/JOptionPane.html (or http://bit.ly/rthGIM).
70. http://java.sun.com/docs/books/tutorial/uiswing/components/textarea.html (or http://bit.ly/aIZb9G).
71. http://download.oracle.com/javase/6/docs/api/javax/swing/JTextPane.html (or http://bit.ly/npKP6R).
72. http://download.oracle.com/javase/6/docs/api/javax/swing/JEditorPane.html (or http://bit.ly/nhGRWU).
73. http://java.sun.com/docs/books/tutorial/uiswing/components/editorpane.html (or http://bit.ly/8X4XoX).
74. http://download.oracle.com/javase/6/docs/api/javax/swing/tree/DefaultTreeModel.html (or http://bit.ly/dpe8qB).
75. http://www.mathworks.com/matlabcentral/newsreader/view_thread/138344 (or http://bit.ly/9CNHVn); http://www.mathworks.com/matlabcentral/newsreader/view_thread/156915 (or http://bit.ly/9jMYOP); and its usage as a Java table: http://www.mathworks.com/matlabcentral/fileexchange/5752 (or http://bit.ly/c4N3c9).

76. http://www.mathworks.com/help/techdoc/matlab_env/bruby4n-1.html#f3-40511 (or http://bit.ly/crY6KJ).

77. http://en.wikipedia.org/wiki/Lucene; http://lucene.apache.org/java/

78. Also used by the *webbot* utility: http://www.mathworks.com/matlabcentral/fileexchange/4023 (or http://bit.ly/d6JYWc) and the compression utility in: http://www.mathworks.com/matlabcentral/fileexchange/8899 (or http://bit.ly/c43CZE), *EZGlobe* utility in: http://www.mathworks.com/matlabcentral/fileexchange/8966 (or http://bit.ly/cgEvYG), and some others.

79. Change notifications are discussed in an official MATLAB technical solution: http://www.mathworks.com/support/solutions/en/data/1-81QJLT/ (or http://bit.ly/bj9y35).

80. http://blogs.mathworks.com/desktop/2009/10/12/the-history-of-keyboard-shortcuts-in-matlab/#comment-6666 (or http://bit.ly/aOAIbd); http://www.mathworks.com/support/solutions/en/data/1-BG4EU1/ (or http://bit.ly/bUDsYD).

81. http://en.wikipedia.org/wiki/TeX

82. For example, http://www.mathworks.com/matlabcentral/fileexchange/11946-tex-editor (or http://bit.ly/gUhTSk).

83. http://UndocumentedMatlab.com/blog/color-selection-components/ (or http://bit.ly/kjBDVn).

84. http://java.sun.com/javase/6/docs/api/javax/swing/JColorChooser.html (or http://bit.ly/9VosZG); http://java.sun.com/docs/books/tutorial/uiswing/components/colorchooser.html (or http://bit.ly/a7RAjb).

85. http://blogs.mathworks.com/desktop/2011/04/18/redesigned-plot-catalog-in-matlab-r2011a/#comment-7698 (or http://bit.ly/fWxHWX).

86. http://www.mathworks.com/matlabcentral/newsreader/view_thread/154144#386877 (or http://bit.ly/a5aH41).

87. http://blogs.mathworks.com/desktop/2009/03/16/click-for-more-information/ (or http://bit.ly/cQnBWP).

88. http://blogs.mathworks.com/desktop/2010/07/19/that-tooltip-looks-familiar/ (or http://bit.ly/92YyYc).

89. http://java.sun.com/docs/books/tutorial/uiswing/components/menu.html (or http://bit.ly/cMm0Ke).

90. http://www.jidesoft.com/javadoc/com/jidesoft/swing/StyledLabel.html (or http://bit.ly/cUzTP4); PDF developer guide: http://www.jidesoft.com/products/JIDE_Common_Layer_Developer_Guide.pdf (or http://bit.ly/9x61e1).

91. http://www.jidesoft.com/javadoc/com/jidesoft/swing/StyledLabelBuilder.html (or http://bit.ly/baFokP).

92. http://www.jidesoft.com/javadoc/com/jidesoft/swing/ClickThroughStyledLabel.html (or http://bit.ly/9yiZ5w).

93. http://en.wikipedia.org/wiki/MIME

94. http://UndocumentedMatlab.com/blog/syntax-highlighted-labels-panels/#comment-13214 (or http://bit.ly/aTWvvy).

95. See http://java.sun.com/docs/books/tutorial/uiswing/components/dialog.html#features (or http://bit.ly/aOzOLw).

96. http://java.sun.com/docs/books/tutorial/uiswing/components/dialog.html (or http://bit.ly/cNtSGs).

97. http://java.sun.com/javase/6/docs/api/javax/swing/JOptionPane.html (or http://bit.ly/aVGRxD).

98. http://java.sun.com/javase/6/docs/api/javax/swing/JDialog.html (or http://bit.ly/d8q0B3).

99. http://publicobject.com/glazedlists/ (or http://bit.ly/bqkgQM).

100. http://www.javaworld.com/javaworld/jw-10-2004/jw-1025-glazed.html (or http://bit.ly/cIY8HA).

101. http://blogs.mathworks.com/desktop/2007/03/29/shortcuts-for-commonly-used-code/#comment-5753 (or http://bit.ly/aRvA9c); http://www.mathworks.com/support/solutions/en/data/1-1BM76/ (or http://bit.ly/9PSVQj); in old MATLAB releases, the relevant code was com.mathworks.ide.cmdline.CommandHistory.deleteAllHistoryForDesktop() — see http://www.mathworks.com/matlabcentral/newsreader/view_thread/ 80617 (or http://bit.ly/9kRTRO).

102. http://www.mathworks.com/support/solutions/en/data/1-8F3C38/ (or http://bit.ly/dpWFSn).

103. http://stackoverflow.com/questions/4405536/history-command-buffer-in-matlab (or http://bit.ly/heUqDG).

104. http://www.mathworks.com/matlabcentral/newsreader/view_thread/278672#734272 (or http://bit.ly/9w7tSW).

105. http://www.mathworks.com/help/releases/R2006a/techdoc/rn/r2006a_v7_2_graphics.html (or http://bit.ly/90tkhY).

106. http://www.mathworks.com/matlabcentral/fileexchange/ (or http://bit.ly/anwdaP).

107. http://blogs.mathworks.com/desktop/2009/09/21/the-front-page-of-the-file-exchange-your-desktop/ (or http://bit.ly/a4KcQN).

108. http://www.jidesoft.com/products/action.htm (or http://bit.ly/bLEUbv); PDF developer guide: http://www.jidesoft.com/products/JIDE_Action_Framework_Developer_Guide.pdf (or http://bit.ly/d8Rrec).

109. http://www.jidesoft.com/products/oss.htm (or http://bit.ly/dyDh22); PDF developer guide: http://www.jidesoft.com/products/JIDE_Common_Layer_Developer_Guide.pdf (or http://bit.ly/9x61e1).

110. https://jide-oss.dev.java.net/ (or http://bit.ly/dg0Myq).

111. http://www.jidesoft.com/products/component.htm (or http://bit.ly/dxGV1m); PDF developer guide: http://www.jidesoft.com/products/JIDE_Components_Developer_Guide.pdf (or http://bit.ly/9BFgg9).

112. http://www.jidesoft.com/products/dialogs.htm (or http://bit.ly/crvSRc); PDF developer guide: http://www.jidesoft.com/products/JIDE_Dialogs_Developer_Guide.pdf (or http://bit.ly/9wGiDE).

113. http://www.jidesoft.com/products/dock.htm (or http://bit.ly/bdI5RA); PDF developer guide: http://www.jidesoft.com/products/JIDE_Docking_Framework_Developer_Guide.pdf (or http://bit.ly/bNvZgP).

114. http://www.jidesoft.com/products/grids.htm (or http://bit.ly/aetw9H); PDF developer guide: http://www.jidesoft.com/products/JIDE_Grids_Developer_Guide.pdf (or http://bit.ly/a88Xzt).

115. http://www.jidesoft.com/javadoc/ (or http://bit.ly/bcevRu).

116. http://www.jidesoft.com/forum/ (or http://bit.ly/9Q0A9j).

117. http://www.jidesoft.com/history/ (or http://bit.ly/b4MEVn); R2008a uses 2.1.2.01 (Aug 2007) and R2010a uses 2.7.1 (Aug 2009).

118. http://www.jidesoft.com/javadoc/com/jidesoft/swing/JideSwingUtilities.html (or http://bit.ly/9ozlvw).

119. http://blogs.mathworks.com/desktop/2007/05/18/do-you-dock-figure-windows-what-does-your-desktop-look-like/#comment-7041 (or http://bit.ly/bIvtgI).

120. http://poi.apache.org/spreadsheet/ (or http://bit.ly/9kyv0m).

121. http://UndocumentedMatlab.com/blog/jide-property-grids/ (or http://bit.ly/bF6LpJ), written by Lebente Hunyadi (http://www.mathworks.com/matlabcentral/fileexchange/authors/60898 or http://bit.ly/9OVCJJ).

122. http://www.mathworks.com/help/techdoc/ref/inspect.html (or http://bit.ly/b23wxl).

123. http://www.mathworks.com/help/techdoc/creating_guis/f7-998368.html (or http://bit.ly/b23wxl).

124. http://www.mathworks.com/help/techdoc/creating_plots/f9-47085.html#f9-43456 (or http://bit.ly/aud12c).

125. http://www.mathworks.com/matlabcentral/fileexchange/28732-property-grid (or http://bit.ly/cn218w).

126. http://UndocumentedMatlab.com/blog/jide-property-grids/comment-page-1/#comment-9476 (or http://bit.ly/971sbb).

127. http://www.jidesoft.com/products/JIDE_Grids_Developer_Guide.pdf (or http://bit.ly/a88Xzt).

128. http://www.jidesoft.com/javadoc/ (or http://bit.ly/bcevRu).

129. http://www.jidesoft.com/javadoc/com/jidesoft/grid/PropertyTable.html (or http://bit.ly/9sVZGe).

130. http://www.jidesoft.com/javadoc/com/jidesoft/grid/PropertyTableModel.html (or http://bit.ly/cJXcYb).

131. http://www.jidesoft.com/javadoc/com/jidesoft/grid/Property.html (or http://bit.ly/9sfINj).

132. http://www.jidesoft.com/javadoc/com/jidesoft/grid/DefaultProperty.html (or http://bit.ly/cgwE2P).

133. http://www.jidesoft.com/javadoc/com/jidesoft/grid/PropertyPane.html (or http://bit.ly/9aG8gN).

134. http://UndocumentedMatlab.com/files/javaclass.m (or http://bit.ly/d1DSbB).

135. http://UndocumentedMatlab.com/blog/advanced-jide-property-grids (or http://bit.ly/a7vaqs), also by Lebente Hunyadi.

136. http://java.sun.com/docs/books/tutorial/uiswing/components/table.html#editrender (or http://bit.ly/coLlLg).

137. http://java.sun.com/docs/books/tutorial/uiswing/components/spinner.html (or http://bit.ly/b1gg56).

138. http://www.jidesoft.com/javadoc/com/jidesoft/grid/SpinnerCellEditor.html (or http://bit.ly/b27yFT).

139. http://java.sun.com/docs/books/tutorial/uiswing/components/button.html#checkbox (or http://bit.ly/9oxWwX).

140. http://www.jidesoft.com/javadoc/com/jidesoft/grid/BooleanCheckBoxCellEditor.html (or http://bit.ly/adJ7K4).

141. http://www.jidesoft.com/javadoc/com/jidesoft/grid/EditorContextSupport.html (or http://bit.ly/cCzYAo).

142. For example, `BooleanCellEditor` is documented in http://www.jidesoft.com/javadoc/com/jidesoft/grid/BooleanCellEditor.html (or http://bit.ly/aN1hmE).

143. http://www.mathworks.com/matlabcentral/fileexchange/26784-property-sheet (or http://bit.ly/bdvcnM).

144. http://sourceforge.net/projects/fraeser/files/matlab_gui_extensions (or http://bit.ly/atM2x3).

145. http://www.mathworks.com/matlabcentral/fileexchange/28732-property-grid (or http://bit.ly/cn218w).

146. http://UndocumentedMatlab.com/blog/date-selection-components/ (or http://bit.ly/bpe1V4).

147. A list of non-JIDE alternatives is presented here: http://UndocumentedMatlab.com/blog/date-selectioncomp onents/#Alternatives (or http://bit.ly/aoTMQv).

148. http://www.jidesoft.com/javadoc/com/jidesoft/combobox/DateChooserPanel.html (or http://bit.ly/cfENEX).

149. http://www.jidesoft.com/javadoc/com/jidesoft/combobox/CalendarViewer.html (or http://bit.ly/ddt08L).

150. http://www.jidesoft.com/javadoc/com/jidesoft/combobox/DateComboBox.html (or http://bit.ly/9S0gA6).

151. http://www.jidesoft.com/javadoc/com/jidesoft/combobox/DateSpinnerComboBox.html (or http://bit.ly/ duNrRv).

152. http://www.jidesoft.com/javadoc/com/jidesoft/combobox/MonthChooserPanel.html (or http://bit.ly/ b7oONr).

153. http://www.jidesoft.com/javadoc/com/jidesoft/combobox/MonthComboBox.html (or http://bit.ly/aDgGib).

154. http://www.mathworks.com/matlabcentral/fileexchange/14583-uicomponent (or http://bit.ly/dbtov3).

155. This was mentioned in http://www.mathworks.com/support/solutions/en/data/1-OVWJ9/ (or http://bit. ly/9GH2Cm).

156. http://java.sun.com/docs/books/tutorial/uiswing/misc/timer.html (or http://bit.ly/bvpQHX).

157. http://commons.apache.org/logging/ (or http://bit.ly/bLoxXv); http://en.wikipedia.org/wiki/Log4j (or http:// bit.ly/ainRgK); http://blogs.mathworks.com/desktop/2009/07/06/calling-java-from-matlab/#comment-6655 (or http://bit.ly/fM0G7S).

158. http://www.jgoodies.com/freeware/forms/ (or http://bit.ly/8XMIwD); https://forms.dev.java.net/ (or http://bit.ly/biUx9g); PDF tutorial: http://www.jgoodies.com/articles/forms.pdf (or http://bit.ly/azLgCf); Presentation: http://www.jgoodies.com/articles/layout.pdf (or http://bit.ly/aVHwkd).

159. http://explodingpixels.wordpress.com/2008/05/02/sexy-swing-app-the-unified-toolbar/ (or http://bit.ly/ deQcsy) — this was written by Ken Orr, who was a lead MATLAB Desktop designer until late 2009, in his personal blog. Note that the blog post does NOT mention MATLAB in any way, and the connection I made here is pure speculation.

160. http://www.mathworks.com/matlabcentral/fileexchange/25862-enhanced-input-dialog-box (or http://bit. ly/9YiiJl).

161. http://www.jgoodies.com/freeware/looks/ (or http://bit.ly/96XUpq); https://looks.dev.java.net/ (or http:// bit.ly/c8Ay7m).

162. http://UndocumentedMatlab.com/blog/jgraph-and-bde/ (or http://bit.ly/hfZUTO).

Customizing
MATLAB® Controls

As noted in Chapter 3, all **MATLAB** uicontrols are based on underlying Java Swing components, or more precisely on internal **MATLAB** classes which extend the standard Swing components. This section will detail these components, show how to access them, and describe how they can be customized in our **MATLAB** application.

The underlying components can be found using the ***findjobj*** utility described earlier:[†]

```
>> hButton = uicontrol('Style','ToggleButton');

>> jButton = findjobj(hButton)
jButton =
    javahandle_withcallbacks.com.mathworks.hg.peer.ToggleButtonPeer$
    hgToggleButton

>> jButton.java
ans =
com.mathworks.hg.peer.ToggleButtonPeer$hgToggleButton[...]
```

The component can also be found and inspected visually, rather than programmatically, using ***findjobj***'s graphic hierarchical component tree display, displayed when ***findjobj*** is called with no output arguments:

```
>> findjobj(hButton); %or: findjobj; to search from the figure root
```

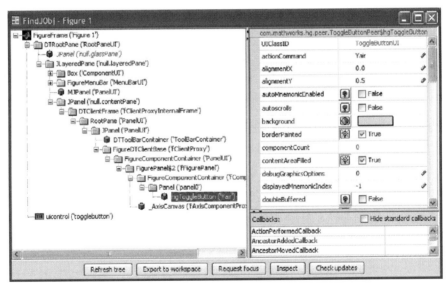

uicontrol **underlying Java component found using** *findjobj*

Once found, the component can be inspected using ***methods/methodsview, get/set, inspect*** and ***uiinspect***, as described in Chapter 1. It is usually best to use the returned MATLAB ***handle()*** wrapper rather than the actual Java object for setting object properties (see a discussion in Section 3.4); however, inspecting the Java object is easier since ***handle()*** reports all arguments

---

[†] ***Findjobj*** is discussed in detail in Section 7.2.

and return values as "MATLAB array", and does not report the constructor methods, super-class, nor static class fields:

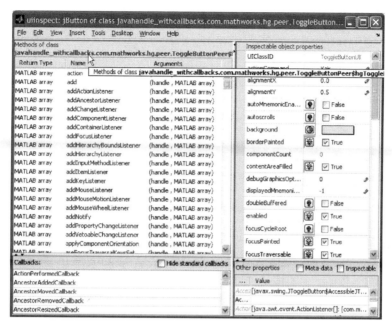

jButton.uiinspect **or:** uiinspect(jButton)

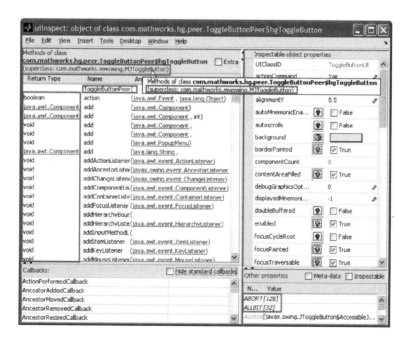

jButton.java.uiinspect **or:** uiinspect(jButton.java) **(See color insert.)**

Of course, the real purpose in finding the underlying Java reference handle is to access the Java component's properties, callbacks and methods, which offer much more extensive functionality than those exposed by the corresponding MATLAB control.

For example, the Java handle can be used to force display of the component tooltip:[†]

```
set(hButton,='TooltipString','This is a tooltip string...');
jButton = findjobj(hButton);
import java.awt.event.ActionEvent;
action = jButton.getActionMap.get('postTip');
actionEvent = ActionEvent(jButton,ActionEvent.ACTION_PERFORMED,'postTip');
action.actionPerformed(actionEvent);

% Or, for an EDT-safe action invocation:
awtinvoke(action, 'actionPerformed(Ljava.awt.event.ActionEvent;)', ...
          actionEvent);
```

Note that the corresponding new way for EDT-safe invocation (see Section 3.2) fails (I believe that this is due to an internal MATLAB bug):

```
>> javaMethodEDT('actionPerformed',action,actionEvent);
??? Error using ==> javaMethodEDT
Java exception occurred:
java.lang.IllegalAccessException: Class
com.mathworks.jmi.AWTUtilities$Invoker$3 can not access a member of
class javax.swing.ToolTipManager$Actions with modifiers "public"
```

Several examples of customizing the Java components were presented in Section 3.3.1. For example, updating a component's mouse-over cursor:

```
jButton = findjobj(hButton);
jButton.setCursor(java.awt.Cursor(java.awt.Cursor.HAND_CURSOR));
```

Let us now look at all the standard MATLAB controls, explore their underlying Java component(s) and describe their added functionality benefits.

[†] http://UndocumentedMatlab.com/blog/spicing-up-matlab-uicontrol-tooltips/#comment-1173 (or http://bit.ly/5oFn8M). Unfortunately, this method does not work consistently — I could not determine why or under which exact circumstances it fails. As an alternative, use Geoffrey Akien's tooltip utility (http://www.mathworks.com/matlabcentral/fileexchange/26283 or http://bit.ly/6ZkYmJ). See Section 6.12 for additional information.

6.1 PushButton

uicontrol(*'Style'*, *'pushbutton'*) is MATLAB's default and simplest uicontrol: a simple push-button. It uses the `com.mathworks.hg.peer.PushButtonPeer$1` class, that extends MATLAB's `com.mathworks.mwswing.MJButton` class, which itself extends Swing's `javax.swing.JButton` class.[1]

Perhaps the simplest and most striking customization of pushbuttons (and most other uicontrols) is using Swing's inherent HTML and CSS support (see Section 3.3.3). Here is an example of an HTML-formatted pushbutton:[2]

```
tooltip ='<html>HTML-aware<br><b>tooltips</b><br><i>supported';
labelTop='<HTML><center><FONT color="red">Hello</Font> <b>world</b>';
labelBot =['<div style="font-family:impact;color:green"><i>What a</i>'...
          '<Font color="blue" face="Comic Sans MS">nice day!'];
hButton = uicontrol('Style','pushbutton', 'tooltip',tooltip, ...
                   'string',[labelTop '<br>' labelBot], 'position',pos);
```

(See color insert.)

`JButton` has many useful properties and methods, which are unavailable when using the MATLAB *uicontrol* handle, and are essentially common to all Java components:

■ **Border** — this property, common to all Java Swing components, sets the border that surrounds the component's interior.[3] Section 3.3.1 showed how to set different borders and how they affect a button's appearance. The related **BorderPainted** property indicates whether or not a component's border is set.

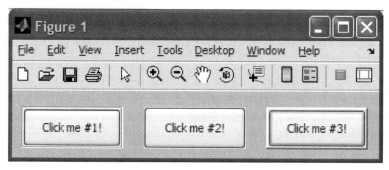

Buttons having different Border values

■ **ContentAreaFilled** — this boolean property (default=true) indicates whether to fill the component's interior with component-specific graphics. On a button such as our jButton, this determines a flat (2D) versus 3D appearance (note: the flat appearance does not distinguish between depressed/undepressed states):[†]

ContentAreaFilled = true   **ContentAreaFilled = false**

This property can be used to display a linkable label in the GUI figure (there are other ways to achieve this, but it seems to me this is one of the easiest). For example, this property is used to display the link in the ***FindJObj*** utility:

```
>> labelStr = '<html><center><a href="">Undocumented<br>Matlab.com';
>> hButton = uicontrol('string',labelStr,'pos',[20,20,100,35]);
>> jButton = findjobj(hButton)
jButton = javahandle_withcallbacks.com.mathworks.hg.peer.PushButtonPeer$1

>> callbackStr = 'web(''http://Undocumentedmatlab.com'');';
>> set(jButton,'ActionPerformedCallback',callbackStr);
>> jButton.setCursor(java.awt.Cursor(java.awt.Cursor.HAND_CURSOR));
>> jButton.setContentAreaFilled(false);
```

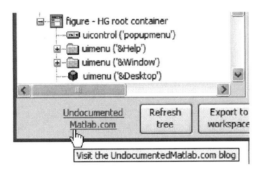

The same visual effect can be achieved by simply clearing the **Border** property:

```
>> jButton.setBorder([]);[4]
>> jButton.setBorderPainted(false); % an alternative
```

Note that Java components have a lighter shade of gray background color than the default MATLAB figure. To set the Java component's background color to the same color as the figure's, we can either set the figure's **Color** property, or the Java component's **Background** property. For example, setting the Java component's **Background** property to the figure's **Color** can be done using the following code segment:

```
>> c = mat2cell(get(gcf,'color'),1,[1,1,1]);
>> jButton.setBackground(java.awt.Color(c{:}));
```

[†] **ContentAreaFilled** is actually the preferred way of setting transparent controls (instead of setting the **Opaque** property): http://java.sun.com/javase/6/docs/api/javax/swing/AbstractButton.html#setContentAreaFilled(boolean) (or http://bit.ly/ce6Ebs). However, this has no effect in MATLAB figure windows, as explained in Section 7.3.3.

Unfortunately, using the *setBackground()* method has a side effect of restoring the default border, something that *setContentAreaFilled()* does not clear. For this reason, it may be better to either use *setBorder([])* rather than *setContentAreaFilled-(false)*, or to set the figure's **Color** property rather than the Java component's **Background** property:

```
>> color = jButton.getBackground.getRGBComponents([]);  % R,G,B,Alpha
>> set(gcf, 'color', color(1:3))
```

- **DebugGraphicsOptions** — (default=0) sets debugging options useful for debugging the graphic appearance.[5]
- **Cursor** — (default=[]) sets the component-specific cursor, as described in Section 3.3.1. Note that this property must be used with the accessor methods (*getCursor()/ setCursor()*) — *get/set* fail here, as described in Section 3.3.1. If the property value is set to [] (the default value), then the container's cursor is used and this component will display no special cursor upon hover.

 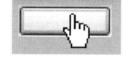 

Standard Cursor   **Nonstandard Cursor**   **Custom Cursor**

- **DisplayedMnemonicIndex** (default=–1) indicates the character position within the text label where the *Mnemonic*[6] (i.e., keyboard shortcut) should be displayed. Mnemonics are similar to, but distinct from, keyboard *accelerators*: Mnemonics only set the focus for a displayed item (e.g., clicking <Alt-F> to set the focus on the File main menu); accelerators activate the action of the associated component, even if it is hidden (e.g., <Control-S> to save the current document). Java components can have both a mnemonic and an accelerator. Associated property **Mnemonic** (default=0) indicates the mnemonic's ASCII code. Also see related property **AutoMnemonicEnabled**, described below. In the following case, **DisplayedMnemonicIndex**=3 (remember that Java indices start at 0) and **Mnemonic**=73 (='r'):[†]

Note that HTML-rendered buttons do not underline the mnemonic character — we need to underline it ourselves: <html><font color='red'>Yai<u>r</u> ...

[†] Sometimes the **MnemonicIndex** should not be the first occurrence of **Mnemonic**. For example, in the File menu, we wish to highlight the second occurrence of **Mnemonic** 'a': 'Save A̲s' rather than 'S̲ave As'; setting **MnemonicIndex** to 5 achieves this.

- **DisabledIcon, DisabledSelectedIcon, Icon, PressedIcon, RolloverIcon, RolloverSelectedIcon, SelectedIcon** — these icons may be set to present a different appearance depending on component state. Refer to Section 4.6.1 for an example. The associated property **IconTextGap** (default=4) determines the gap in pixels between the icon and the button text label.

  Associated properties **HorizontalTextPosition** and **VerticalTextPosition** specify the label text's alignment relative to the label icon. These properties accept `javax.swing.SwingConstants` integer values: **HorizontalTextPosition** accepts LEFT (=2), CENTER (=0, default), RIGHT (=4), LEADING (=10) or TRAILING (=11);[†] **VerticalTextPosition** accepts TOP (=1), CENTER (=0, default), or BOTTOM (=3). Users should normally prefer to use `javax.swing.SwingConstants` rather than their numeric values. The reason is that although unlikely, numeric values may change between JVM versions, platforms and implementations. Also, enumerated constants are more readable and maintainable than numbers.

  For example, let us display an icon to the right and upward of the text:

```
myIcon = fullfile(matlabroot,'/toolbox/matlab/icons/warning.gif');
jButton.setIcon(javax.swing.ImageIcon(myIcon));
jButton.setHorizontalTextPosition(javax.swing.SwingConstants.LEFT);
jButton.setVerticalTextPosition(javax.swing.SwingConstants.BOTTOM);
```

HorizontalTextPosition = `SwingConstants.LEFT`	CENTER	CENTER
VerticalTextPosition = `SwingConstants.BOTTOM`	BOTTOM	CENTER

- **HorizontalAlignment** — default=0) sets the component's label (text and icon) alignment relative to the component's horizontal center. Like **HorizontalTextPosition** above, **HorizontalAlignment** accepts the following values: `javax.swing.SwingConstants.RIGHT, LEFT, CENTER, LEADING, TRAILING`.

 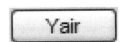 

HorizontalAlignment = `SwingConstants.LEADING`	**HorizontalAlignment =** `SwingConstants.CENTER` (=default)	**HorizontalAlignment =** `SwingConstants.TRAILING`

---

† For a Latin locale, LEADING and LEFT have the same effect, as do TRAILING and RIGHT. This may be different for LTR (Hebrew/Arabic) or Far-Eastern locales.

The Java **HorizontalAlignment** property is particularly important since the MATLAB **HorizontalAlignment** *uicontrol* property has no effect on non-text Windows controls (i.e., all controls except editboxes and text labels).[7]

■ **VerticalAlignment** — (default=0) similar to **HorizontalAlignment**: This property sets the component's label alignment relative to the component's vertical center. Like **VerticalTextPosition** above, **VerticalAlignment** accepts any of the following integer values: `javax.swing.SwingConstants.TOP` (=1), `BOTTOM` (=3), or `CENTER` (=0, default):

VerticalAlignment =   **VerticalAlignment =**   **VerticalAlignment =**
`SwingConstants.TOP`   `SwingConstants.CENTER`   `hSwingConstants.BOTTOM`
                                        (=default)

The **HorizontalAlignment** and **VerticalAlignment** properties can be combined. For example,

HorizontalAlignment = `LEFT` and **VerticalAlignment =** `TOP`

■ **Margin** — (default = `java.awt.Insets(2,-1,2,-1)`) this property sets the margin insets[8] (top, left, bottom, right) between the button's borders and the internal label. Setting **Margin** to [] will revert to the default margin, a 2-pixel vertical and −1 pixel horizontal insets:

```
jButton.setMargin(java.awt.Insets(8,8,8,8));
```

Default Margin   **8-pixel Margin†**

---

† For some unknown reason, only the vertical margin has any visible effect on pushbutton labels, at least on Windows XP. I tend to believe that this is a JVM bug, rather than a MATLAB one, but perhaps I am missing something here.

- **FocusCycleRoot** (default=false), **FocusPainted** (default=true), **FocusTraversalKeys** (default=[]), **FocusTraversalPolicy** (default=[]), **FocusTraversalPolicyProvider** (default=false), **FocusTraversalPolicySet** (default=false), **Focusable** (default=true), **ManagingFocus** (default=false), **NextFocusableComponent** (default=[]), **RequestFocusEnabled** (default=true) and **VerifyInputWhenFocusTarget** (default=true) all relate to the component's focus cycle, that is, selecting (setting the focus on) the component using the keyboard.[9] MATLAB documentation calls the focus cycle *"tab-order"* but only allows selecting the focus cycle order, using the ***uistack*** function — for all the extra functionality we need to use these Java properties (also see Section 3.3.4, as well as the **FocusTraversable** property below).

- **MultiClickThreshhold** — (default=0) sets the number of milliseconds between subsequent processed user mouse clicks on the button. Any clicks that occur within the specified number (e.g., fast double-clicks) will be considered by the component as only a single click.[10] The default value of 0 means that all clicks will be processed separately; this is often undesirable in GUI applications. Note that the property value is in milliseconds, not seconds.

- There are quite a few other properties, but I personally find them less useful. However, by all means feel free to inspect them and test alternative values.

In addition to the property accessor (getter/setter/checker) methods, JPushButton has the following useful method:

- *doClick()* — programmatically simulate a button click. This has the same effect as an interactive user button click, and the corresponding callback(s) is/are invoked. An optional numeric input parameter to this method specifies the time in milliseconds in which the button should remain visibly "pressed" before popping back to its normal state.

In addition to the standard Swing component callbacks (see Section 3.4), JPushButton defines several non-standard callbacks:

- **ActionPerformedCallback** — fired when the button is clicked.
- **CaretPositionChangedCallback** — unused.
- **InputMethodTextChangedCallback** — unused.
- **ItemStateChangedCallback** — fired when the button state (depressed or not) has changed. This can only be done programmatically, via the *setSelected()* method, or by setting the **Selected** property on the Java handle, or by setting the **Value** property on the MATLAB handle.[†] It cannot be done interactively since interactive (mouse) clicking immediately bounces the button to its undepressed state.

[†] Note the potentially confusing difference in property names between the MATLAB and Java objects: MATLAB's **Value** (a boolean 'on'/'off' property) is equivalent to the Java object's **Selected** property (a boolean true/false, 0/1 property); MATLAB's **Selected** property has no Java object equivalent — its boolean value controls highlighting of the uicontrol with a unique black border.

- **StateChangedCallback** — Similar to **ItemStateChangedCallback**, but also fired when the mouse enters and leaves the button bounds thereby modifying the button appearance with a special highlight border.

In addition to all the regular JPushButton properties, methods and callbacks, MATLAB's PushButtonPeer$1 adds the following public properties:

- **AutoMnemonicEnabled** — boolean flag (default=false); if set, then setting the button label (via *setText()*) with a string containing an ampersand (&) will automatically process the string and use the & location as an indication of a requested mnemonic, saving programmers the trouble of also setting the **DisplayedMnemonicIndex** and **Mnemonic** properties. For example,

```
jButton.setAutoMnemonicEnabled(true)
jButton.setText('&Yair')
```

| **AutoMnemonicEnabled** = true (responds to <Alt-Y> key-clicks) | **AutoMnemonicEnabled**=false (=default) |

- **FlyOverAppearance** — boolean flag (default=false); if set, the button appearance is changed to a flat (2D) appearance with a special 3D border effect displayed on mouse hover. This appearance is useful for toolbar buttons:

| **FlyOverAppearance** = false | **FlyOverAppearance** = true |

- **FocusTraversable** — boolean flag (default=true); if set then the component participates in the figure's focus-traversal round,[†] meaning that clicking <Tab> or <Control-Tab>[‡] will eventually come around to selecting the component (=setting the focus on it). If this flag is false, the component cannot be selected with <Tab> — only with the mouse (also see Section 3.3.4).

[†] MATLAB does not enable bypassing a component altogether, only setting its tab-order; we need the Java property to achieve this.

[‡] On Windows systems; the keyboard event is of course different for other platforms.

And the following public methods:

■ *hasFlyOverAppearance* — returns a boolean flag indicating the **FlyOverAppearance** property value;

■ *hideText* — a convenience (equivalent) form for *setText([]);*

■ *isAutoMnemonicEnabled* — returns a boolean flag indicating the **AutoMnemonic Enabled** property value;

■ *setAutoMnemonicEnabled(flag)* — sets the **AutoMnemonicEnabled** property;

■ *setFlyOverAppearance*(flag) — sets the **FlyOverAppearance** property value;

■ *setFocusTraversable(flag)* — sets the **FocusTraversal** property value;

■ *setBackgroundPainter()* — sets a painter object that is used to paint the button before the standard *paint()* is called, effectively painting the button's background, possibly modifying the button appearance (I never tried this). The method accepts `com.math-works.mwswing.Painter`-compliant objects where `Painter` is an interface that declares only the standard Swing *paint()* method.

...while not surprisingly removing the *JButton()* constructors from public view.

Like all other MATLAB uicontrols, pushbuttons obey Java's look-and-feel, as explained in Section 3.3.2. Here is how the pushbutton uicontrol appears with different L&Fs on R2011a (JVM 1.6) on the Windows XP platform:[†]

```
import javax.swing.UIManager
originalLnF = UIManager.getLookAndFeel;

% Modify the platform Look-and-Feel
UIManager.setLookAndFeel('javax.swing.plaf.metal.MetalLookAndFeel');
    % or: com.sun.java.swing.plaf.motif.MotifLookAndFeel
    % or: com.sun.java.swing.plaf.windows.WindowsLookAndFeel
    % or: com.sun.java.swing.plaf.windows.WindowsClassicLookAndFeel
    % or: com.jgoodies.looks.plastic.Plastic3DLookAndFeel
hButton = uicontrol('Style','pushbutton', 'String','Yair');

% Restore the standard platform L&F
UIManager.setLookAndFeel(originalLnF);

% Update the uicontrol to the newly updated L&F
jButton = findjobj(hButton);
jButton.updateUI;
```

Unselected (underpressed) / Selected (depressed) rows of pushbuttons labeled "Yair" shown in Windows L&F, Win Classic, Metal L&F, Motif L&F, Plastic L&F

Note that a bug existed in MATLAB versions R14 (7.0) through R2006a (7.2) in which the background color of buttons could not be set on Windows XP machines;[11] to fix this problem on those WinXP MATLAB versions, do one of the following:

- Set the Windows Scheme to 'classic';[12]
- Update the L&F to some other temporary value before creating the button;
- Add the `-Dswing.noxp = true` option to the *java.opts* file (see Section 1.9).[13]

6.2 ToggleButton

uicontrol(*'Style'*,*'togglebutton'*) is similar to the pushbutton uicontrol described in the previous section, except that it has a dual state (depressed and undepressed). A togglebutton can remain in a depressed (selected) state, unlike pushbuttons which automatically pop back up to the undepressed state. The togglebutton uicontrol uses the `com.mathworks.hg.peer.Toggle-ButtonPeer$hgToggleButton` class that extends MATLAB's `com.mathworks.mwswing.MJToggleButton` class, which itself extends Swing's `javax.swing.JToggleButton` class.[14]

`JToggleButton` and `hgToggleButton` have the same appearance, properties, and methods as the pushbutton *uicontrol*. Togglebuttons also expose the following public constructor method:

- *ToggleButtonPeer$hgToggleButton* — Java object constructor method. Accepts 0–3 optional arguments: text string, icon image, and depressed state flag.

On the other hand, togglebuttons do not expose the *setBackgroundPainter* method that the regular pushbuttons expose.

Some nonstandard callbacks behave slightly differently than in pushbuttons:

- **ActionPerformedCallback** — fired when the button is clicked <u>in either button state</u> (depressed and undepressed).
- **ItemStateChangedCallback** — fired when the button state (depressed or not) has changed, either interactively (button click) or programmatically (via the *setSelected()* method or by setting the **Selected** property).

6.3 RadioButton

uicontrol(*'Style'*,*'radiobutton'*) is also similar to the pushbutton uicontrol described earlier. It uses the `com.mathworks.hg.peer.RadioButtonPeer$1` class that extends MATLAB's `com.mathworks.mwswing.MJRadioButton` class, which itself extends Swing's `javax.swing.JRadioButton` class.[15] `JRadioButton` actually extends `JToggleButton`, so it inherits all of `JToggleButton`'s properties, methods, and callbacks.

Radiobuttons also support HTML and CSS formatting, like most other uicontrols. Here is an example of an HTML-formatted radiobutton:

```
tooltip = '<html>HTML-aware<br><b>tooltips</b><br><i>supported';
labelTop= '<HTML><center><FONT color="red">Hello</Font> <b>world</b>';
labelBot=['<div style="font-family:impact;color:green"><i>What a</i>'...
```

```
                    '<Font color="blue" face="Comic Sans MS">nice day!'];
    hButton = uicontrol('Style', 'radiobutton', 'tooltip',tooltip, ...
                    'string', [labelTop '<br>' labelBot], 'position',pos);
```

(See color insert.)

JRadioButton and RadioButtonPeer$1 have similar useful properties and methods as the pushbutton uicontrol, except the **FlyOverAppearance** and **FocusTraversable** features (properties and associated methods). Like the pushbutton and togglebutton uicontrols, radiobutton also supports the **AutoMnemonic** feature which the superclass JRadioButton does not.

JRadioButton is basically a simple button with a pre-defined radio icon. This means that the circle **Icon** image can be changed to any other icon:

```
    jButton = findjobj(hButton);
    myIcon = fullfile(matlabroot,'/toolbox/matlab/icons/csh_icon.png');
    jButton.setIcon(javax.swing.ImageIcon(myIcon));
```

If we change the default radio icon, we also need to specify the **SelectedIcon** property: **Icon** will then display in the unselected state (empty radio circle by default), while **SelectedIcon** will display in the selected state (filled radio circle by default).

A corollary of the radio icon issue is that the **HorizontalTextPosition** and **VerticalTextPosition** properties (and related getter/setter methods) are very relevant to radio buttons, controlling the text position in relation to the adjacent radio icon. As explained in the pushbutton Section 6.1, **HorizontalTextPosition** accepts any of javax.swing.SwingConstants.RIGHT (=4), LEFT (=2), CENTER (=0), LEADING (=10) or TRAILING (=11, default), whereas **VerticalTextPosition** accepts javax.swing.SwingConstants.TOP (=1), BOTTOM (=3), or CENTER (=0, default):[†]

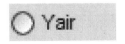      

HorizontalTextPosition = Swing Constants.TRAILING **(=default)**[‡]   **HorizontalTextPosition** = SwingConstants.CENTER   **HorizontalTextPosition** = SwingConstants.LEADING

[†] Users should only use one of the javax.swing.SwingConstants values. The numeric values may change between JVM versions, platforms and implementations (although this is unlikely). Using the constants is also much more readable and maintainable than using the numeric values.

[‡] Note that in radiobutton's case for a Latin locale, TRAILING and LEFT have the same effect, as do LEADING and RIGHT. This is different than the case for pushbuttons and togglebuttons (where TRAILING⇔RIGHT and LEADING⇔LEFT). This may also be different for LTR (Hebrew/Arabic) or Far-Eastern locales.

Another icon-related property which is particularly useful for radiobuttons is **IconTextGap** (default=4 pixels), which determines the gap in pixels between the icon and the text label.† Gap values may be negative, in which case the text overlaps the icon. Let us compare different gap sizes:

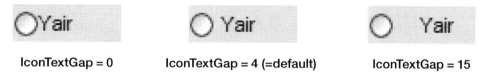

| IconTextGap = 0 | IconTextGap = 4 (=default) | IconTextGap = 15 |

Here is how the radiobutton uicontrol appears with different L&Fs on R2007b (JVM 1.6) on the Windows XP platform, as explained in Sections 3.3.2 and 6.1:

| | Windows L&F | Windows Classic L&F | Metal L&F | Motif & Plastic L&F |

6.4 Checkbox

uicontrol('Style','checkbox') is very similar to the radiobutton uicontrol described earlier. It uses the `com.mathworks.hg.peer.CheckboxPeer$1` class that extends MATLAB's `com.math-works.mwswing.MJCheckBox` class, which itself extends Swing's `javax.swing.JCheckBox` class.[16] Like `JRadioButton`, `JCheckBox` extends `JToggleButton`, so it inherits all its properties, methods and callbacks. The entire discussion about radio icons presented above therefore also applies to checkboxes.

Like most other uicontrols, checkboxes also support HTML and CSS formatting:

```
tooltip = '<html>HTML-aware<br><b>tooltips</b><br><i> supported';
labelTop= '<HTML><center><FONT color="red">Hello</Font> <b>world </b>';
labelBot=['<div style="font-family:impact;color:green"><i>What a</i> '...
          '<Font color="blue" face="Comic Sans MS">nice day!'];
hButton = uicontrol('Style','checkbox', 'tooltip',tooltip, ...
                    'string', [labelTop '<br>' labelBot], 'position',pos);
```

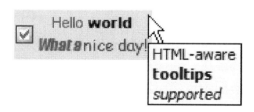

(See color insert.)

---

† **IconTextGap** was already introduced in Section 6.1 above, and also exists for pushbutton and togglebutton.

Here is how the checkbox uicontrol appears with different L&Fs on R2007b (JVM 1.6) on the Windows XP platform:

Unselected

Selected

Windows L&F   Windows Classic L&F   Metal L&F   Motif & Plastic L&F

JCheckBox is selectable just like JRadioButton and JToggleButton, and can therefore participate in a javax.swing.ButtonGroup[17] for mutually exclusive behavior. This is the standard manner of programming a list of mutually exclusive radio buttons, but a ButtonGroup can contain a mixture of JRadioButton, JCheckBox, and JToggleButton. However, the MATLAB **uibuttongroup** allows only radio and toggle buttons, not checkboxes, to participate in a group. This limitation is probably intended to prevent unintentional usability mistakes, in which checkboxes (which are non-exclusive by well-known convention) behave exclusively. This is a typical example of how MATLAB attempts to simplify programming at the expense of customizability. If we specifically wish to include checkboxes in a mutually exclusive button group (warning: this violates the well-known convention), then we therefore have the following options:

- Use pure Java components (via **javacomponent**) and the Swing ButtonGroup.
- Use MATLAB checkbox uicontrols without **uibuttongroup** and programmatically select/deselect the other checkboxes whenever some checkbox is selected (using the **Callback** property).
- Use MATLAB radiobutton uicontrols and **uibuttongroup**, then replace the radio icons with checkbox icons via **findjobj** and the component's *setIcon()* method.
- Use MATLAB radiobutton uicontrols and **uibuttongroup**, then update the radio **uicontrol**'s style to 'checkbox' (!!!). Note that the controls must first be created as radio buttons for the mutual exclusive behavior to work. Also note that while this behavior is undocumented and as such is subject to change in future MATLAB versions without prior notice, it does work as of MATLAB 7.13 (R2011b).

A useful customization that can be done to checkboxes is to set a tri-state (mixed) mode. This could indicate, for example, a yes/no/maybe situation, or empty/full/partial. Luckily, MATLAB already uses this tri-state mode in its CheckBoxTree component, described at the end of Section 5.2.2. Looking at the com.mathworks.mwswing.checkboxtree package, we note the TriStateButtonUI class that is used by the CheckBoxTree component to set a mixed-state checkbox appearance. The CheckBoxTreeCellRenderer class contains an *updateClient-Property(SelectionState)* method that apparently tells TriStateButtonUI which checkbox state to display. Acting on hunch, I tried a few property names, until I hit on "selectionState" Here is an end-to-end example:

```
import com.mathworks.mwswing.checkboxtree.*
jButton = findjobj(hButton);
jButton.setUI(TriStateButtonUI(jButton.getUI));
jButton.putClientProperty('selectionState', SelectionState.MIXED);
jButton.repaint;
```

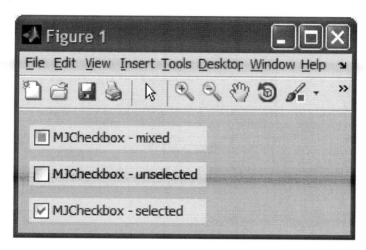

Tri-state checkbox (yes/no/maybe)

Adventurous readers can use `TriStateButtonUI` and `SelectionState` also on radio-button uicontrols. It works, but the visual appearance would probably not be understood by users. Therefore, I advise not to use it in practice.

Note that there is another alternative for tri-state checkboxes — using the `com.mathworks.mwt.MWCheckbox` component, as explained in Section 5.3. Finally, we can also use JIDE's `TristateCheckBox`[18] to achieve a similar result.

MWT's `MWCheckbox` vs. MWSwing's `MJCheckBox`

We sometimes need to design option forms with right-aligned checkboxes, that is, with the label to the left of the checkbox icon. This can easily be achieved by setting the checkbox's **HorizontalTextPosition** and **HorizontalAlignment** properties:

```
jButton.setHorizontalTextPosition(jButton.java.LEFT);
jButton.setHorizontalAlignment(jButton.java.TRAILING);
```

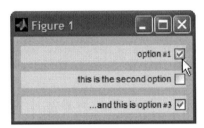

Right-aligned checkboxes

Right-aligned checkboxes appear to be useful for multi-option (form) dialog window. However, if we need to also align different controls (e.g., combo or edit boxes), it becomes more difficult to correctly align the labels and the controls. In such a case, consider using FormPanel (see Section 5.5.1) or the JGoodies Forms package, which is even more powerful and flexible than FormPanel (see Section 5.8.2).

6.5 Editbox

uicontrol('Style','edit') and the rest of the uicontrols are different from the uicontrols presented so far, in that they are not buttons. Still, they share many similarities with the button uicontrols described earlier. There are two distinct uicontrols called "editbox" in MATLAB: a single-line editbox and a multi-line editbox. MATLAB automatically uses the single-line control if the **Max** property is set to 1 (the default value, backward compatible with early MATLAB versions). If **Max** > 1, the multi-line editbox is used.

Editboxes do not share other control's support for HTML content. However, multi-line editboxes have their own extensive internal HTML support, described below.

Beware of a possible pitfall using MATLAB uicontrols: when switching styles, including switching between the single-line and multi-line editbox versions, MATLAB replaces the underlying Java component with a new component that has default properties. Therefore, if we need any customizations to the uicontrol, then we should ensure that they are done after setting the final uicontrol style, otherwise they will be forgotten.

Here is how the editbox uicontrol appears with different L&Fs on R2007b (JVM 1.6) on the Windows XP platform:[†]

---

† Plastic L&F looks like Motif for single-line editboxes and like Metal for multi-line editboxes.

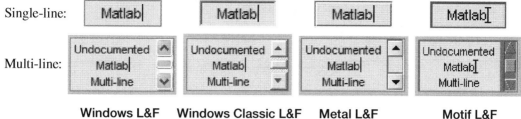

Single-line:

Multi-line:

Windows L&F   Windows Classic L&F   Metal L&F   Motif L&F

Note that a bug existed in MATLAB versions R14 (7.0) through R2006a (7.2) in which the background color of single-line editboxes could not be set on Windows XP machines;[19] to fix this problem on those WinXP MATLAB versions, we must do one of these:

- Set the editbox's **Max** property to 2, thereby modifying to a multi-line editbox with the same dimensions. You may wish to hide the scrollpane's scrollbars;[†]
- Set the Windows Scheme to "classic";
- Update the L&F to some other temporary value before creating the editbox;
- Add the `-Dswing.noxp=true` option to the *java.opts* file (see Section 1.9).[20]

6.5.1 Single-Line Editbox

The simple default single-line editbox uicontrol uses the `com.mathworks.hg.peer.EditTextPeer$hgTextField` class that extends MATLAB's `com.mathworks.mwswing.MJTextField` class, which itself extends Swing's `javax.swing.JTextField` class.[21]

`JTextField` has many useful properties and methods missing from MATLAB's *uicontrol* handle, in addition to the standard pushbutton ones presented in Section 6.1:

- **Border** — this property was already described for pushbuttons but merits special mention here, since the editbox border is such a pronounced feature of this component.[22] A CSSM user recently requested[23] to remove this border altogether. This can be done by simply setting this property to []:

```
jEditbox.setBorder([]);
```
[24]

Default border   No border

Now let us set a relatively complex border: a raised bevel border (giving the appearance of a pushbutton) surrounded by a thick rounded red border:[‡]

```
import javax.swing.BorderFactory java.awt.Color
outer = javax.swing.border.LineBorder(Color.red,4,true);
```

---

† set(hEditbox,'Max',2); jScrollPane=findjobj(hEditbox); jScrollPane.setVerticalScrollBarPolicy(21); %21 = never.

‡ This border is only given as an example of setting complex multi-level border — I am not assuming for even a moment that this specific border is useful in any practical GUI application.

```
inner = BorderFactory.createRaisedBevelBorder();
border = BorderFactory.createCompoundBorder(outer,inner);
jEditbox.setBorder(border);
```

Complex multi-layered border

- **Editable** — (default=true) a boolean flag indicating whether or not the editbox text can be modified. Note that the MATLAB HG handle (hEditbox) only allows setting the **Enable** property (its jEditbox Java counterpart is called **Enabled**), but not to set an enabled yet uneditable control — this can only be done using the Java **Editable** property.
- **Caret** — this property, common to all Java Swing data-entry components, sets a javax.swing.text.DefaultCaret[25] object that controls the text caret appearance (relevant for editable editboxes only).

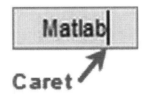

Specific Caret properties and callbacks include:
- **BlinkRate** (default=530 [milliseconds] on Windows XP). Set to 0 to stop blinking altogether.
- **Dot** (default=end of text) — indicates the caret's character position, with 0 indicating the position before the first character. Setting the **Dot** position automatically moves the **Mark** position (see below), effectively cancelling any text selection. The **Dot** position is also reflected in JTextField's **CaretPosition** and **SelectionEnd** properties.
- **Mark** (default=**Dot**) — indicates the character position of the start of the selection range. This property is read-only. To set the **Mark** position, use the *moveDot()* method, which moves the **Dot** position while leaving the **Mark** position unchanged. For example, if the text is "MATLAB" then *setDot(1)* followed by *moveDot(5)* will select the "atla" substring leaving the caret blinking at **Dot** position 5, following the selection, between the "a" and the "b" characters:

setDot(1) followed by *moveDot(5)*

By default, MATLAB's editbox uicontrol (`EditTextPeer$hgTextField`) sets the Mark to 0 and the **Dot** to the end of the text, thereby selecting the entire text. However, this only becomes visible when the editbox receives focus (programmatically, or via keyboard or mouse).

Associated read-only properties **X** and **Y** hold the **Mark** position in pixels; **CenterX** and **CenterY** hold the selection's midpoint position.

Editboxes also support bidi (RTL) text, via a `javax.swing.text.Position.Bias`[26] object attached to both **Dot** and **Mark**. Mixed bidi text can create odd-looking selections (this is normal for non-Latin languages), and is fully supported:†

Non-Latin RTL characters (note the odd selection behavior)

```
>> jEditbox.getCaret
ans =
Dot = (1, Forward) Mark = (4, Backward)
```

The **Mark** position is also reflected in the `JTextField`'s **SelectionStart** property. However, unlike **Mark**, the **SelectionStart** property is settable. It is usually easier to set `JTextField`'s **SelectionStart** and **SelectionEnd** properties (possibly in a single MATLAB *set* command) than to get the **Caret** reference, set the **Dot** and then use *moveDot()*. However, both methods are equivalent.

■ **Bounds** — indicates the pixel bounds of the selection (if any) and the caret. If Mark=Dot (i.e., no selection), **Bounds** will still indicate a minimal 10-pixel-wide boundary. Associated read-only properties **MinX**, **MinY**, **MaxX** and **MaxY** hold the **Bounds** edge points, which can also be obtained from the **Bounds** object.

■ **SelectionVisible** — (default=true) a boolean flag which enables hiding the selection — this property is reverted to true each time the editbox gains focus. To overcome this, do not clear this property directly but only in the editbox's **FocusGainedCallback**:

```
cbStr = 'set(get(gcbo,''Caret''),''SelectionVisible'',''off'')';
set(jEditbox,'FocusGainedCallback',cbStr);
```

■ **Visible** — (default=true) a boolean flag which enables hiding the blinking caret. Like the **SelectionVisible** property, this property constantly reverts to true, so it needs to be cleared in the editbox's **FocusGainedCallback**.

---

† MATLAB has many serious bugs in bidi support, including a serious bug that hangs the MATLAB desktop when RTL text is entered in the Command Window. However, the basic editbox control usually behaves nicely with RTL text. On some past MATLAB releases, non-Latin characters were lost if the uicontrol style was modified or the editbox was converted into multi-line (Max > 1). However, this is fixed as of R2011a.

- **UpdatePolicy** — indicates whether the caret position should change automatically when the text is modified (=default, `caret.UPDATE_WHEN_ON_EDT=0`), or should remain in place regardless of text changes (=`caret.NEVER_UPDATE=1`).[†]
- **StateChangedCallback** — this callback is fired whenever the caret (**Dot**) position changes, either programmatically or interactively. The callback's EventData will contain information of the new **Dot** position.

- As noted above, the associated `JTextField` properties **SelectionStart**, **SelectionEnd** and **CaretPosition** mimic the Caret's **Dot** and **Mark** properties. The *moveCaretPosition()* method behaves just like the caret's *moveDot()* method. Another alternative is to use the *select(selectionStart,selectionEnd)* method to select the requested text range.
- **SelectedText** is a read-only property holding the text currently selected, between the **Mark** and **Dot** (or **SelectionStart** and **SelectionEnd**) positions. Associated property **Text** holds the entire text within the editbox. Note that both these properties hold a `java.lang.String` object, which should be cast to a MATLAB string via MATLAB's built-in *char* function.
- **SelectionColor** and **SelectedTextColor** ought to change the foreground and background colors of the selected text. These properties too are overridden whenever the editbox gains focus, and so need to be overridden in the editbox's **FocusGained Callback**:

```
cbStr = ['set(gcbo,''SelectionColor'',java.awt.Color.red,' ...
           '''SelectedTextColor'',java.awt.Color.blue)'];
set(jEditbox, 'FocusGainedCallback', cbStr);
```

Non-standard selection colors and FocusGainedCallback

- **CaretColor** property controls caret color. Remember that Java colors are set differently from MATLAB colors. Unlike **SelectionColor** and **SelectedTextColor**, this **CaretColor** property is not automatically overridden and can therefore be set outside the **FocusGainedCallback**. For example, to set a red caret:

```
jEditbox.setCaretColor(java.awt.Color(1.0,0,0));
jEditbox.setCaretColor(java.awt.Color.red);    % an alternative
```

- **DisabledTextColor** controls the text color (default=gray) when the editbox is disabled. This property can also be set outside the **FocusGainedCallback**.

[†] There is also an ALWAYS_UPDATE option — read the `DefaultCaret` documentation (http://java.sun.com/javase/6/docs/api/javax/swing/text/DefaultCaret.html or http://tinyurl.com/cdg5ta) for more information.

■ **ScrollOffset** — sets the positive pixel offset (default=0=leftmost character) of the left-most character to display in the visible editbox rectangle. The leftmost character chosen is always rounded to the nearest character pixel edge. If the remaining length to the end of the text is shorter than the editbox width, then the actual **ScrollOffset** used will be smaller than the required value.

Note that the **Caret** position is NOT modified, and might become hidden to the right or left of the displayed text. For example, for the text "Yair 01234567890 abcdef" with **Caret** position set to 4 in a 60-pixel-wide editbox:

Yair|12345€	234567890	7890 abcde	890 abcdef
ScrollOffset t 0	**ScrollOffset = 30**	**ScrollOffset = 60**	**ScrollOffset = 90**

■ **HorizontalVisibility** — this read-only property provides a `javax.swing.BoundedRangeModel`[27] object that specifies the content text's visible and total width in pixels. Note that the object's **Extent** property is smaller than the editbox's **Width**, to account for the editbox's **Border** and internal **Margins**; the object's **Value** property is the same as the editbox's **ScrollOffset**:

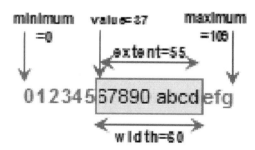

■ **Document** — holds a `javax.swing.text.PlainDocument`[28] object that contains the editbox content document metadata. The **Document** object is a pivotal element of the editbox component.[29] Interesting things possible with this object include setting callbacks upon text insert, update or removal (delete); inserting text at a specific position; replacing or removing (deleting) text; retrieving a specific paragraph (relevant for multi-line editboxes) or text sub-range;[30] adding undo/redo support to the editbox[31] and installing a `DocumentFilter`[32] to process the text.[33] Users could theoretically set the editbox's **Document** property to a `StyledDocument`[34] object (e.g., `javax.swing.text.html.HTMLDocument`[35]), but I have not tried this myself. As an alternative, consider using a `SyntaxTextPane` component (see Section 5.5.1).

■ **DragEnabled** — (default=false) a boolean flag indicating whether the editbox contents can be mouse-dragged externally as a DND source (for example, onto an editor, command line or any DND target). The **DropMode**,[†] **DropLocation**, **DropTarget** and **TransferHandler** properties enable the editbox act as a DND target, accepting externally dragged data as input sources. Refer to Section 3.7 for details.

- **FocusAccelerator** — (default=***char***(0)) sets the keyboard accelerator sequence that will cause the receiving text component to get the focus. The accelerator will be the key combination of the <Alt> key and the specified character, converted to upper-case. Any previous key accelerator setting, including menu-bar accelerators, will be superseded. A ***char***(0) key has the effect of turning off the focus accelerator. By default, there is no focus accelerator key (i.e., an accelerator of \0=***char***(0)). For example, let us set the accelerator to <Alt>-E, overriding the menu-bar's default accelerator for the Edit menu:

```
>> jEditbox.setFocusAccelerator('e');
>> jEditbox.getFocusAccelerator          % let us check...
ans =
E      <= 'e' converted to 'E', meaning an <Alt>-E accelerator
```

Special methods not discussed above, which may be useful for MATLAB programmers (all of these are standard `JTextField` methods, not MATLAB extensions) are the following:

- *cut(), copy() & paste()* — do the corresponding clipboard action.
- *enableInputMethods(flag)* — enable inputting non-standard multi-stroke international unicode characters. Input methods are an important aspect of Java i18n (internationalization) and well beyond the scope of this book.[36]
- *replaceSelection(string)* — replaces the selected text with the specified string. If no sub-text is currently selected, this method has the effect of inserting the specified string at the caret position.
- *selectAll()* — selects the entire editbox text, equivalent to *select(0,intmax)*
- *getScrollableBlockIncrement, getScrollableUnitIncrement* — returns the integer scrollbar increments that correspond to scrollbar block or single-unit movement in any direction. This roughly corresponds to the HG handle's **SliderStep** property, although **SliderStep** holds double values equivalent to the **Block/Unit** increments divided by the (**Max–Min**) value.
- *viewToModel(point)* — returns the character position of the specified pixel position. This is useful in mouse callbacks, in which the mouse position is known from the callback's EventData and we wish to get the corresponding text position. The reverse transformation is done via *modelToView(position)*.

[†] For backward compatibility, the default `DropMode` value is `DropMode.USE_SELECTION`. Usage of `DropMode.INSERT` is recommended to get the expected user experience of inserting DND drops at the **Caret** position.

Non-standard callbacks include:

- **ActionPerformedCallback** — fired when <Enter> is clicked in the editbox.
- **CaretUpdateCallback** — fired when the caret position has changed.
- **CaretPositionChangedCallback** — relevant for Input Methods only.
- **InputMethodTextChangedCallback** — relevant for Input Methods only.

`EditTextPeer$hgTextField` exposes two extra properties not available in `JTextField`:

- *setSelectAllOnFocus(flag)* — (settable-only; default=true) selects entire text and sets the caret at its end whenever the control gains focus, regardless of prior selection.[37] This may seem like a regular setter method for a **SelectAllOnFocus** property, but we must be careful to always use this method and <u>never use the alternative *set(jEditbox, 'selectAllOnFocus', flag)* because it</u> **crashes MATLAB**,[†] whereas using *setSelect AllOnFocus(flag)* is entirely safe.
- **TipWhenTruncatedEnabled** — (default=false) overrides the tooltip processing to automatically display the full editbox content text whenever the box is too short to visually present the entire text. When the box is larger than the visual appearance of the text, the tooltip is deleted. Setting this property to true overrides any previous tooltip that might have been set for the editbox. This property has two accessor methods: *isTipWhenTruncatedEnabled()* and *setTipWhenTruncatedEnabled(flag)*.

6.5.2 *Multi-Line Editbox*

Multiline editboxes, created when the **Max** property is set or updated to a higher value than 1, use the `EditTextPeer$hgTextEditMultiline` class, which extends `com.mathworks.hg. peer.utils.MJMultilineText`, `com.mathworks.mwswing.MJTextPane` and `javax.swing. JTextPane`[38] (a `JEditorPane`-derived class). So, this simple-looking textbox is actually a very powerful HTML-aware editor component, although it is pre-configured as non-HTML-aware by default (see below).[39]

Multi-line editboxes are a compound component, composed of a container (a `com.math-works.hg.peer.utils.UIScrollPane` object) which includes three children, as expected from any `ScrollPane`:[40] a `javax.swing.JViewport`[41] that contains the `EditTextPeer$hg-TextEditMultiline` component, and horizontal/vertical scrollbars.[‡]

The scrollbars are simple `javax.swing.JScrollPane.ScrollBar` instances of `javax. swing.JScrollBar` that shall be described shortly (Section 6.7, *Slider **uicontrol***).

[†] At least on MATLAB 7.5 (R2007b) through 7.13 (R2011b) running JVM 1.6 on a Windows XP PC. I suspect this may be due to a missing *isSelectAllOnFocus()* method in MATLAB's implementation.

[‡] `com.mathworks.hg.peer.utils.UIScrollPane$1` and `$2`, which directly extend `javax.swing. JScrollPane.ScrollBar`: http://java.sun.com/javase/6/docs/api/javax/swing/JScrollPane.ScrollBar.html (or http:// tinyurl.com/dzxtmr).

Since most of the multi-line editbox customizations would naturally be needed for the editing component, we need to dig within the scrollpane container to get its reference:

```
>> jScrollPane = findjobj (hEditbox);

>> jScrollPane.list
com.mathworks.hg.peer.utils.UIScrollPane[...]
 javax.swing.JViewport[...]
   com.mathworks.hg.peer.EditTextPeer$hgTextEditMultiline[...]
 com.mathworks.hg.peer.utils.UIScrollPane$1[...]
   com.sun.java.swing.plaf.windows.WindowsScrollBarUI$WindowsArrowButton[...]
   com.sun.java.swing.plaf.windows.WindowsScrollBarUI$WindowsArrowButton[...]
 com.mathworks.hg.peer.utils.UIScrollPane$2[...]
   com.sun.java.swing.plaf.windows.WindowsScrollBarUI$WindowsArrowButton[...]
   com.sun.java.swing.plaf.windows.WindowsScrollBarUI$WindowsArrowButton[...]

>> jViewport = jScrollPane.getViewport;†
>> jEditbox = jViewport.getView;
```

Before diving into the editbox component's customization, let us investigate some of the interesting ones available in its container jScrollPane:[42]

- **HorizontalScrollBarPolicy** — controls the appearance of the horizontal (bottom) scrollbar. As explained for **uitable** (Section 4.1.3) and **uitree** (Section 4.2.1) above, the possible values for this property are jScrollPane. HORIZONTAL_SCROLLBAR_AS_ NEEDED (=30), HORIZONTAL_SCROLLBAR_NEVER (=31, default‡) and HORIZONTAL_ SCROLLBAR_ALWAYS (=32).

  Note that if our handle is a MATLAB *handle* wrapper (as returned from the *findjobj* utility), rather than a naked Java reference, then the policies need to be accessed via the *java* builtin function, as explained in Section 3.4:

  ```
  >> scrollPolicy = jScrollPane.HORIZONTAL_SCROLLBAR_NEVER
  ??? No appropriate method, property, or field
  HORIZONTAL_SCROLLBAR_NEVER for class
  javahandle_withcallbacks.javax.swing.JScrollPane.

  >> scrollPolicy=jScrollPane.java.HORIZONTAL_SCROLLBAR_NEVER
  scrollPolicy =
      31
  ```

  Also note that setting a non-default **HorizontalScrollBarPolicy** requires using the *setWrapping(false)* method. See the discussion of *setWrapping()* below for more details and a usage example.

---

† Using jScrollPane.*getViewport()* is preferable to *getComponent(0)* in this case, since the sub-somponents order in the ScrollPane might be different on some JVM implementations. For the same reason we use jViewport.*getView()* rather than *getComponent(0)* to get the editbox object within the Viewport.

‡ The default **HorizontalScrollBarPolicy** on non-Windows platforms may be different.

Finally, note that updating the HG handle (`hEditbox`) **Position** property (either programmatically or by resizing its container) has a side-effect of automatically reverting the scrollbar policies to their default values (`HORIZONTAL_SCROLLBAR_NEVER` and `VERTICAL_SCROLLBAR_ALWAYS/NEVER`). It is therefore advisable to set `jScrollPane`'s **ComponentResizedCallback** to "unrevert" the policies:

```
hjScrollPane = handle(jScrollPane,'CallbackProperties');
scrollPolicy = hjScrollPane.java.HORIZONTAL_SCROLLBAR_AS_NEEDED;
callback = @(h,e)set(h,'HorizontalScrollBarPolicy',scrollPolicy);
set(hjScrollPane,'ComponentResizedCallback',callback);
```

- **VerticalScrollBarPolicy** — similarly to **HorizontalScrollBarPolicy**, controls the appearance of the vertical (right) scrollbar. Accepted values are: `VERTICAL_SCROLLBAR_AS_NEEDED` (=20), `VERTICAL_SCROLLBAR_NEVER` (=21) and `VERTICAL_SCROLLBAR_ALWAYS` (=22). The default **VerticalScrollBarPolicy** is `VERTICAL_SCROLLBAR_ALWAYS` for editboxes taller than 19 pixels[†] and `VERTICAL_SCROLLBAR_NEVER` otherwise. This default setting causes the vertical scrollbar to appear even when unneeded (i.e., when the entire editbox content is visible). It is therefore useful to set both scrollbars to `*_AS_NEEDED`:

`VERTICAL_SCROLLBAR_ALWAYS` **(default)**   `VERTICAL_SCROLLBAR_AS_NEEDED`

In some cases, users may wish to specifically set a `VERTICAL_SCROLLBAR_NEVER` policy. For example, after converting a single-line editbox to a multi-line one (by setting its **Max** property) to solve the editbox background problem on WinXP for MATLAB versions 7.0-7.2 (see discussion at the top of Section 6.5).

As above, note that updating `hEditbox`'s **Position** automatically reverts **VerticalScrollBarPolicy** to its default value of `VERTICAL_SCROLLBAR_ALWAYS/NEVER`.

- **WheelScrollingEnabled** — (default=true) is a boolean flag controlling whether the mouse wheel should enable vertical scrolling within the editbox.
- **Background** and **Foreground** — these have no effect since the opaque viewport, scrollbars and `hgTextEditMultiline` objects occlude the scrollpane. Set these properties individually for the requested sub-components.

[†] Recall that the default uicontrol height is 20 pixels, thereby making VERTICAL_SCROLLBAR_ALWAYS the default policy.

- **Enabled** — (default=true) indicates whether the entire scroll-pane component is enabled. To disable editing yet enable scrolling, use the `hgTextEditMultiline`'s **Editable** property instead.
- **ToolTipText** — this string property can be set for the entire scroll-pane, or individually for the scrollbars and `hgTextEditMultiline` sub-components.
- **AdjustmentValueChangedCallback** — this callback is fired continuously whenever the scrollbar value is modified, via resizing, dragging, or clicking.
- *anchorToBottom()* — this useful method, added by MATLAB's `MJScrollPane`'s extension of the standard Swing `JScrollPane`, ensures that whenever the caret position is at the editbox bottom, then if the editbox is resized (for example, by resizing its figure window or updating its **Position** property) or its contents are updated, then the visible viewport will update to show the bottom portion. This is useful for logger-type editboxes, where data is constantly updated at the bottom of the box and the scroll-pane needs to keep showing the last item.
- *getViewport(), getVerticalScrollBar(), getHorizontalScrollBar()* — convenience methods for retrieving the scroll-panes sub-components.

Multiline editboxes share many of the properties, methods and callbacks as single-line editboxes. Readers who wish to customize multi-line editboxes are therefore encouraged to read previous sections (Section 6.1 for pushbuttons; Section 6.5.1 for single-line editboxes).

Some of these shared customizations may be even more useful on multi-line than on single-line editboxes. For example, several CSSM readers have requested to use a multi-line editbox to present an activity log which would be constantly updated. However, when setting the ***uicontrol***'s **String** property with the modified log data, the caret position always reverts to the starting position (top line), rather than the end (bottom line, with the most recent log data). In order to fix this, use one of the following:[43]

```
jEditbox.setCaretPosition(jEditbox.getDocument.getLength);
jEditbox.setCaretPosition(intmax);   % alternative
```

Caret at top row (default)

Caret at bottom (via Java)

A related functionality of the containing `MJScrollPane` enables automatic scrolling to the bottom of the scroll-box whenever the editbox contents change. This is done using the *anchorToBottom()* method described above.

The default MATLAB implementation of the editbox uicontrol simply enables a multi-line vertical-scrollable text area using the system font. However, the underlying `JTextPane` object enables many important customizations, including the ability to specify <u>different</u> font attributes (size/color/bold/italic, etc.) and paragraph attributes (alignment, etc.) for text segments (called *style runs*) and the ability to embed images, HTML† and other controls. There are several alternative methods of doing this. From easiest to hardest these involve:

■ *setFont(), setForeground()* and other similar methods (or HG handle properties) affect the entire content pane, not individual style runs.
■ Use the *setPage(url)* method to load a text page from the specified URL (any preexisting editbox content will be erased). The page contents may be plain text, HTML‡ or RTF. The content type will automatically be determined and the relevant `StyledEditorKit`[44] and `StyledDocument`[45] will be chosen for that content. Additional `StyledEditorKit` content parsers can be registered to handle additional content types.[46] Here is an example loading an HTML page:

```
>> jEditbox.setPage('http://tinyurl.com/c27zpt');
```

where the URL's contents are:[47]

```
<html><body >
<img src="images/dukeWaveRed.gif" width="64" height="64">
This is an uneditable <code>JEditorPane</code>, which was
<em>initialized</em> with <strong>HTML</strong> text <font
size=-2> from</font> a <font size=+2>URL</font>.
<p>An editor pane uses specialized editor kits to read, write,
display, and edit text of different formats. The Swing text
package includes editor kits for plain text, HTML, and RTF. You
can also develop custom editor kits for other formats. <script
language="JavaScript" src="/js/omi/jsc/s_code_remote.js">
</script> </body></html>
```

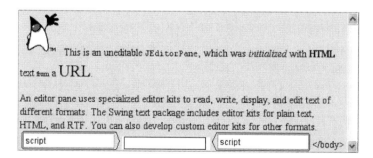

MATLAB's editbox *uicontrol* displaying a webpage (note how <script> tags are not supported)

---

† Note that some non-simple HTML/CSS features are not supported by this Swing component. If we need to display them, then we can use a browser control as explained in Section 8.3.2.
‡ JVM 1.6 supports HTML 3.2 and partially CSS, but not JavaScript. Future JVM versions will support HTML 4 & CSS 2.

- Set a specific `StyledEditorKit` (via *setEditorKit*) or **ContentType** properties, then use *setText()* to set the text, which should be of the appropriate content type. Note that setting **EditorKit** or **ContentType** clears any existing text and left-aligns the contents (hgTextEditMultiline is center aligned by default). Also note that HTML <div>s get their own separate lines and that <html> and <body> opening and closing tags are accepted but unnecessary. For example,

```
>> jEditbox.setEditorKit(javax.swing.text.html.HTMLEditorKit);
>> % alternative: jEditbox.setContentType('text/html');
>> jEditbox.setText('<b><div style="font-
family:impact;color:green">Matlab</div></b> GUI is <i><font
color="red">highly</font></i> customizable')
```

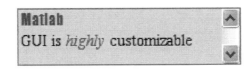

(See color insert.)

Let us show another usage example, of an event log file, spiced with icons and colored text based on event severity. First, define the logging utility function:

```
function logMessage(jEditbox,text,severity)

    % Ensure we have an HTML-ready editbox
    HTMLclassname = 'javax.swing.text.html.HTMLEditorKit';
    if ~isa(jEditbox.getEditorKit,HTMLclassname)
        jEditbox.setContentType('text/html');
    end

    % Parse the severity and prepare the HTML message segment
    if nargin < 3, severity = 'info'; end
    switch lower(severity(1))
        case 'i',   icon = 'greenarrowicon.gif'; color = 'gray';
        case 'w',   icon = 'demoicon.gif';       color = 'black';
        otherwise,  icon = 'warning.gif';        color = 'red';
    end
    icon = fullfile(matlabroot,'toolbox/matlab/icons',icon);
    iconTxt = ['<img src="file:///',icon,'" height=16 width=16>'];
    msgTxt = [' <font color=',color,'>',text,'</font>'];
    newText = [iconTxt,msgTxt];
    endPosition = jEditbox.getDocument.getLength;
    if endPosition > 0, newText = ['<br/>' newText]; end

    % Place the HTML message segment at the bottom of the editbox
    currentHTML = char(jEditbox.getText);
    jEditbox.setText(strrep(currentHTML,'</body> ',newText));
    endPosition = jEditbox.getDocument.getLength;
    jEditbox.setCaretPosition(endPosition); % end of content

end  % logMessage
```

Now, let us use this logging utility function to log some messages:

```
logMessage(jEditbox, 'a regular info message...');
logMessage(jEditbox, 'a warning message...',     'warn');
logMessage(jEditbox, 'an error message!!!',      'error');
logMessage(jEditbox, 'a regular message again...', 'info');
```

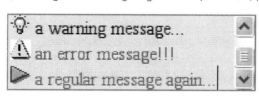

(See color insert.)

HTML editboxes are normally editable, images included. In actual applications, we may wish to prevent editing the display log. To do this, simply call jEditbox.*setEditable(false)*.

Setting a hyperlink handler is easy: first we need to ensure that we are using an HTML content-type document. Next, set the editbox to be uneditable (hyperlinks display correctly when the editbox is editable, but are unclickable), using jEditbox.*setEditable(false)*. Finally, set the callback function in the editbox's **HyperlinkUpdateCallback** property. As per Section 3.4, we set the callback on the editbox's *handle*, not the base reference:

```
jEditbox.setContentType('text/html');
jEditbox.setText('link: <a href="http://UndocumentedMatlab.
com">UndocumentedMatlab.com');
jEditbox.setEditable(false); % hyperlinks require non-editable
hjEditbox = handle(jEditbox,'callbackproperties');
set(hjEditbox,'HyperlinkUpdateCallback',@linkCallbackFcn);

function linkCallbackFcn(src,eventData)
   url = eventData.getURL;     % java.net.URL object
   description = eventData.getDescription; % URL string
   jEditbox = eventData.getSource;
   switch char(eventData.getEventType)
      case char(eventData.getEventType.ENTERED)
            disp('link hover enter');
      case char(eventData.getEventType.EXITED)
            disp('link hover exit');
      case char(eventData.getEventType.ACTIVATED)
            jEditbox.setPage(url);
   end
end % linkCallbackFcn†
```

(See color insert.)

---

† Additional methods of displaying hyperlinks are discussed in Sections 3.3.1, 5.5.1, 6.9, 8.3.1, and 8.3.2.

■ Setting the styles programmatically, one style run after another. This can be done via the text-pane's **Document** object (introduced in the single-line editbox section above). Individual character ranges can be set using the Document's *setCharacterAttributes-(startPos,endPos,attributeSet,replaceFlag)* method,[48] or entire style runs can be inserted via *insertString(startPos,text,attributeSet).*[49] Attributes are updated using the static methods available in `javax.swing.text.StyleConstants`.[50] These methods include setting character attributes (font/size/bold/italic/strike-through/underline/superscript/subscript and foreground/background colors), paragraph attributes (indentation/spacing/tab-stops/bidi), image icons and any Swing Component (buttons, etc.).

For example, let us adapt Sun's official `JTextPane` example[51] to MATLAB:

```
initString = {'This is an editable ',... %regular
              'JTextPane', ... %italic
              ' - a styled ', ... %bold
              'colored ', ... %color
              'text ', ... %small
              'component, ', ... %large
              ['which supports embedded components',10], ...
              [' ',10], ... %button
              ['...and embedded images...',10], ... %regular
              ' ', ... %icon
              [10,'JTextPane is a subclass of JEditorPane', ...
               ' that uses a StyledEditorKit and ' ...
               'StyledDocument, and provides cover ' ...
               'methods for interacting with those objects.']};
initStyles = {'regular','italic','bold','color','small',...
              'large','regular','button','regular','image','regular'};

import javax.swing.text.* % StyleContext, StyleConstants

defaultContext = StyleContext.getDefaultStyleContext();
defaultAttrs = defaultContext.getStyle(StyleContext.DEFAULT_STYLE);
doc = jEditbox.getStyledDocument();
regular = doc.addStyle('regular', defaultAttrs);

style = doc.addStyle('italic', regular);
StyleConstants.setItalic(style, true);

style = doc.addStyle('bold', regular);
StyleConstants.setBold(style, true);

style = doc.addStyle('color', regular);
StyleConstants.setForeground(style, java.awt.Color.red);
StyleConstants.setBackground(style, java.awt.Color.cyan);

style = doc.addStyle('small', regular);
StyleConstants.setFontSize(style, 10);

style = doc.addStyle('large', regular);
StyleConstants.setFontSize(style, 16);
```

```
style = doc.addStyle('image', regular);
StyleConstants.setAlignment(style,StyleConstants.ALIGN_CENTER);

pigUrl = 'http://tinyurl.com/calqqu';⁵²
pigImage = javax.swing.ImageIcon(java.net.URL(pigUrl));
StyleConstants.setIcon(style,pigImage);

style = doc.addStyle('button', regular);
StyleConstants.setAlignment(style,StyleConstants.ALIGN_CENTER);

icon = fullfile(matlabroot,'/toolbox/matlab/icons/matlabicon.gif');
jButton = javax.swing.JButton(javax.swing.ImageIcon(icon));
hjButton = handle(jButton,'callbackproperties');
set(hjButton,'ActionPerformedCallback','beep')
StyleConstants.setComponent(style, jButton);

for styleRunIdx = 1 : length(initString)
   doc.insertString(doc.getLength(),initString{styleRunIdx},...
                    doc.getStyle(initStyles{styleRunIdx}));
end
```

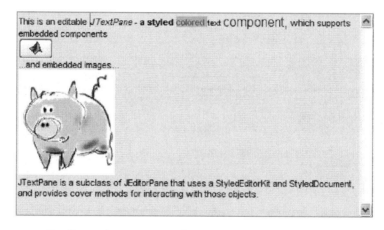

Programmatically setting separate Document style runs (See color insert.)

Note that if a styled multi-line editbox is converted to a single-line editbox (by setting hEditbox's **Max** property to 1), then it loses all style information, embedded images and components. Returning to multi-line mode will therefore show only the plain-text.

Multi-line properties which do not exist in single-line editboxes include:

- **CharacterAttributes** — determines the style attributes of a specified character range. Read the reference⁵³ for details.
- **Container** — a reference to the editbox's grandparent scrollpane.
- **ContentType** — the automatically inferred content-type of the text. Determines the parsing **EditorKit** and the type of **Document**. 'text/plain', 'text/html', and 'text/rtf' are supported by default.

- **EditorKit** — the editor parsing the text (see above).
- **InputAttributes** — the default attribute set for the entire edit-pane.
- **LogicalStyle** — the default style object for the paragraph at the current caret position. Any paragraph element without an explicit style will use this style.
- **Page** — enables setting the editbox's contents from a specified URL.
- **ParagraphAttributes** — determines the style attributes of a specified paragraph element. Read the reference[54] for details.
- **StyledDocument** — holds a version of the **Document** with style information.
- **ScrollableTracksViewportWidth, ScrollableTracksViewportHeight** — (default=true) read-only flags that indicate the scrollability of the editbox's content-pane in the corresponding direction.[55] These properties also exist in single-line editboxes (with a default value of false), but are more useful in multi-line editboxes. **ScrollableTracksViewportWidth** is related to the *setWrapping* method described below.
- **HyperlinkUpdateCallback** — see a description of hyperlink support above.

Single-line editbox properties/methods/callbacks unavailable in Multi-line editboxes:

- **Columns** — this property enables setting the `JTextField`'s **PreferredWidth** based on the specified number of columns, roughly equivalent to the number of "m"-width characters fitting in the visible editbox rectangle. Since editbox uicontrols are effectively controlled by their container and the HG handle's position array, the **Columns** property is of no effective use in MATLAB.†
- **HorizontalAlignment** — while this property does not exist per-se, MATLAB's implementation (`hgTextEditMultiline`) added the *setHorizontalAlignment()* method that enables setting the horizontal text alignment: 0=left, 1=center (default), 2=right.‡ This can also be set directly via the HG handle:

```
set(hEditbox,'HorizontalAlignment','right');
```

- **SelectAllOnFocus** — when a multiline editbox gains focus, the previous selection and caret position are maintained, unlike the default behavior of single-line editboxes.
- **ScrollOffset** — use the viewport's properties to set the displayed text range.
- **TipWhenTruncatedEnabled** — because of the viewport scrollbars, there is no need for this feature in a multi-line editbox.
- **ActionPerformedCallback** — clicking <Enter> in a multi-line editbox simply moves the caret to the next editbox row, beneath the current row. Therefore, this callback is not necessary in a multi-line editbox.

---

† Users who need to specify the uicontrol's width in term of characters, can set the HG handle's **Units** property to "characters" before setting the handle's **Position** property.

‡ Note that these values are different than the `SwingContants` accepted by the single-line editbox. Be careful NOT to use hgTextEditMultiline's static constant fields LEFT_ALIGNMENT (=0), CENTER_ALIGNMENT (=0.5) and RIGHT_ALIGNMENT (=1) since these values are half of the actual values accepted by the *setHorizontalAlignment* method.

Note that due to being contained in a scrollpane, the space that is actually taken by the editing component `EditTextPeer$hgTextEditMultiline` is smaller than the size of the uicontrol. Whereas the default uicontrol size is 60 × 20 pixels, the contained `EditTextPeer$hgText-EditMultiline` only uses 41 × 18: a one-pixel margin is used for the scrollpane border at the top, bottom and left, and 18 pixels are used by the vertical scrollbar on the right (the bottom horizontal scrollbar is hidden by default in MATLAB's multi-line *uicontrol*). In comparison, single-line editboxes use the entire 60 × 20 space.

Interesting `hgTextEditMultiline` methods include:

- *setWrapping(flag)* — this may seem like a regular setter method for a **Wrapping** property, but we must be careful to always use this method and <u>never use the alternative *set(jEditbox,'wrapping',flag)* because it</u> **crashes MATLAB**,[†] whereas using *setWrapping(flag)* is entirely safe.

By default, line-wrapping is turned on, effectively disabling horizontal scrolling. For this reason, MATLAB set the **HorizontalScrollBarPolicy** to `HORIZONTAL_SCROLLBAR_NEVER`. However, in some cases, it may be more useful to turn line-wrapping off and horizontal scrolling on. Here is a usage example:

```
jEditbox.setWrapping(false);
newPolicy = jScrollPane.HORIZONTAL_SCROLLBAR_AS_NEEDED;
jScrollPane.setHorizontalScrollBarPolicy(newPolicy)
```

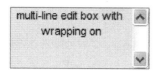 

Note that this method only works for the default `EditorKit`, and fails for `HTMLEditorKit` — This is due to HTML's inherent wrapping behavior, as can easily be seen in any browser webpage.

- *scrollToReference(refString)* — for HTML documents that contain a reference, calling this method will scroll the content-pane so that the requested reference will become visible, preferably at the top of the content pane. For example, if the document contains a tag <div name="123">, then *scrollToReference('123')* will scroll this <div> into view. A similar effect is seen when using *setPage('#123')*, since the *setPage* method calls *scrollToReference* internally. This method has no effect on non-HTML document.

- *replaceSelection(string)* — inserts the specified text string into the content pane as a replacement for the currently-selected content. If no content is selected then a simple insertion takes place.

[†] At least on MATLAB 7.5 (R2007b) through 7.13 (R2011b) running JVM 1.6 on a Windows XP PC. I suspect this may be due to a missing *isWrapping()* method in MATLAB's implementation. Oddly, the crash only affects *handle*(jEditbox), which accepts a *true/false* value; on the other hand, using *set(jEditbox,'wrapping','on'/'off')* on the naked Java reference is entirely safe. odd …

Note that the text is HTML-encoded if the current content-type and `EditorKit` are HTML-based, so *replaceSelection('123')* will actually insert "<b> 123</b>" instead of the requested HTML code — use *setText* to solve this problem. `HTMLEditorKit` also contains several methods[56] to insert HTML tags (e.g., '123') at specific `HTML.Tag`[57] locations (e.g., 'HTML.Tag.Body') using the `HTMLEditorKit.`*insertHTML()* method and the `HTMLEditorKit.InsertHTML-TextAction`[58] inner class.

- *insertComponent(Component)* — inserts the specified Swing component (button/ table, etc.) into the content pane as a replacement for the currently selected content. If no content is selected then a simple insertion takes place.

- *insertIcon(ImageIcon)* — insert the specified image (a object) into the content pane as a replacement for the currently-selected content. If no content is selected then a simple insertion takes place.

```
pigUrl = 'http://tinyurl.com/calqqu';[59]
pigImage = javax.swing.ImageIcon(java.net.URL(pigUrl));
jEditbox.insertIcon(pigImage);
```

- *registerEditorKitForContentType(contentTypeString, editorKitClassName)* — associates a particular `EditorKit` class with the specified content type. The specified class should extend `StyledEditorKit` or one of its descendants (e.g., `HTMLEditorKit` or `RTFEditorKit`). An `EditorKit` object is not created at this time, only a registry binding. The actual `EditorKit` will automatically be created at run-time, when the specified content-type is detected.

- *getEditorKitForContentType(contentTypeString)* — returns the `EditorKit` object responsible for parsing the specified content type: 'text/plain', 'text/html', and 'text/rtf' are supported by default; the default `javax.swing.text.StyledEditorKit` is returned for unregistered content types.

- *setEditorKitForContentType(contentTypeString,editorKit)* — sets the `EditorKit` object responsible for parsing the specified content type.

- *getEditorKitClassNameForContentType* — similar to the *getEditorKitForContent-Type* method described above, but returns a class name, not an `EditorKit` object. This method returns a `java.lang.String` object that should be cast to a MATLAB string using the built-in **char** function.

- *createEditorKitForContentType(contentTypeString)* — creates an `EditorKit` object that should be associated with a specific content type, from the list of registered

EditorKits. It is usually unnecessary to call this method directly: *getEditorKitFor-ContentType* always calls it if it fails to find an association.

- *addStyle(name,baseStyle)* — adds a named style to the list of known logical styles. This style can then be customized independently of its baseStyle, and uses to insert a style run into the document using doc.*insertString* (see above).
- *getStyle(name)* — returns the requested named style.
- *removeStyle(name)* — removes the specified style from the list of known styles.

6.5.3 *The JEditorPane Alternative*

As a side-note, we can always use JEditorPane directly in our MATLAB code to display HTML, without having to go through the ***uicontrol*** route.[60] This was the solution that Mikhail posted to a StackOverflow forum query:[61]

```
mytext = ['<html><body><table border="1">' ...
          '<tr><th>Month</th><th>Savings</th></tr>' ...
          '<tr><td>January</td><td>$100</td></tr>' ...
          '</body></html>'];

% Create a figure with a scrollable JEditorPane
jEdit = javax.swing.JEditorPane('text/html', mytext);
jPanel = javax.swing.JScrollPane(jEdit);
[hcomponent, hcontainer] = javacomponent(jPanel, [], gcf);
set(hcontainer, 'units', 'normalized', 'position', [0,0,1,1]);

% Turn anti-aliasing on (R2006a, Java 5.0)
java.lang.System.setProperty('awt.useSystemAAFontSettings', 'on');
jEdit.setFont(java.awt.Font('Arial', java.awt.Font.PLAIN, 13));
honorDisplayPropName = javax.swing.JEditorPane.HONOR_DISPLAY_PROPERTIES;
jEdit.putClientProperty(honorDisplayPropName, true);
```

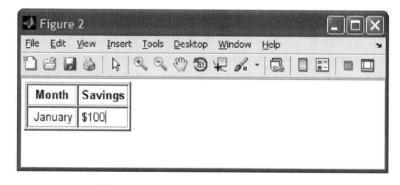

Editable **HTML-aware** JEditorPane

Mikhail's code included setting SwingUtilities2's **AA_TEXT_PROPERTY_KEY** property for anti-aliasing. Unfortunately, SwingUtilities2 was an unsupported and undocumented internal class in Java 1.5[62] (undocumented/unsupported by Sun, not MathWorks for a change…)

and completely disappeared in Java 1.6 (which is bundled with MATLAB R2007b onward). Therefore, SwingUtilities2 can only be used on MATLAB releases R14 SP2 (7.0.4) through R2007a (7.4) — on any other MATLAB version this will throw an error. For newer releases, use JIDE's AA_TEXT_PROPERTY_KEY alternative (JIDE is bundled with MATLAB and this is supported even on new MATLAB releases — see Section 5.7):

```
try
    % This only works on Java 1.5 (Matlab R14SP2 to R2007a):
    propName = com.sun.java.swing.SwingUtilities2.AA_TEXT_PROPERTY_KEY;
catch
    % This works for Java 1.6 (Matlab R2007b onward):
    propName = com.jidesoft.swing.JideSwingUtilities.AA_TEXT_PROPERTY_KEY;
end
jEdit.putClientProperty(propName, true);
```

Alternatively, add the following switch to your *java.opts* file (see Section 1.9):

```
-Dswing.aatext = true
```

With this switch, we no longer need to set anti-aliasing separately for each component. It is entirely harmless to set this switch even on MATLAB/Java versions that do not support it (the switch is simply ignored in these cases).

As a final note regarding font anti-aliasing, take a look at the static methods supplied by the built-in com.mathworks.services.AntialiasedFontPrefs class. Specifically, we can use the following code snippet:

```
try
    import com.mathworks.services.AntialiasedFontPrefs
    AntialiasedFontPrefs.setDesktopFontAntialiased(true);
catch
    % Never mind...
end
```

Note that while JEditorPane's support for HTML is extensive, it is incomplete. It also does not contain a JavaScript engine or other web-related features we have come to expect in a browser. For the more complex browser support see Section 8.3.2.

Another related alternative is to use the builtin com.mathworks.webintegration.start-page.framework.view.ResourcePane class that extends JEditorPane with some useful methods for setting HTML title, subtitles, text, hyperlinks, and image resources.

6.6 Listbox

uicontrol('Style','listbox') is similar to the multi-line editbox uicontrol described earlier. It uses the com.mathworks.hg.peer.ListboxPeer$UicontrolList class that extends MATLAB's com.mathworks.mwswing.MJList class, which itself extends Swing's javax.swing.JList class.[63]

Like multi-line editboxes, listboxes are actually composed of a container (a `com.mathworks.`
`hg.peer.utils.UIScrollPane` object) which includes three children, as expected from any
ScrollPane:[64] a `javax.swing.JViewport`[65] that contains the `ListboxPeer$UicontrolList`
component, and horizontal/vertical scrollbars.[†] Readers are referred to the multi-line editbox
section for a detailed description of scrollpanes.

```
>> hListbox = uicontrol('Style','List','String',{'item #1','item #2'});

>> jScrollPane = java(findjobj(hListbox))
jScrollPane =
com.mathworks.hg.peer.utils.UIScrollPane[...]

>> jListbox = jScrollPane.getViewport.getView
jListbox =
com.mathworks.hg.peer.ListboxPeer$UicontrolList[...]
```

The default initial value of the `UIScrollPane`'s **HorizontalScrollBarPolicy** is `HORIZONTAL_`
`SCROLLBAR_AS_NEEDED` (=30) for listboxes wider than 35 pixels and `HORIZONTAL_SCROLLBAR_`
`NEVER` (=31) for narrower listboxes, a setting which is usually satisfactory. The default
VerticalScrollBarPolicy is `VERTICAL_SCROLLBAR_ALWAYS` (=22) for listboxes taller than 25
pixels and `VERTICAL_SCROLLBAR_NEVER` (=21) for shorter listboxes; users will probably wish
to change this value to `VERTICAL_SCROLLBAR_AS_NEEDED` (=20), at least for tall listboxes.

Note that as with the multi-line editbox's case, MATLAB's scroll-pane implementation
automatically reverts the policy to the default configuration whenever the listbox is resized,
losing any user specification. It is therefore advisable to set `jScrollPane`'s **Component-
ResizedCallback** in order to "unrevert" the policies:

```
hjScrollPane = handle(jScrollPane,'CallbackProperties');
scrollBarPolicy = hjScrollPane.java.VERTICAL_SCROLLBAR_AS_NEEDED;
callback = @(h,e) set(h,'VerticalScrollBarPolicy',scrollBarPolicy);
set(hjScrollPane,'ComponentResizedCallback',callback);
```

Listboxes share `JTable` and `JTree`'s use of a separate **CellRenderer**,[66] **Model**,[67] and
SelectionModel.[68] Readers are referred to Sections 4.1.1 and 4.2.4 for a description of these
concepts and how they can be used in MATLAB applications. For a detailed description of
`JList` customization, read the Swing documentation.[69] Some of the interesting customizations
will be presented below.

Many of the customizations presented earlier in this chapter (borders, cursors, margins
etc.) are also relevant for listboxes. Perhaps the simplest and most striking customization of

---

[†] `com.mathworks.hg.peer.utils.UIScrollPane$1` and `$2`, which directly extend `javax.swing.`
`JScrollPane.ScrollBar`: http://java.sun.com/javase/6/docs/api/javax/swing/JScrollPane.ScrollBar.html (or
http://tinyurl.com/dzxtmr).

listbox is using its inherent HTML and CSS support, shared with most other uicontrols (see Section 3.3.3):

HTML-enriched listbox (See color insert.)

Presented here is how the listbox uicontrol appears with different L&Fs on R2007b (JVM 1.6) on the Windows XP platform:

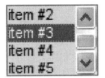 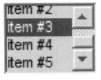 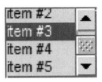 

    Windows L&F   **Windows Classic**   **Metal & Plastic L&F**   **Motif L&F**

`JList` has many useful properties and methods missing from MATLAB's ***uicontrol*** handle, in addition to the standard pushbutton ones presented Section 6.1:

- **SelectionMode** — this property holds the listbox selection mode: `javax.swing.`
  `ListSelectionModel.SINGLE_SELECTION` (=0), `SINGLE_INTERVAL_SELECTION` (=1)
  or `MULTIPLE_INTERVAL_SELECTION` (=2). `SINGLE_SELECTION` is the default when the
  HG handle (hListbox)'s **Max** property value[†] =1, which is **Max**'s default value;
  `MULTIPLE_INTERVAL_SELECTION` is the default when **Max** > 1. When applicable,
  interval selection is done by holding down the <Shift> key, and separate (multiple)
  selections by holding down the <Ctrl> key.[‡]

 Note that using a non-`SINGLE_SELECTION` **SelectionMode** requires the HG handle (hListbox)'s **Max** property to be > 1, otherwise a warning message will be displayed in the MATLAB Command Window and the listbox will be hidden from view. This is also MATLAB's behavior for the equivalent *set(hListbox,'**Max**',1,'**Value**',[1:4])*:

[†] Actually, **Max–Min** is the relevant value being checked, but by default **Min** value=0 and there is really no reason to modify this default value in listboxes.

[‡] On Windows platforms only — the standard selection process is somewhat different on other platforms (e.g., using the <Command> key on Mac OS).

```
>> import javax.swing.ListSelectionModel
>> selectionMode = ListSelectionModel.MULTIPLE_INTERVAL_SELECTION;
>> jListbox.setSelectionMode(selectionMode);

>> % Now use <Shift> to select several list items
Warning: single-selection listbox control requires a scalar Value
Control will not be rendered until all its parameter values are valid
(Type "warning off MATLAB:hg:uicontrol:ParameterValuesMustBeValid" to
suppress this warning.)

>> get(hListbox,'Value')
ans =
     2    4    5    6

>> set(hListbox,'Max',2);    % Now ok - no more warning messages!
```

 Also note that changing the HG handle's **Max** property causes the underlying `jListbox` to be recreated, and so it must again be found using ***FindJObj***. The previous `jListbox` handle will still exist but will NOT update the visible object, which may cause many hard-to-diagnose bugs. This unfortunate effect happens for all ***uicontrol***s. Therefore, always re-retrieve the Java handle after modifying the HG **Min/Max** property.

SINGLE_SELECTION   SINGLE_INTERVAL_   MULTIPLE_INTERVAL_
                              SELECTION   SELECTION

- **SelectedIndex** — holds the topmost (smallest) index of the selected listbox items. For the `ListSelectionModel.SINGLE_SELECTION` selection model, this is simply the index of the single selected item. Index values start at 0 for the topmost listbox item, and have increasing integer values for items lower in the listbox. If nothing is selected, **SelectedIndex** holds −1. Setting **SelectedIndex** value has the effect of programmatically selecting the corresponding item, firing relevant callbacks just as selecting the item by mouse or keyboard clicks. Setting the value to −1 is accepted but does not deselect the current selection (use the *clearSelection()* method to clear the current selection); setting lower negative numbers throws an error.
- **SelectedValue** — holds the string value of the **SelectedIndex** item, or [] if nothing is selected. Setting this property can only be done via the *setSelectedValue(itemString, scrollFlag)*, in which *scrollFlag* indicates whether the item should scroll into view once selected; if the specified *itemString* does not exist, then no error is thrown and the current selection is retained.

- **SelectedIndices**—holds the list of selected item indices. For the `ListSelectionModel.SINGLE_SELECTION` model this is equal to **SelectedIndex**; for `SINGLE_INTERVAL_SELECTION` and `MULTIPLE_INTERVAL_SELECTION` it may contain a column array of sorted int32 index values (e.g. [0;1;2]). Note that when setting **SelectedIndices**, we need not worry about array orientation or numeric class: `jListbox.`*setSelectedIndices([0,2:4,6])* works as expected.

  This property roughly corresponds to the MATLAB HG handle (`hListbox`)'s **Value** property (whose topmost index is 1 and holds a row array of indices). As noted above, a multi-valued `jListbox`.**SelectedIndices** (or the equivaluent `hListbox`.**Value**) require `hListbox`.**Max** > 1, otherwise a warning message will appear and the listbox will be hidden from view.

- **SelectedValues** — holds an array of the selected item values (strings). To convert to a MATLAB cell array, use MATLAB's built-in *cell* function. Note that this property holds an array even if only one or even no item is selected:

```
>> values = jListbox.getSelectedValues.cell;
>> values = cell(jListbox.getSelectedValues) % equivalent form
values =
    'Undocumented'
    'Matlab'
```

- **MinSelectionIndex, MaxSelectionIndex** — read-only properties holding the index of the topmost (smallest) and bottommost (largest) selected listbox items. **MinSelectionIndex** is the same as **SelectedIndex** or min (**SelectedIndices**); **MaxSelectionIndex** is the same as max(**SelectedIndices**).

- **FirstVisibleIndex, LastVisibleIndex** — read-only properties holding the index values of the topmost (smallest) and bottommost (largest) visible listbox items. The HG handle (`hListbox`) **ListboxTop** property corresponds to **FirstVisibleIndex** but starts at 1 not 0; **LastVisibleIndex** has no HG equivalent.

- **AnchorSelectionIndex, LeadSelectionIndex** — read-only properties holding the index values of the most recently updated **SelectionInterval**. The most recent index0 is considered the "anchor" and the most recent index1 is considered the "lead". Some interfaces display these indices specially, for example, Windows95 displays the lead index with a dotted yellow outline. Using these properties enable us to distinguish between selection of indices 1-to-4 and 4-to-1.

- **SelectionEmpty** — a read-only boolean flag (*isSelectionEmpty*) indicating whether there is any active selection. See related method *clearSelection()*.

- **SelectionForeground, SelecionBackground** — sets the foreground and background colors (should be a `java.awt.Color` object) of selected items.

- **Container** — a read-only MATLAB extension to `JList`, which returns a reference to the container `ScrollPane` object (same as jEditbox.*getParent.getParent*).

- **LayoutOrientation** — (default=`jListbox.VERTICAL`=0) sets the layout of listbox items within the viewport. The default **LayoutOrientation** (`jListbox.VERTICAL`=0) indicates regular top-to-bottom arrangement; jListbox.VERTICAL_WRAP=1 sets a horizontal item layout, wrapping to a new row as necessary for the maximum number of rows determined by the **VisibleRowCount** property (default=8); jListbox. HORIZONTAL_WRAP=2 sets a vertical item layout, wrapping to a new column at row number **VisibleRowCount**. For example,

 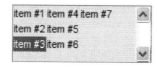 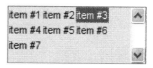

LayoutOrientation = VERTICAL	VERTICAL_WRAP	HORIZONTAL_WRAP
VisibleRowCount is irrelevant	**VisibleRowCount** = 3	**VisibleRowCount** = 3

- **FixedCellHeight, FixedCellWidth** — holds the listbox's cells height (default=13 pixels[†]) and width (default=−1). A −1 value means that the actual size is determined by the default platform-dependent **CellRenderer** size:

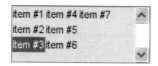 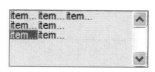 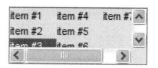

FixedCellHeight = −1	**FixedCellHeight** = 10	**FixedCellHeight** = 16
FixedCellWidth = −1	**FixedCellWidth** = 30	**FixedCellWidth** = 50

- **PrototypeCellValue** — (default=[]) holds a "typical" item value (string) which helps the **CellRenderer** to optimize **CellHeight** and **CellWidth**.
- **CellViewerEnabled** — (default=false) this boolean property was added by MATLAB's MJList listbox implementation (does not exist in Swing's standard JList). If set to true, a special tooltip containing the entire item text is displayed if the item under the current mouse cursor position is truncated (does not fit within the listbox width). Here are some examples:

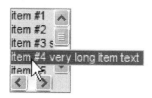 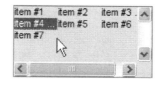 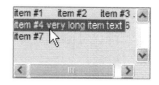

	LayoutOrientation = HORIZONTAL_WRAP
LayoutOrientation = VERTICAL	before & after mouse over the truncated item

---

[†] Actually, the default JList CellHeight is −1, but this translates into relatively large margins, so MathWorks apparently decided to specifically set a 13-pixel CellHeight in the Matlab listbox *uicontrol* implementation.

- **RightSelectionEnabled** — (default=false) a boolean flag indicating whether items can be selected by right-clicking. By default this is disabled, to enable presentation of a uicontextmenu upon right-click. This is a MATLAB MJList extension, not available in Swing's standard JList.

- **DragSelectionEnabled** — (default=false) a boolean flag indicating whether to move/ extend the selection interval[†] when the mouse drags (i.e., clicks an item and then moves) in the upward or downward direction. This is also a MATLAB MJList extension, not available in Swing's standard JList.

- **ValueIsAdjusting** — (default=false) a boolean flag indicating whether the contents are modified (e.g., during a drag operation). Programmers seldom need to set this value, but may use it as explained in the documentation[70] and in the **ValueChangedCallback** description below.

- **ValueChangedCallback** — fired when the SelectedIndex value has changed, pro-grammatically or interactively. The callback is actually fired *twice* for each update: once at the beginning (when **ValueIsAdjusting**=true) and once at the end (**ValueIs-Adjusting**=false). This callback is the Java equivalent of the MATLAB HG handle (hListbox)'s **Callback** property.[71]

- **CellRenderer** — holds a ListCellRenderer object[‡] (a JLabel implementation) which is responsible for displaying the listbox's cell items. To leverage the full power of listbox **CellRenderers**, create a separate Java class that implements the ListCellRenderer[72] interface, and then set the **CellRenderer** property to an instance of it. Use the documentation,[73] the descriptions in the *uitable* and *uitree* sections (4.1.1 and 4.2.4, respectively), or the example in Section 7.6.1.

- **ListData** — a settable property specifying the listbox items. **ListData** accepts a cell array of strings (or any Java Object e.g. IconImages or JLabels with both icons and text). We shall see the use of this property in Section 6.6.2.

Interesting special methods include:

- *addSelectionInterval(anchorIndex,leadIndex)* — adds listbox items *anchorIndex* through *leadIndex* (or vice versa if *leadIndex* < *anchorIndex*) to the current selection. This is especially useful for a MULTIPLE_INTERVAL_SELECTION model; for SINGLE_ SELECTION only *leadIndex* is selected; for SINGLE_INTERVAL_SELECTION the requested interval is merged with the existing interval if they overlap or are adjacent, otherwise the new interval replaces the existing selection interval.

[†] Depending on the **SelectionMode**, a multi-item interval or only a single item can be selected.

[‡] Actually, an object of class javax.swing.DefaultListCellRenderer.UIResource: http://java.sun. com/javase/6/docs/api/javax/swing/DefaultListCellRenderer.UIResource.html (or http://tinyurl.com/csvc79).

- *setSelectionInterval(anchorIndex,leadIndex)* — sets the selection to specified interval in a `SINGLE_INTERVAL_SELECTION` or `MULTIPLE_INTERVAL_SELECTION` SelectionModel; for the `SINGLE_SELECTION` model only *leadIndex* is selected.

- *removeSelectionInterval(anchorIndex,leadIndex)* — deselects any item in the specified interval.

- *clearSelection()* — deselects all listbox items. **SelectionEmpty** will become true.

- *ensureIndexIsVisible(index)* — scrolls the listbox as necessary to ensure the requested item index is visible in the displayed viewport.

- *getCellBounds(anchorIndex,leadIndex)* — returns the bounding `Rectangle`[74] (pixels relative to the top-left listbox corner) of the specified cell interval. This can be used for special highlighting of the selection bounds.

- *getNextMatch(prefixString,startIndex,searchDirection)* — returns the index of the next item whose toString starts with the specified *prefixString*, starting the search at the *startIndex* item, in the specified *searchDirection*.[†] This enables us to program an easy keyboard navigation/selection based on user key-clicks.

- *indexToLocation(index)* — returns the pixel position of the specified index. This may be useful for moving the mouse cursor to point at a specified listbox item. The corresponding *locationToIndex(Point)* returns the item index, useful in mouse callbacks.

- *isSelectedIndex(index)* — returns a boolean flag indicating whether the specified item index is selected.

- *setSelectionAppearanceReflectsFocus(flag)* — this may seem like a regular setter method for a **SelectionAppearanceReflectsFocus** property, but be careful to always use this method and <u>never use the alternative *set(jListbox, 'SelectionAppearance-ReflectsFocus',flag)*</u> that **crashes MATLAB** R2008a and earlier,[‡] whereas using *setSelectionAppearanceReflectsFocus(flag)* is safe.

This settable-only boolean property (default=true) is another MATLAB extension. It has the apparent effect of using a different (grayish) selection background whenever the listbox loses the focus, reverting to the standard selecting background when focus is regained.[75]

Note that *setSelectionAppearanceReflectsFocus* "freezes" the current selection color, so if the listbox is not in focus when we run it, the gray selection color will remain even when the listbox is back in focus...

[†] *searchDirection* is either `javax.swing.text.Position.Bias.Forward` or `Backward`. See: http://java. sun.com/javase/6/docs/api/javax/swing/text/Position.Bias.html (or http://tinyurl.com/cranzr).

[‡] I suspect that this may be due to a missing *isSelectionAppearanceReflectsFocus* method in MATLAB's implementation. This bug was fixed in MATLAB 7.7 (R2008b).

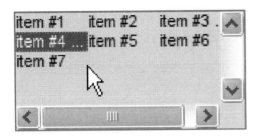

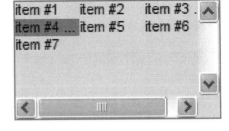

Focus gained   Focus lost

6.6.1 The Listbox Data Model

The listbox model, retrieved via the *getModel* method or the **Model** property (*get(jListbox, 'Model')*), enables item manipulation. Using the getter methods or callback properties is always safe. However, when updating simple listbox string items, it is better to update the HG handle's **String** property rather than the **Model**, since **Model** update breaks the connection between Java and MATLAB items. The **Model**'s callbacks and methods include the following:

- **ContentsChangedCallback** — fired after listbox content items have changed.
- **IntervalAddedCallback** — fired after one or more items were added. The EventData object contains the start and end indices of the added interval.
- **IntervalRemovedCallback** — fired after one or more items are removed. The EventData object contains the start and end indices of the removed interval.

- *getElementAt(index)* — returns the item at the specified index. Also: *get(index)*.
- *firstElement()* — returns the topmost listbox item.
- *indexOf(item)* — returns the index of the first occurrence of the specified item in the listbox; if not found, returns −1. *indexOf(item,searchIndex)* starts the search at the specified *searchIndex*. *lastIndexOf(item)* returns the index of the last occurrence and similarly for *lastIndexOf(item,searchIndex)*.
- *elements()* — returns the enumeration object of all the listbox items.
- *insertElementAt(item,index)* — inserts the specified item at the specified index. Also: *add(index,item)* which is equivalent, but which some consider better because it conforms to the List Collections interface.
- *setElementAt(item,index)* — replaces the specified item index with a new *item*. Also: *set(index,item)* which is equivalent but may be better (such as *add* above).
- *removeElementAt(index)* — removes the specified item from the listbox. Also: *remove(index)* which is equivalent but which some consider better (such as *add* above).
- *removeElement(item)* — removes the first (lowest index) occurrence of item.
- *removeRange(minIndex,maxIndex)* — removes the specified index interval.
- *removeAllElements()* — empties the listbox. Also; *clear()*, equivalent but better
- *getSize()* — returns the number of listbox items. Also: *capacity()*.

■ *setSize(int)* — sets the listbox size but may cause many usability problems if mis-used, so I suggest not using this method. Instead, one needs to update the HG String.

6.6.2 *Customizing the Appearance of Listbox Items*

Customizing listbox items requires combined use of several of the properties/methods introduced above, particularly **ListData** and **CellRenderer**. For example, let us present icon images rather than text (string) items: A simple solution is to use HTML images,[76] but this looks bad due to the narrow row height of listbox items (setting the HTML image height attribute does not improve this situation):

```
imgSrc = 'http://www.google.com/intl/en_ALL/images/logo.gif';
uicontrol('Style','listbox','Units','pixel','Pos',[0 0 250 100], ...
          'String', {'a', ['<HTML><IMG SRC="' imgSrc '">'], 'c'});
```

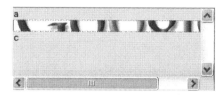

A better solution is to use the **ListData** property, as follows:

```
% Prepare the list of ImageIcon objects
iconsFolder = fullfile(matlabroot,'toolbox/matlab/icons');
imgs = dir(fullfile(iconsFolder,'*.gif'));
for iconIdx = 1 : length(imgs)
    iconFilename = fullfile(iconsFolder,imgs(iconIdx).name);
    icons{iconIdx} = javax.swing.ImageIcon(iconFilename);
end

% Display the ImageIcon list in 18x18 cells within the listbox
jScrollPane = findjobj(hButton);
jListbox = jScrollPane.getViewport.getView;
jListbox.setLayoutOrientation(jListbox.HORIZONTAL_WRAP)
jListbox.setVisibleRowCount(4)
jListbox.setFixedCellWidth(18)      % icon width = 16 + 2px margin
jListbox.setFixedCellHeight(18)     % icon height = 16 + 2px margin
jListbox.setListData(icons)
```

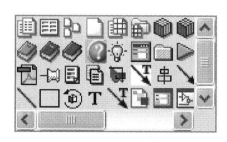

(See color insert.)

Note that after setting the **ListData** property, the item data become different from the HG handle's **String** property, possibly causing programming bugs. Also, the HG **String** property cannot be modified after updating **ListData**.[†] It is therefore advised to update **ListData** only if we wish to list non-string items; for string items, only use HG's **String** property. If we do update **ListData**, then we must first ensure that there are exactly as many items in the HG **String** property as the number of items we will supply **ListData;** otherwise, the listbox might disappear if we select a **ListData** item beyond HG's range.

The default listbox **CellRenderer** knows how to render (display) text and icons; other objects are rendered by converting them to text strings, using their *toString* method. This looks bad. For example, let us try to display labels having both text <u>and</u> icons:

```
% Prepare the list of JLabel objects
iconsFolder = fullfile(matlabroot,'toolbox/matlab/icons');
imgs = dir(fullfile(iconsFolder,'*.gif'));
for idx = 1 : length(imgs)
   iconFname = imgs(idx).name;
   iconFname = fullfile(iconsFolder, iconFname);
   jLabels{idx} = javax.swing.JLabel;
   jLabels{idx}.setIcon(javax.swing.ImageIcon(iconFname));
   jLabels{idx}.setText(iconFname);
   jLabels{idx}.setToolTipText(['Item-specific tooltip: ' iconFname]);
end

% Set the JLabel objects in the model
set(hListbox,'String',{imgs.name}); % ensure consistent HG size

%jListbox.setListData(jLabels); % easy but bad - see note above
model = javax.swing.DefaultListModel;'
for idx = 1:length(imgs)
   model.addElement(jLabels{idx});
end
jListbox.setModel(model);
```

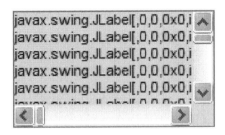

[†] This later problem can be solved by replacing the call to *setListData(dataArray)* with the following:
```
model = javax.swing.DefaultListModel;
for idx = 1:length(dataArray),  model.addElement(dataArray{idx});   end
jListbox.setModel(model);
```
The String property can now be updated, although it still remains unrelated to the model.

[‡] The role of the Model in Swing components is explained here: http://java.sun.com/products/jfc/tsc/articles/architecture/ (or http://tinyurl.com/atggc).

Therefore, to display non-icon objects we need a custom **CellRenderer**. First place the following code in a file called *LabelListBoxRenderer.java*:

```java
import java.awt.*;
import javax.swing.*;
public class LabelListBoxRenderer extends JLabel
                                  implements ListCellRenderer
{
    public LabelListBoxRenderer() {
        setOpaque(true);
        setHorizontalAlignment(LEFT);
        setVerticalAlignment(CENTER);
    }

    // return a label displaying both text and image.
    public Component getListCellRendererComponent(
                                  JList list,
                                  Object value,
                                  int index,
                                  boolean isSelected,
                                  boolean cellHasFocus) {

        try {
            // Try assuming the object is a JLabel
            JLabel jLabel = (JLabel) value;
            setIcon(jLabel.getIcon());
            setText(jLabel.getText());
            list.setToolTipText(jLabel.getToolTipText());
        } catch (Exception e) {
            // Oops... the object is probably not a JLabel
            setIcon(null);
            setText(value.toString());
            list.setToolTipText(null);
        }
        if (isSelected) {
            setBackground(list.getSelectionBackground());
            setForeground(list.getSelectionForeground());
        } else {
            setBackground(list.getBackground());
            setForeground(list.getForeground());
        }
        setEnabled(list.isEnabled());
        setFont(list.getFont());
        setOpaque(true);
        return this;
    }
}
```

Next, compile this file and place the generated *LabelListBoxRenderer.class* file in our MATLAB's Java classpath (see Section 1.6 for details). Now use this LabelListBoxRenderer class in MATLAB:

```
jListbox.setCellRenderer(LabelListBoxRenderer);
jListbox.setFixedCellHeight(16); % give the icons some space...
```

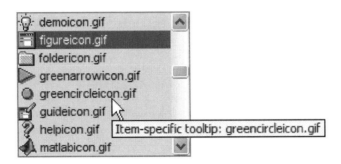

(See color insert.)

 Note: The following Section 6.6.3 shows an entirely different alternative for presenting item-specific tooltips.

Since we have ensured setting the HG **String** property with the label texts, we can now work at the MATLAB HG level almost as usual:

```
>> listboxStrs = get(hListbox, 'string');
>> index = get(hListbox, 'value');
>> disp(listboxStrs{index})
figureicon.gif
```

However, setting the HG **String** property, while now allowed,[†] causes a side effect of replacing the existing listbox model with a new string-based model:

```
listboxStrs{index} = 'Undocumented Matlab';
set(hListbox, 'string', listboxStrs);
```

---

† Since we have used the `model.addElement()` loop instead of the simple `setListData()` call.

To overcome this, set the labels via the model, not the HG **String** property:[†]

```
jListbox.getModel.setElementAt('Undocumented Matlab',index-1);
```

The **ListData** property is settable-only, and so cannot be used to get the list of all existing items. Instead, use jListbox.*getModel.toArray.cell* to get the list as a MATLAB cell array. This list can also be gotten as an enumerated list by using jListbox.*getModel.elements*. See Section 2.1.3 for enumeration usage details.

As another example of customizing listbox items, consider the request I once received to set a dedicated behavior (e.g., display a dedicated text) for selected listbox items:

```
import java.awt.Component;
import javax.swing.*;
public class LabelListBoxRenderer extends JLabel
                                  implements ListCellRenderer
{
    public LabelListBoxRenderer() {
        setOpaque(true);
        setHorizontalAlignment(LEFT);
    }

    // return a label displaying both text and image.
    public Component getListCellRendererComponent(
                                  JList list,
                                  Object value,
                                  int index,
                                  boolean isSelected,
                                  boolean cellHasFocus)
    {
        setText(value.toString());
        if (isSelected) {
            if (selectedItemText != null)
                setText(selectedItemText); // override label text
            setBackground(list.getSelectionBackground());
            setForeground(list.getSelectionForeground());
```

[†] This method has the disadvantage that now the HG String values are inconsistent with the model (displayed) values. To fix this, set the HG String and then update the model with JLabels for all the non-updated values. This is admittedly awkward.

```
        } else {
            setBackground(list.getBackground());
            setForeground(list.getForeground());
        }
        setEnabled(list.isEnabled());
        setFont(list.getFont());
        setOpaque(true);
        return this;
    }

    public void setSelectedItemText(String text) {
        selectedItemText = text;
    }

    public String getSelectedItemText() {
        return selectedItemText;
    }
}
```

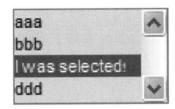

6.6.3 *Dynamic (Item-Specific) Context-Menus and Tooltips*

Item-specific context-menus and tooltips can be achieved with pure-MATLAB code, without the necessity for a custom Java **CellRenderer** as discussed above.[77] The trick is to trap the **MousePressedCallback** (for item-specific context-menu) and **MouseMovedCallback** (for item-specific tooltip) events. Below are sample implementations for both of these.

The benefit of using this approach, compared to the Java **CellRenderer** approach, is that no Java knowledge or programming is necessary — it can be achieved by any proficient MATLAB programmer. However, it should be realized that this pure-MATLAB approach has limitations: Custom **CellRenderer** enables customization of listbox item appearance (for example, icons, etc.), something that (to the best of my knowledge) cannot be done in pure MATLAB.

We start by defining a basic context menu, then trap **MousePressedCallback** to update it based on the current mouse position, and then display the updated menu. There are several ways in which we could pass the basic jmenu object to the **MousePressedCallback** function — in this case we choose to simply pass it as an extra (4th) run-time input argument. Alternatively, we could store the jmenu object in the UserData or elsewhere accessible from within the callback.

Note that for our purposes here, it is better to trap **MousePressedCallback** rather than **MouseClickedCallback**, since **MousePressedCallback** fires immediately when the mouse button is pressed, without waiting for its release as **MouseClickedCallback** does:

```
% Prepare the context menu (note the use of HTML labels)
menuItem1 = javax.swing.JMenuItem('action #1');
menuItem2 = javax.swing.JMenuItem('<html><b>action #2');
menuItem3 = javax.swing.JMenuItem('<html><i>action #3');

% Set the menu items' callbacks
set(menuItem1,'ActionPerformedCallback',@myFunc1);
set(menuItem2,'ActionPerformedCallback',{@myfunc2,data1,data2});
set(menuItem3,'ActionPerformedCallback','disp ''action #3 ...''');

% Add all menu items to context menu (with internal separator)
jmenu = javax.swing.JPopupMenu;
jmenu.add(menuItem1);
jmenu.add(menuItem2);
jmenu.addSeparator;
jmenu.add(menuItem3);

% Convert to a callback-able reference handle
jListbox = handle(jListbox, 'CallbackProperties');

% Set the mouse-click event callback
set(jListbox, 'MousePressedCallback', ...
              {@mousePressedCallback,hListbox,jmenu});

% Mouse-click callback
function mousePressedCallback(jListbox,jEventData,hListbox,jmenu)

    if jEventData.isMetaDown  % right-click is like a Meta-button

        % Get the clicked list-item
        %jListbox = jEventData.getSource;
        mousePos = java.awt.Point(jEventData.getX,jEventData.getY);
        clickedIndex = jListbox.locationToIndex(mousePos) + 1;
        listValues = get(hListbox,'string');
        clickedValue = listValues{clickedIndex};

        % Modify the context menu or some other element
        % based on the clicked item. Here is an example:
        itemStr = ['<html><b><font color="red">' clickedValue];
        item = jmenu.add(itemStr);

        % Remember to call jmenu.remove(item) in item callback
        % or use the timer hack shown here to remove the item:
        timerFcn = {@removeItem,jmenu,item};
        start(timer('TimerFcn',timerFcn,'StartDelay',0.2));

        % Display the (possibly-modified) context menu
        jmenu.show(jListbox, jEventData.getX, jEventData.getY);
        jmenu.repaint;

    else

        % Left-click - do nothing (do NOT display context-menu)
    end
end   % mousePressedCallback
```

```
% Remove the extra context menu item after display
function removeItem(hObj,eventData,jmenu,item)
    jmenu.remove(item);
end    % removeItem


% Menu items callbacks must receive at least 2 args:
% hObject and eventData: user-defined args follow these two
function myfunc1(hObject, eventData)
    % ...


function myFunc2(hObject, eventData, myData1, myData2)
    % ...
```

For dynamic item-specific tooltips, similarly trap **MouseMovedCallback**:

```
% Convert to a callback-able reference handle
jListbox = handle(jListbox, 'CallbackProperties');


% Set the mouse-movement event callback
set(jListbox, 'MouseMovedCallback', {@mouseMovedCallback,hListbox});


% Mouse-movement callback
function mouseMovedCallback(jListbox, jEventData, hListbox)

    % Get the currently-hovered list-item
    mousePos = java.awt.Point(jEventData.getX, jEventData.getY);
    hoverIndex = jListbox.locationToIndex(mousePos) + 1;
    listValues = get(hListbox,'string');
    hoverValue = listValues{hoverIndex};

    % Modify the tooltip based on the hovered item
    msgStr = sprintf('<html>item #%d: <b>%s</b></html>', ...
                    hoverIndex, hoverValue);
    set(hListbox, 'Tooltip',msgStr);
end    % mouseMovedCallback
```

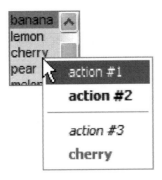

Dynamic context-menu (See color insert.)

Dynamic tooltip

6.7 Popup Menu (a.k.a. Drop-Down, Combo-Box)

uicontrol(*'Style','popupmenu'*) is similar to the listbox uicontrol described in the previous section. It uses the `com.mathworks.hg.peer.ComboboxPeer$MLComboBox` class that extends MATLAB's `com.mathworks.mwswing.MJComboBox` class, which itself extends Swing's `javax.swing.JComboBox` class.[78]

Unlike listboxes, `JComboBox` is not embedded within a scroll-pane. Instead, it is a simple container for a text field, an arrow button and the popup (drop-down) window.

Many of the pushbutton properties, methods and callbacks presented in Section 6.1 also apply to popup windows. This includes, of course, HTML formatting:

```
tooltip = '<html>HTML-aware<br><b>tooltips</b><br><i>supported';
hPopup = uicontrol('Style', 'popup', 'tooltip',tooltip, 'string',{ ...
    '<HTML><FONT color="red">Hello</Font></html>', 'world', ...
    '<html><div style="font-family:impact;color:green"><i>What a', ...
    '<Html><Font color="blue" face="Comic Sans MS">nice day!'});
```

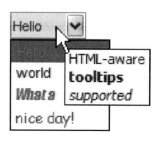

(See color insert.)

Presented here is how the dropdown uicontrol appears with different L&Fs on R2007b (JVM 1.6) on the Windows XP platform:

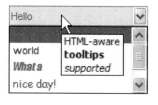

Windows L&F

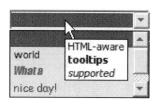

Windows Classic L&F

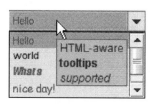

Metal L&F

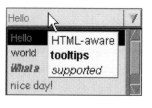

Motif L&F

Plastic L&F

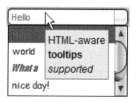

Nimbus L&F

- **MaximumRowCount** — (default=20) sets the maximum number of drop-down items to display together, before requiring a scrollbar. In effect this value controls the maximal popup height:

MaximumRowCount = 20   **MaximumRowCount = 3**   **MaximumRowCount = 2**
(default)

- **Editable** — (default=false) a boolean flag which controls whether the text field that displays the currently-selected item is editable or not. Unfortunately, this looks ok (if we ignore the missing left border, which is fixable — see below) only for text items, not HTML-rendered items:[†]

Editable text item string   **Editable HTML item string**

A more serious problem is that after editing an entry, the popup control disappears, displaying the following error in the Command Window:

```
Warning: popupmenu control requires that Value be an integer within String
range
Control will not be rendered until all of its parameter values are valid.
```

The reason for this behavior is that when the combo-box object detects that the text field's content match none of the popup list items, it automatically sets the **SelectedIndex** to −1 and therefore MATLAB's HG **Value** property to 0. At this point the MATLAB implementation kicks in, hiding the uicontrol since it considers 0 an invalid value for the **Value** property. This is similar to the check being done to test for an empty HG **String** value (=no items):

```
>> set(hPopup,'string',[])
popupmenu control requires a non-empty String
Control will not be rendered until all of its parameter values are valid.
```

[†] This happens since the underlying `JTextField` component does not support HTML rendering — see Section 6.5.1.

Bruno Luong on CSSM has suggested[79] clearing these particular warnings:

```
warning('off','MATLAB:hg:uicontrol:ParameterValuesMustBeValid')
```

Unfortunately, as far as I could see this has an effect only on MATLAB R2008a onward and in any case does not prevent the control from being hidden — it just prevents the warning from showing on the Command Window.

It therefore appears that the only easy way to really implement an editable popup menu is NOT to use MATLAB's ***uicontrol*** but rather Swing's standard JComboBox, which has none of these problems/limitations:

```
items = {'option #1','option #2','option #3'};
model = javax.swing.DefaultComboBoxModel(items);
jPopup.setModel(model);
jPopup.setEditable(true);
jPopup = javacomponent('javax.swing.JComboBox',position,hFig);
```

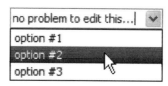

- **PopupVisible** — (default=false) a boolean flag which controls whether the popup window is currently (or should be) displayed. If this property is updated, then the focus is automatically transferred to the popup window for easy item selection using the keyboard (up/down/enter keys). There are also equivalent convenience methods *showPopup()/hidePopup()*.

  On a Windows platform the **PopupVisible** property is toggled, thereby showing/hiding the popup window, whenever the user clicks <Alt-Up> or <Alt-Down> when the combo-box has focus.

- **PopupWidthConstrained** — (default=false) a boolean flag which is another MATLAB MJComboBox extension to the standard Swing JComboBox. It is apparently used to constrain the width of the drop-down list to the width of the text field. MathWorks took the trouble to add this feature because Swing JComboBox's width is constrained, causing a difficulty in distinguishing between popup values when the control is relatively narrow; MATLAB's MJComboBox's default unconstrained behavior is much more user-friendly:

PopupWidthConstrained
= false (default)

PopupWidthConstrained
= true

Note that the **PopupWidthConstrained** property's read accessor methods is the expected *isPopupWidthConstrained()*, thereby also enabling the expected MATLAB-standard format of *get('PopupWidthConstrained')*. However, the property update accessor method is not the expected *setPopupWidthConstrained(flag)* but rather a nonstandard *setConstrainPopupWidth(flag)*. For this reason, it is impossible to set this property using *set('PopupWidthConstrained',...)*, but only via *setConstrainPopupWidth()*:

```
>> set(jPopup,'PopupWidthConstrained',true)
??? Changing the 'PopupWidthConstrained' property of javahandle_
withcallbacks.com.mathworks.hg.peer.ComboboxPeer$MLComboBox is not allowed.

>> jPopup.setPopupWidthConstrained(true)
??? No appropriate method or public field
setPopupWidthConstrained for class javahandle_withcallbacks.com.
mathworks.hg.peer.ComboboxPeer$MLComboBox.
```

- **PrototypeDisplayValue** — (default = []) holds a "typical" item value (string or Object) which helps the Renderer to optimize the text field's height and width.
- **SelectedIndex** — holds the index of the selected drop-down item. Index values start at 0 for the topmost listbox item, and have increasing integer values for items lower in the listbox. Setting **SelectedIndex** value has the effect of programmatically selecting the corresponding item, firing relevant callbacks just as selecting the item by mouse or keyboard clicks. This property roughly corresponds to the MATLAB HG handle (hPopup)'s **Value** property (whose topmost index is 1).
- **SelectedItem** — holds the currently-selected drop-down value. Note that this value may be HTML-encoded if this was how the combo-box was initially set:

```
>> jPopup.getSelectedItem
ans =
<HTML><FONT color="red">Hello</Font></html>
```

  SelectedItem can be updated, but has no effect in the MATLAB implementation since it requires a com.mathworks.hg.peer.ComboboxPeer$ComboBoxElement object. Instead, update the **SelectedIndex** property.
- **TipWhenTruncatedEnabled** — (default=false) a very useful boolean flag that overrides the tooltip processing to automatically display the full text field content whenever it is too short to visually present the entire text. When the box is larger than the visual appearance of the text, the tooltip is deleted. Setting this property to true overrides any previous tooltip that might have been set for the popup uicontrol. **TipWhenTruncatedEnabled** has no effect on regular non-editable popup uicontrols, only on editable ones.

Truncated-text tooltip
(editable popup only)

■ **Model** — (default=javax.swing.DefaultComboBoxModel[80]) The popup's data model[†] is an object that implements the ComboBoxModel[81] interface, which is an extension of the basic ListModel.[82] The standard DefaultComboBoxModel is used by the MATLAB implementation, and there is not much reason to modify it. The model can be used to access/modify presented items, as described below.

Using the model's getter methods or callbacks is always safe. However, as with listboxes (see Section 6.6.1), when updating simple popup list string items, it is better to update the HG handle's **String** property than the **Model**, since **Model** update breaks the connection between Java and MATLAB items.

Interesting **Model** methods and properties include:

■ **SelectedItem** — see the **SelectedItem** property described above.
■ **Size** — a read-only property holding the number of popup list items.
■ **ContentsChangedCallback** — fired after popup contents (items) have changed.
■ **IntervalAddedCallback** — fired after one or more items were added. The EventData object contains the start and end indices of the added interval.
■ **IntervalRemovedCallback** — fired after one or more items were removed. The EventData object contains the start and end indices of the removed interval.
■ *getElementAt(index)* — returns the item at the specified index. Also: *get(index)*.
■ *getIndexOf(item)* — returns the index of the specified item in the popup list; if not found, returns –1.
■ *insertElementAt(item,index)* — inserts the specified item at the specified index position within the popup list.
■ *setElementAt(item,index)* — replaces the specified item index with a new *item*.
■ *removeElementAt(index)* — removes specified item from the popup list.
■ *removeElement(item)* — removes the specified item from the popup list.
■ *removeAllElements()* — empties the popup list.

---

† The role of the Model in Swing components is explained here: http://java.sun.com/products/jfc/tsc/articles/architecture/ (or http://tinyurl.com/atggc).

- **ItemCount** — this read-only property returns the number of popup items. This information can also be retrieved from the popup Model.
- **EditorColumnCount** — this settable-only property is a MATLAB `MJComboBox` extension to the standard Swing `JComboBox`. I assume that it helps determine the edit-field's width, like Swing's JTextField's **Columns** property.[83]

 Note: While this may seem like a regular property, be careful to always use the *setEditorColumnCount(int)* method and <u>never use the alternative</u> *set(jPopup,"EditorColu mnCount",number)* since it **crashes MATLAB**,[†] whereas using *setEditorColumnCount(int)* is entirely safe.

- **Editor** — this property holds a reference to the Java class that is responsible for rendering (displaying) and processing (editing) the selected item in the textfield. **Editor** is only relevant when the popup control is editable (see the **Editable** property above); otherwise, the **Renderer** is used. The default uicontrol **Editor** is a `com.mathworks.mwswing.MJComboBox$DefaultEditor` object, which is a MATLAB extension of Swing's generic `javax.swing.plaf.basic.BasicComboBoxEditor`.[‡]

```
>> jPopup.setEditable(true)
>> editor = jPopup.getEditor
editor =
com.mathworks.mwswing.MJComboBox$DefaultEditor@55b768
```

The main use of the editor object is in customization of its internal text field component, which can be retrieved via the Editor's read-only **EditorComponent** property or the *getEditorComponent()* method. MATLAB's editor is similar to Swing's, except that it uses an `MJTextField` component,[§] whereas Swing uses `JTextField`. Refer to Section 6.5.1 (Single-line editboxes) for a discussion of `MJTextField` and its possible customizations.

```
>> textField = jPopup.getEditor.getEditorComponent
textField =
com.mathworks.mwswing.MJComboBox$DefaultEditor$BorderTextField[...]

% Let's fix the missing left border that's due to bad x location
>> textField.getBounds
ans =
```

[†] At least on MATLAB 7.5 (R2007b) through 7.13 (R2011b) running JVM 1.6 on a Windows XP PC. I suspect this may be due to a missing *getEditorColumnCount* method in MATLAB's implementation.

[‡] http://java.sun.com/javase/6/docs/api/javax/swing/plaf/basic/BasicComboBoxEditor.html (or http://tinyurl.com/cubnfs); in R2009b MathWorks reverted to using the simple standard Swing component, rather than a MATLAB extension.

[§] Actually, a `com.mathworks.mwswing.MJComboBox$DefaultEditor$BorderTextField` object, that extends `MJTextField`, which in turn extends Swing's `JTextField`. In R2009b this was reverted to a simple `JTextField`, and the associated hidden left border problem disappeared so there does not seem to be reason for much customization anyway.

```
java.awt.Rectangle[x=-1,y=1,width=41,height=20] %note negative x
>> textField.setLocation(java.awt.Point(1,1))
```

Before: hidden left border　　　After: fixed left border

In addition to **EditorComponent**, the editor contains an **Item** property (and associated *getItem/setItem* methods) that holds the current text-field contents:

```
>> jPopup.getEditor.getItem
ans =
<HTML><FONT color="red">Hello</Font></html>
```

The editor has a single callback property, **ActionPerformedCallback**, which is fired when the editing is finalized (by clicking <Enter>).

If we investigate the editor object (e.g. using *uiinspect*), then we will note that it has the public methods *focusGained(),focusLost()*. However, for some reason, the MATLAB implementation does not expose these focus event methods as callbacks. Instead, use jPopup's **FocusGainedCallback**, **FocusLostCallback**.

The editor's *selectAll()* method can be used to select the entire text-field's contents. This is simply a convenience method for the JTextField.*selectAll()* method of the editor's internal text-field component.

In practice, I can see some use for customizing the editor's text-field component, but not much use for replacing the editor or text-field components with some other user-provided objects.

 Note: When setting the **Editor** property, the specified editor must be a Java class implementing the javax.swing.ComboBoxEditor[84] interface. Nonimplementing objects will be accepted without error or warning, but the editor simply will not be replaced, causing hard-to-trace bugs.

■ **Renderer** — exactly like listbox controls, the **Renderer** property of popup (combobox) controls holds a ListCellRenderer object[†] (a JLabel implementation) which is responsible for displaying the control's items. To leverage the full power of **Renderers**, create a separate Java class that implements the ListCellRenderer[85] interface, and then set the **Renderer** property to an instance of the class. For sample usages refer to the documentation,[86] or the *uitable* and *uitree* in Sections 4.1.1 and 4.2.4, respectively.

[†] Actually, an object of class com.mathworks.mwswing.MJComboBox$CorrectedBorderRenderer, that extends Swing's defaulr javax.swing.plaf.basic.BasicComboBoxRenderer: http://java.sun.com/javase/6/docs/api/javax/swing/plaf/basic/BasicComboBoxRenderer.html (or http://tinyurl.com/cg5xmw).

Here is a simple example mimicking the listbox example of the previous section (as can be seen, except for very minor differences the code looks the same):

```
% Prepare the list of JLabel objects
iconsFolder = fullfile(matlabroot,'toolbox/matlab/icons');
imgs = dir(fullfile(iconsFolder,'*.gif'));
for idx = 1 : length(imgs)
    iconFilename = fullfile(iconsFolder, imgs(idx).name);
    iconTooltip = ['Item-specific tooltip: ' imgs(idx).name];
    jLabels{idx} = javax.swing.JLabel;
    jLabels{idx}.setIcon(javax.swing.ImageIcon(iconFilename));
    jLabels{idx}.setText(imgs(idx).name);
    jLabels{idx}.setToolTipText(iconTooltip);
end

% Set the JLabel objects in the model
set(hPopup,'String',{imgs.name}); % ensure consistent HG size
model = javax.swing.DefaultComboBoxModel; %not DefaultListModel
for idx = 1:length(imgs)
    model.addElement(jLabels{idx});
end
jPopup.setModel(model);

% Set the display Renderer
% Note: LabelLisitBoxRenderer was presented in section 6.6 above
jPopup.setRenderer(LabelListBoxRenderer); %not setCellRenderer()
jPopup.setFixedCellHeight(16); % give the icons some space...
jPopup.setMaximumRowCount(8); % override the default 20
```

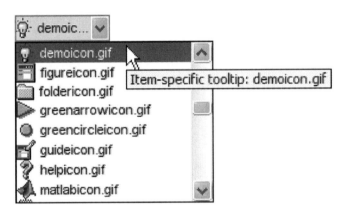

Like listboxes, since we have ensured setting the HG **String** property with the label texts, we can now work at the MATLAB HG level almost as usual:

```
>> listboxStrs = get(hPopup,'string');
>> index = get(hPopup,'value');

>> disp(listboxStrs{index})
demoicon.gif
```

However, like listboxes, setting the HG **String** property has a side-effect of replacing the existing listbox model with a new string-based model:

```
listboxStrs{index} = 'Undocumented Matlab';
set(hPopup,'string',listboxStrs);
```

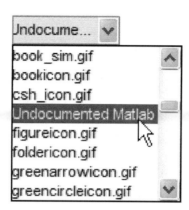

To overcome this, set the labels via the Java **Model**, not the HG **String** property. In listboxes we did this via the *setElementAt()* method. Unfortunately, the `javax.swing.DefaultComboBoxModel` class does not expose such a public method, so we use *insertElementAt(item,index)* followed by *removeElementAt(index)*:[†]

```
% The following is not allowed - argh!
%model.setElementAt('Undocumented Matlab',index-1);

% ...instead, we use a combination of insert + remove:
model.insertElementAt('Undocumented Matlab',index-1);
model.removeElementAt(index);
```

[†] This method has the disadvantage that now the HG String values are inconsistent with the model (displayed) values. To fix this, set the HG String and then update the model with JLabels for all the non-updated values. This is admittedly awkward.

Note: If the popup control is **Editable**, then the **Renderer** only controls the items presented in the popup list, while the selected item in the text-field uses the **Editor** (not **Renderer**) for rendering. If the control is uneditable, **Renderer** is used for both the popup list and the text-field.

The popup control exposes several non-standard callbacks, as follows:

- **ActionPerformedCallback** — fired when the selected item has changed.
- **ItemStateChangedCallback** — fired twice when the selected item has changed: first deselecting the previous item, then for selecting the new item.
- **CaretPositionChangedCallback** — unused.
- **InputMethodTextChangedCallback** — unused.
- **PopupMenuCanceledCallback** — fired when the popup menu is closed. This callback is very similar to **PopupMenuWillBecomeInvisibleCallback** and they usually fire together (**PopupMenuCanceledCallback** before **PopupMenuWillBecomeInvisible-Callback**).
- **PopupMenuWillBecomeInvisibleCallback** — fired when the popup menu is about to be hidden (closed). Also see **PopupMenuCanceledCallback** above.
- **PopupMenuWillBecomeVisibleCallback** — fired when the popup menu is about to be displayed.

And here are several interesting methods exposed by the jPopup Java peer object:

- *removeAll(), removeAllItems(), removeItem(item), removeItemAt(index), addItem(item), insertItemAt(item,index), getItemAt(index)* — these are simply convenience methods for the corresponding model methods (see below).
- *configureEditor(editor,item)* — sets the control's editor component with a default initial selected item presented for editing in the control's text-field.
- *selectWithKeyChar(char)* — selects the nearest item corresponding to the specified character. This can be used in **KeyPress** callbacks for easy keyboard navigation/selection based on user key-clicks.
- *hidePopup(), showPopup()* — these are pretty much self-explanatory.[87]

6.8 Slider

uicontrol('Style','slider') is actually implemented as a Java Swing `JScrollBar` rather than a `JSlider`.[†] It uses the `com.mathworks.hg.peer.SliderPeer$MLScrollBar` class that extends

---

[†] I do not know the reason for naming this control "slider" rather than "scrollbar". It would seem that this historic misnomer now prevents MATLAB from implementing an actual slider control, which is pretty common in modern GUIs. In this section, the terms "slider" and "scrollbar" will be used interchangeably. See Section 3.3.1 for `JSlider` examples.

MATLAB's `com.mathworks.mwswing.MJScrollBar` class, which itself extends Swing's `javax.swing.JScrollBar` class.[88]

Like `JComboBox`, `JScrollBar` is not embedded within a scroll-pane. Instead, it is a simple container for the central part (knob/thumb and track/trough), and the two scroll buttons (which are instances of `javax.swing.plaf.basic.BasicArrowButton`):

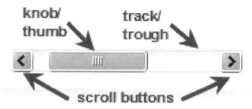

Many of the pushbutton properties, methods and callbacks presented in Section 6.1 also apply to sliders. Note that HTML formatting is generally irrelevant for this control, which does not contain a text field — it is only relevant for the tooltip.

```
tooltip='<html>HTML-aware<br><b>tooltips</b><br><i>supported';
hSlider=uicontrol('style','slider', 'tooltip',tooltip);
```

Here is how the slider *uicontrol* appears with different L&Fs on R2007b (JVM 1.6) on the Windows XP platform (Plastic L&F is similar to Metal, with an added tooltip drop-shadow; Windows L&F is similar to the Classic with slightly altered arrows):

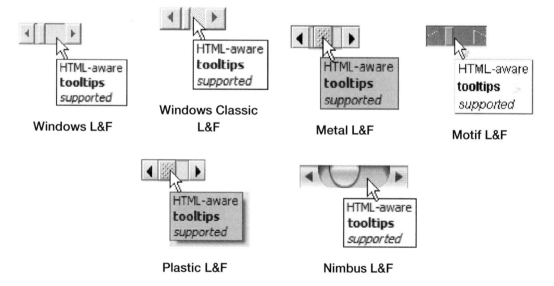

`MJScrollBar` has many useful properties and methods missing in MATLAB's *uicontrol* handle, in addition to the standard pushbutton ones presented in Section 6.1:[†]

---

† All of these are standard Swing `JScrollBar` properties and methods — MATLAB's extension does not really extend much.

- **BlockIncrement** — (default=100,000[†]) sets the change in the scrollbar value whenever the scroll-bar track is clicked, moving the knob a full "block". The corresponding MATLAB property is **SliderStep** (2nd value), that has a default value of 0.1. Note that while Java defines increments as integers, MATLAB defines them as floating-point portions. An internal algorithm (which changed in R2008b) preserves the correct arithmetic ratios between **BlockIncrement**, **UnitIncrement**, **VisibleAmount**, **Value**, **Maximum**, and **Minimum**.

  Here is a simple example, showing corresponding values in MATLAB and Java:

	Min	Max	Unit Increment	Block Increment	Visible Amount	Value
MATLAB	2	4	0.01	0.1	0.2	2.5
Java 2008a	200000000	420000000	2000000	20000000	20000000	250000000
Java 2008b +	– 100000	1000000	10000	100000	100000	150000

- **UnitIncrement** — (default=10,000[‡]) sets the change in the scrollbar value whenever an arrow button is clicked. The corresponding MATLAB property is **SliderStep** (1st value), which has a default value of 0.01. Note that while Java defines increments as integers, MATLAB defines them as floating-point portions.
- **VisibleAmount** — (default=**BlockIncrement**) sets the size of the scrollbar knob. In MATLAB, this cannot be different from the block increment value, but using this Java property we can specify different values for the knob size.
- **Value** — (default=**Minimum**) sets the scrollbar value and knob position. The corresponding MATLAB property is also **Value**.
- **Maximum** — (default=1,000,000[§]) sets the maximal scrollbar value. The corresponding MATLAB property is **Max**, which has a default value of 1.
- **Minimum** — (default=–**BlockIncrement**[¶]) sets the minimal scrollbar value. The corresponding MATLAB property is **Min** that also has a default value of 0.

 Note: In pre-2008 Matlab releases, whenever the MATLAB slider control's **Min** or **Max** properties were updated, the control was re-created and a new Java handle had to be retrieved.[††] Apparently this does not happen when updating the related MATLAB **SliderStep** or **Value** properties, so in these cases the Java handle does not need to be retrieved again. It also does not happen on modern MATLAB releases.

[†] 10,000,000 up to MATLAB 7.6 (R2008a); 100,000 starting with MATLAB 7.7 (R2008b).

[‡] 1,000,000 up to MATLAB 7.6 (R2008a); 10,000 starting with MATLAB 7.7 (R2008b).

[§] 110,000,000 up to MATLAB 7.6 (R2008a); 1,000,000 starting with MATLAB 7.7 (R2008b).

[¶] 0 up to MATLAB 7.6 (R2008a); -**BlockIncrement** starting with MATLAB 7.7 (R2008b).

[††] This appears to be the standard case for *uicontrol*s — also see for example in editboxes: http://www.mathworks.com/matlabcentral/newsreader/view_thread/244383 (or http://tinyurl.com/dzhk5c).

■ **Orientation** — (default=0) sets the scrollbar orientation: 0=horizontal; 1=vertical. There is no corresponding MATLAB property: MATLAB automatically determines the slider orientation based on the size of the *uicontrol*: if the control is more wide than tall, then a horizontal orientation is used, otherwise a vertical orientation is used. Using the Java **Orientation** property, we can override this behavior to specify wide vertical or narrow horizontal sliders:

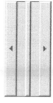

Wide vertical slider   **Narrow horizontal slider**

Sliders (or rather, `JScrollBars`) have a single non-standard callback:

■ **AdjustmentValueChangedCallback** — fired continuously when the scrollbar track or an arrow button is clicked, or when the knob is dragged. Compare this with the standard MATLAB **Callback**, which is only fired when the mouse button is released. This has been an ongoing complaint from MATLAB users for many years (e.g., here[89]), which can easily be solved by using the Java callback.

Note: MathWorks is indeed aware of this issue, as evidenced by their addition of an internal *ActionEvent* event to the slider's *schema.class* object, as an alternative to using **AdjustmentValueChangedCallback**. *ActionEvent* is the method used by several built-in MATLAB functions, such as *imscrollpanel*:

```
>> hSlider = uicontrol('Style','slider', ...);
>> hcSlider = classhandle(handle(hSlider));
>> hcSlider.Events.get
                   Name: 'ActionEvent'
     EventDataDescription: 'Action Event'

>> handle.listener (hSlider, 'ActionEvent', @myCallbackFunction);'
```

6.9 Text Label

uicontrol('Style','text') uses a `com.mathworks.hg.peer.LabelPeer$1` class that extends MATLAB's `com.mathworks.hg.peer.utils.MultilineLabel` class, which itself extends Swing's standard `javax.swing.JComponent` class.[90] It has no sub-components.

---

† In R2008b we can also use the *addlistener* function, but in earlier MATLAB releases *addlistener* only worked for Java objects.

`LabelPeer$1` shares some formatting properties with pushbuttons, described in Section 6.1 (specifically, **Border, HorizontalAlignment, VerticalAlignment**). In addition, `LabelPeer$1` has the following interesting property:

- **LineWrap** — (default=true) a flag indicating whether the label's string should wrap onto a new line if the control's width is smaller than the string label's extent. There is no corresponding MATLAB property.

MATLAB's text ***uicontrol*** is pretty bland "out-of-the-box" because it is created as a simple borderless label. In some cases, it may be advisable to make the text label noticeable (e.g., for displaying alerts or error messages).[91] This can be achieved by changing the font (which can be done via MATLAB properties), background color, and/or by adding a **Border**, as explained in Sections 3.3.1, 6.1, and 6.5.1.

As an alternative to the limited text ***uicontrol***, we can use the ***text*** function (which supports Tex/Latex formatting), a borderless button, or a ***javacomponent*** with `JLabel`.

HTML formatting is unfortunately **NOT** supported by `LabelPeer$1`'s text field — it is only relevant for the tooltip. As noted in Section 3.3.3, we can overcome this limitation by using a standard Java Swing `JLabel`, which does support HTML:[92]

```
%show the 'for all' and 'beta' symbols and other HTML formatting
str = '<html>&#8704;&beta; <b>bold</b><i><font color="red">label';
jLabel = javaObjectEDT('javax.swing.JLabel',str);
[hcomponent,hcontainer] = javacomponent(jLabel,[100,100,80,20],gcf);
```

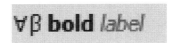

HTML label

Note: for styled labels consider the other alternatives presented in Section 5.5.1.

In many cases, it is useful to use a `com.mathworks.mwswing.MJLabel` rather than its `JLabel` superclass, because of `MJLabel`'s **TipWhenTruncatedEnabled** property:

```
str = 'This is an MJLabel string';
jLabel = javaObjectEDT('com.mathworks.mwswing.MJLabel',str);
jLabel.setTipWhenTruncatedEnabled(true);
[hcomponent,hcontainer] = javacomponent(jLabel,[10,50,50,30],gcf);
```

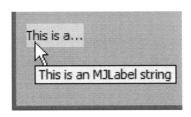

Note that MJLabel also support HTML, but we need to duplicate the initial "&" in the text string because MJLabel removes it as a mnemonic indicator by default (unless a second optional argument is passed to *setText(text,mnemonicFlag)* with a *false* value). Also note that HTML text always wraps and does not truncate:

```
str = '<html>&&#8704;&beta; <b>bold</b><i> <font color="red">label';
jLabel.setText(str);
```

HTML label before leading "&" duplication... **...and after "&" duplication**

This default behavior of *setText(text)* is misleading — one would have expected compatibility with JLabel's *setText(text)*, meaning a default value of *false* for the optional *mnemonicFlag*, rather than *true*. For this reason, the following would give the same results (note that the leading "&" is not duplicated; also note the extra *mnemonicFlag = false* input argument to *setText*):

```
% Note: the '&' is not duplicated; extra setText flag = false
str = '<html>&#8704;&beta; <b>bold</b><i> <font color="red">label';
jLabel.setText(str, false);
```

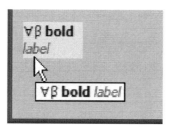

Same effect with *mnemonicFlag = false* instead of "&" duplication (See color insert.)

The mnemonic functionality is a very useful feature when combined with the *setLabelFor()* method: Labels can be associated with another component (e.g., an adjacent editbox) such that pressing the label's mnemonic hot-key will bring the associated component into focus:

```
% Create the editbox
hEdit = uicontrol('style','edit','pos',[70,10,100,20]);
set(hEdit,'string','Initial value');
jEdit = findjobj(hEdit);
```

```
% Create the label
str = 'My &data:'; % mnemonic hotkey = <Alt> -D
jLabel = javaObjectEDT('com.mathworks.mwswing.MJLabel',str);
[hcomponent,hcontainer] = javacomponent(jLabel,[10,10,50,20],gcf);

% Synchronize the label and figure background colors
color = get(gcf,'Color');
colorCells = mat2cell(color,1,[1,1,1]);
jLabel.setBackground(java.awt.Color(colorCells{:}));

% Associate the label with the editbox
jLabel.setLabelFor(jEdit.java);
```

Editbox selected after pressing <Alt>-D (the label's mnemonic)

The mnemonic hot-key functionality can be turned on and off at will, using MJLabel's *setArmed(flag)* method:

```
jLabel.setArmed(false);
```

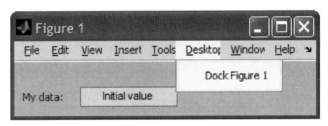

Same figure with label disarmed (note the missing underscore in the label);
<Alt>-D now activates the Desktop menu item

com.mathworks.widgets.HyperlinkTextLabel is another HTML-aware multi-line label component, which is explained in detail in Section 5.5. This component is particularly advantageous for displaying hyperlinks, as explained there. Additional methods of displaying hyperlinks are discussed in Sections 3.3.1, 6.5.2, 8.3.1, and 8.3.2.

6.10 Frame

uicontrol('Style','frame') uses the com.mathworks.hg.peer.FramePeer$1 class that extends MATLAB's com.mathworks.mwswing.MJPanel class, which itself extends Swing's standard javax.swing.JPanel class.[93] It has no sub-components.

HTML formatting is generally irrelevant for this control — it is only relevant (and supported) for the tooltip.

LabelPeer$1 is actually a simple JPanel that has a simple **Border** (a javax.swing.border.LineBorder object), which can be customized as explained in Sections 3.3.1, 6.1, and 6.5.1 (Single-line editbox). As such, it does not have any interesting formatting properties, methods or callbacks (except the **Border**). For this and many other reasons, it is advisable to use MATLAB *uipanel* wherever frames were considered for use — read the following section for more details.

6.11 Uipanel

uipanel is considered a UI control although it is created using a dedicated function (*uipanel*) rather than the *uicontrol* function. *uipanel* uses a standard java.awt.Panel with a single com.mathworks.hg.peer.LabelPeer$1 sub-component for the panel's title (this is the same object used for text *uicontrol*s — see Section 6.9). Since text labels do not support HTML formatting, the *uipanel*'s title similarly does not support HTML. However, HTML is indeed supported for the *uipanel*'s tooltip.

The panel's title handle can be retrieved by inspecting the panel's children (be careful not to confuse the title with other possible panel children). The title handle can also be retrieved directly, using the undocumented MATLAB property **TitleHandle**:

```
hTitle = findall(hPanel, 'parent',hPanel, 'style','text');
hTitle = get(hPanel,'TitleHandle')   % a direct alternative
```

We can use this handle to modify the panel's title to any control. For example,

```
titlePos = get(hTitle,'position');
titlePos(3) = 70;
set(hTitle,'style','checkbox','string','All options','pos',titlePos);
```

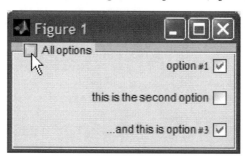

Note that if *uipanel* is created with no title (or an empty one), no java.awt.Panel nor com.mathworks.hg.peer.LabelPeer$1 sub-component are created (see Section 7.3.3). Only when the panel's title is updated are these two objects created. If the title is then reset to ""or [], these Java objects are made invisible (but not deleted).

Unfortunately, it seems that only the panel's title object is customizable at the Java level: Unlike other *uicontrols*, *uipanel* itself does not have any accessible Java peer object, and so cannot be customized at the Java level. It appears that the HG information is stored elsewhere and that the panel is painted directly onto the figure's canvas.

If we need customizable panel layout[94] or borders (see Sections 3.3.1, 6.1, and 6.5.1), then we have to use a JPanel object, and this will force us to use Java controls for all the components contained within the panel (see Section 3.8). We cannot present a MATLAB plot axes within rounded panel borders, for example. This is indeed a great pity.

6.12 Tooltips[†]

All **uicontrols**, just like all Swing components, have tooltips. In standard documented/supported MATLAB, the only thing which can be customized in tooltips is their text contents. We have already shown in Section 3.3.3, that tooltips support HTML contents, which can lead to very innovative and informative tooltips, which can even display images.

But HTML formatting is not the only thing that can be customized in tooltips. In this section, we shall present several other usages, which can easily be extended by interested readers.

6.12.1 Displaying a Tooltip on Disabled Controls[95]

One issue with the stock MATLAB **uicontrol** tooltips is that if we turn the uicontrol's **Enable** property to "inactive" or "off", its tooltip no longer displays. This is the behavior that we normally want, but occasionally we wish to display a tooltip on a disabled control, for example, to explain why the control is disabled.

We can use the *findjobj* utility (see Section 7.2.2) to find the Java handle for the **uicontrol**. This handle can then be used to set the tooltip text. The tooltip will display if we disable the control using its Java handle's **Enabled** property rather than the MATLAB handle's **Enable** property:

```
hButton = uicontrol('String','Button');
jButton = findjobj(hButton);
set(jButton,'Enabled',false);
set(jButton,'ToolTipText','This is disabled for a reason');
```

As any Java object, properties can also be set using corresponding accessor methods:

```
javaMethodEDT('setEnabled',jButton,false);
javaMethodEDT('setToolTipText',jButton,'Button is disabled for a reason');
```

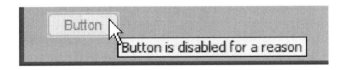

Tooltip on a disabled uicontrol

[†] The bulk of this section was contributed by Matthew (Matt) Whitaker (http://bit.ly/8YFX7W).

Unfortunately, this hack does not work for "inactive" controls. There is no direct Java analogy for inactive controls — it is a MATLAB extension. It appears that MATLAB somehow intercepts mouse events associated with inactive controls. Section 6.12.4 explains how event callback can be used to display tooltips for such controls.

As an alternative for inactive edit-box controls, we can simulate the inactive behavior by setting the Java object's **Editable** property (or by using its *setEditable()* accessor method), then setting the tooltip. Note that the extremely useful Java **Editable** property is unavailable in the MATLAB handle, for some inexplicable reason:

```
hEditbox = uicontrol('String','Edit Text','Style','edit');
jEditbox = findjobj(hEditbox);
set(jEditbox,'Editable',false);
set(jEditbox,'ToolTipText','Text is inactive for a reason');
```

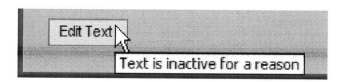

Tooltip on a non-editable editbox

6.12.2 *Displaying a Tooltip on Truncated Text*

If we want to conditionally display a tooltip for an editbox ***uicontrol*** when the text exceeds the control's width, we can use the **TipWhenTruncatedEnabled** property (or its corresponding *setTipWhenTruncatedEnabled()* method). This will display a tooltip with the editbox contents if the string is shown truncated. This saves the user from having to scroll through the control to see its contents. I often use this for edit controls that may contain long path names:

```
hEditbox(1) = uicontrol('Style','edit','Units','norm','Pos', ...
[0.1,0.8,0.4,0.05], 'String','Too Short');

hEditbox(2) = uicontrol('Style','edit','Units','norm','Pos', ...
[0.1,0.7,0.2,0.05], 'String','Long Enough to Display a Tool Tip');

jEditbox1 = findjobj(hEditbox(1));
jEditbox2 = findjobj(hEditbox(2));

% property-based alternative
set(jEditbox1,'TipWhenTruncatedEnabled',true);


% method-based alternative
javaMethod('setTipWhenTruncatedEnabled',jEditbox2,true);
```

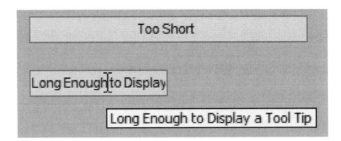

TipWhenTruncatedEnabled tooltip

The **TipWhenTruncatedEnabled** property is also available for multi-line editboxes, but has (obviously) no effect when scrollbars are present. Also note that setting the **TipWhenTruncatedEnabled** property to true overrides any previous tooltip that might have been set for the editbox.

Finally, note that the **TipWhenTruncatedEnabled** property can also be set for the editbox component of popup-menu (aka drop-down) controls, <u>after</u> they have been set to be editable using their Java **Editable** property (note that both properties are false by default for MATLAB uicontrols). In the following screenshot, the drop-down's editbox component contained an HTML snippet that is shown unformatted within the edit-box and HTML-formatted in the de-truncated tooltip:

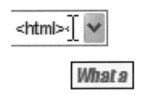

De-truncated HTML-format tooltip

6.12.3 Controlling Tooltip Timing

As you have probably noticed, there is a slight delay between the time the mouse enters the control and when the tooltip actually appears. If we display a tooltip over a control for sufficiently long, the tooltip will then disappear. Sometimes the default delays are too slow or fast for our application. These delay times can be controlled using the `javax.swing.ToolTipManager`.[96] `ToolTipManager` sets these parameters globally (including for MATLAB desktop components), but they are not persistent between MATLAB sessions.

Some examples using the `ToolTipManager`:

```
btn = uicontrol('String','Button','Tooltip', ...
        'This is a button.','Pos',[100,100,75,25]);

txt = uicontrol('Style','edit','String','Edit Text', ...
        'Tooltip','This is editable text','Pos',[100,50,75,25]);
```

```
% Use a static method to get ToolTipManager object
tm = javax.swing.ToolTipManager.sharedInstance;

% Get the delay before display in milliseconds (=750 on my system)
initialDelay = javaMethodEDT('getInitialDelay',tm);

% Set tooltips to appear immediately
javaMethodEDT('setInitialDelay',tm,0);

% Get delay before tooltip disappears (=10000 (10 sec) on my system)
dismissDelay = javaMethodEDT('getDismissDelay',tm);

% Set the dismiss delay to 2 seconds
javaMethodEDT('setDismissDelay',tm,2000);

% Turn off all tooltips in system (including the Matlab desktop)
javaMethodEDT('setEnabled',tm,false);
javaMethodEDT('setEnabled',tm,true); % ...now turn them back on
javaMethodEDT('setInitialDelay',tm,initialDelay);
javaMethodEDT('setDismissDelay',tm,dismissDelay);
```

Note the extensive use of the ***javaMethodEDT*** function to execute Java Swing methods on the Swing Event Dispatch Thread (EDT — see Section 3.2).

6.12.4 *Displaying a Tooltip on Inactive Controls*[97]

Section 6.12.2 explained that displaying tooltips on inactive controls is problematic since MATLAB appears to intercept mouse events to these inactive controls, so even setting the tooltip on the underlying Java object will not work: The Java object appears not to receive the mouse-hover event and therefore does not "know" that it is time to display the tooltip.

There is an undocumented Java technique[98] (Java also has some…) for forcing a tooltip to appear using the **ActionMap** of the *uicontrol*'s underlying Java object to get at a *postTip* action. A **WindowButtonMotionFcn** callback could be used to check if the mouse was above the inactive control, then triggering the forced tooltip display. We will need to chain existing **WindowButtonMotionFcn** callbacks and handle ModeManagers that override them. All this is admittedly difficult to implement.

The Image Processing Toolbox has the nice pair of ***iptaddcallback*** and ***iptremovecallback*** functions that largely handle these issues. But for general MATLAB, there seemed to be no alternative until I[†] remembered that events trigger callbacks. I decided to use a listener for the *WindowButtonMotion* event to detect the mouse motion. The advantage of using an event listener is that we do not disturb any existing **WindowButtonMotionFcn** callback. We still need to be somewhat careful that our listeners do not do conflicting things, but it is a lot easier than trying to manage everything through the single **WindowButtonMotionFcn**.

[†] Matt Whitaker.

A demonstration of this appears below with some comments following (note that this code uses the ***FindJObj*** utility — see Section 7.2):

```matlab
% Illustrates how to make a tooltip appear on an inactive control
function inactiveBtnToolTip
   h = figure('WindowButtonMotionFcn',@motionFcn);
   col = get(h,'color');
   lbl = uicontrol('Style','text', 'Pos',[10,160,120,20], ...
                   'Background',col, 'HorizontalAlignment','left');
   btn = uicontrol('Parent',h, 'String','Button', ...
                   'Enable','inactive', 'Pos',[10,40,60,20]);
   uicontrol('Style','check', 'Parent',h, ...
             'String','Enable button tooltip', ...
             'Callback',@chkTooltipEnable, 'Value',1, ...
             'Pos',[10,80,180,20], 'Background',col);

   % Create the tooltip and postTip action
   jBtn = findjobj(btn);
   import java.awt.event.ActionEvent;
   javaMethodEDT('setToolTipText',jBtn,'This button is inactive');
   actionMap = javaMethodEDT('getActionMap',jBtn);
   action = javaMethodEDT('get',actionMap,'postTip');
   actionEvent = ActionEvent(jBtn, ActionEvent.ACTION_PERFORMED, ...
                             'postTip');

   % Get control's extents +2 pixels to compare to the mouse position
   margin = [-2,-2,4,4]; % define a narrow band around the control
   btnPos = getpixelposition(btn) + margin;
   left = btnPos(1);
   right = sum(btnPos([1,3]));
   btm = btnPos(2);
   top = sum(btnPos([2,4]));

   % Add a listener on mouse movement events
   tm = javax.swing.ToolTipManager.sharedInstance; %tooltip manager
   pointListener = handle.listener(h,'WindowButtonMotionEvent', ...
                                   @figMouseMove);

   % inControl is a flag to prevent multiple postTip action triggers
   % while mouse remains in the button
   inControl = false;

   function figMouseMove(src,evtData) %#ok

      %get the current point
      cPoint = evtData.CurrentPoint;
      if cPoint(1) >= left && cPoint(1) <= right &&...
         cPoint(2) >= btm && cPoint(2) <= top
         if ~inControl %we just entered
            inControl = true;
            action.actionPerformed(actionEvent); %show the tooltip
         end %if
```

```
        else
            if inControl %we just existed
                inControl = false;
                %toggle to make it disappear when leaving button
                javaMethodEDT('setEnabled',tm,false);
                javaMethodEDT('setEnabled',tm,true);
            end %if
        end %if
    end %figMouseMove

    %illustrate we can still do regular window button motion callback
    function motionFcn(varargin)
        str = sprintf('Mouse position: %d, %d',get(h,'CurrentPoint'));
        set(lbl,'String',str);
        drawnow;
    end %motionFcn

    function chkTooltipEnable(src,varargin)
        if get(src,'Value')
            set(pointListener,'Enable','on');
        else
            set(pointListener,'Enable','off');
        end %if
    end %chkTooltipEnable

end %inactiveBtnToolTip
```

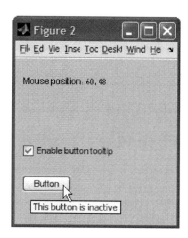

Tooltip on an inactive button

A few comments on the code:

1. The code illustrates that we can successfully add an additional event listener to listen for mouse motion events, while still invoking the original **WindowButtonMotionFcn** callback. This makes chaining callbacks easier.

2. *handle.listener* objects have an **Enable** property that can be used to turn the listener on and off. This can be seen in the *chkTooltipEnable()* callback for the check box in the code above. If we wanted to permanently remove the listener we would simply use *delete(pointListener)*. Note that *addlistener* adds a hidden property to the object being listened to, so that the listener is tied to the object's lifecycle. If we create a listener directly using *handle.listener* we are responsible for its disposition. Unfortunately, *addlistener* fails for HG handles on R2008a and earlier MATLAB releases, so we use *handle.listener* directly. On the other hand, *addlistener* does provide some extra sanity checks that *handle.listener* does not: it checks whether the listened handle actually has the requested event (evoking an error if not); also, starting in R2009a, it checks that the event name conforms to HG2 naming conventions (evoking a warning if not).[†] See Appendix B for additional information on *handle.listener*.

3. The code illustrates a good practice when tracking rapidly firing events like mouse movement of handling reentry into the callback while it is still processing a previous callback. Here, we use the inControl flag to prevent the *postTip* action being continuously fired during mouse hover over the control.

4. I was unable to determine if there is any corresponding action for the *postTip* to dismiss tips so I resorted to using the ToolTipManager to toggle its own **Enable** property to cleanly hide the tooltip as the mouse leaves the control.

5. Extensive use of the *javaMethodEDT* function was made to execute Java methods that affect swing components on Swing's EDT (see above).

Each MATLAB callback has an associated event with it. Some of the ones that might be immediately useful at the figure-level are **WindowButtonDown**, **WindowButtonUp**, **WindowKeyPress**, and **WindowKeyRelease**. They can all be accessed through *handle.listener* or *addlistener* as in the code above. Unfortunately, events do not always have names that directly correspond to the callback names. To see the list of available events for a particular MATLAB object, use the following code, which relies on another undocumented function — *classhandle*. Here we list the events for *gcf*:

```
>> get(get(classhandle(handle(gcf)),'Events'),'Name')
ans =
    'SerializeEvent'
    'FigureUpdateEvent'
    'ResizeEvent'
    'WindowKeyReleaseEvent'
    'WindowKeyPressEvent'
    'WindowButtonUpEvent'
    'WindowButtonDownEvent'
    'WindowButtonMotionEvent'
    'WindowPostChangeEvent'
```

[†] HG2 event names do not have an "Event" suffix.

References

1. http://java.sun.com/javase/6/docs/api/javax/swing/JButton.html (or http://tinyurl.com/2tnxxc).
2. http://UndocumentedMatlab.com/blog/button-customization/ (or http://tinyurl.com/yzaze7x).
3. http://java.sun.com/docs/books/tutorial/uiswing/components/border.html (or http://bit.ly/afp0gT); http://UndocumentedMatlab.com/blog/customizing-uicontrol-border/ (or http://bit.ly/cX76rK).
4. http://www.mathworks.com/matlabcentral/newsreader/view_thread/284619#754494 (or http://bit.ly/c3PpcU).
5. http://java.sun.com/javase/6/docs/api/javax/swing/JComponent.html#setDebugGraphicsOptions(int) (or http://bit.ly/4B3vX2).
6. http://java.sun.com/docs/books/tutorial/uiswing/components/menu.html#mnemonic (or http://tinyurl.com/c3fdu8).
7. http://www.mathworks.com/help/techdoc/ref/uicontrol_props.html#bqxoims (or http://bit.ly/aE3NCi).
8. http://java.sun.com/javase/6/docs/api/java/awt/Insets.html (or http://tinyurl.com/38calj).
9. http://java.sun.com/docs/books/tutorial/uiswing/misc/focus.html (or http://tinyurl.com/5curo); also read the very informative http://java.sun.com/javase/6/docs/api/java/awt/doc-files/FocusSpec.html (or http://tinyurl.com/cqom4d).
10. http://java.sun.com/javase/6/docs/api/javax/swing/AbstractButton.html#setMultiClickThreshhold(long) (http://bit.ly/95L42L).
11. http://www.mathworks.com/matlabcentral/newsreader/view_thread/145846 (or http://tinyurl.com/db5bjf).
12. http://www.mathworks.com/matlabcentral/newsreader/view_thread/163279#438778 (or http://tinyurl.com/cnx52y); http://www.mathworks.com/matlabcentral/newsreader/view_thread/161302#416409 (or http://tinyurl.com/dfxqz3); http://www.mathworks.com/matlabcentral/newsreader/view_thread/99285 (or http://tinyurl.com/d8c64v).
13. http://www.mathworks.com/support/bugreports/194025 (or http://tinyurl.com/mjoatj).
14. http://java.sun.com/javase/6/docs/api/javax/swing/JToggleButton.html (or http://tinyurl.com/4aufxh).
15. http://java.sun.com/javase/6/docs/api/javax/swing/JRadioButton.html (or http://tinyurl.com/czz2or).
16. http://java.sun.com/javase/6/docs/api/javax/swing/JCheckBox.html (or http://tinyurl.com/34lppe).
17. http://java.sun.com/javase/6/docs/api/javax/swing/ButtonGroup.html (or http://tinyurl.com/d9k36z).
18. http://www.jidesoft.com/javadoc/com/jidesoft/swing/TristateCheckBox.html (or http://bit.ly/9LcWOG).
19. http://www.mathworks.com/matlabcentral/newsreader/view_thread/71720 (or http://tinyurl.com/d2bsyk).
20. http://www.mathworks.com/support/bugreports/194025 (or http://tinyurl.com/mjoatj).
21. http://java.sun.com/javase/6/docs/api/javax/swing/JTextField.html (or http://tinyurl.com/2tpnjg).
22. http://UndocumentedMatlab.com/blog/customizing-uicontrol-border/ (or http://bit.ly/cX76rK).
23. http://www.mathworks.com/matlabcentral/newsreader/view_thread/244383 (or http://tinyurl.com/dzhk5c).
24. http://www.mathworks.com/matlabcentral/newsreader/view_thread/284619#754494 (or http://bit.ly/c3PpcU).
25. http://java.sun.com/javase/6/docs/api/javax/swing/text/DefaultCaret.html (or http://tinyurl.com/cdg5ta).
26. http://java.sun.com/javase/6/docs/api/javax/swing/text/Position.Bias.html (or http://tinyurl.com/cranzr).
27. http://java.sun.com/javase/6/docs/api/javax/swing/BoundedRangeModel.html (or http://tinyurl.com/ckvxff).
28. http://java.sun.com/javase/6/docs/api/javax/swing/text/PlainDocument.html (or http://tinyurl.com/de6479); also read; http://java.sun.com/javase/6/docs/api/javax/swing/text/Document.html (or http://tinyurl.com/33zp32).
29. http://java.sun.com/docs/books/tutorial/uiswing/components/generaltext.html#document (or http://tinyurl.com/5h38a).
30. http://java.sun.com/products/jfc/tsc/articles/text/element_interface/ (or http://tinyurl.com/c8p99u).
31. http://java.sun.com/docs/books/tutorial/uiswing/components/generaltext.html#undo (or http://tinyurl.com/cmv8wk).
32. http://java.sun.com/javase/6/docs/api/javax/swing/text/DocumentFilter.html (or http://tinyurl.com/cctt56).
33. http://java.sun.com/docs/books/tutorial/uiswing/components/generaltext.html#filter (or http://tinyurl.com/dackdp).
34. http://java.sun.com/javase/6/docs/api/javax/swing/text/StyledDocument.html (or http://tinyurl.com/d2ylgx).

35. http://java.sun.com/javase/6/docs/api/javax/swing/text/html/HTMLDocument.html   (or   http://tinyurl.com/cnqpcj).

36. Read introductions here: http://java.sun.com/j2se/1.4.2/docs/guide/imf/index.html (or http://bit.ly/aIf3jo) and here: http://java.sun.com/products/jfc/tsc/articles/InputMethod/inputmethod.html (or http://bit.ly/cgNcQo). Additional info: http://java.sun.com/javase/technologies/desktop/articles.jsp#I18N (or http://bit.ly/9nlIXG).

37. http://www.mathworks.com/matlabcentral/newsreader/view_thread/285437#757783   (or   http://bit.ly/b5utRD).

38. http://java.sun.com/javase/6/docs/api/javax/swing/JTextPane.html (or http://tinyurl.com/2sqmm6).

39. http://java.sun.com/docs/books/tutorial/uiswing/components/text.html (or http://bit.ly/7duW9X).

40. http://java.sun.com/javase/6/docs/api/javax/swing/JScrollPane.html   (or   http://tinyurl.com/2hp2we); http://java.sun.com/docs/books/tutorial/uiswing/components/scrollpane.html   (or   http://tinyurl.com/o64g).   `com.mathworks.hg.peer.utils.UIScrollPane` extends `com.mathworks.mwswing.MJScrollPane` which directly extends Swing's `javax.swing.JScrollPane`.

41. http://java.sun.com/javase/6/docs/api/javax/swing/JViewport.html (or http://tinyurl.com/2r3a4s).

42. For an overview of how to use scroll-panes, read http://java.sun.com/docs/books/tutorial/uiswing/components/scrollpane.html (or http://tinyurl.com/o64g).

43. http://UndocumentedMatlab.com/blog/setting-line-position-in-edit-box-uicontrol (or http://tinyurl.com/dhe8yo).

44. http://java.sun.com/javase/6/docs/api/javax/swing/text/StyledEditorKit.html (or http://tinyurl.com/cq5zxx).

45. http://java.sun.com/javase/6/docs/api/javax/swing/text/StyledDocument.html   (or   http://tinyurl.com/d2ylgx).

46. http://java.sun.com/javase/6/docs/api/javax/swing/JEditorPane.html#registerEditorKitForContentType (java.lang.String,%20java.lang.String) (or http://tinyurl.com/dzhrg2).

47. http://java.sun.com/docs/books/tutorial/uiswing/examples/components/TextSamplerDemoProject/src/components/TextSamplerDemoHelp.html (or http://tinyurl.com/c27zpt).

48. Example usage: http://javatechniques.com/blog/setting-jtextpane-font-and-color/ (or http://tinyurl.com/dx673x).

49. Example usage: http://java.sun.com/docs/books/tutorial/uiswing/components/generaltext.html (or http://tinyurl.com/482pk9).

50. http://java.sun.com/javase/6/docs/api/javax/swing/text/StyleConstants.html   (or   http://bit.ly/cCu6bY). Also look at the internal classes: `StyleConstants.CharacterConstants`, `StyleConstants.ColorConstants`, `StyleConstants.FontConstants`, and `StyleConstants.ParagraphConstants`.

51. http://java.sun.com/docs/books/tutorial/uiswing/components/editorpane.html (or http://tinyurl.com/b7elr); Javasourcecode:http://java.sun.com/docs/books/tutorial/uiswing/examples/components/TextSamplerDemoProject/src/components/TextSamplerDemo.java (or http://tinyurl.com/yvmf7r).

52. http://java.sun.com/docs/books/tutorial/uiswing/examples/components/TextSamplerDemoProject/src/components/images/Pig.gif (or http://tinyurl.com/calqqu).

53. http://java.sun.com/products/jfc/tsc/articles/text/element_interface/#changingCharacterAttributes (or http://bit.ly/c8NBqZ).

54. http://java.sun.com/products/jfc/tsc/articles/text/element_interface/#changingParagraphAttributes (or http://bit.ly/chdAaN).

55. http://java.sun.com/javase/6/docs/api/javax/swing/JEditorPane.html#getScrollableTracksViewport Width() (or http://bit.ly/a8qDy7).

56. A detailed description and Tic-Tac-Toe usage example can be found here: http://java.sun.com/products/jfc/tsc/articles/tictactoe/index.html (or http://tinyurl.com/chg3ob).

57. http://java.sun.com/javase/6/docs/api/javax/swing/text/html/HTML.Tag.html (or http://tinyurl.com/cdo5fo).

58. http://java.sun.com/javase/6/docs/api/javax/swing/text/html/HTMLEditorKit.InsertHTMLTextAction.html (or http://bit.ly/aODmK8).

59. http://java.sun.com/docs/books/tutorial/uiswing/examples/components/TextSamplerDemoProject/src/components/images/Pig.gif (or http://tinyurl.com/calqqu).

60. http://UndocumentedMatlab.com/blog/gui-integrated-html-panel/ (or http://bit.ly/7cNXYM).

61. http://stackoverflow.com/questions/1903516/matlab-displaying-markup-html-or-other-format/1903990#1903990 (or http://bit.ly/5aYj7d).

62. http://www.jroller.com/gfx/entry/be_ready_for_java_se (or http://bit.ly/4PKz2l).

63. http://java.sun.com/javase/6/docs/api/javax/swing/JList.html (or http://tinyurl.com/3x52m2).

64. http://java.sun.com/javase/6/docs/api/javax/swing/JScrollPane.html (or http://tinyurl.com/2hp2we); http://java.sun.com/docs/books/tutorial/uiswing/components/scrollpane.html (or http://tinyurl.com/o64g). `com.mathworks.hg.peer.utils.UIScrollPane` extends `com.mathworks.mwswing.MJScrollPane` which directly extends Swing's `javax.swing.JScrollPane`.

65. http://java.sun.com/javase/6/docs/api/javax/swing/JViewport.html (or http://tinyurl.com/2r3a4s).

66. http://java.sun.com/javase/6/docs/api/javax/swing/DefaultListCellRenderer.html (or http://tinyurl.com/clhxet). For a usage example read http://java.sun.com/javase/6/docs/api/javax/swing/ListCellRenderer.html (or http://tinyurl.com/28qs5u).

67. http://java.sun.com/javase/6/docs/api/javax/swing/ListModel.html (or http://tinyurl.com/4tv9ob) and its default implementation http://java.sun.com/javase/6/docs/api/javax/swing/DefaultListModel.html (or http://tinyurl.com/dhkco7).

68. http://java.sun.com/javase/6/docs/api/javax/swing/ListSelectionModel.html (or http://tinyurl.com/38b2tk) and its default implementation http://java.sun.com/javase/6/docs/api/javax/swing/DefaultListSelectionModel.html (or http://bit.ly/ciLKj8).

69. http://java.sun.com/docs/books/tutorial/uiswing/components/list.html (or http://tinyurl.com/5h7mx); http://java.sun.com/products/jfc/tsc/tech_topics/jlist_1/jlist.html (or http://tinyurl.com/2hf2qu).

70. http://java.sun.com/javase/6/docs/api/javax/swing/ListSelectionModel.html#setValueIsAdjusting(boolean) (or http://bit.ly/bHDCQJ).

71. See related: http://www.mathworks.com/matlabcentral/newsreader/view_thread/169102 (or http://bit.ly/eKeEA7).

72. http://java.sun.com/javase/6/docs/api/javax/swing/ListCellRenderer.html (or http://tinyurl.com/28qs5u).

73. http://java.sun.com/javase/6/docs/api/javax/swing/JList.html#renderer (or http://tinyurl.com/c9nyac); http://java.sun.com/javase/6/docs/api/javax/swing/ListCellRenderer.html (or http://tinyurl.com/28qs5u); http://java.sun.com/javase/6/docs/api/javax/swing/DefaultListCellRenderer.UIResource.html (or http://tinyurl.com/csvc79); http://java.sun.com/docs/books/tutorial/uiswing/components/combobox.html#renderer (or http://tinyurl.com/86ljz).

74. http://java.sun.com/javase/6/docs/api/java/awt/Rectangle.html (or http://tinyurl.com/ccnqxq).

75. http://www.mathworks.com/matlabcentral/newsreader/view_thread/277385 (or http://bit.ly/9jJ9ic).

76. http://www.mathworks.com/matlabcentral/newsreader/view_thread/278059#731584 (or http://bit.ly/afSMmT).

77. http://UndocumentedMatlab.com/blog/setting-listbox-mouse-actions/ (or http://tinyurl.com/ylpfcxa).

78. http://java.sun.com/javase/6/docs/api/javax/swing/JComboBox.html (or http://tinyurl.com/3cfejq).

79. http://www.mathworks.com/matlabcentral/newsreader/view_thread/173313#445679 (or http://tinyurl.com/d9gl9f).

80. http://java.sun.com/javase/6/docs/api/javax/swing/DefaultComboBoxModel.html (or http://tinyurl.com/2lrpxu).

81. http://java.sun.com/javase/6/docs/api/javax/swing/ComboBoxModel.html (or http://tinyurl.com/cbsjhs).

82. http://java.sun.com/javase/6/docs/api/javax/swing/ListModel.html (or http://tinyurl.com/4tv9ob).

83. http://java.sun.com/javase/6/docs/api/javax/swing/JTextField.html#setColumns(int) (or http://tinyurl.com/c9hlcw).

84. http://java.sun.com/javase/6/docs/api/javax/swing/JComboBox.html#setEditor(javax.swing.ComboBoxEditor) (or http://bit.ly/b4UasM).

85. http://java.sun.com/javase/6/docs/api/javax/swing/ListCellRenderer.html (or http://tinyurl.com/28qs5u).

86. http://java.sun.com/docs/books/tutorial/uiswing/components/combobox.html#renderer (or http://tinyurl.com/86ljz).

87. http://www.mathworks.com/matlabcentral/newsreader/view_thread/292881#784678 (or http://bit.ly/aWBrPc).

88. http://java.sun.com/javase/6/docs/api/javax/zsing/JScrollBar.html (or http://tinyurl.com/32zssr).

89. http://www.mathworks.com/matlabcentral/newsreader/view_thread/144980 (or http://tinyurl.com/b99jnb).

90. http://java.sun.com/javase/6/docs/api/javax/swing/JComponent.html (or http://tinyurl.com/yqp2u6).

91. http://www.mathworks.se/matlabcentral/newsreader/view_thread/246424 (or http://tinyurl.com/ybnua54).

92. http://www.mathworks.com/matlabcentral/newsreader/view_thread/265569 (or http://tinyurl.com/yjhgcns).

93. http://java.sun.com/javase/6/docs/api/javax/swing/JPanel.html (or http://tinyurl.com/39qfmh).

94. See a discussion in http://java.sun.com/docs/books/tutorial/uiswing/components/panel.html

95. http://UndocumentedMatlab.com/blog/additional-uicontrol-tooltip-hacks/ (or http://bit.ly/appIpJ).

96. http://java.sun.com/javase/6/docs/api/javax/swing/ToolTipManager.html (or http://bit.ly/9fG6sV).

97. http://UndocumentedMatlab.com/blog/inactive-control-tooltips-event-chaining/ (or http://bit.ly/c8voeY).

98. http://UndocumentedMatlab.com/blog/spicing-up-matlab-uicontrol-tooltips/#comment-1173 (or http://bit.ly/5oFn8M).

The Java Frame

MATLAB figures are nowadays basically Java Swing objects, proxy derivatives of the Swing `JFrame` container. This has not always been so: in MATLAB releases prior to 7, figures were coded using native platform functionality. In MATLAB 7, this has apparently been recoded using Java Swing, probably in an attempt to increase cross-platform compatibility and to reduce code maintenance and development costs. The MATLAB code wraps all the Java code to ensure a smooth transition, and so MATLAB programmers may remain oblivious of the underlying Java infrastructure.

However, we can take full advantage of the capabilities offered by Java if we step outside MATLAB's documented boundaries into the undocumented realms of the Swing `JFrame`. The door into these realms is using the MATLAB figure's undocumented **JavaFrame** property. This is a hidden property, which cannot be seen when typing *get(hFig)* or *set(hFig)*.[†] However, it is fully accessible, just like any other familiar figure property such as **Tag, UserData**, or **Name**. Note that this is a read-only (un-settable) property, just like the **Type, FixedColors**, or **BeingDeleted** properties:

```
>> jFrame = get(gcf,'JavaFrame')
jFrame =
com.mathworks.hg.peer.FigurePeer@14b3c93
```

Starting with MATLAB R2008a (7.6), a warning message appears whenever the **JavaFrame** property is accessed:

```
Warning: figure JavaFrame property will be obsoleted in a future release. For
more information see the JavaFrame resource on the MathWorks website.[‡]
```

For the moment, at least as of release R2011b (7.13), these warnings are harmless, and can easily be turned off:[§]

```
warning('off','MATLAB:HandleGraphics:ObsoletedProperty:JavaFrame');
```

Alternatively, retrieving the **JavaFrame** property of the figure's *handle()*-ed handle does not display any warning message:

```
jFrame = get(handle(gcf),'JavaFrame');        % no warning displayed
```

However, the warning may be a bad omen for things breaking down in some future MATLAB release. I shall be on the lookout for this and try to find workarounds if and when it happens. The first place to look for answers if and when this happens should be this book's website.[1]

MathWorks is currently evaluating whether to discontinue the **JavaFrame** property (causing an application error if used), or to continue supporting it for backward compatibility. I encourage

---

[†] At least not normally, it can be seen, together with other undocumented properties, after typing. *set(0 'HideUndocumented', 'off')*. See http://UndocumentedMatlab.com/blog/getundoc-get-undocumented-object-properties/ (or http://bit.ly/ns3Cog).

[‡] The hyperlink links to http://www.mathworks.com/javaframe

[§] This warning workaround is even used by MATLAB itself, in functions such as *javacomponent.m* that use the **JavaFrame** property.

all users of this property to let MathWorks know how and why it is important to them, so that they may be more inclined to preserve it. Users can do so by using the official feedback form: http://www.mathworks.com/javaframe

7.1 Java Frame Properties and Methods

Once retrieved, the jFrame handle can be used to set several interesting properties which have not been exposed as regular properties in the MATLAB figure.

7.1.1 Window Minimization and Maximization

The JavaFrame's **Minimized** and **Maximized** properties affect the window state. Note that these properties are Java booleans which accept true/false (or 1/0), not the regular MATLAB 'on'/'off':

```
% Three alternative possibilities of setting Minimized/Maximized:
jFrame.setMinimized(true);
set(jFrame,'Minimized',true); % note interchangeable 1⇔true, 0⇔false
jFrame.handle.Minimized = 1;
```

jFrame follows Java convention: the method that retrieves boolean values is called *is<Propname>()* instead of *get<Propname>*. In our case: *isMaximized()* and *isMinimized()*:

```
flag = jFrame.isMinimized;          % Note: isMinimized, not getMinimized
flag = get(jFrame,'Minimized');
flag = jFrame.handle.Minimized;
```

All MATLAB releases from the past years have the **Minimized** and **Maximized** properties in jFrame, but some old releases do not.[2] I therefore advise to always prefer using the corresponding jFrameProxy properties, rather than the jFrame ones, as explained in Section 7.3.7.

7.1.2 Docking and Undocking

All MATLAB figures, and most other MATLAB windows, dialogs, and so on, are dockable Clients within enclosing Groups, which can themselves be docked into enclosing Groups (see Section 8.1 for details). Using the Java Desktop object, we can control the docking state and target of these client windows and their Group containers.

Regular MATLAB figures have two fully documented properties that affect docking:[3]

- **WindowStyle** — sets the docking window's state (normal/modal/docked)
- **DockControls** — controls the display of docking controls in the figure

Using these properties presents several limitations: we cannot hide the **DockControls** when a figure is docked. Also, we cannot define a docking target — the Figures group is always automatically used. Both limitations can be overcome using the **JavaFrame** object handle — the first limitation is overcome in Section 7.3.6; we can overcome the second limitation using the **JavaFrame**'s **GroupName** property.

Setting nondefault docking groups can be useful in applications that open multiple figure windows that need to be semantically grouped. For example, a monitoring application that opens several real-time graphs may wish to have them grouped in a separate group from the main UI controls; alternatively, graphs of input channels may be separated from output channels by using separate groups. Control panels with separate internal resizable panels can also be implemented this way.

Before setting the figure's group, we need to be acquainted with another read-only **JavaFrame** property, **Desktop**, which returns the reference of the main MATLAB desktop.[†] Chapter 8 describes this reference and its uses in detail:

```
>> jFrame.getDesktop
ans =
com.mathworks.mde.desk.MLDesktop@146b111
```

The JavaFrame's **GroupName** property affects the group to which the figure belongs when it is docked (programmatically or by clicking its ⬦ docking icon — see Section 7.3.6).[4] If the group name does not exist, it needs to be created using jFrame.*getDesktop.addGroup(groupName)*. These actions are combined in jFrame's *setDesktopGroup(jFrame.getDesktop,groupName)* method that automatically creates a new group if necessary. By default, frames belong to the "Figures" group. Specifying any other *groupName* enables creating a new docking group or docking in existing desktop groups. The list of existing group names can be retrieved via the jFrame.*getDesktop.getGroupTitles* method (other desktop-related and group-related methods are described in Section 8.1) or **GroupTitles** read-only property.[‡]

For example, let us dock our figure into the Editor group:

```
>> get(jFrame.getDesktop,'GroupTitles')
ans =
    'Editor'
    'Figures'
    'Web Browser'
    'Array Editor'§         % 'Variable Editor' on new releases
    'File Comparisons'¶      % 'File and Directory Comparisons'...
>> jFrame.setGroupName('Editor');
>> jFrame.setDesktopGroup(jFrame.getDesktop,'Editor');  % alternative
```

[†] This reference can also be retrieved by other means — see Chapter 8 (The MATLAB Desktop).

[‡] The difference between them is that *getGroupTitles* returns an array of java.lang.String objects, while get(...,'Grouptitles') returns a MATLAB cell array of strings (char) which is easier to use in MATLAB applications.

[§] This is called "Variable Editor" on R2008a (MATLAB 7.6) onward.

[¶] This is called "File and Directory Comparisons" on R2008a (MATLAB 7.6), later renamed "Comparison Tool" in R2011a (MATLAB 7.12).

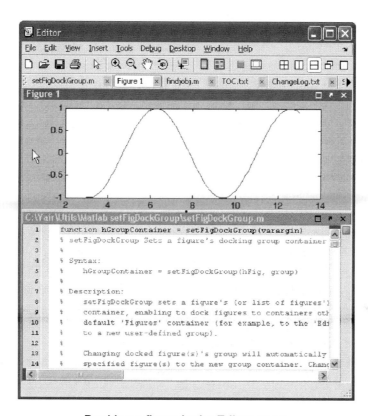

Docking a figure in the Editor group

I have created a convenience utility for setting the figure docking based on the properties and methods above. This utility, called ***setFigDockGroup***, is available on the MathWorks File Exchange.[5]

Additional docking-related methods which might be of interest are:[†]

- jFrame.*getGroupName()* or ***get***(jFrame, 'GroupName')
- jDesktop.*addClient(clientComponent, Name, . . .)*
- jDesktop.*getGroupMembers(groupName)*
- jDesktop.*setGroupDocked(groupName, stateFlag)*
- jDesktop.*setClientDocked(figureName, groupName, stateFlag)*

Also see some additional docking-related customizations in Section 7.3.6.

7.1.3 UI-Related JavaFrame Properties

UIControlBackgroundCompatibilityMode property affects the appearance of MATLAB ***uicontrol*** backgrounds. It is an odd property: it is gettable as usual, via the three alternatives above (jFrame.*getUI...*, ***get***(jFrame, 'UI...'), and jFrame.handle.UI...). However,

---

[†] Used in %matlabroot%\toolbox\matlab\plottools\plottools.m.

it can only be set via the first of these alternatives, using jFrame.*setUIControlBackground-CompatibilityMode*. It also does not appear as a property in *get*(jFrame), although *get*(jFrame, 'UI...') works as expected ....

The value of **UIControlBackgroundCompatibilityMode** is normally 0 (=jFrame. UICONTROLBACKGROUND_OS), which lets the platform-dependent OS (Operating System) decide on the uicontrol's background appearance. It can also be set to 1 (=jFrame. UICONTROLBACKGROUND_COMPATIBLE), which will cause all subsequent uicontrols in the figure to have a background appearance which is compatible across platforms. This can easily be seen in button uicontrols on Windows, where the default XP OS background appearance for buttons does not enable a background color, unlike older (aka *Classic*) Windows system.[6] Note that in order to solve the XP issue, MATLAB 7.5 (R2007b)[†] was fixed to automatically switch to Compatible appearance mode for uicontrols which have their **BackgroundColor** property set (right-most button in the following screenshot):

```
% Left-most button: normal (OS) background appearance
uicontrol('string','click me!','position',[10,10,100,20]);

% Middle button: Compatible-mode background appearance
bgMode = jFrame.UICONTROLBACKGROUND_COMPATIBLE;
jFrame.setUIControlBackgroundCompatibilityMode(bgMode);
uicontrol('string','click me!','position',[130,10,100,20]);

% Right-most button: back in OS mode, but automatically switched to
% Compatible mode because of the BackgroundColor
bgMode = jFrame.UICONTROLBACKGROUND_OS;
jFrame.setUIControlBackgroundCompatibilityMode(bgMode);
uicontrol('string','click me!','background','y');
```

Windows L&F; different UIControlBackgroundCompatibilityModes values

Compare this to other look-and-feels (L&Fs):

Motif L&F

Metal L&F[†]

Windows Classic L&F

Of course, MATLAB programmers always have the alternative option of using Swing controls instead of MATLAB uicontrols. The Swing controls enable a very high degree of flexibility in defining the control's background appearance.[7]

The jFrame.*showTopSeparator(flag)* method controls the display of the divider line that separates the figure content from the top menu/toolbar:

```
jFrame.showTopSeparator(false)
```

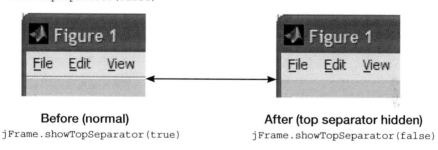

Before (normal)	After (top separator hidden)
jFrame.showTopSeparator(true)	jFrame.showTopSeparator(false)

Another useful method is *getCurrentEdit()* (despite the method name, there is no corresponding **CurrentEdit** property). This method returns the current editing component peer. This can be useful, for example, when setting a figure-wide keyboard action callback and wishing to determine which of several editing components (editboxes, etc.) triggered the callback.

7.1.4 Miscellaneous Other JavaFrame Properties

UserLastMethodID apparently reports the integer ID of the latest graphical user-interface (GUI) method used in this jFrame. This ID can be translated into a text description via jFrame's *getUserMethodDescription(id)* method:

```
>> disp(jFrame.getUserLastMethodID)
   23
```

[†] The Metal L&F has a hover effect on the central button: a bluish border.

```
>> disp(jFrame.getUserMethodDescription(jFrame.getUserLastMethodID))
setVisual
```

Like any other Java object, the full list of jFrame properties and methods can be explored using MATLAB's ***get, inspect***, or ***methodsview*** functions, or by using my *findjobj* or *uiinspect* utilities, as explained above. I have presented those properties and methods that I felt were of special interest; readers are welcome to investigate the others.

The **NativeWindowHandle** and **NativeChildWindowHandle** read-only jFrame properties return OS-native handles of the window. Native-code programmers (e.g., using C/C++ code) can access these handles in order to add graphics, modify the window decorations (the blue frame surrounding the content), etc.[8] A similar native handle is returned by jFrame. fHG1Client.getWindow.*getHWnd.*[†] This latest HWND belongs to the top-level window frame,[9] whereas **NativeWindowHandle** appears to be the handle ID of the window's axis canvas (see the description below). **NativeChildWindowHandle** is a superposition of the OpenGL child-pane handle (in the listing below, the handle value would be 0x00D116A400D116A4):

```
HWND = int32(jFrame.getNativeChildWindowHandle/2^32);
HWND = bitshift(jFrame.getNativeChildWindowHandle,-32);   %alternative
```

This is what the Spy++ application[‡] reported for a top-level window frame (**HWND**) and its axis canvas (**NativeWindowHandle**):

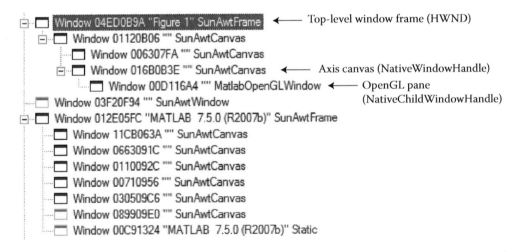

Spy++ listing for a figure window (top tree) and the MATLAB desktop (bottom tree)

[†] Or com.mathworks.util.NativeJava.getHWnd(jFrame.fHG1Client.getWindow). In R2007b and earlier, use fFigureClient rather than fHG1Client.

[‡] Spy++ is bundled with Microsoft Visual Studio. A good alternative is the free Winspector utility (http://www.softpedia.com/get/Security/Security-Related/Winspector.shtml or http://bit.ly/bfQJYg).

When we click any of these handles in Spy++, a small window is displayed with relevant information. Here are screenshots for **HWND** and **NativeWindowHandle**:

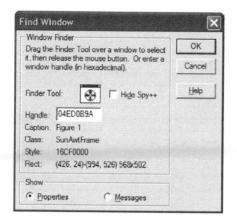

Top-level window frame (HWND)   Axis canvas (NativeWindowHandle)

7.2 FindJObj and the Java Frame Components Hierarchy

The jFrame handle can be used to access (traverse) the entire Java component hierarchy in the figure. This is done using several possible alternatives, as shown below for R2008a (other MATLAB releases return different results, as explained below):

```
>> jContainer1 = jFrame.getAxisComponent.getParent.getParent'
jContainer1 =
com.mathworks.hg.peer.FigureComponentContainer[,0,0,560x420,layout=java
.awt.BorderLayout,alignmentX=0.0,alignmentY=0.0,border=,flags=9,
maximumSize=minimumSize=,preferredSize=]

>> jContainer2=jFrame.fHG1Client.getContentPane'
jContainer2 =
com.mathworks.hg.peer.FigureComponentContainer[,0,0,560x420,...]

>> isequal(jContainer1,jContainer2)
ans =
     1
```

The GUI subcomponents hierarchy can be listed for any Java Swing container by using its *list()* method, which dumps the listing in text format onto the Command Window. If we invoke *list()* on the top-level figure ancestor, a Java container proxy of class com.mathworks. hg.peer.FigureFrameProxy$FigureFrame, then we get a listing of the entire figure-frame GUI hierarchy.

Alternatively, simply click <Ctrl> + <Shift> + F1 in any MATLAB window (including the Desktop) to see the same frame hierarchy output. This is a built-in Java diagnostic tool,

---

† See, for example, the code within ***getfigurefordesktopclient.m***.
‡ In R2007b and earlier, use fFigureClient rather than fHG1Client.

available since very early Java GUIs,[10] and by extension in all MATLAB releases since 6.0 (R12), since all MATLAB GUI since R12 is Java-based.†

Here is the list generated by the top-level figure ancestor in R2008a, edited for clarity. Note that different MATLAB releases and/or different platforms might generate slightly different hierarchy lists:

```
>> jFrame.fFigureClient.getWindow.list

com.mathworks.hg.peer.FigureFrameProxy$FigureFrame
 com.mathworks.mwswing.desk.DTRootPane
  javax.swing.JPanel
  javax.swing.JLayeredPane
   javax.swing.Box
    javax.swing.Box$Filler
    com.mathworks.mwswing.desk.DTTitleButton
    javax.swing.Box$Filler
   com.mathworks.hg.peer.FigureMenuBar
    com.mathworks.hg.peer.MenuPeer$FigureMJMenu
    com.mathworks.hg.peer.MenuPeer$FigureMJMenu
    com.mathworks.hg.peer.MenuPeer$FigureMJMenu
    ...

  com.mathworks.mwswing.MJPanel
   com.mathworks.mwswing.desk.DTClientFrame
    com.mathworks.mwswing.desk.DTInternalFrame$RootPane
     javax.swing.JPanel
      com.mathworks.mwswing.desk.DTToolBarContainer
       com.mathworks.mwswing.MJPanel
        com.mathworks.mwswing.MJToolBar
         com.mathworks.mwswing.MJButton
         com.mathworks.mwswing.MJButton
         com.mathworks.mwswing.MJButton
         com.mathworks.mwswing.MJButton
         com.mathworks.mwswing.MJButton
         com.mathworks.mwswing.MJToolBar$VisibleSeparator
          javax.swing.JSeparator
         com.mathworks.mwswing.MJToggleButton
         com.mathworks.mwswing.MJToolBar$VisibleSeparator
         ...
      com.mathworks.hg.peer.FigureClientProxy$FigureDTClientBase
       com.mathworks.hg.peer.FigureComponentContainer
        com.mathworks.hg.peer.FigurePanel$2
         com.mathworks.hg.peer.FigureComponentContainer
          com.mathworks.hg.peer.ActiveXCanvas
          com.mathworks.hg.peer.FigureAxisComponentProxy$_AxisCanvas
```

Docking icon

Main figure menu bar

Main figure tool bar

Main figure content area

---

† http://www.mathworks.com/support/solutions/en/data/1-9SFV31/ (or http://tinyurl.com/yje3edc); R12 had native (non-Java) figures, so in R12 we can only list the Editor or the Desktop which are Java-based; in later MATLAB releases, the MATLAB figure also became a listable Java frame.

In MATLAB 7.7 (R2008b), the content-pane subcomponent hierarchy has changed slightly:

```
  ...
  com.mathworks.hg.peer.FigureClientProxy$FigureDTClientBase
   com.mathworks.mwswing.MJPanel
    com.mathworks.hg.peer.HeavyweightLightweightContainerFactory
$FigurePanelContainerHeavy
      com.mathworks.hg.peer.FigureComponentContainer
      com.mathworks.hg.peer.FigureAxisComponentProxy$_AxisCanvas
```

7.2.1 FindJObj

Earlier in this book, I explained that all MATLAB GUIs (except the axes plotting engine[†]) are based on Java components. I have mentioned the *findjobj* utility as a means to access the underlying Java components to enable customizations that are unavailable in standard MATLAB, as well as to display the component hierarchy of complex MATLAB containers.

The time has now come for a formal introduction of *findjobj*, explaining its uses and internal mechanism. Of course, readers are welcome to continue using *findjobj* as a black-box utility, but I believe that important insights can be gained from understanding its inner details. *findjobj*'s code is available for free download on the MathWorks File Exchange.[11] It is one of my favorite submissions and is apparently well liked by users, being highly reviewed and highly downloaded.

findjobj has two main purposes:

1. Find the underlying Java object reference of a given MATLAB handle, to enable programmatic object customization. Historically this was the original purpose,[12] hence the utility's name. *findjobj* was meant to extend MATLAB's standard *findobj* function, which does not find Java components.[13]
2. Display a container's internal components hierarchy in a graphical user interface, to facilitate visualization of complex containers. This was later extended to also display and allow modification of the subcomponents' properties and callbacks.[14]

7.2.2 Finding the Underlying Java Object of a MATLAB Control

findjobj's heart is finding a control's underlying Java handle. Unfortunately, this is not exposed by MATLAB except in very rare cases. I could not find a way to directly access the underlying Java-peer handle. Therefore, I resorted to getting the control's enclosing Java frame (window) reference, and then working down its subcomponents hierarchy until finding the Java object(s) which satisfy the position and/or class criteria. To get the enclosing Java frame (aka **TopLevelAncestor**), I use the MATLAB figure's undocumented **JavaFrame** property (see Section 7.1). Since MATLAB releases R2008a onward issue a standard warning when using this property, I have turned off this warning in *findjobj*'s code (see discussion above).

[†] See Section 7.3.1.

Traversing the frame's hierarchy presents several challenges: Main-menu items are accessed using different functions than other Swing components or subcontainers and are not programmatically accessible until first displayed. I overcome this latter challenge by simulating a menu-open action (menu.*doClick()*, see Section 4.6.4) when menus should be searched (this feature is off by default since it takes several seconds and also changes the GUI focus). For "regular" subcontainers, sometimes we need to loop over *getComponent(...)* and in some other cases over *getChildAt(...)*.

Another challenge was presented by the fact that Java positions start at (0,0) in the top-left corner increasing rightward and downward, rather than starting at (1,1) in the bottom-left and increasing upward as in MATLAB (see Section 3.1.1). Moreover, Java positions are always pixel-based and relative to their parent container, which is different from MATLAB (if the MATLAB units is 'pixels', then the value is absolute; if 'normalized', then it returns a nonpixel value). To further complicate matters, some MATLAB controls have a different size than their Java counterparts: some controls have 5-pixel margins, while others do not; some controls are shifted by a pixel or two from their container's border (for a total offset of up to 7 pixels), while some controls (such as popup-menus) have an entirely different reported size. In theory, we could use the MATLAB component's undocumented **PixelBounds** property (much faster than ***getpixelposition***), but unfortunately **PixelBounds** turns out to be unreliable, sometimes returning erroneous values. Finally, different Java containers/components return their position differently: for some, it is a *getLocation()* method; for others, it is *getX()/getY()*, and for others it is the **X** and **Y** properties (without any corresponding *getX()/getY()* accessor methods!).

Having finally overcome all these challenges (and quite a few smaller ones, documented within the source code), I have wrapped the algorithm in a function interface that tries to emulate *findobj*'s. Using *findjobj* can now be as easy as

```
% Modify the mouse cursor when over the button
hButton = uicontrol('string','click me!');
jButton = findjobj(hButton);
jButton.setCursor(java.awt.Cursor(java.awt.Cursor.HAND_CURSOR))
```

Modified *uicontrol* Cursor—a Java property

... or as complex as (for other examples of using *findjobj*, see Sections 4.2.5 and 4.7.3):

```
% Find all non-button controls with the specified label
jControls = findjobj('property',{'text','click me!'}, 'not','class','button');
```

Note that when using *findjobj* in a GUIDE-generated m-file, we must place the call to *findjobj* in the m-file's *_OutputFcn()* function rather than the *_OpeningFcn()*. The reason is that *_OpeningFcn()* is invoked <u>before</u> the figure is made visible and the Java peers are created. *findjobj* cannot find the non-existent Java peers at this point. On the other hand, *_OutputFcn()* is invoked immediately <u>after</u> it is made visible, when the peers are in place so *findjobj* can find them.

7.2.3 GUI for Displaying Container Hierarchy, Properties, and Callbacks

When *findjobj* is called with no output arguments, the function infers that the user requests to see the GUI version, rather than to get the control's Java handle:

```
>> findjobj(gcf); % or: findjobj(gcf)
```

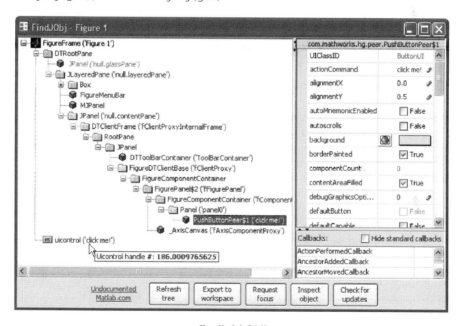

findjobj GUI

There are several noteworthy aspects in this graphical hierarchy presentation:

The hierarchy tree is displayed using the internal `com.mathworks.hg.peer.UITreePeer` Java object. This is the object that underlies the semi-documented *uitree* function (see Section 4.2). The hierarchy subcomponents are presented as tree nodes, each having a separate icon based on the component type. Where possible (toolbar buttons, for example), the component's icon image is used for its corresponding tree node. A `javax.swing.JProgressBar` is presented while the tree is being populated, an action that can take a few seconds depending on the target figure's complexity. Some tree branches that are normally uninteresting are automatically collapsed: hidden containers (these are also grayed-out), menu-bars, tool-bars, and scroll-bars. In parallel to the Java container hierarchy, a separate tree branch is presented with the corresponding MATLAB Handle-Graphics (HG) hierarchy.

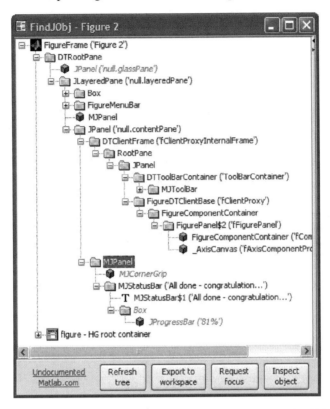

Another *findjobj* GUI example — note the hidden (gray) items, the HG tree branch, and the auto-collapsed `MJToolBar` **container**

Each node item gets a unique tooltip (see first GUI screenshot above), as well as a unique context-menu (right-click menu) with actions that are relevant for that node:

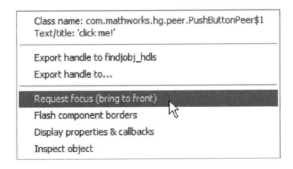

Item-specific context-menu

Labels presented a particular challenge: Java labels do not store their text string in any accessible property. So, while we could see the label handle in the GUI hierarchy, its contents string could not be presented as for other controls. This was confusing in GUI figures that had many text labels, which could not be distinguished. The solution was to associate Java labels with the figure's MATLAB (HG) label handles by size and position — once a Java label was associated, I got the text string from its HG handle.

Finally, a node-selection callback is attached to the tree that will flash a red border around the GUI control when its corresponding Java node-item is clicked/selected:

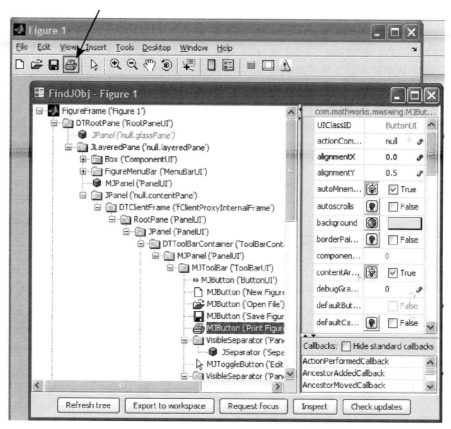

Flashing red border around a selected toolbar icon using *findjobj* (See color insert.)

Once the tree was done, I set out to display and enable modifications of component properties and callbacks in separate adjacent panels. An internal `com.mathworks.mlwidgets.inspector.PropertyView` component is used to display the properties. This is the JIDE component (see Section 5.7.3) that underlies the built-in ***inspect*** function. To prevent a JIDE run-time alert, `com.mathworks.mwswing.MJUtilities.`*initJIDE()* is called before the JIDE component is first used. A label is added to the table's header, displaying the currently selected subcomponent's class (e.g.,`'javax.swing.JButton'`), and a tooltip with a color-coded list of its properties.

The callbacks table was implemented using com.jidesoft.grid.TreeTable to enable easy column resizing, but this is otherwise used as a simple data table. A checkbox was added to filter out the 30-odd standard Swing callbacks, which are non-unique to the selected subcomponent (tree node). All the panels — tree, properties, and callbacks — are then placed in resizable javax.swing.JSplitPanes and presented to the user.

Space here is limited and *findjobj* is some 3000 lines long, so I have obviously not covered everything. I encourage readers to download the utility and explore the code.

7.2.4 *The Java Frame Container Hierarchy*

Here is *findjobj*'s view of the textual Java Frame listing that was presented earlier. As indicated by *findjobj*, each subcomponent has a separate set of properties, methods, and callbacks. I labeled some important hierarchy components in this screenshot. Each numbered label will now be discussed separately in the following sections:

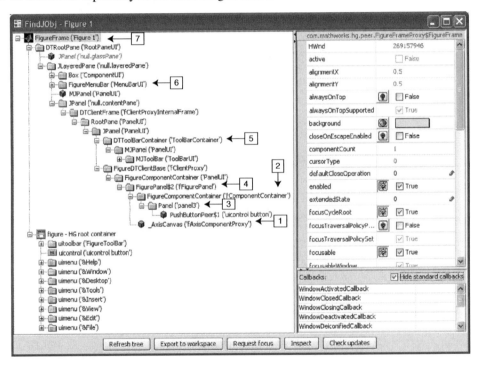

findjobj presentation of the Java Frame object hierarchy

7.3 Important Java Frame Containers

7.3.1 *AxisCanvas*

AxisCanvas is the graphic java.awt.Canvas where MATLAB plots all axes and graphs as graphic elements. Unfortunately, I am unaware of a direct access to internal elements within this graphic area — only via MATLAB's HG handles.

```
>> jAxisCanvas = jFrame.getAxisComponent
jAxisCanvas =
com.mathworks.hg.peer.FigureAxisComponentProxy$_AxisCanvas[fAxisComponentProxy,
0,0,560x420]
```

The AxisCanvas area overlaps the entire content pane area, but is <u>always</u> displayed beneath the FigureComponentContainer which includes all uicontrols.[15] This behavior is due to their common parent FigurePanel$2's OverlayLayout.[16] This layout can be reversed, hiding uicontrols and ActiveXes beneath the axes, by changing the component's Z-order within their FigurePanel$2, as follows:[†]

```
>> jFigPanel2 = jAxisCanvas.getParent;
>> awtinvoke(jFigPanel2,'setComponentZOrder',jAxisCanvas,0); % 1 = >0
```

Note that all uicontrols will now disappear since the AxisCanvas is opaque and extends over the entire content pane. This can be used to temporarily hide figure uicontrols (e.g., when printing GUIs or during heavy application processing).

Other interesting properties and methods of AxisCanvas:

The **Enabled** boolean flag property (default = true; settable via jAxisCanvas. *setEnabled(flag)* or *set(jAxisCanvas,'Enabled','on/off')*; gettable via *isEnabled* or *get*) controls whether the figure axes are clickable or not. This may be useful, for example, to temporarily prevent zoom-in/out until some calculation completes. Related flag property **Focusable** (default = true) controls the ability to focus on the axes (e.g., by using the Tab key).

The **Visible** flag property (default = true; set via *setVisible*, etc.) controls the visibility of all figure axes. Setting this property should only be done on the EDT, using *awtinvoke* or *javaMethodEDT*. If set to 0 (*awtinvoke(jAxisCanvas, 'setVisible',0)*[‡]), all figure axes disappear at once leaving only the uicontrols visible, and vice versa.

NativeWindowHandle reports the axes canvas handle ID, enabling external programs to directly draw on this canvas. This is also the handle returned by jAxisCanvas.getPeer. *getHWnd* and jFrame.*getNativeWindowHandle*. This handle is presumably used by MATLAB to draw on the canvas using JAWT[§] via external libraries such as *hg.dll* on Windows. Note that it is a different handle ID than the HWND returned for the top-level figure frame. Related read-only (nonsettable) property **NativeWindowHandleValid** is a boolean flag reporting whether or not the handle ID is valid.[¶] For example, after a figure window is closed, **NativeWindowHandle** = 0 and **NativeWindowHandleValid** = false.

[†] This must be done on the EDT, hence the use of *awtinvoke()*; we can also use *javaMethodEDT()* in MATLAB R2008a+.

[‡] We cannot use *set(jAxisCanvas,'Visible',0)* here because it does not use the EDT, unless *javaObjectEDT(jAxisCanvas)* was previously invoked. Alternatively, we could have used the equivalent method *awtinvoke(jAxisCanvas,'hide')*.

[§] http://en.wikipedia.org/wiki/Java_AWT_Native_Interface (or http://tinyurl.com/5omllu); http://java.sun.com/j2se/1.5.0/docs/guide/awt/1.3/AWT_Native_Interface.html (or http://tinyurl.com/dhnbtp). The reason for using JAWT may be performance, or perhaps MathWorks' wish to reuse the existing code from MATLAB 6 and earlier.

[¶] Note that as a boolean Java flag, its accessor method is *isNativeWindowHandleValid()*, not *getNativeWindowHandleValid()*.

Cursor can be used to get or set the cursor (mouse pointer) shape. Other MATLAB functions also do this (*getptr, setptr*) at the figure level, so this property is not very useful here. However, it is very useful for specific components, for hover effects.

AxisCanvas includes all the standard Swing callbacks mentioned in Chapter 3, including focus, keyboard, and mouse actions. In fact, figure-level callbacks are only effective when set on the AxisCanvas, not on any other figure hierarchy container.[17]

AxisCanvas is used in several places within this book: Section 3.7 used AxisCanvas for trapping drag-and-drop (DND) events onto a MATLAB figure; Section 3.8 used AxisCanvas to reparent a MATLAB figure's axes onto a new Java container; and Section 10.1 explained that the *UISplitPane* utility uses AxisCanvas for setting mouse callbacks that behave better than similar callbacks at the figure level.

7.3.2 *FigureComponentContainer*

FigureComponentContainer is a transparent container for all the figure's controls, including MATLAB *uicontrol*s, Java Swing controls (added with *javacomponent*), and ActiveX controls (added with *actxcontrol*). It normally overlaps AxisCanvas such that its contained controls are always painted on top of MATLAB axes/plots. This is a consequence of its default Z-order of 0, as explained above. This container can be directly accessed as follows:

```
>> jControlsPanel = jFrame.getFigurePanelContainer.getComponent(0)
jControlsPanel =
com.mathworks.hg.peer.FigureComponentContainer[fComponentContainer,...]
```

FigureComponentContainer is a com.mathworks.mwswing.MJPanel object (that extends javax.swing.JPanel). This object shares many of the properties, methods, and callbacks of AxisCanvas (or most other Swing components for that matter), so these shall not be repeated. Perhaps the single interesting property in this object is **ComponentCount** (read-only), which holds the number of controls within the container (there is also a corresponding *getComponentCount()* method).

7.3.3 *Component's Private Container*

Each uicontrol and ActiveX that is added to the figure will have a separate java.awt.Panel container within the FigureComponentContainer. These containers cannot be accessed directly, since they are dynamic in nature. However, they can be accessed as separate children of the FigureComponentContainer, as follows:[†]

```
>> jControlsPanel = jFrame.getFigurePanelContainer.getComponent(0);
>> jButtonHGPanel = jControlsPanel.getComponent(0);
```

---

† Or by using the *findjobj* utility, as explained below.

```
>> jButton = jButtonHGPanel.getComponent(0)
jButton =
com.mathworks.hg.peer.PushButtonPeer$1[,0,0,60x20,alignmentX=0.0,...,
text=uicontrol button,defaultCapable=false]
```

Since `java.awt.Panel` is a heavyweight container (panel.*isLightweight()* = false), it cannot be made transparent. Since all **uicontrol**s, ActiveXes, and **javacomponent**s are automatically contained in such a `Panel` by MATLAB,[†] this means that even if the control or component is made transparent, this would still have no effect since their containing `Panel` is opaque. This prevents plot axes from showing beneath transparent controls and is also the reason for MATLAB controls being Java peers, not simple Swing components.[18] Bill York, MATLAB GUI development manager, explained there were technical reasons for choosing heavyweight `Panel` over the lightweight `JPanel`.[‡]

Note that jFrame.*getActiveXCanvas()* does <u>not</u> return the handle of an existing ActiveX Canvas, as its name would imply. Rather, **actxcontrol.m**[§] uses it to create a <u>new</u> ActiveX Canvas within `FigureComponentContainer`, into which the ActiveX control is then embedded using the native handle ID and the **NativeWindowHandle** and **NativeWindowHandleValid** properties (and their corresponding accessor methods).

7.3.4 FigurePanel or ContainerFactory

`FigurePanel$2` is the common container panel for `FigureComponentContainer` and `AxisCanvas`. Its `OverlayLayout` manager is responsible for the axes/controls overlap issue discussed above. It can be directly accessed as follows:

```
>> jFigPanel2 = jFrame.getFigurePanelContainer
jFigPanel2 =
com.mathworks.hg.peer.FigurePanel$2[fFigurePanel,0,0,560x420,layout=
javax.swing.OverlayLayout]
```

In MATLAB 7.9 (R2009b), com.mathworks.hg.peer.FigurePanel$2 was modified to com.mathworks.hg.peer.HeavyweightLightweightContainerFactory$2, and once again in MATLAB 7.12 (R2011a) to ...$FigurePanelContainerHeavy. All these objects extend java.awt.Panel and appear to be very similar to each other.

[†] **javacomponent**s are contained within a com.mathworks.hg.peer.HGPanel, which extends java.awt.Panel; ActiveXes are contained within a com.mathworks.hg.peer.ActiveXCanvas, which extends the heavyweight java.awt.Canvas via com.mathworks.hg.peer.AxisCanvas; **uicontrol**s used a simple java.awt.Panel until R2009b, when MATLAB started using com.mathworks.hg.peer.UIComponentHeavyweightContainer which is simply a java.awt.Panel extension.

[‡] http://www.mathworks.com/matlabcentral/newsreader/view_thread/268556#702864 (or http://bit.ly/7VNzo0). Note that while individual controls are opaque, the entire figure window can be made fully or partially transparent — see Section 7.3.7.

[§] Feel free to look at its code: **edit**('*actxcontrol*').

7.3.5 *DTToolBarContainer*

`DTToolBarContainer` is the container for all figure toolbars. Each toolbar is contained within a `com.mathworks.mwswing.MJPanel`, which is contained in the parent `DTToolBarContainer`. This parent can be directly accessed as follows:

```
>> jToolBar = jFrame.fHG1Client.getFrameProxy.getToolBarContainer'
jToolBar =
com.mathworks.mwswing.desk.DTToolBarContainer[ToolBarContainer,0,...]

>> % an alternative:
>> jFigPanel2 = jFrame.getFigurePanelContainer;
>> jControlsPanel = jFigPanel2.getComponent(0);
>> jToolBar = jControlsPanel.getTopLevelAncestor.getToolBar;
```

The toolbars themselves are accessible from their `DTToolBarContainer` parent using `jToolBar.getComponent(toolbarIndex).getComponent(0)`.[‡] The standard figure toolbars only contain togglebuttons and separators:

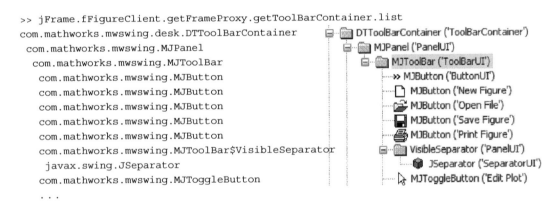

```
>> jFrame.fFigureClient.getFrameProxy.getToolBarContainer.list
com.mathworks.mwswing.desk.DTToolBarContainer
 com.mathworks.mwswing.MJPanel
  com.mathworks.mwswing.MJToolBar
   com.mathworks.mwswing.MJButton
   com.mathworks.mwswing.MJButton
   com.mathworks.mwswing.MJButton
   com.mathworks.mwswing.MJButton
   com.mathworks.mwswing.MJButton
   com.mathworks.mwswing.MJToolBar$VisibleSeparator
    javax.swing.JSeparator
   com.mathworks.mwswing.MJToggleButton
   . . .
```

MATLAB figure toolbars are typically `com.mathworks.mwswing.MJToolBar` objects.[§] We can add our own `JToolBar` objects to their `DTToolBarContainer` parent with a simple `jToolBar.add()` command (of course, this can also be done with MATLAB's standard *uitoolbar* function).

Toolbars, like all other Swing containers, may also contain other Swing components, just like the MATLAB Desktop's Workspace toolbar:

[†] In R2007b and earlier, use `fFigureClient` rather than `fHG1Client`.

[‡] Remember that Java indices start at 0, so the first (topmost) toolbar will have an index of 0.

[§] `com.mathworks.mwswing.MJToolBar` extends `javax.swing.JToolBar`.

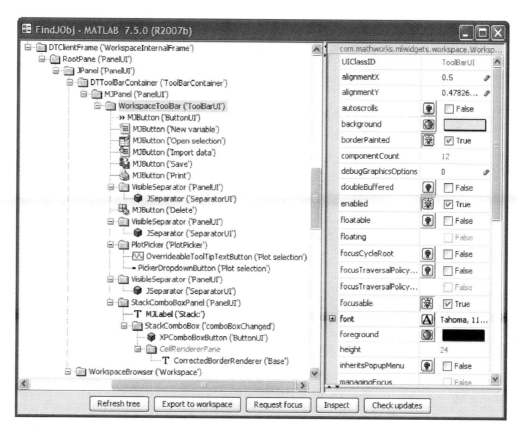

findjobj **presentation of a toolbar with non-standard controls (See color insert.)**

Readers are referred to Section 4.5, where toolbar creation and customization is extensively discussed.

7.3.6 *FigureMenuBar and Docking Controls*

FigureMenuBar contains the figure's menu. It can be accessed directly as so:

```
>> jMenuBar = jFrame.fHG1Client.getMenuBar†
jMenuBar =
com.mathworks.hg.peer.FigureMenuBar[,0,0,543x21,layout = javax.swing.plaf.basic.
DefaultMenuLayout,alignmentX = . . .,paintBorder = true]
```

FigureMenuBar is basically just a container for MJMenuItems and MJMenus. MJMenus are displayed as cascading submenus, as shown below (the appearance of menus, menu items, and submenus is greatly affected by the platform and chosen L&F):

---

† In R2007b and earlier, use fFigureClient rather than fHG1Client.

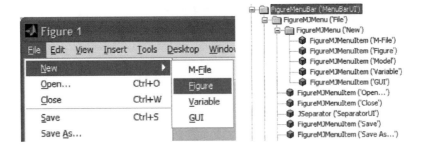

Cascading submenus

Additional menus, submenus, and menu items may be added to their parent, using the parent's *add()* method. Section 4.6 and many online resources[19] describe how to set up and customize menus, and this shall not be repeated here.

A sibling container of the FigureMenuBar and the content pane is a `javax.swing.Box` object that contains the window docking controls, which are `com.mathworks.mwswing. desk.DTTitleButton` objects. Docking is described in detail in Section 7.1.2. Using these `DTTitleButton` handles, we can hide/display the docking controls, even when this is not allowed in regular MATLAB:

```
>> set(gcf, 'WindowStyle','docked', 'DockControls','off')
??? Error using ==> set
Cannot set DockControls to 'off' while WindowStyle is 'docked'
```

Unfortunately, in the R2008a compiler, MathWorks removed the docking controls and disabled the figure docking properties.[20] This can be easily overcome:[†]

```
jFrame = get(handle(hFig),'JavaFrame');
jFrame.fHG1Client.setClientDockable(true);    %R2007b-: use .fFigureClient
```

7.3.7 FigureFrame

FigureFrame is the topmost component in the frame's hierarchy. It represents the `jFrame` window itself, although it is not itself a valid Swing container or component, but just a proxy reference of it. It can be accessed directly as follows (in R2007b and earlier, use `fFigureClient` rather than `fHG1Client`):

```
>> jFrameProxy=jFrame.fHG1Client.getWindow()
jFrameProxy=
com.mathworks.hg.peer.FigureFrameProxy$FigureFrame[fClientProxyFrame,
227,25,568x502,invalid,layout=java.awt.BorderLayout,title=Figure 1,resizable,
normal,defaultCloseOperation=DO_NOTHING_ON_CLOSE,rootPane=com
.mathworks.mwswing.desk.DTRootPane[,4,30,560x468,...],...]
```

---

† http://UndocumentedMatlab.com/blog/docking-figures-in-compiled-applications/ (or http://bit.ly/l0GCDj). In R2007b and earlier, use fFigureClient rather than fHG1Client — refer to the blog article for additional details.

jFrameProxy can also be retrieved via other alternatives:

- `jFrame.fHG1Client.getContentPane.getTopLevelAncestor()`
- `jFrame.getFigurePanelContainer.getComponent(0).`
  `getTopLevelAncestor()`
- `jControlsPanel.getRootPane.getParent()` — jControlsPanel can be
  replaced by any component reference[†]
- `javax.swing.SwingUtilities.getWindowAncestor(jFrame.`
  `getAxisComponent)`[‡]

 Beware of the potential pitfall of using the seemingly innocent `jFrame.fFigureCli-`
ent.*getFrameProxy()*: this will return a `FigureFrameProxy` object, that is an instance
of the `FigureFrameProxy$FigureFrame`'s container class, instead of the requested
`FigureFrameProxy$FigureFrame` itself.

While individual figure components cannot be made **transparent** (see Section 7.3.3),
FigureFrame can be used to set the entire figure's transparency level, ranging from 0.0
(fully-transparent) to 1.0 (fully-opaque, the default figure value):[§]

```
com.sun.awt.AWTUtilities.setWindowOpacity(jFrameProxy,0.8);
jFrameProxy.repaint;
```

A semi-transparent window can be used for another interesting effect, of **blurring**.[21] This
blurring effect is typically used to visually indicate a disabled or inactive window. The trick is
to overlay the main (blurred) window with an empty semi-transparent overlay window that has
the same size and position.

FigureFrame contains several useful properties and methods for MATLAB applications:

HWnd, described above, is the native window handle ID that could probably enable modifying the window frame and similar aspects, although I have never tried this myself.

AlwaysOnTop is a boolean flag (default = false) specifying whether or not the figure window
should always remain on top of all other windows.[22] This top-most state is only applicable when
the window is shown (i.e., not minimized or made invisible). If two overlapping windows have this
flag set, then the last window whose flag was set or that was selected will be shown on top of the
other window. Note that clearing **AlwaysOnTop** does not hide or minimize the window — the

---

[†] This method can be used to retrieve `FigureFrameProxy` without directly using the **JavaFrame** property. Basically,
javacomponent is ran with an invisible/transparent. `JLabel`, which is then used to get the `RootPane` and the `Frame` references: http://www.mathworks.com/matlabcentral/newsreader/view_thread/104129#275318 (or http://tinyurl.com/dkfzn5).

[‡] This is the method used in *editmenufcn*, for example.

[§] Available on Java 6 update 10 and later, pre-bundled in MATLAB R2009b (7.9) and later, or as a retrofit (see Section 1.8). See
additional details in http://UndocumentedMatlab.com/blog/transparent-matlab-figure-window/ (or http://bit.ly/fykuiH).

window will remain in its previous (top-most) position, but it could now be overlapped by other windows.

Not all MATLAB platforms support **AlwaysOnTop**. This depends on the Operating System's capabilities. To check whether the target platform supports this functionality, check the **AlwaysOnTopSupported** property (jFrameProxy.*isAlwaysOnTopSupported*). If **AlwaysOnTopSupported** = false, try alternative solutions on the File Exchange.[23]

The **CloseOnEscapeEnabled** flag (default = false) controls whether the window can be closed by clicking the <Esc> keyboard button. This behavior is typically expected in popup dialogs and message-boxes, but not in complex GUI windows. For this reason, setting this flag to true is mostly useful for small MATLAB figures that act as popup notification windows.

DefaultCloseOperation — when a figure window is closed, in reality it is NOT disposed/deleted, but merely hidden from view, thereby preserving an active reference to the frame object and all its contained objects. This is because MATLAB's default frame-closing action is not DISPOSE_ON_CLOSE (=2), but rather DO_NOTHING_ON_CLOSE (=0). Internal MATLAB listener code does the actual cleanup, in effect mimicking Java's default JFrame closing action (HIDE_ON_CLOSE =1).[24] After a MATLAB figure closes, it is still accessible via java.awt.Frame.*getFrames()*. Therefore, to prevent potential memory leaks when multiple MATLAB figures are opened and closed, set the figure's Frame **DefaultCloseOperation** property to DISPOSE_ON_CLOSE:

```
jFrame.setDefaultCloseOperation(jFrame.DISPOSE_ON_CLOSE);
% or: set(jFrame,'DefaultCloseOperation',2);
```

Using EXIT_ON_CLOSE (=3) can be a nice way to exit MATLAB when the user application window closes[25] (it also reportedly solves an issue of premature exit in compiled applications[26]):

```
jFrame.setDefaultCloseOperation(jFrame.EXIT_ON_CLOSE);
% or: set(jFrame,'DefaultCloseOperation',3);
```

Undecorated — this flag property always has a true value, which cannot be modified. It indicates that the frame window has a title bar and border which are set by the current L&F's so-called decoration. Unfortunately, we cannot modify this property. The reason for this is that Java prevents removing window decorations after a window in displayed and we cannot access a figure's Java components before the window is displayed, so that is a bit of a catch-21. One can try playing around with his/her Java L&F's properties (see Section 3.3.2), but this is not for the faint-hearted.

As suggested in Section 3.8, we could theoretically create another undecorated pure-Java JFrame and transfer the entire MATLAB figure's content onto this new (undecorated) JFrame. In practice, implementing this causes numerous problems.[27]

Another possible alternative, which I have not been able to successfully implement, is to invoke jFrameProxy.*dispose()*, then *setUndecorated(true)*, *pack()*, and finally *show(true)*. This removes the decoration, but unfortunately all the content as well …

WindowDecorationStyle in `jFrameProxy`'s single child (`DTRootPane`) does not appear to make any difference, so this is also a dead-end with regard to window decoration.

The **Enabled** flag property (default = true), already discussed above for axes, is useful at this (figure window) level: clearing this flag (setting to false or 'off') disables any interactive action on the figure window, including clicks anywhere in the window frame (including the minimize/maximize/close icons in the window's decoration [frame]; main menu; toolbars and content), as well as selecting the window (Alt-Tab in Windows). This is very useful in preventing user actions while some complicated computation is done. We must take special care to handle programming exceptions; otherwise, the window might stay in its inactive state and cannot even be closed.

In effect, using this property can be an alternative to MATLAB's built-in (yet undocumented) *uisuspend* and *uirestore* functions. Note that the fully-documented functions *uiwait* and *uiresume* are similar in functionality and in name to the undocumented *uisuspend* and *uirestore* functions. We should not confuse between these two function sets, and most importantly, should not confuse between *uiresume* and *uirestore* after using their counterparts earlier in the code — odd things might happen if we do so. Also, note that *uiresore* has a minor bug in that it overwrites the figure's **WindowScrollWheelFcn** callback with the **WindowKeyPressFcn** value.[†]

I created a convenience utility called *enableDisableFig*, available on the MATLAB File Exchange,[28] for setting figure modality based on `jFrameProxy`'s **Enabled** property.

Setting the **Modal** flag, or the figure handle's **WindowStyle** property to 'modal' is safer than using **Enabled** (because the window can be closed), and can often answer the requirements.[29] While a modal window is less powerful than a disabled window in some aspects (e.g., modal windows enable selection and closing), it has the advantage of safety and of locking the MATLAB desktop for input or selection. Users who wish can set both properties, thereby gaining an added level of window modality, but if there is a bug somewhere, then they will need the Operating System's task-manager to kill MATLAB, since neither the window nor the MATLAB desktop will be accessible.

MaximumSize and **MinimumSize** control the limits to which the window may be resized, either interactively or programmatically:[‡]

```
% Enable window sizes between 300x200 and 400x300:
newMaxBounds = java.awt.Dimension(400,300);    % width, height
jFrameProxy.setMaximumSize(newMaxBounds);
jFrameProxy.setMinimumSize(java.awt.Dimension(300,200))

% Clear all limitations
set(jFrameProxy,'MaximumSize',[],'MinimumSize',[]);
```

Minimized and **Maximized** flags control the window state accordingly. This feature has long been a requested feature of MATLAB figures, which is still not exposed by its HG handle,

---

[†] http://www.mathworks.com/support/bugreports/646025 (or http://bit.ly/drhhqa); apparently fixed in R2011a (MATLAB 7.12).

[‡] Even when maximized, windows appear no larger than their stated **MaximumSize**.

but can easily be done via Java.[†] Note that these properties are Java booleans which accept true/false (or 1/0), not the regular MATLAB 'on'/'off':

```
% Three alternative possibilities of setting Minimized:
jFrameProxy.setMinimized(true);
set(jFrameProxy,'Minimized',true);
jFrameProxy.handle.Minimized = true;
```

When either the **Maximized** or **Minimized** properties are changed back to false, the window is restored to regular mode (its **RestoredLocation** and **RestoredSize**).

Maximized and **Minimized** are mutually-exclusive, meaning that no more than one of them can be 1 (or true) at any time. This is automatically handled by jFrameProxy — users only need to be aware that a situation in which a window is both maximized and minimized at the same time is impossible (duh!).

Some MATLAB releases have the **Maximized** and **Minimized** properties in jFrame itself. However, because not all releases have this property,[30] it is advisable to always use the corresponding properties on jFrameProxy — this should be much more portable since these are standard java.awt.Window properties, whereas jFrame is an internal unsupported MATLAB object that is prone to change between releases.

The **RestoredLocation** (a java.awt.Point(x,y) object) and **RestoredSize** (a java.awt.Dimension(width,height) object) properties control the position and size of the figure window after un-maximizing or un-minimizing.

The **FocusableWindowState** flag (default = true) controls whether the window can become the active window. Clearing this flag is the standard mechanism for windows used as a floating palette or toolbar and thus should be nonfocusable.[31] Setting the flag on a visible window can have a delayed effect on some platforms: the actual change may happen only when the window becomes hidden and then visible again. To ensure consistent behavior across platforms, set the window's focusable state when it is invisible and then show it. Note that the similarly-named property **Focusable**, inherited from java.awt.Component, appears to have no effect at the window level.

The read-only **Focused** flag can be read[‡] in run-time (e.g., within callback functions) to determine whether or not the window is currently in focus. Actions such as requesting focus (jFrameProxy.*requestFocus()*) can then be applied accordingly.

StatusBar, **StatusBarText**, and **StatusBarVisible** were presented in detail in Section 4.7 and shall not be repeated here.

jFrameProxy also contains several properties which are reflected in the HG handle: **Title** (equivalent to HG's **Name** property), **Modal** (equivalent to **WindowStyle** = 'modal'), **Resizable** (equivalent to **Resize**), **Bounds** (equivalent to **Position**), and so on.

[†] Spawning several submissions on the File Exchange that do exactly this. Also see http://www.mathworks.com/matlabcentral/newsreader/view_thread/82958 (or http://tinyurl.com/cfa4wt).

[‡] Like all Java flags, using *get*(jFrameProxy,*'Focused'*) or jFrameProxy.*isFocused()*, not jFrameProxy.*getFocused()*.

Among the nonproperty-related methods, we should note *transferFocus()* and *transferFocusBackward()* that act like interactive Alt-Tab and Alt-Ctrl-Tab, transferring focus to the next/previous window. Also note *resize(width,height)* and *requestFocus()*.

Additional nonstandard yet self-explanatory callbacks exposed by `jFrameProxy` are:

- **WindowActivatedCallback,**   **WindowDeactivatedCallback**
- **WindowOpenedCallback,**   **WindowClosedCallback,**
  WindowClosingCallback
- **WindowIconifiedCallback,**   **WindowDeiconifiedCallback**
- **WindowGainedFocusCallback,** **WindowLostFocusCallback**
- **WindowStateChangedCallback**

7.4 BeanAdapters

As explained in Section 3.4, MATLAB automatically generates Java-bean adapter wrapper objects for MATLAB handles. For example,

```
>> h = uicontrol('string','click me!'); % a simple button
>> class(java(handle(h)))
ans =
uicontrolBeanAdapter0
```

As mentioned in Section 3.4, these wrapper objects should not be confused with the actual Java peer objects described earlier in this chapter. The adapter objects expose little more than the basic functionality of the MATLAB handles. On the other hand, the peer objects (obtainable using the *findjobj* utility) are much more powerful, exposing significantly more properties, callbacks, and methods.

Let us take a look at a simple button, for example:[32]

```
h = uicontrol('string','click me!');  % a simple button

% BeanAdapter:
hj1 = java(handle(h));
disp(hj1.class);              % => uicontrolBeanAdapter0
m1 = methods(hj1);            % => 140 methods
p1 = fieldnames(get(hj1));    % => 52 properties

% Java HG peer:
hj2 = findjobj(h);
disp(hj2.class);              % =>
      javahandle_withcallbacks.com.mathworks.hg.peer.PushButtonPeer$1
m2 = methods(hj2);            % => 347 methods
p2 = fieldnames(get(hj2));    % => 152 properties
```

References

1. http://www.UndocumentedMatlab.com/blog/JavaFrame (or http://bit.ly/97M1Cb).
2. For example, http://www.mathworks.com/matlabcentral/newsreader/view_thread/151005#379698 (or http://bit.ly/cCUjzk); http://www.mathworks.com/matlabcentral/newsreader/view_thread/155823 (or http://bit.ly/dmzSK2).

3. http://www.mathworks.com/help/techdoc/creating_plots/f5-41409.html (or http://bit.ly/8d5xi2).
4. http://www.mathworks.com/matlabcentral/newsreader/view_thread/126957#321453 (or http://bit.ly/8zJaYi).
5. http://www.mathworks.com/matlabcentral/fileexchange/16650 (or http://tinyurl.com/cjw5la).
6. http://www.mathworks.com/matlabcentral/newsreader/view_thread/161302 (or http://tinyurl.com/c8x42g); http://www.mathworks.com/matlabcentral/newsreader/view_thread/145846 (or http://tinyurl.com/db5bjf); http://www.mathworks.com/matlabcentral/newsreader/view_thread/99285 (or http://tinyurl.com/d8c64v); http://www.mathworks.com/matlabcentral/newsreader/view_thread/94240 (or http://tinyurl.com/d8kbz4).
7. For example, http://www.mathworks.com/matlabcentral/newsreader/view_thread/147025 (or http://tinyurl.com/c6hmhw).
8. A few examples on the File Exchange: http://www.mathworks.com/matlabcentral/fileexchange/2041 (or http://bit.ly/inId6a); http://www.mathworks.com/matlabcentral/fileexchange/31437 (or http://bit.ly/jxWkk6); http://www.mathworks.com/matlabcentral/fileexchange/3434 (or http://bit.ly/lNwF97).
9. http://www.mathworks.com/matlabcentral/newsreader/view_thread/73110 (or http://tinyurl.com/c93vno).
10. http://java.sun.com/developer/TechTips/1998/tt0909.html#tip1 (or http://tinyurl.com/yapzb74).
11. http://www.mathworks.com/matlabcentral/fileexchange/14317 (or http://tinyurl.com/bnprwc).
12. http://www.mathworks.com/matlabcentral/newsreader/view_thread/143275 (or http://bit.ly/5XPiv1).
13. http://UndocumentedMatlab.com/blog/findjobj-find-underlying-java-object/ (or http://bit.ly/8YFRnE).
14. http://UndocumentedMatlab.com/blog/findjobj-gui-display-container-hierarchy/ (or http://bit.ly/aZnrW9).
15. http://www.mathworks.com/support/solutions/en/data/1-3Y3C84/ (or http://bit.ly/7AD4QO).
16. http://java.sun.com/javase/6/docs/api/javax/swing/OverlayLayout.html (or http://tinyurl.com/ddwvf2).
17. http://UndocumentedMatlab.com/blog/uicontrol-callbacks/#comment-22139 (or http://bit.ly/bexuYP).
18. The technical differences between heavyweight and lightweight Java components are explained here: http://java.sun.com/products/jfc/tsc/articles/mixing/index.html (or http://tinyurl.com/2cztz7); http://java.sun.com/developer/technicalArticles/GUI/mixing_components/index.html (or http://bit.ly/6S4TjE).
19. http://java.sun.com/docs/books/tutorial/uiswing/components/menu.html (or http://tinyurl.com/dsxgl); or in www.java2s.com: http://www.java2s.com/Tutorial/Java/0240__Swing/0400__JMenuBar.htm (or http://tinyurl.com/c4k43v), http://www.java2s.com/Tutorial/Java/0240__Swing/0380__JMenu.htm (or http://tinyurl.com/cjk7wg) and http://www.java2s.com/Tutorial/Java/0240__Swing/0420__JMenuItem.htm (or http://tinyurl.com/dacdce).
20. http://www.mathworks.com/support/solutions/en/data/1-A6XFET/ (or http://bit.ly/dbK2MU).
21. http://UndocumentedMatlab.com/blog/blurred-matlab-figure-window/ (or http://bit.ly/hT4e86).
22. Sample MATLAB usage: http://www.mathworks.com/matlabcentral/newsreader/view_thread/128026 (or http://bit.ly/bFzTju); http://www.mathworks.com/matlabcentral/newsreader/view_thread/155685 (or http://bit.ly/cbb54A).
23. For example, http://www.mathworks.com/matlabcentral/fileexchange/8642 (or http://tinyurl.com/cj362s) or http://www.mathworks.com/matlabcentral/fileexchange/20694 (or http://tinyurl.com/cwrtmc). Note that some other submissions (e.g., http://www.mathworks.com/matlabcentral/fileexchange/17166, http://www.mathworks.com/matlabcentral/fileexchange/14103 or http://www.mathworks.com/matlabcentral/fileexchange/11684) are simply wrappers for the **AlwaysOnTop** Java property.
24. http://java.sun.com/docs/books/tutorial/uiswing/components/frame.html#windowevents (or http://bit.ly/cH93xG).
25. http://java.sun.com/javase/6/docs/api/javax/swing/JFrame.html#setDefaultCloseOperation(int) (or http://bit.ly/aefKo9).
26. http://www.mathworks.com/matlabcentral/newsreader/view_thread/293841#849882 (or http://bit.ly/nIVQV2).
27. http://www.mathworks.com/matlabcentral/newsreader/view_thread/284932#755936 (or http://bit.ly/cB9l2d).
28. http://www.mathworks.com/matlabcentral/fileexchange/15895 (or http://bit.ly/gR25Ok).
29. http://www.mathworks.com/matlabcentral/newsreader/view_thread/247071 (or http://tinyurl.com/ddxn9c).
30. For example, http://www.mathworks.com/matlabcentral/newsreader/view_thread/151005#379698 (or http://bit.ly/cCUjzk); http://www.mathworks.com/matlabcentral/newsreader/view_thread/155823 (or http://bit.ly/dmzSK2).
31. http://java.sun.com/javase/6/docs/api/java/awt/Window.html#setFocusableWindowState(boolean) (or http://bit.ly/dp4XJI).
32. http://www.mathworks.com/matlabcentral/newsreader/view_thread/270589#709652 (or http://bit.ly/5vVnYt).

Chapter 8

The MATLAB® Desktop

The MATLAB integrated development environment (IDE), commonly known as the MATLAB Desktop, has continuously evolved since its first Java-based MATLAB 6.0 (R12) appearance back in 2000. The Desktop now includes an integrated Editor, Command Window, Command History, Profiler, Help Browser, and other supporting tools.

In this chapter, I will discuss several customizations that can be done to these windows and tools. It should be noted that the Desktop code-base is very extensive, so there may be many customizations still undiscovered. Moreover, the Desktop remains under constant MathWorks development, each new release adding new features. We should also keep in mind that undocumented aspects sometimes change in new releases, and this is especially true for Desktop-related features.

On the brighter side, the entire Desktop and related tools is nowadays Java-based. This means that with entry hooks, that will be described shortly, we have a very wide range of possible customizations that can be achieved, as the following pages will show. We are only limited by the time we can spare for investigation and trials.

We start by retrieving the Desktop's Java handle. This can be done using several alternatives, which are used by several built-in MATLAB functions:[†]

```
>> jDesktop = com.mathworks.mde.desk.MLDesktop.getInstance
jDesktop =
com.mathworks.mde.desk.MLDesktop@e80d28

>> jDesktop = com.mathworks.mlservices.MatlabDesktopServices.getDesktop;

>> jDesktop = get(get(handle(gcf),'JavaFrame'),'Desktop');'
```

Note that all these methods only work in MATLAB 7.0 and later releases. For MATLAB 6 versions, we need to use a slightly different approach:[‡]

```
>> jDesktop = com.mathworks.ide.desktop.MLDesktop.getMLDesktop;
```

In this chapter, I will mainly use the MATLAB 7 object, as the MATLAB 6 object is much less versatile and customizable. MATLAB 6 users can use the ideas presented in this chapter to find corresponding functionality in their system, if it is available.

8.1 Desktop Functionality and Layout

8.1.1 The Java Desktop Object

All MATLAB windows, panels, figures, and Editor documents, are called *Clients*. These Clients are all part of a containing *Group*, into which they can be docked. The default Groups are 'Editor', 'Figures', 'Web Browser', 'Variable Editor', 'File and Directory Comparisons'

[†] For example, %matlabroot%/toolbox/matlab/general/desktop.m and %matlabroot%/toolbox/matlab/uitools/uiopen.m.

[‡] For a discussion of the **JavaFrame** property, see Chapter 7.

(or: 'Comparison Tool'), and 'special figures'. Even when undocked, Clients remain part of their Group. This enables group-wide actions to be performed, such as closing the entire group, or bringing the group's last used Client into focus.

Some Clients are special — they can only exist once within a group. Example of this is the Desktop's Command Window, or the Figure group's plot-editing panels. These special Clients are called *Singletons* and have dedicated handling methods.

These are some of `jDesktop` object's methods, many of which are Client and/or Group-related:

- *activate()* — brings the MATLAB Desktop into focus. It does not change the Desktop's currently active window, unlike *showCommandWindow()* et al.

- *addGroup(...), addClient(...)* — discussed in Section 7.1.2.

- *attemptClose(), attemptMainFrameClose()* — close the main Desktop window; equivalent to **exit/quit**.

- *canClose()* — returns a flag (normally true) indicating whether or not the main Desktop can be closed.

- *canHaveMainFrame()* — returns a flag (normally true) indicating whether or not the main Desktop is displayable.

- *cascadeDockedDocuments(groupName)* — causes all docked documents within the specified group (e.g., 'Editor' or 'Figures') to become cascaded (as opposed to maximized, for example).

- *closeClient(client)* — closes the specified client window (an Editor document or a Figure window), if it is found. *client* is the Java Frame/Document's handle, or its name (e.g., the figure title, Editor document filename or panel name[†]). *closeClient(clientName, groupName)* only closes the window if it is contained in the specified docking group, thereby limiting accidental closures. There are also similar *removeClient(...)* methods.

- *closeGroup(groupName)* — closes the specified group, if it is currently open (displayed), and all figures attached to it (not necessarily docked).[2] *closeGroup(groupName, dockedOnlyFlag, flag)* — *dockedOnlyFlag* specifies whether to close undocked group figures/documents; I do not know the purpose of the second *flag*. There are also similar *removeGroup(...)* methods.

- *closeGroupSingletons(groupName)* — closes singleton clients belonging to the specified group. Used by the built-in **plottools** function to close plot-editting panels. Also see: *getSingletonTitles(), isSingleton(), restoreGroupSingletons()*.

- *closeCommandHistory(), closeCommandWindow(), closeFileBrowser(), closeFileExchange(),[‡] closeHelpBrowser(), closeMainFrame(), closeProfiler(), closeProjectExplorer(),[§] closeWorkspaceBrowser()* — closes the corresponding window/panel, if it

---

[†] 'Command Window', 'Command History', 'Current Directory', 'Workspace', 'Help,' 'Profiler', 'Figure Palette', 'Plot Browser', or 'Property Editor'.

[‡] *closeFileExchange()* was added in R2009b (MATLAB 7.9).

[§] *closeProjectExplorer()* was removed in R2008b (MATLAB 7.7).

is currently open (displayed). *closeMainFrame(flag)* indicates whether or not to save the desktop layout.

- *disableThreadSafeGetMethods()* — disable Command Window warnings and prevention when trying to invoke non-thread-safe (EDT) methods. See Section 3.2 (The Event Dispatch Thread) for a detailed discussion. Corresponding *enableThreadSafeGet-Methods()* reverses this thread-unsafe behavior, and *areThreadSafeGetMethods-Enabled()* returns the current state (default=false).

- *getClearCommandAction( ), getClearHistoryAction( ), getClearWorkspaceAction( )* — return a `javax.swing.Action`[3] object handle. Accelerator keystrokes can be set via *setAccelerator, setAcceleratorSequence.*

- *getClient(clientName, groupName)* — returns a handle to the requested client window, optionally specifying its containing group; *getTitle(clientHandle)* and *getShort-Title(clientHandle)* return the client's corresponding window name in the Editor, with and without the file path, respectively.

- *getClientByName(internalName)* returns the handle to the client that has this *internal-Name* (the name returned by its *getName()* method or **Name** property).

- *getClientGroup(clientName)* returns the client's containing group name.

- *getClientLocation(clientHandle)* — returns a `com.mathworks.mwswing.desk.DTLocation` object specifying the client's location, docking state, restored locations, and other similar location data (see Section 8.1.3); *getGroupLocation(groupName)* returns a similar object for the specified docking group; *getLastUndockedLocation (clientHandle)* returns a similar object of the client's location when undocked (or [] if it was never undocked).

- *getClientShortTitles()* — returns an array of all the open client names; *getClientTitles()* returns the same list with the clients' full names (this includes path for open Editor documents and the Current Directory). **ClientShortTitles** and **ClientTitles** are the corresponding read-only (un-settable) properties.

- *getCommandWindowHWND()* — returns the Desktop window's HWND value, possibly useful for interfacing MATLAB and other applications.[†] This is the same value as returned from `jDesktop`.*getMainFrame().getHWnd().*[‡]

- *getDocumentArrangement(groupName)* — returns a numeric value indicating how client windows/documents are arranged within the specified group. Possible values are 1 (maximized), 2 (tiled) or 3 (floating/cascaded). *setDocumentArrangement(group-Name,value,size)*[§] updates the setup:

```
% Set a 2x3 titled Editor arrangement
```

[†] Similar to the figure window's HWND value, discussed in Section 7.1.

[‡] Note the different spelling: HWND vs. HWnd — Java is case sensitive so be careful to use the correct spelling in each case.

[§] *Size* is a `java.awt.Dimension` object specifying table in tiles: `java.awt.Dimension (numCols,numRows)`. It is only relevant for tiled documents, so set *size* = [] for maximized (1) or floating/cascaded (3) arrangement values. See http://www.mathworks.com/matlabcentral/newsreader/view_thread/155225#411327 (or http://bit.ly/6sjxUg).

```
jDesktop.setDocumentArrangement('Editor',2,java.awt.Dimension(2,3));
jDesktop.setDocumentArrangement('Editor',2,[]); %reuse previous size
```

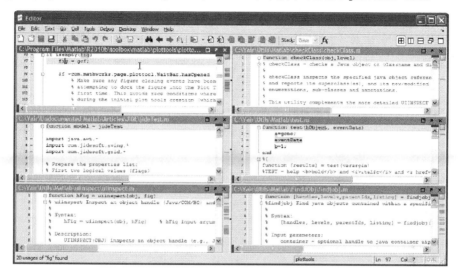

2 × 3 tiled document arrangement

- *getDocumentTiledDimension()* returns the Dimension of the group tiles table when arranged as titled, or a 1 × 1 dimension otherwise:

```
>> jDesktop.getDocumentTiledDimension('Editor')
ans =
java.awt.Dimension[width=2,height=3]
```

- *getDocumentBarPosition(groupName)* — returns a numeric value indicating how the client document bar, which contains buttons with the names of all the clients contained in the group, is arranged within the specified group. Possible values are 1 (top), 3 (right), 5 (bottom), 7 (left), and −1 (hidden). *setDocumentBarPosition(groupName, value)* updates the bar position.

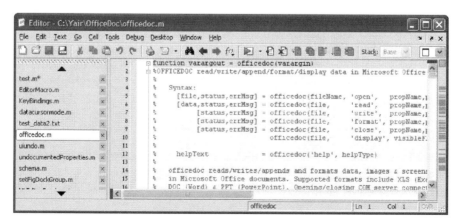

Left document bar position

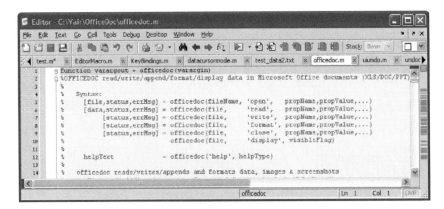

Top document bar position

- *getDocumentContainment()* — This returns the number 2. I am uncertain what this represents, but I suspect that it may be related to jDesktop's *PER_FRAME* (=1) and *PER_GROUP* (=2) static fields. There is also a corresponding *setDocument-Containment(value)* method and a **DocumentContainment** property.

- *getContainingFrame(clientHandle)* — returns a handle to the client's Java container, a com.mathworks.hg.peer.FigureClientProxy$FigureDTClientBase object. This object is actually mid-level in the Frame hierarchy, as shown in Section 7.2. It can be used to access the figure's Java sub-components in lieu of the **JavaFrame** property (see Chapter 7 — The Java Frame).

- *getFrameContainingGroup(groupName)* — returns a handle to the Java Frame that contains the specified group. To get the Editor frame's handle, for example, run jDesktop.*getFrameContainingGroup('Editor')*, which is equivalent to: jDesktop.*getGroupContainer('Editor').getTopLevelAncestor*.

- *getGroup(clientHandle)* — returns the group name of the group that contains the specified client window/panel.

- *getGroupContainer(groupName)* — returns an object handle similar to that returned by *getFrameContainingGroup(groupName)*, only lower down the Frame hierarchy.

- *getGroupContainerInsets(groupName)* — returns a java.awt.Insets[4] object that contains the internal margins of the group container. For example, setting the **DocumentBarPosition** (see above) value to 1 (top) will set the top inset to 21 and the rest to 0; setting the value to 7 (left) will set only the left inset; *getInternalFrame-Insets(clientHandle)* returns the corresponding client insets.

- *getGroupLocation(groupName)* — see *getClientLocation()* discussed above.

- *getGroupMembers(groupName)* — returns an array of handles of the client windows/panels contained within the specified group. If the group name is not found, then the Desktop's client handles are returned:

```
>> jDesktop.getGroupMembers('Editor')
ans =
```

```
java.awt.Component[]:
    [com.mathworks.mde.editor.EditorViewContainer]
    [com.mathworks.mde.editor.EditorViewContainer]
    [com.mathworks.mde.editor.EditorViewContainer]

>> % Remember: group names are case-sensitive !!!
>> jDesktop.getGroupMembers('editor') % not found, use 'Desktop'
ans =
java.awt.Component[]:
    [com.mathworks.mde.cmdwin.CmdWin              ]
    [com.mathworks.mde.cmdhist.CmdHistoryWindow  ]
    [com.mathworks.mde.workspace.WorkspaceBrowser]
    [com.mathworks.mde.help.HelpBrowser          ]
```

- *getGroupName()* — returns the Desktop's docking group name (='Desktop'). The corresponding read-only (unsettable) property is **GroupName**.

- *getGroupTitles()* — returns the titles of all available groups. There is a corresponding read-only (un-settable) **GroupTitles** property.† The titles are returned as a Java array of Strings; to convert to a MATLAB cell-array we can simply use the *get()* mechanism, or use the built-in *cell* function to cast the Java Strings array into a MATLAB char cell array:

```
>> jDesktop.getGroupTitles          % returns a Java String object array
ans =
java.lang.String[]:
    'Editor'
    'Figures'
    'Web Browser'
    'Variable Editor'
    'File and Directory Comparisons'

>> get(jDesktop,'GroupTitles')      % returns a Matlab char cell-array
ans =
    'Editor'
    'Figures'
    'Web Browser'
    'Variable Editor'
    'File and Directory Comparisons'

>> groupNames = cell(jDesktop.getGroupTitles);      % equivalent
>> groupNames = jDesktop.getGroupTitles.cell;       % equivalent
```

- *getInternalFrameInsets(clientHandle)* — see *getGroupContainerInsets* above.

- *getLastDocumentSelectedInGroup(groupName)* — returns the handle of the latest client that was selected (accessed) in the specified group. This would normally indicate the currently active (in-focus) document/figure/panel.

- *getLastUndockedLocation(clientHandle)* — see *getClientLocation()* above.

---

† Note that some of these titles were changed between MATLAB releases. For example, 'Array Editor' => 'Variable Editor'.

- *getLayoutSavePolicy()* — returns a `com.mathworks.mwswing.desk.Desktop$LayoutSavePolicy` object representing when Desktop layout changes are saved; *setLayoutSavePolicy(layout)* updates the policy to one of `LayoutSavePolicy`'s static (predefined) layouts: *NEVER, UPON_EXIT* or *UPON_CHANGE* (=default). There is also a corresponding read–write (settable) **LayoutSavePolicy** property. See *initMainFrame(...), restoreLayout(...)* and Section 8.1.3 for additional details.

- *getMainFrame()* — returns a handle to the Desktop's Java Frame, a `com.mathworks.mde.desk.MLMainFrame` object. We can use this object to access specific Desktop components. This is described in detail in Section 8.1.2.

- *getMajorVersion(), getMinorVersion()* — returns the internal Desktop version (3 and 0 respectively on my system, regardless of the MATLAB release).

- *getSelected()* — returns a handle to the currently selected client:

```
>> jDesktop.getSelected
ans =
com.mathworks.mde.cmdwin.CmdWin[cw_DTClientBase,0,0,1074x731,...]
```

- *getSelectedGroup()* — returns the name of the currently selected group, or [] if the selected group is the Desktop; *getSelectedInGroup(groupName)* returns the client handle currently selected within the specified group.

- *getShortTitle(clientHandle), getTitle(clientHandle)* — see *getClient()*

- *getSingletonShortTitles(), getSingletonTitles()* — returns the list of titles of all singleton clients. There are also corresponding properties. See also: *isSingleton(), closeGroupSingletons(), restoreGroupSingletons().*

```
>> jDesktop.getSingletonTitles
ans =
java.lang.String[]:
    'Command Window'
    'Command History'
    'Current Directory'
    'Workspace'
    'Help'
    'Profiler'
    'Figure Palette'
    'Plot Browser'
    'Property Editor - Figure'
```

- *getToolBarRegistry()* — returns an object containing information about the current configuration of the Main ('MATLAB') and Editor ('Editor' and 'Cell mode'/'Codepad') toolbars and their contained buttons/controls. Also see *showToolBarCustomization-Panel()* below.

- *getWindowRegistry()* — returns the list of currently open non-Desktop client windows (i.e., Editor documents and figure windows).

- *groupToFront(groupName)* — brings the specified group into focus.

- *hasClient(client)* — returns a flag indicating whether or not the specified client handle or name exists; *hasClient(clientName,groupName)* limits the check to those clients contained within the specified group.

- *hasGroup(groupName)* — returns a flag indicating whether a group exists.

- *hasMainFrame()* — returns a flag indicating whether or not the Desktop is currently in use. This is what the built-in ***desktop('-inuse')*** returns internally.

- *hideClient(client)* — hides (closes) the specified client (by handle or name) if it exists; *hideClient(clientName,groupName)* limits the action to a client contained within the specified group. Compare: *showClient()* below.

- *initMainFrame(splashFlag, restoreLayoutFlag, minimizedFlag)* — starts the main Desktop frame, if it is not already shown; *splashFlag* controls whether or not the splash screen should appear;[†] optional *restoreLayoutFlag* (default =true) indicates whether to reuse the last-used layout or the default one; optional *minimizedFlag* (default=false) controls whether the Desktop should start in minimized mode. The built-in ***desktop*** and ***desktop('-norestore')*** actually call *initMainFrame* internally. See *getLayoutSavePolicy(..)*, *restoreLayout(...)* and Section 8.1.3 for other related layout methods.

- *isClientDocked(), isClientHidden(), isClientMaximized(), isClientMinimized(), isClientSelected(), isClientShowing(), isClientUnfurled()* — all these check a client handle or name for the specified state.[‡] These methods also accept an optional second *groupName* argument. See *plottools.m* for sample usage.

- *isGroupDocked(groupName), isGroupMaximized(...), isGroupMinimized(...), isGroupSelected(...), isGroupShowing(...), isGroupUnfurled(...)* — similar to the corresponding Client methods but only accepting a single *groupName* parameter. For some reason there is no corresponding *isGroupHidden* method.

- *isSingleton(clientName)* — returns a flag indicating whether or not the client is a singleton; *isDocument(clientName)* apparently returns the reverse flag. See also: *getSingletonTitles(), closeGroupSingletons(), restoreGroupSingletons()*.

- *minimizedDockedDocuments(groupName)* — minimizes all clients which are currently docked in the specified group.[§]

- *removeClient(client), removeGroup(groupName)* — see the corresponding *closeClient(), closeGroup()* methods.

- *restoreGroupSingletons(groupName)* — opens the singleton clients attached to the specified group. Used by the built-in ***plottools*** function to open plot-editting panels. See also: *getSingletonTitles(), closeGroupSingletons()*.

- *restoreLayout(layoutName)* — restores the specified layout; *saveLayout(layoutName)* saves the current layout under the specified name; *restorePreviousLayout()* is pretty much self-explanatory. See Section 8.1.3 for additional information regarding desktop layouts.

[†] This corresponds to the -nosplash startup option; note that the splash screen only appears after a predefined timeout.

[‡] *isClientHidden()* only accepts a client handle; all the others accept client handle, clientName or clientName with groupName.

[§] *minimizedDockedDocuments* appears to be a naming typo: should be *minimizeDockedDocuments*.

- *setClientDocked(client, groupName, dockedFlag)* — sets the specified client (*handle*, or *clientName* with optional *groupName*) docking state.
- *setClientLocation(client, groupName, location)* — sets the specified client (*handle*, or *clientName* with optional *groupName*) location (a com.mathworks.mwswing. desk.DTLocation object — see Section 8.1.3).
- *setClientMaximized(client, groupName, maximizedFlag)* — sets the specified client (*handle*, or *clientName* with optional *groupName*) window maximization state (true=maximized; false=regular).
- *setClientMinimized(client, groupName, minimizedFlag, dockPosition)* — sets the specified client (*handle*, or *clientName* with optional *groupName*) window minimization state (true=minimized; false=regular). The optional *docPosition* value indicates the minimized position when the group is docked: 1 (top), 3 (right=default), 5 (bottom), 7 (left) or 0 (center — fills entire container group).
- *setClientSelected(client, groupName, selectedFlag)* — sets the specified client (*handle*, or *clientName* with optional *groupName*) selection state, effectively bringing the client into active focus if *selectedFlag* = true.
- *setDefaultDesktop(), setDefaultLayout() setCommandAndHistoryLayout(), setCommandOnlyLayout()* — see Section 8.1.3 for details.
- *setDocumentColumnWidths(groupName, relativeWidths)* — if the group's document arrangement (see *getDocumentArrangement* above) is tiled, this method sets the relative widths of the tile columns; *setDocumentRowHeights* does a similar action in the vertical direction, spanning rows. For example,[5]

```
jDesktop.setDocumentArrangement('Editor',2,java.awt.Dimension(3,4));
jDesktop.setDocumentColumnWidths('Editor',[.6,.2,.1,.1]);
jDesktop.setDocumentRowHeights('Editor',[.3,.5,.2]);
jDesktop.setDocumentColumnSpan('Editor',1,1,2);
```

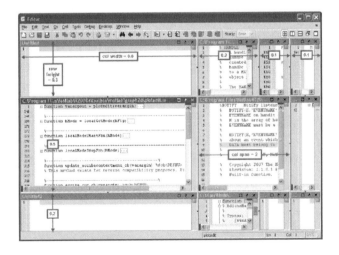

Document cell width, height and span example (See color insert.)

- *setDocumentColumnSpan(groupName, rowNum, colNum, spanNum)* — if the group's document arrangement (see *getDocumentArrangement* above) is tiled, this method sets the tile at the specified row/column to span the specified number of tile columns to the right. Note that row and column numbers start at 0 (top left) and the minimal *spanNum* is 1.

- *setGroupDocked(groupName, flag), setGroupMaximized(groupName, flag), setGroupMinimized(groupName, flag), setGroupLocation(groupName, location)* — sets the requested state flag or *location* value for the specified group. Groups are always docked to their parent (the Desktop). Restoring the group's appearance is done by setting *flag* = false. Like *setClientMinimized*, *setGroupMinimized* also accepts a third optional numeric value indicating the minimized position when the group is docked: 1 (top), 3 (right = default), 5 (bottom), 7 (left), or 0 (center — fills entire container group).

- The top-level window of docked groups is the Desktop. So, unless we undock the group, operations done to its window will affect the entire desktop:[6]

```
container = jDesktop.getGroupContainer('Figures').getTopLevelAncestor;
container.setSize(width,height); % e.g., (500,300)
container.setAlwaysOnTop(true); % or false to return to normal
container.setMaximized(true); % or false to return to normal
container.setMinimized(true); % or false to return to normal
container.setVisible(true); % or false to hide: ignore the Java error
```

- *setStatusText(text)* — displays the specified text in the Desktop's status bar. There is also a write-only (un-gettable) **StatusText** property:

```
jDesktop.setStatusText('testing 123...');
set(jDesktop, 'StatusText', 'testing 123...'); % an alternative
```

Note that the statusbar text is only settable if the Desktop frame is visible. It is therefore prudent to test this in advance, as the built-in *uiopen* function does:

```
dt = javaMethod('getInstance', 'com.mathworks.mde.desk.MLDesktop');
if dt.hasMainFrame
    dt.setStatusText(message);
else
    disp(message);
end
```

- *showClient(clientName, groupName, location, selectFlag)* — shows the requested client; optional *groupName* limits the action to clients in the specified group; optional *location* indicates the client's new DTLocation ([] indicates "do not change"; see Section 8.1.3); optional *selectFlag* indicates whether the client should be selected (brought into focus) after display. Note related method *showClient(clientHandle, location, selectFlag)* that accepts a *clientHandle* rather than a *clientName*; *toFront(clientHandle)*; and *showClientHidden(clientHandle,location,selectFlag)*.

- *showCommandWindow()* — brings the MATLAB Desktop's Command Window into focus, displaying it if it was hidden or minimized. This is what the built-in **command-window** function does internally. As an alternative, run **uimenufcn(0,'WindowComm-andWindow').**[†]

- *showCommandHistory()* — brings the Command History panel into focus. Alternatively, use the following (used by the builtin **commandhistory** function):

```
com.mathworks.mde.cmdhist.CmdHistoryWindow.invoke;
```

- *showFileBrowser()* — brings the File Browser panel/window into focus. Alternatively, use the following (used by the built-in **filebrowser** function):

```
com.mathworks.mde.filebrowser.FileBrowser.invoke;
```

- *showHelpBrowser()* — brings the Help Browser panel/window into focus. Alternatively, use the following (used by the built-in **helpbrowser** function):

```
com.mathworks.mlservices.MLHelpServices.invoke;
```

- *showProfiler()* — brings the Profiler panel/window into focus. Alternatively, use the following (used by the built-in **profile** function):

```
com.mathworks.mde.profiler.Profiler.invoke;
```

- *showWorkspaceBrowser()* — brings the Workspace Browser panel/window into focus. Alternatively, use the following (used by the builtin **workspace** function):

```
com.mathworks.mlservices.MLWorkspaceServices.invoke;
```

- *showGroup(groupName, flag)* — brings the specified group into focus. *flag* must apparently be true for *showGroup* to have any effect. *showGroup* is used by many built-in MATLAB functions (**plottools** for example).
- *showToolBarCustomizationPanel(toolbarName)* — displays the Preferences panel responsible for toolbars configuration. Also see *getToolBarRegistry()*.

Some of these methods can also be invoked directly from `com.mathworks.mlser-vices.MatlabDesktopServices`, using static internal methods. For example,

```
com.mathworks.mlservices.MatlabDesktopServices.showCommandWindow;⁷
com.mathworks.mlservices.MatlabDesktopServices.closeCommandWindow;
com.mathworks.mlservices.MatlabDesktopServices.setCommandOnlyLayout;
```

[†] This alternative is used by **commandwindow** when Java and/or the desktop are not present, in a MATLAB session started with the '-nojvm' or '-nodesktop' option. **uimenufcn** is discussed in section 4.6.1. *showCommandWindow()* can also be invoked directly from `com.mathworks.mlservices.MatlabDesktopServices.show...()`. See a discussion of alternatives here: http://www.mathworks.com/matlabcentral/newsreader/view_thread/269716 (or http://bit.ly/6g7dX2).

8.1.2 *The Desktop Frame*

As seen in Section 8.1.1, the jDesktop.*getMainFrame()* method returns a handle to the Desktop Java Frame window (a com.mathworks.mde.desk.MLMainFrame object):

```
>> jDesktopFrame = jDesktop.getMainFrame;
jDesktopFrame =
com.mathworks.mde.desk.MLMainFrame[MainDesktopFrame,...]
```

Since this is also the first window created when MATLAB starts,[†] we can also retrieve this handle by other means:

```
jWindows = java.awt.Window.getOwnerlessWindows;
jDesktopFrame = jWindows(1);
jFrames = java.awt.Frame.getFrames;
jDesktopFrame = jFrames(1);
```

Note that it is not always assured that MLMainFrame would be the first Frame. For example, in R2011a it is actually the second.

This handle enables access to the entire Desktop GUI hierarchy. Before showing specific usage examples of internal hierarchy components, let us explore useful aspects of the Desktop Frame itself, at the window level.

Most of the Desktop Frame's functionality closely follows the figure frame's functionality discussed in Chapter 7 (The Java Frame). Therefore, only the highlights will be presented; interested readers are referred to Chapter 7 for additional details, or to investigate further using the ***uiinspect*** and ***findjobj*** utilities discussed above.

We begin with a long-demanded request to programmatically hide/restore the Desktop. This can be achieved by several alternatives using the Frame handle:[8]

```
jDesktopFrame.show;
jDesktopFrame.hide;

jDesktopFrame.show(flag);        % true = show, false = hide

jDesktopFrame.setVisible(flag); % or: set(jDesktopFrame,'Visible',...)
```

Similarly, we can minimize/maximize the entire Desktop window:[9]

```
jDesktopFrame.setMinimized(flag); % true = minimize; false = restore
jDesktopFrame.setMaximized(flag); % true = maximize; false = restore
```

Or hide its menu bar:[10]

```
jDesktopFrame.getJMenuBar.setVisible(flag); % true = show, false = hide
jDesktopFrame.getJMenuBar.repaint;
```

[†] Unless we used the -nojvm, -noawt and/or -nodesktop startup (command-line) options, in which case we have no Desktop Frame anyway... Those using -nodesktop, then starting the Desktop via the *initMainFrame* method, deserve their punishment.

Other useful methods (most of which have corresponding properties):

```
jDesktopFrame.getDesktop();
jDesktopFrame.getDesktopMenu();
jDesktopFrame.getJMenuBar();
jDesktopFrame.getMenuBar();
jDesktopFrame.getHWnd();
jDesktopFrame.getHeight();
jDesktopFrame.getWidth();
jDesktopFrame.getLocationOnScreen();
jDesktopFrame.getX();
jDesktopFrame.getY();
jDesktopFrame.resize(pixelWidth,pixelHeight);
jDesktopFrame.reshape(x,y, pixelWidth,pixelHeight);
jDesktopFrame.repaint();
jDesktopFrame.requestFocus();
jDesktopFrame.setAlwaysOnTop(flag);
jDesktopFrame.setBackground(java.awt.Color);
jDesktopFrame.setForeground(java.awt.Color);
jDesktopFrame.setCloseOnEscapeEnabled(flag);
jDesktopFrame.setCursor(value or java.awt.Cursor);
jDesktopFrame.setEnabled(flag);
jDesktopFrame.setFont(javaFont);
jDesktopFrame.setLocation(x,y);
jDesktopFrame.setRestoredLocation(x,y);
jDesktopFrame.setMaximumSize(java.awt.Dimension);
jDesktopFrame.setMinimumSize(java.awt.Dimension);
jDesktopFrame.setPreferredSize(java.awt.Dimension);
jDesktopFrame.setRestoredSize(java.awt.Dimension or x,y);
jDesktopFrame.setSize(java.awt.Dimension or x,y);
jDesktopFrame.setModal(flag);
jDesktopFrame.setResizable(flag);
jDesktopFrame.setSelected(flag);
jDesktopFrame.setStatusBar(com.mathworks.mwswing.MJStatusBar);
jDesktopFrame.setStatusBarVisible(flag);
jDesktopFrame.setStatusText(text);
jDesktopFrame.setTitle(text);
jDesktopFrame.toBack();
jDesktopFrame.toFront();
```

For example, the *setTitle()* method (or the corresponding **Title** property) was used by a CSSM user to set different titles to different MATLAB sessions that run concurrently on the same computer — setting separate titles enables easy distinction between these separate sessions;[11] The statusbar can be updated using the *setStatus*()* methods, as explained in Section 4.7; The desktop frames position and size can be controlled using the *set*Location()* and *set*Size()* methods; and so on.

In addition to these methods/properties, and the 30-odd standard callbacks (see Section 3.4), the Desktop frame has the following non-standard callbacks:

- **DragEnterCallback, DragExitCallback, DragOverCallback, DropCallback, DropActionChangedCallback** — used for Drag-&-Drop ops.
- **WindowIconifiedCallback, WindowDeiconifiedCallback** — these refer to minimization and unminimization operations.
- **WindowActivatedCallback, WindowDeactivatedCallback.**
- **WindowGainedFocusCallback, WindowLostFocusCallback.**
- **WindowOpenedCallback, WindowClosedCallback, WindowClosingCallback.**
- **WindowStateChangedCallback.**

The Desktop components hierarchy, accessible via the `jDesktopFrame` handle, should be handled with more care than the Figure Frame, for two reasons:

- The Figure Frame hierarchy is much less prone to change between MATLAB releases. There is much attention in recent releases to improvement of the Desktop functionality and usability, and this is expected to impact the internal Desktop hierarchy and sub-components, much more than for the Figure.
- The Desktop hierarchy is highly dependent on the Desktop state and layout, to a larger extent than the Figure frame hierarchy.

Having thus been fairly warned, there are quite a few interesting things we can do with internal Desktop components, which are discovered down the hierarchy tree.

Here is an example: In answer to a CSSM user request to hide the Desktop's lower edge (the panel that contains the "Start" button and status bar), we locate the status bar's parent panel and simply hide it:[12]

```
jDesktop.getMainFrame.getStatusBar.getParent.setVisible(false);
```

Further examples of using the Desktop object and its contained sub-components are shown in the rest of this Section 8.1.

8.1.3 *Organizing the Desktop Clients*

The MATLAB Desktop enables users to switch between different presentation layouts of the Desktop panels (Command Window, Workspace, etc.).[13] This has been supported as far back as MATLAB 6 (R12), with newer MATLAB releases adding improved functionality such as the ability to save user-defined layouts, as Kristin Thomas explained in the official MATLAB Desktop blog.[14]

The only supported way to save and switch layouts is to use the Desktop's main menu. Since Kristin has posted her write-up, a few people have posted unanswered follow-up comments requesting to know how to programmatically save and switch layouts.[15] I will now show how this can be done.

First, we need to get the Java handle of the MATLAB desktop. We can then investigate this handle using the built-in ***methodsview*** function or my ***uiinspect*** utility which was described above. We quickly see the relevant layout-related functions, which we can put to good use:

```
% Get the desktop's Java handle (Matlab 7 only)
jDesktop = com.mathworks.mde.desk.MLDesktop.getInstance;

% Inspect the available desktop functions
methodsview(jDesktop);
uiinspect(jDesktop);

% Get and set the layout-saving policy
oldPolicy = jDesktop.getLayoutSavePolicy;
set(jDesktop,'LayoutSavePolicy',
jDesktop.getLayoutSavePolicy.NEVER);

% Save the current layout
jDesktop.saveLayout('Yair');

% Switch between different layouts
jDesktop.restoreLayout('Yair');
jDesktop.restoreLayout('Default');
jDesktop.restoreLayout('History and Command Window');

% Switch between pre-defined layouts
jDesktop.setDefaultLayout();
jDesktop.setDefaultDesktop();
jDesktop.setCommandOnlyLayout();
jDesktop.setCommandAndHistoryLayout();
```

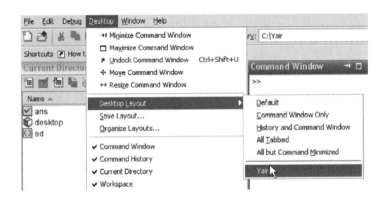

Desktop layout menu in MATLAB 7

The new "Yair" layout is stored as an XML file (*YairMATLABLayout.xml*) in the ***prefdir*** folder.[†] This layout file can be copied between users and MATLAB installations, with some care when copying onto an earlier-release installation.[16]

[†] For example, C:\Documents and Settings\Yair\Application Data\MathWorks\MATLAB\R2008a\YairMATLABLayout.xml on a Windows platform. It can be edited via: **edit**(fullfile(prefdir,'YairMATLABLayout.xml')).

Note that trying to restore an invalid layout name simply does nothing (without throwing an error). This may be misleading, so if we plan to change layouts programmatically (e.g., in startup scripts), then we should first check the existence of the layout file (using the ***dir*** or ***exist*** built-in functions).

Also note that the layout functionality relies heavily on unsupported and undocumented internal implementation, which may change without prior notice between MATLAB releases. The code snippet above works on several MATLAB 7 releases. But for MATLAB 6.0 (R12), for example, it needs to be modified, as follows:

```
% Get the desktop's Java handle (Matlab 6 only)
jDesktop = com.mathworks.ide.desktop.MLDesktop.getMLDesktop;

% Inspect the available desktop functions
% Note: in Matlab 6, methodsview() did not accept object handles
methodsview('com.mathworks.ide.desktop.MLDesktop');
%uiinspect(jDesktop); % UIINSPECT does not work on Matlab 6

% Save the current layout
% saving the desktop is not possible in Matlab 6

% Switch between different layouts
jDesktop.set5PanelLayout;
jDesktop.setTallLayout;
jDesktop.setShortLayout;
jDesktop.setDefaultDesktop;
jDesktop.setMolerMode;'       % ='Command window only'
```

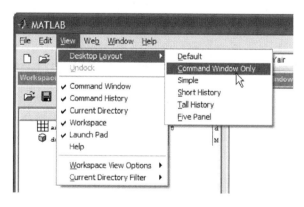

Desktop layout menu in MATLAB 6

Note that MATLAB 6 did not have a generic *restoreLayout()* function, instead using a few pre-defined *setXXX()*. Also note that MATLAB 6 did not have the *saveLayout()* function — it did however have *saveDesktop(string,string)* and *restoreDesktop(string,string)*, which I leave as an exercise to readers.

[†] This is presumably an internal-MathWorks reference to MATLAB's founder Cleve Moler.

Perhaps the main lesson to be learned from this sample customization is that essentially anything that can be done via the menu can also be done programmatically.

Using desktop layouts is easy and interactive, but if we want finer-grained control over the clients' location we need to use other methods, which in general require the use of a com. mathworks.mwswing.desk.DTLocation object:

```
com.mathworks.mwswing.desk.DTLocation.create('NW');
```

or of one of DTLocation's derivative classes: DTNestedLocation, DTFloating-Location, DTTiledLocation, DTBorderLocation.

8.1.4 Customizing the Desktop Toolbars

The MATLAB Desktop has two toolbars displayed by default beneath the main menu: the MATLAB Toolbar, and the Shortcuts Toolbar:

The top ("MATLAB") toolbar is a standard Java toolbar that can be customized similarly to figure toolbars (see Section 4.5).

Here is one particular use-case that is specific to the MATLAB Toolbar: A reader has approached me with a request to trap current-directory changes in order to update project-specific settings. To achieve this, we first locate the Desktop's "Current Directory" drop-down in the MATLAB Toolbar (this is done using the *findjobj* utility, which searches from jDesktopFrame downward). We then set this component's **ActionPerformedCallback** property:[†]

```
% First get a reference to the 'Current Directory' drop-down component
% Note that we use handle 0 = desktop
hCombo = findjobj(0,'class','CwdComponentSet$CustomComboBox');

% Set the component's ActionPerformedCallback property
set(hCombo, 'ActionPerformedCallback', 'disp([''New dir: '' pwd])');

% The following is typed in the Command Window
>> cd c:\          % or via the 'Current Directory' drop-down
New dir: c:\              <= this is presented by the callback
>>
```

[†] A different mechanism for monitoring current-folder changes is to use com.mathworks.jmi.MatlabPath (see Section 9.2.2), but apparently this requires writing some Java code to handle 'CWD_CHANGE' events and cannot be done with pure MATLAB code as in here.

Other button controls in the Desktop toolbar can be accessed via the toolbar container's reference handle:[17]

```
hMainFrame=com.mathworks.mde.desk.MLDesktop.getInstance.getMainFrame;
hToolbar = hMainFrame.getContentPane.getComponent(0).getComponent(0);
```

The bottom ("Shortcuts") toolbar is a reflection of the contents of the [*prefdir* '/shortcuts. xml'] file, which can be textually edited. The toolbar, which is also a Java toolbar, has special access functions in the com.mathworks.mlwidgets.shortcuts.ShortcutUtils class. For example, to add a new shortcut to the toolbar:[18]

```
name = 'My New Shortcut';
cbstr = 'disp(''My New Shortcut'')'; % will be eval'ed when clicked
iconfile = 'c:\path\to\icon.gif'; % default icon if it is not found
isEditable = 'true';
scUtils = com.mathworks.mlwidgets.shortcuts.ShortcutUtils;
category = scUtils.getDefaultToolbarCategoryName; % = 'Toolbar Shortcuts'
scUtils.addShortcutToBottom(name,cbstr,iconfile,category,isEditable);
```

To add a shortcut to the Help Browser (also known as a "Favorite"), simply set the category to scUtils.getDefaultHelpCategoryName (='Help Browser Favorites' on English-based MATLAB installations); to add the shortcut to the "Start" button, set the category to "Shortcuts".

To remove a shortcut, use the *removeShortcut(category, shortcutName)* method (this method does not complain if the specified shortcut does not exist):[19]

```
scUtils.removeShortcut('Toolbar Shortcuts', 'My New Shortcut');
```

The *addShortcutToBottom()* method does not override existing shortcuts. Therefore, to ensure that we do not add duplicate shortcuts, we must first remove any possibly existing shortcut using *removeShortcut()*, before adding it. Alternatively, we could loop over all existing category shortcuts checking their label, and adding a new shortcut only if it is not already found, as follows:

```
>> category = scUtils.getDefaultToolbarCategoryName;
>> scVector = scUtils.getShortcutsByCategory(category);
>> scArray = scVector.toArray
ans =
java.lang.Object[]:
    [com.mathworks.mlwidgets.shortcuts.Shortcut]
    [com.mathworks.mlwidgets.shortcuts.Shortcut]
    [com.mathworks.mlwidgets.shortcuts.Shortcut]
    . . .

>> char(scArray(1))
ans =
How to Add

>> foundFlag = 0;
   for scIdx = 1:length(scArray)
      if strcmp(char(scArray(scIdx)), 'My New Shortcut')
         foundFlag = 1; break;
```

```
        end
    end

>> if ~foundFlag, scUtils.addShortcutToBottom(...); end
```

Following my advice on the StackOverflow forum,[20] Richie Cotton wrapped the code above for adding and removing toolbar shortcuts, in a user-friendly utility that can now be found on the MATLAB File Exchange[21] and on his blog.[22]

We are not limited to the above-mentioned three default categories ('Toolbar Shortcuts', 'Help Browser Favorites', and 'Shortcuts'). In fact, we can add new categories as follows:

```
scUtils.addNewCategory('New category name');
```

Removing a category is also easy: Simply use the afore-mentioned *removeShortcut* method with an empty shortcut name — this will immediately remove the entire shortcuts category, along with all its contents:

```
scUtils.removeShortcut('category name to remove',[]);
```

Shortcuts are normally visible in the toolbar and in the MATLAB start menu. However, using com.mathworks.mlwidgets.shortcuts.ShortcutTreePanel they can also be displayed in any user GUI:

```
jShortcuts = com.mathworks.mlwidgets.shortcuts.ShortcutTreePanel;
[jhShortcuts,hPanel] = javacomponent(jc,[10,10,300,300],gcf);
```

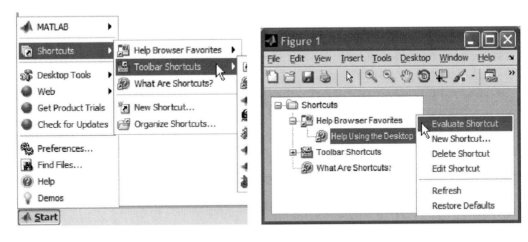

MATLAB start-menu shortcuts   Shortcuts presented in a user GUI

8.2 System Preferences[23]

MATLAB's user preferences are stored in the matlab.prf text file,[24] which is stored in the user's MATLAB preferences folder (**prefdir**):[25]

```
edit(fullfile(prefdir,'matlab.prf'));
```

Each preference appears to be on a separate line in the following format: <pref-name>=<pref-type><pref-value> where <pref-type> appears to be one of these:

- B => boolean/logical flag
- C => color (RGB numeric value)
- F => font (type,size,name)
- I => int16
- J => int64
- R => rectangular area (x,y,h,w)
- S => string/char

Some examples:

```
EditorShowLineNumbers = Btrue
EditorMaxCommentWidth = I120
```

We can read the preference names from this MATLAB.prf file and then use the following (you-guessed-it) Java calls to get/set the values:[26]

```
com.mathworks.services.Prefs.get<type>Pref(<pref-name>)
com.mathworks.services.Prefs.set<type>Pref(<pref-name>, newValue);
```

where <type> is one of: Boolean, Color, RGBColor, Font, Integer, Rectangle, String, Double (I believe Doubles get converted to int64 — possibly a bitwise[†] casting since both use 64 bits). For example,

```
com.mathworks.services.Prefs.getBooleanPref('LayoutSnapToGrid')
com.mathworks.services.Prefs.setIntegerPref('LayoutGridWidth', 25)
... Prefs.setStringPref('HelpSelectedProducts','MATLAB')
```

Preference values can also be retrieved using the undocumented built-in functions *feature* or *system_dependent* (unfortunately, there is no corresponding *set* feature):

```
>> NumericFormat = feature('getpref','GeneralNumFormat2')
NumericFormat =
Slong g
```

Adding a second argument to *get<type>Pref()* indicates a default value that is returned if <pref-name> is not already set or defined:

```
>> disp(com.mathworks.services.Prefs.getIntegerPref('xxxx',123))
   123
```

We can programmatically set any preference key we like — we are not limited to MATLAB's built-in set. I used this feature in my ***cprintf*** utility (described in the following section), to set

† Suggested by MathWorks: http://www.mathworks.com/support/solutions/en/data/1-OVWJ9/ (or http://bit.ly/9GH2Cm).

user-defined colors for later use by the desktop's UI syntax-highlighting engine. The relevant code segment is this:

```
% Convert Matlab RGB vector into a known style name, e.g. '[255,37,0]'
function styleName = getColorStyle(rgb)

   % Convert Matlab RGB array into a Java Color object
   intColor = int32(rgb*255);
   javaColor = java.awt.Color(intColor(1), intColor(2), intColor(3));

   % Preference key name format: '[RRR,GGG,BBB]'
   styleName = sprintf('[%d,%d,%d]',intColor);

   % Set/update the preference with this Java Color
   com.mathworks.services.Prefs.setColorPref(styleName,javaColor);
end % getColorStyle
```

...which in turn adds entries such as the following to my matlab.prf file:

```
[12,34,67] = C-15982013
```

Note that -15982013 = 0xFF0C2243, which is the RGB value [12, 34, 67] with an opaque alpha value (0xFF). This color value can later be retrieved using:

```
>> disp(com.mathworks.services.Prefs.getColorPref('[12,34,67]'))
java.awt.Color[r = 12,g = 34,b = 67]
```

After modifying the preferences, we need to notify all the components that use them to update themselves, standard practice in Java:[†]

```
com.mathworks.services.ColorPrefs.notifyColorListeners('ColorsText')
com.mathworks.services.ColorPrefs.notifyColorListeners('ColorsBackground')
```

 Warning: I published much of this information on the CSSM forum[27] back in 2007. Ben Steiner then shared his experience on that thread, as follows:

"For anyone else that's playing with this: I don't advise trying to edit the matlab.prf via MATLAB(!). I created a situation that made MATLAB unworkable. I did find that deleting the matlab.prf completely (in frustration) solved it."

8.3 Command Window

There are several ways by which we can get a direct handle to the Command Window edit pane (which contains the editable *Document*). We can start with the Desktop's Java Frame handle and work our way down the extremely complex hierarchy tree. A better alternative is to use this shortcut:[28]

```
jDesktop = get(get(handle(gcf),'JavaFrame'),'Desktop');
jTextArea = get(getMainFrame(jDesktop), 'FocusOwner');
jTextArea = jDesktop.getMainFrame.getFocusOwner; % an alternative
```

---
† More information on changing the Command Window colors can be found in the following section.

We can also start with the handle to the Command Window itself, and find our way down its scroll-pane hierarchy tree:

```
jDesktop = com.mathworks.mde.desk.MLDesktop.getInstance;
cmdWin = jDesktop.getClient('Command Window');
jTextArea = cmdWin.getComponent(0).getViewport.getView
```

Yet another alternative is to get the corresponding Document listener target:

```
cmdWinDoc = com.mathworks.mde.cmdwin.CmdWinDocument.getInstance;
listeners = cmdWinDoc.getDocumentListeners;

% Loop over all listener objects until we find the required JTextArea
% Note: jTextArea's actual position in the listeners list may vary'
jTextArea = [];
for listenerIdx = 1 : length(listeners)
    comp = listeners(listenerIdx);
    if comp.isa('javax.swing.JTextArea$AccessibleJTextArea')
        jTextArea = comp.getAccessibleParent.getComponent(0);
        break;
    end
end
```

The returned `jTextArea` is an object of class `com.mathworks.mde.cmdwin.XCmdWndView`, which derives from the standard Java Swing `JTextArea` component.[29]

Using the `jTextArea` handle, we can directly access (read/modify) the Command Window edit pane text. Note that the retrieved text is a `java.lang.String` object, so in order to use it in MATLAB it is better to immediately convert it using the ***char*** function:

```
cwText = char(get(jTextArea,'Text')); % or: jTextArea.getText.char;
```

To get the reference handle of the Command Window's containing Frame, do the following:[‡]

```
try
    % Matlab 7
    jDesktop = com.mathworks.mde.desk.MLDesktop.getInstance;
    cmdWin = jDesktop.getClient('Command Window');
    cmdWinFrame = cmdWin.getTopLevelAncestor;
catch
    % Matlab 6
    jDesktop = com.mathworks.ide.desktop.MLDesktop.getMLDesktop;
    cmdWin = jDesktop.getClient('Command Window');
    cmdWinFrame = cmdWin.getTopLevelWindow;
end
```

[†] The `JTextArea` position in the listeners array may change. In fact, it is the 3rd item in R2008a, but the 4th item in R2008b.

[‡] When the Command Window is docked in the Desktop (as it normally is), the containing Frame can also be retrieved via jDesktop.*getMainFrame()*, as described in Section 8.1.2.

8.3.1 *Controlling Command Window Colors*

A very common request over the years[30] has been to enable programmatic customization of the Command Window colors. This need has two subrequirements:

- Customizing the entire Command Window foreground/background colors.
- Customizating specific text segments outputted to the Command Window.

In the preceding section, I have shown how the MATLAB preferences can be modified programmatically. The specific customization for setting the Command Window colors (the first sub-requirement) is:[31]

```
% Do not use system color
import com.mathworks.services.*
Prefs.setBooleanPref('ColorsUseSystem',0);

% Use the specified colors for foreground/background
% (instead of the default black on white)
Prefs.setColorPref('ColorsBackground', java.awt.Color.yellow);
ColorPrefs.notifyColorListeners('ColorsBackground');
```

Programmatically controlling the Command Window colors is important, for example, when we have two MATLAB applications open at the same time and wish to visually distinguish between them. Unfortunately, this has the side-effect of setting the colors even for future MATLAB sessions, since the corresponding preferences have changed. It also has the effect of changing the color in all MATLAB text panes — not just the Command Window (e.g., the Command History pane).

So, if we only wish to set the colors temporarily and/or only update the Command Window, we should not modify the system preferences. Instead, simply use the jTextArea handle directly. This also has the benefit of applying immediately (no need for notifications as in the preceding section). Several alternatives are presented in the following code snippet (we may need to tweak it for particular MATLAB versions):

```
jTextArea.setBackground(java.awt.Color.yellow);
jTextArea.setBackground(java.awt.Color(1,1,0));
set(jTextArea,'Background','yellow');
set(jTextArea,'Background',[1,1,0]);
```

We can do the same with the **Foreground** property:

```
jTextArea.setForeground(java.awt.Color(0,0,1)); % =blue
```

This can be used, for example, to flash the Command Window, alerting the user to some event:

```
for idx = 1 : 5
    jTextArea.setBackground(java.awt.Color.red); pause(0.2);
    jTextArea.setBackground(java.awt.Color.white); pause(0.2);
end
```

Let us now turn to the more difficult task of customizing the color of specific text segments within the Command Window. For this we use `jTextArea`'s dormant syntax highlighting capabilities. In fact, these capabilities are not entirely dormant: they are used by MATLAB in two very specific cases:

- Errors (or rather: output to STDERR) are displayed in a **red** color.
- Hyperlinks are displayed as **<u>underlined blue.</u>**

We can use this within our programs as follows:[†]

```
% Use hyperlink style:
disp('<a href="">my text</a>');

% Use STDERR (fid = 2) style:
fprintf(2,'my text\n');
```

We would perhaps have expected `jTextArea` to support HTML formatting like the rest of the *uicontrols*. Unfortunately, `jTextArea` (like Swing's standard `JTextArea` of which `jTextArea` is an instance) does not automatically support HTML formatting. In fact, `jTextArea`'s default Document object, which holds the text-area's text and font style information, is an extension of Swing's `javax.swing.text.PlainDocument`,[32] which does not allow any text style formatting. And the `jTextArea` object itself is a simple `JTextArea`, which does not enable using a styled Document object. Perhaps in a future version MathWorks would be willing to use the almost identical (syntactically wise) `JTextPane`,[33] which does enable styled text runs. Instead of using `JTextPane`, MATLAB apparently implemented their support for STDERR and hyperlink styles using a custom `com.mathworks.mde.cmdwin.CmdWinSyntaxUI` class that extends `javax.swing.plaf.basic.BasicTextAreaUI`. Unfortunately, these are internal classes that we users cannot customize.

After many trials and errors, frustrations and blind alleys, an idea occurred to me:[34] Perhaps we could fool the UI class to think that our text should be syntax highlighted? We would then have a few more colors with which to play (comments=green, strings=purple, etc.). So I took a look at the `jTextArea`'s **Document** component (that holds all text and style info) and there I saw that MATLAB uses several custom attributes with the style and hyperlink information:

- **SyntaxTokens** attribute holds style color strings such as 'Colors_M_Strings' for strings, or 'CWLink' for hyperlinks.
- **LinkStartTokens** attribute holds the segment start offsets for hyperlinks (–1 for non-hyperlinked, 0+ for hyperlink).
- **HtmlLink** attribute holds the URL target (java.lang.String object) for hyperlinks, or null ([]) for non-hyperlink.

I played a hunch and modified the style of a specific text segment and lo-and-behold, its Command Window color changed! Unfortunately, I found out that I cannot just *fprintf(text)* and

---

[†] http://UndocumentedMatlab.com/blog/changing-matlab-command-window-colors-part2/(orhttp://tinyurl.com/y97mhsx). Additional methods of displaying hyperlinks are discussed in Sections 3.3.1, 5.5.1, 6.5.2, 6.9, and 8.3.2.

then modify its style — for some unknown reason MATLAB first needs to place the relevant segment in a "styled" mode (or something similar). I tried to *fprintf(2,text)* to set the red (error) style, but this did not help. But when I prepended a simple hyperlink space character I got what I wanted — I could now modify the subsequent text to any of the predefined syntax highlighting colors/styles.

But is it possible to use any user-defined colors, not just the predefined syntax highlighting colors? I then remembered my earlier discovery that "Colors_M_Strings" and friends are simply system preference color objects that can be updated:

```
import com.mathworks.services.*
Prefs.setColorPref('Colors_M_Strings',java.awt.Color(...));
```

So I played another hunch and tried to set a new custom preference:

```
>> Prefs.setColorPref('yair',java.awt.Color.green);
>> Prefs.getColorPref('yair')
ans =
java.awt.Color[r=0,g=255,b=0]
```

So far so good. I now played the hunch and changed the text element's style name from 'Colors_M_Strings' to 'yair' and luckily the new green color took effect!

So we can now set any style color (and underline it by enclosing the text in a non-target-url hyperlink), as long as we define a style name for it using Prefs.*setColorPref()*. How can we ensure the color uniqueness for multiple colors? The answer was to simply use the integer RGB values of the requested color, something such as [47,0,255].

But we still have the hyperlinked (underlined) space before our text — how do we get rid of it? I tried to set the relevant LinkStartTokens entry to –1 but could not: unlike **SyntaxTokens** which are modifiable Java objects, **LinkStartTokens** is an immutable numeric vector. I could, however, set its URL target to null ([]) to prevent the mouse cursor from changing when hovering over the space character, but could not remove the underline. I then had an idea to simply hide the underline by setting the character style to the Command Window's background color. The hard part was to come up with this idea — implementation was then relatively easy:

```
% Get a handle to the Command Window component
mde = com.mathworks.mde.desk.MLDesktop.getInstance;
cmdWin = mde.getClient('Command Window');
xCmdWndView = cmdWin.getComponent(0).getViewport.getView;

% Store the Command Window background color as a special color pref
% This way, if the Command Window background color changes (via
% File/Preferences), it will also affect existing rendered strings
cwBgColor = xCmdWndView.getBackground;
com.mathworks.services.Prefs.setColorPref('CW_BG_Color',cwBgColor);

% Now update the space character's style to 'CW_BG_Color'
% See within the code: setElementStyle(docElement,'CW_BG_Color',...)
```

Having thus completed the bulk of the hard detective/inductive work, I now had to contend with several other obstacles before the code could be released to the public:

- Older MATLAB versions (e.g., 7.1 R14) use the **Document** style elements slightly differently and I needed to find a solution that would work well on all MATLAB 7 versions (this took quite some time…).[†] I even succeeded in implementing most features on MATLAB 6 — this again was quite an effort.

- If the text is not newline ('\n')-terminated, then sometimes it is not rendered properly. Adding a forced Command Window *repaint()* helped solve much of this problem, but some quirks still remain (see *cprintf*'s help section).

- Multi-line text ('abra \n kadbra') creates several style elements which needed to be processed separately.

- Adding exception handling, argument processing, and so on, to ensure foolproof behavior. For example, accepting case-insensitive and partial style names.

- Debugging the code was very difficult because whenever the debugger stopped at a breakpoint, "k>>" is written to the Command Window thereby ruining the displayed element! I had to devise nontrivial instrumentation and post-processing (see within the code).

Bottom line: we now have a very simple and intuitive utility that is deceivingly simple, but took many hours of investigation to develop. I never imagined it would be so difficult when I started, but this just makes the engineering satisfaction greater. 😊

The *cprintf* utility, which is the outcome of these extensive labors, is now available for download from the MATLAB File Exchange.[35] Based on the reviews/ratings it received and the number of times it was downloaded, it appears to be a very popular utility that answers a real user need.

```
>>
>>
>>    cprintf('text',     'regular black text');
      cprintf('hyper',    'followed %s ','by');
      cprintf('k',        '%d colored', 4);
      cprintf('-comment','& underlined');
      cprintf('err',      'elements\n');
      cprintf('cyan',     'cyan');
      cprintf('-green',   'underlined green');
      cprintf(-[1,0,1],  'underlined magenta');
      cprintf([1,0.5,0],'and multi-\nline orange\n');

 regular black text followed by 4 colored & underlined elements
 cyan underlined green underlined magenta and multi-
 line orange
 >> |
```

cprintf — display styled formatted text in the Command Window (See color insert.)

As in other cases, it is hoped that some future MATLAB version will have this capability built into the language. MathWork's Ken Orr of the Desktop development team has indeed publically commented that:[36]

[†] It appears that MATLAB 7.13 (R2011b) modified the underlying **Document** in a manner that required a major fix.

"We would definitely like to allow users to change the color of their text in the Command Window (this is a frequent request). We're thinking about more robust highlighting mechanisms now."

8.3.2 *Help Popup and Integrated Browser Controls*

The Command Window's `jTextArea` can also be used to programmatically display a popup window with user-specified text or HTML, on MATLAB releases that support the help popup (R2007b onward):[37]

```
jDesktop = com.mathworks.mde.desk.MLDesktop.getInstance;
jTextArea = jDesktop.getMainFrame.getFocusOwner;
jClassName = 'com.mathworks.mlwidgets.help.HelpPopup';
jPosition = java.awt.Rectangle(0,0,400,300);
helpTopic = 'surf';
javaMethodEDT('showHelp',jClassName,jTextArea,[],jPosition,helpTopic);
```

Notes:

1. `jPosition` sets the popup's pixel size and position (X, Y, Width, Height). Remember that Java counts from the top down (contrary to MATLAB) and is 0-based (see Section 3.1.1). Therefore, `Rectangle(0,0,400,300)` is a 400×300 window at the screen's top-left corner.

2. On R2007b we must use the equivalent but more cumbersome ***awtinvoke*** function instead of ***javaMethodEDT*** (see Sections 1.1 and 3.2).

3. On R2011b, the code snippet above can be simplified by this replacement: `helpUtils.errorDocCallback('surf')`

4. `helpTopic` is the help topic of our choice (the output of the ***doc*** function). To display arbitrary text, create a simple .m file that only has a main help comment with the arbitrary text, which will be presented in the popup.

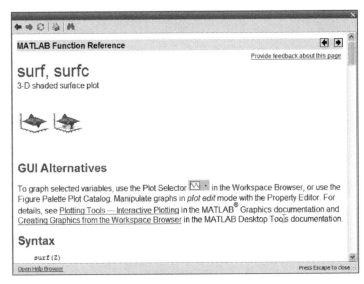

(See color insert.)

So if we had a *sample.m* file with the following contents:

```
function sample
% The text in this function's main comment will be presented in the
% help popup. <a href="http://UndocumentedMatlab.com">Hyperlinks</a>
% are supported, but unfortunately not full-fledged HTML.
```

Then we would get the following popup displayed (the m-file help text is converted into HTML using the undocumented internal function ***help2html***):

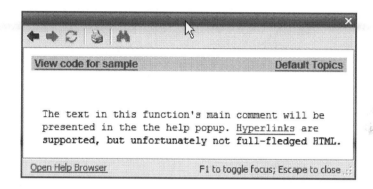

Help popup with user-created arbitrary text

Note that MATLAB's embedded web-browser control (detailed later in this section), accepts the "matlab:" protocol for hyperlinks. When matlab: hyperlinks are clicked, the control invokes the target in the MATLAB Command Prompt as an actual MATLAB command. For example, "matlab:dir(pwd)" will run the command ***dir(pwd)***.[38]

Well, this does get the message across, but looks rather dull. It would be nice if this could be improved to provide full-scale HTML support. Unfortunately, MATLAB documentation says this cannot be done:[39]

> *The **doc** function is intended only for reference pages supplied by The MathWorks. The exception is the **doc** UserCreatedClassName syntax. **doc** does not display HTML files you create yourself. To display HTML files for functions you create, use the **web** function*

Luckily for us, there is a back door: The idea is to search all visible Java windows for the HelpPopup (a modeless undecorated MJDialog Java window[40]). For some reason, MATLAB does not reuse existing HelpPopup windows but always creates a new instance. In any case, we search for the single visible HelpPopup, and then move down its component hierarchy to the internal web browser (a com.mathworks.mlwidgets.html.HTMLRenderer object), then update its content with HTML text or a webpage URL:

```
% Find the Help popup window
jWindows = com.mathworks.mwswing.MJDialog.getWindows;
```

```
jPopup = [];
for idx=1 : length(jWindows)
    if strcmp(get(jWindows(idx),'Name'),'HelpPopup')
        if jWindows(idx).isVisible
            jPopup = jWindows(idx);
            break;
        end
    end
end

% Update the popup with selected HTML
html=['Full HTML support: <b><font color=red>bold</font></b>, '...
      '<i>italic</i>, <a href="matlab:dir">hyperlink</a>, ' ...
      'symbols (&#8704;&#946;) etc.'];
if ~isempty(jPopup)
    browser = jPopup.getContentPane.getComponent(1).getComponent(0);
    browser.setHtmlText(html);
end
```

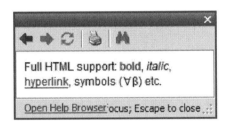

Help popup with HTML content

We can display HTML content and highlight certain keywords using the *setHtmlText-AndHighlightKeywords()* method:

```
browser.setHtmlTextAndHighlightKeywords(html,{'support','symbols'});
```

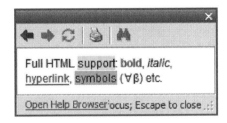

Help popup with HTML content and highlighting (See color insert.)

Instead of specifying the HTML content, we can point this browser to a URL webpage location (no need for the "http://" prefix) using *setCurrentLocation()*:

```
browser.setCurrentLocation('UndocumentedMatlab.com');
```

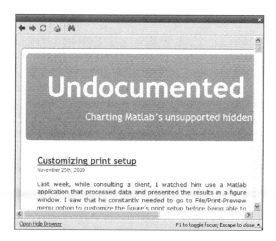

Help popup browser displaying a URL webpage

The `HTMLRenderer` includes a full-fledged browser (which may be different across MATLAB releases and platforms). This browser supports HTML, CSS, JavaScript and other web-rendering aspect that we would expect from a modern browser. Being a full-fledged browser, we have some control over its appearance using multiple internal methods of either this object or one of its children.[†] Interested readers may use my ***uiinspect*** utility to explore these options (see Section 1.3):

```
>> browser.list
com.mathworks.mlwidgets.html.HTMLRenderer[HTMLRenderer,...]
 com.mathworks.mlwidgets.html.WebRenderer[...]
  com.webrenderer.windows.MozillaBrowserCanvas[canvas0,0,0,792x587]
 com.mathworks.mwswing.MJPanel[HTMLRendererInfoMessageBarPanel,...]
  com.mathworks.mwswing.MJPanel[...]
   com.mathworks.mwswing.MJLabel[...]
   com.mathworks.mwswing.MJScrollPane[...]
   ...
```

Technically, `HTMLRenderer` is actually just a `JPanel` containing the actual browser. Luckily for us, MathWorks extended this panel class with the useful methods presented above, that forward the user requests to the actual internal browser. This way, we do not need to get the actual browser reference (although we can, of course).

My ***popupPanel*** utility, downloadable on the MATLAB File Exchange,[41] encapsulates all the above, displaying MATLAB doc pages, arbitrary text, HTML or webpages.

An interesting exercise left for the readers, is adapting the main heavy-weight documentation window (Help Browser) to display user-created HTML help pages. This can be achieved by means very similar to those shown in this section.

[†] http://www.mathworks.com/matlabcentral/newsreader/view_thread/243727#670555 (or http://bit.ly/csoohl); note that some internal `HTMLBrowserPanel` were removed or have changed between MATLAB releases, so test carefully.

Another option for displaying webpages in a stand-alone pop-up window is to use the `com.mathworks.mlservices.MLHelpServices` class, described in Section 5.6:

```
docRoot = char(com.mathworks.mlservices.MLHelpServices.getDocRoot);
url = ['jar:file:///' docRoot '/techdoc/help.jar!/ref/lasterror.html'];
com.mathworks.mlservices.MLHelpServices.cshDisplayFile(url);
com.mathworks.mlservices.MLHelpServices.cshSetSize(600,400);
com.mathworks.mlservices.MLHelpServices.cshSetLocation(500,250);
```

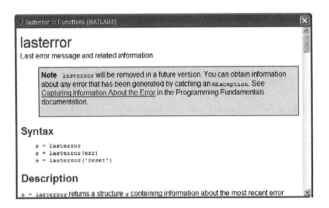

Although meant for displaying help pages, `MLHelpServices` can be used for any URL (the text:// protocol is apparently not supported, but regular HTML webpages are):

```
com.mathworks.mlservices.MLHelpServices.cshDisplayFile('google.com');
```

Of course, as the official documentation states, we can always use the fully supported *web* function to display HTML or URLs.[†] Under the hood, *web* uses the same `HTMLRenderer` as our `HelpPopup`. The benefit of using the methods shown here is the use of a lightweight popup window that is well-integrated with existing MATLAB help.

[†] Urs Schwarz (aka "us") has uploaded *sweb* to the File Exchange, a highly recommended extension to *web*: http://www.mathworks.com/matlabcentral/fileexchange/26034 or http://bit.ly/7VJQUC.

If we do wish to display a full-blown browser window, then we should consider using the internal WebBrowser object returned as the second output parameter from the *web()* function. WebBrowser enables fine-grained programmatic control using its supplied methods:[†]

```
% Create a new browser window and point it to a webpage URL
jBrowser = com.mathworks.mde.webbrowser.WebBrowser.createBrowser;
jBrowser.setCurrentLocation('www.UndocumentedMatlab.com');
% or: [status,jBrowser,url] = web('www.UndocumentedMatlab.com');

% Wait for the contents to be available, then close the browser window
pause(1); s = {};
while isempty(s)
    s = char(jBrowser.getHtmlText);
    pause(0.1);
end
jDesktop = com.mathworks.mde.desk.MLDesktop.getInstance;
jDesktop.removeClient(jBrowser);
```

HelpPopup's browser component is actually a stand-alone component that we can embed in our MATLAB GUI applications as a component (unlike WebBrowser, which creates a full-blown window).[42] In fact, MATLAB's browser object predates PopupPanel by many years and quite a few releases.

Here is a simple example in which a MATLAB listbox selects an adjacent webpage. This simple example shows how the Java browser object can easily be controlled by MATLAB. Specifically, we use two browser states: first we present an HTML text message ('Loading www.cnn.com — please wait...'), then replace it with a webpage:

```
% Prepare the figure window
f = figure('Name','Browser GUI demo', 'Number','off', 'Units','norm');

% Set up the browser panel
jObject = com.mathworks.mlwidgets.html.HTMLBrowserPanel;
[browser,container] = javacomponent(jObject, [], f);
set(container, 'Units','norm', 'Pos',[0.3,0.05,0.65,0.9]);

% Set up the URLs listbox
urls = {'www.cnn.com', 'www.bbc.co.uk', 'myLocalwebpage.html', ...
        'www.Mathworks.com', 'UndocumentedMatlab.com'};
hListbox = uicontrol('style','listbox', 'string',urls, ...
        'units','norm', 'pos',[0.05,0.05,0.2,0.9], 'userdata',browser);

% Set the listbox callback to load selected URL in the browser panel
cbStr = ['strs = get(gcbo,''string''); url = strs{get(gcbo,''value'')};'...
        'browser = get(gcbo,''userdata'');' ...
        'msg = [''<html><h2>Loading '' url '' - please wait...''];' ...
        'browser.setHtmlText(msg); pause(0.1); drawnow;' ...
        'browser.setCurrentLocation(url);'];
set(hListbox, 'Callback',cbStr);
```

[†] http://stackoverflow.com/questions/1311106/running-a-javascript-command-from-matlab-to-fetch-apdf-file (or http://bit.ly/7iLJU3). This example could of course be replaced with a simple *urlread()* — it is not intended to provide a real-life solution but rather demonstrate the WebBrowser usage.

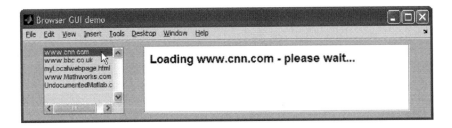

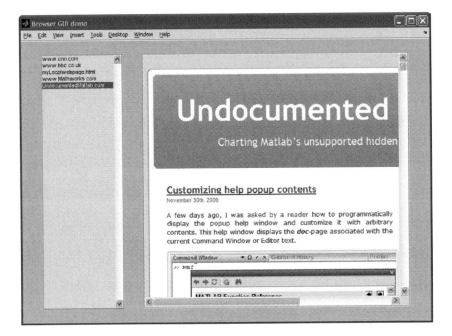

Browser object integrated in a MATLAB GUI (valid webpage)

If the webpage is inaccessible, an error message is displayed:

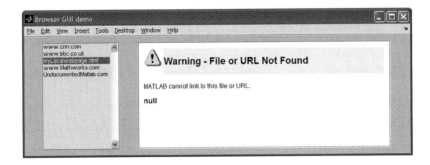

Browser object integrated in a MATLAB GUI (missing webpage)

Note that we can also use a `JEditorPane` component to display HTML (a simpler subset than the browser control), as discussed in Sections 3.3.3 and 6.5.2.[43]

Also note that MATLAB's built-in `WebBrowser` object accepts the nonstandard protocol text:// for displaying HTML contents (<html> and that closing tags are unnecessary here):

```
>> [status,jBrowser,url] = web('text://this is <b><i>a test');
status =
     0
jBrowser =
com.mathworks.mde.webbrowser.WebBrowser[Web Browser,0,0,924x635,...]
url =
text://this is <b><i>a test
```

We can easily expand this simple example to display any HTML message or webpage, in a seamless integration within our GUI.

Several years ago, a CSSM reader asked whether it is possible to set a callback function on the web browser's closure.[44] There are two alternatives for doing this:

1. *set*(jBrowser.getParent,**'ComponentRemovedCallback'**,@myFcn).
2. embed a `com.mathworks.mlwidgets.html.HTMLBrowserPanel` component in the GUI as described above, and then set the figure's **CloseRequestFcn** property like any other regular figure.

Note that MATLAB ships with an additional, completely different, ICE web browser: a commercial product from http://www.icesoft.com. MathWorks themselves recommend switching to ICE in some cases of problems with the standard browser.[45] In most cases I suggest sticking to the built-in browser and not the ICE browser, as I have seen some problems when using the ICE browser in MATLAB. To enable the ICE browser, run the following, after which MATLAB will use ICE by default:

```
com.mathworks.mlwidgets.html.HTMLRenderer.setUseWebRenderer(false);
```

8.3.3 *Modifying the Command Window Prompt*

A reader of the UndocumentedMatlab.com blog emailed me a challenge:[46] Modify the standard MATLAB Command-Window prompt from ">>" to some other string, preferably a dynamic prompt with the current timestamp. At first thought this cannot be done: The Command-Window prompts are hard-coded and to the best of my knowledge cannot be modified via properties or system preferences.

So the prompt can (probably) not be modified in advance, but what if it could be modified after being displayed? It is true that my *cprintf* utility modifies the Command-Window contents to display colored text. But this case is different since *cprintf* runs once synchronously (user-invoked), whereas the prompt appears asynchronously multiple times.

There are two ways of handling multiple asynchronous events in MATLAB: The first approach, which I have eventually used, involves setting a callback on the object. This is a well-known MATLAB practice, although we shall see that it uses an undocumented callback and functionality.

A possible alternative is to set a PostSet ***handle.listener*** on the relevant object property (see Appendix B). This approach is entirely undocumented and not well known. Interested readers can try this approach, rather than the callbacks approach that I have taken.

The solution involved getting the Command-Window reference, then setting its **CaretUpdateCallback**. This callback is fired whenever the desktop text is modified, which is an event we trap to replace the displayed prompt:

```
% Get the reference handle to the Command Window text area
jDesktop = com.mathworks.mde.desk.MLDesktop.getInstance;
try
    cmdWin = jDesktop.getClient('Command Window');
    jTextArea = cmdWin.getComponent(0).getViewport.getView;
catch
    commandwindow;
    jTextArea = jDesktop.getMainFrame.getFocusOwner;
end

% Instrument the text area's callback
if nargin && ~isempty(newPrompt) && ~strcmp(newPrompt,'>> ')
    set(jTextArea,'CaretUpdateCallback',{@setPromptFcn,newPrompt});
else
    set(jTextArea,'CaretUpdateCallback',[]);
end
```

Now that we have the Command-Window object callback set, we need to set the logic of prompt replacement — this is done in the internal MATLAB function *setPromptFcn*. Here is its core code:

```
% Does the display text end with the default prompt?
% Note: catch a possible trailing newline
cwText = char(jTextArea.getText);
pos = strfind(cwText(max(1,end-3):end),'>> ');
if ~isempty(pos)
    % Short prompts need to be space-separated
    if length(newPrompt) < 3
        newPrompt(end+1:3) = ' ';
    elseif length(newPrompt) > 3
        fprintf(newPrompt(1:end-3));
    end
    newLen = jTextArea.getCaretPosition;

    % The Command-Window text should be modified on the EDT
    awtinvoke(jTextArea.java,'replaceRange(Ljava.lang.String;II)', ...
            newPrompt(end-2:end), newLen-3, newLen);
    awtinvoke(jTextArea.java,'repaint()');
end
```

In this code snippet, note that we space-pad prompt string that are shorter than three characters: this is done to prevent an internal-MATLAB mixup when displaying additional text — MATLAB "knows" the Command-Window's text position and it gets mixed up if it turns out to be shorter than expected.

Also note that I use the semi-documented ***awtinvoke*** function (Section 1.1) to replace the default prompt (and an automatically appended space) on the Event Dispatch Thread (Section 3.2). Since MATLAB R2008a, I could use the more convenient ***javaMethodEDT*** function, but I wanted my code to work on all prior MATLAB 7 versions, where ***javaMethod-EDT*** was not yet available.

The callback snippet above would enter an endless loop if not changed: whenever the prompt is modified the callback would have been re-fired, the prompt re-modified and so on endlessly. There are many methods of preventing callback re-entry — here is the one that I chose:[47]

```matlab
function setPromptFcn(jTextArea,eventData,newPrompt)
    % Prevent overlapping reentry due to prompt replacement
    persistent inProgress
    if isempty(inProgress)
        inProgress = 1;  %#ok unused
    else
        return;
    end

    try
        % *** Prompt modification code goes here ***
        pause(0.02);  % force the prompt-change callback to fizzle-out
    catch
        % Never mind - ignore errors...
    end

    % Enable new callbacks now that the prompt has been modified
    inProgress = [];
end   % setPromptFcn
```

Handling both static prompt strings (e.g., '[Yair]') and dynamic prompts (e.g., '[25-Jan-2010 01:00:51]') is done by accepting string-evaluable strings/functions:

```matlab
% Try to evaluate the new prompt as a function
try
    origNewPrompt = newPrompt;
    newPrompt = feval(newPrompt);
catch
    try
        newPrompt = eval(newPrompt);
    catch
        % Never mind - probably a string...
    end
end
```

```
if ~ischar(newPrompt) && ischar(origNewPrompt)
    newPrompt = origNewPrompt;
end
```

I then added some edge-case error handling and wrapped everything in a single utility called *setPrompt*, now available on the File Exchange.[48] Some usage examples:

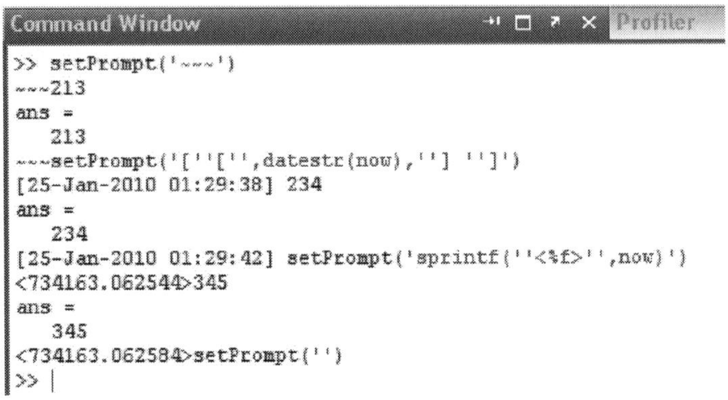

setPrompt usage examples

However, the displayed timestamp is somewhat problematic in the sense that it indicates when the prompt was <u>created</u>, not when the associated Command-Window command was <u>executed</u>. In the screenshot above, the **234** command was executed on [25-Jan-2010 01:29:42], instead of the displayed [25-Jan-2010 01:29:38].

This is somewhat misleading. It would be better if the last (current) timestamp was continuously updated and would therefore always display the latest command's execution time. I use a predetermined *setPrompt* argument of 'timestamp' to indicate that this should be done. *setPrompt* implements this using a MATLAB *timer* as follows:

```
% This is entered in the main function before setting the prompt:
stopPromptTimers;
if nargin && strcmpi(newPrompt,'timestamp')

    % Update initial prompt & prepare a timer to continuously update it
    newPrompt = @()(['[',datestr(now),'] ']);
    start(timer('Tag','setPromptTimer', ...
                'Name','setPromptTimer', ...
                'ExecutionMode','fixedDelay', ...
                'ObjectVisibility','off',...
                'Period',0.99, ...
                'StartDelay',0.5, ...
                'TimerFcn',{@setPromptTimerFcn,jTextArea}));
end

% Stop & delete any existing prompt timer(s)
function stopPromptTimers
```

```
  try
     timers = timerfindall('tag','setPromptTimer');
     if ~isempty(timers)
        stop(timers);
        delete(timers);
     end
  catch
     % Never mind...
  end
end  % stopPromptTimers


% Internal timer callback function
function setPromptTimerFcn(timerObj,eventData,jTextArea)
  try
     try jTextArea = jTextArea.java; catch, end  %#ok
     pos = getappdata(jTextArea,'setPromptPos');
     newPrompt = datestr(now);
     awtinvoke(jTextArea,'replaceRange(Ljava.lang.String;II)', ...
               newPrompt, pos, pos+length(newPrompt));
     awtinvoke(jTextArea,'repaint()');
  catch
     % Never mind...
  end
end  % setPromptTimerFcn
```

8.3.4 Tab Completions

Tab completions have become a standard feature of development environment in recent years. MATLAB has not fallen too far behind, and has repeatedly introduced support and improvements for tab completions in the Desktop Command Window and the MATLAB Editor.[49] When a CSSM reader asked[50] whether it is possible to customize MATLAB tab-completion for user-defined functions. I searched for an answer and found a similar question on StackOverflow that provided the necessary clue:[51]

Apparently, MATLAB has a file called *TC.xml* in its [matlabroot '/toolbox/local/'] folder that contains the definitions of the tab-completable functions and their arguments. In order for a user-defined function's arguments to support tab-completion, a new entry needs to be added to this XML file.

8.3.4.1 TC.xml and TC.xsd   The full syntax of the *TC.xml* file can be found in the *TC.xsd* file, which is located in the same folder as *TC.xml*. Here are some sample definitions from an R2008a *TC.xml* file (which might vary across MATLAB releases):

```
<binding name="addpath" ctype="DIR"/>;
<binding name="help"    ctype="FUN SUBFUN"/>;
<binding name="clear"   ctype="FUN VAR"/>;

<binding name="whos"    ctype="VAR">
  <arg previous="-file" ctype="MATFILE"/>
```

```
</binding>
<binding name="open">
  <arg argn="1" ctype="VAR MATFILE FIGFILE MFILE MDLFILE FILE"/>
</binding>
<binding name="openfig">
  <arg argn="1" ctype="FIGFILE"/>
  <arg argn="2" ctype="VAR" value="new visible invisible reuse"/>
</binding>
<binding name="mlint" ctype="FUN">
  <arg argn="2:10" ctype="VAR" value="-struct -string -id"/>
</binding>
```

The first example defines that an unlimited number of ***addpath*** arguments, all of type DIR. So, when completing any argument of this function in the Command-Window, MATLAB presents only relevant sorted DIR (folder) elements in the pop-up window:

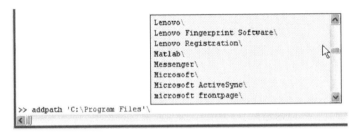

Tab-completion of type DIR (See color insert.)

Similarly, ***help*** defines all its arguments to be a function or sub-function type, so the popup will only be populated with function names currently visible in the desktop:

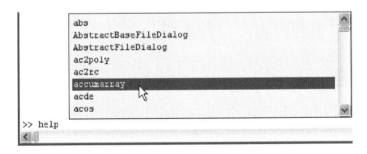

Tab-completion of types FUN & SUBFUN

Similarly, ***clear*** defines all its arguments as function names or variables. Note that the list of available functions/variables may change depending on the current execution stack position. The full list of supported types as defined in *TC.xsd* is: VAR, FUN, SUBFUN, DIR, FILE, MFILE, MATFILE, FIGFILE, MDLFILE, MCOSPCG, MCOSCLASS (the last two available since R2010a), and MESSAGEID (since R2011b).

The **whos** function defines all its arguments as VAR, except a single MATFILE arg that follows a '-file' argument (**whos**'s help page[52] explains why); the **open** function defines tab completion only for its first argument (with plenty of possible types). Likewise, **openfig** is defined as accepting a FIGFILE followed by a VAR with a few extra special-purpose strings[53] that are added to the popup-up menu.

Finally, the **mlint** example shows that multiple arguments can be defined using a single XML definition element. In this case, args #2-10 are defined as VAR (with three extra special-purpose strings), while args #1 and #11+ are defined as FUN.

Careful users can edit the *TC.xml* file (I strongly suggest saving a backup first):

```
edit(fullfile(matlabroot,'toolbox/local/TC.xml'))
```

User-defined functions can easily be added to *TC.xml*, and we can even add/modify the built-in MATLAB functions that are already defined. Note that changes to *TC.xml* only take effect after a MATLAB restart. From then on, all future MATLAB sessions will use the modification, so a really simple one-time edit can improve our workflow for a long time — at least until we upgrade MATLAB, when we will need to redo our edits.

8.3.4.2 TabComplete Utility
To facilitate *TC.xml* editing, I have created a utility called **TabComplete**, which is available on the MATLAB File Exchange.[54] The use of this utility is very simple:

```
tabcomplete test file 'DIR +data -data no_data' VAR
```

defines a user-defined function **test** that accepts a FILE argument, followed by a DIR argument with three special-purpose strings, followed by any number of VAR arguments.

```
<binding name="test" ctype="VAR">
<arg argn="1" ctype="FILE"/>
<arg argn="2" ctype="DIR" value="+data -data no_data"/>
</binding>
```

To define specific argument types without any default type, use

```
tabcomplete test file 'DIR +data -data no_data' ''
```

```
>> % Note: in a previous Matlab session I did:
>> % tabcomplete('test','file','+data -data no_data','');
>>
>>
>>
>>                        +data
>>                        -data
>>                        no_data
>>
>> test('234.jpg','
```

Using *TabComplete* for user-defined functions

Note: MATLAB releases 7.10 (R2010a) through 7.13 (R2011b) have a bug that causes MATLAB to enter an endless loop (full CPU load) whenever tab-completion is requested for an argument that has possible values that are not simple terms. In the case above, "+data" and "-data" both cause this abnormal behavior. In such cases, the only remedy is to kill the MATLAB process via the OS's task manager. Until this bug is fixed, I suggest using only simple values.

TabComplete can retrieve the current list of tab-completion definitions:

```
>> definitions = tabcomplete;
>> definitions(1)
ans =
    functionName: 'addpath'
     defaultType: 'DIR'
     extraValues: ''
        platform: ''
    functionArgs: []

>> definitions(54)
ans =
    functionName: 'openfig'
     defaultType: ''
     extraValues: ''
        platform: ''
    functionArgs: [1x2 struct]

>> definitions(54).functionArgs(1)
ans =
    previousArg: ''
        argType: 'FIGFILE'
    extraValues: ''

>> definitions(54).functionArgs(2)
ans =
    previousArg: ''
        argType: 'VAR'
    extraValues: 'new visible invisible reuse'
```

TabComplete has a few limitations: it does not support the -previous option described above (we can do this by manually editing *TC.xml*). There are also some inherent limitations in MATLAB's TC functionality: changes take effect only after a MATLAB restart (there might be a way to reload the definitions in the current MATLAB session, but I do not know of any); the list of standard types cannot be modified; and the default type does not support extra special-purpose strings as do the numbered args.

8.3.4.3 Additional Aspects of Tab Completion   There is another annoying limitation of MATLAB's tab-completion mechanism: *TC.xml* only supports lowercase function names although MATLAB supports function names with UPPERCASE characters. This limitation can be solved by editing the *TC.xsd* file (not the *TC.xml* file). Instead of:

```
<xsd:simpleType name="tcBindingNameType">
  <xsd:restriction base="xsd:token">
    <xsd:pattern value='[A-Za-z_0-9]+(/[a-z_0-9]+)?'/>
  </xsd:restriction>
</xsd:simpleType>
```

Change the **xsd:pattern** definition element to:[†]

```
<!-- Yair 21/2/2010: added A-Z -->
<xsd:pattern value='[A-Za-z_0-9]+(/[A-Za-z_0-9]+)?'/>
```

(note the way that comments can be added to the XSD/XML files).

An entirely different customization, for user-defined class members, was presented by Michal Kutil.[55] Unfortunately, Michal's hacks only work for user-defined classes, and not for regular MATLAB functions and scripts, either built-in or user-defined.

A related undocumented aspect of tab completions, is the 'tabcompletion' feature.[56] For some unknown reason, this feature cannot be activated via the *feature* function but only via the older *system_dependent* one (in most other cases, these two functions are interchangeable):

```
system_dependent('tabcompletion',false) % use true to turn back on
```

8.3.5 Additional Command Window Uses

The most important addition of `XCmdWndView` compared to the standard `JTextArea` is the support for incremental search/replace, defined in the `com.mathworks.widgets.incSearch.IncSearchInterface` interface, that defines the following methods: *find(string,forwardFlag)*, *findBack(findEventObj)*, *findForward(findEventObj)*, *incSearch(string,forwardFlag)*, *incSearchEOL(string)*, *incSearchNextWord(string)*, *endIncSearch()*, *endIncSearchMoveCaret()*, *replace(findEventObj)*, *replaceAll(findEventObj)*, *startIncSearch(forwardFlag)*, and *clearSearch()*.

Another aspect of the Command Window's `jTextArea`, is its callback support for keyboard and mouse events. We use this feature in Section 8.5, where key-bindings support is discussed in detail. For the purpose of this section, we just show how key-click events in the Command Window can be trapped and processed:[57]

```
jDesktop = com.mathworks.mde.desk.MLDesktop.getInstance;    % Matlab 7
%jDesktop = com.mathworks.ide.desktop.MLDesktop.getMLDesktop;    % Matlab 6
```

[†] This has been fixed in R2011a, as I have suggested.

```
cmdWin = jDesktop.getClient('Command Window');
jTextArea = cmdWin.getComponent(0).getViewport.getView;
h_cw = handle(jTextArea,'CallbackProperties');
set(h_cw, 'KeyPressedCallback', @myMatlabFunction);

% Alternatively, to pass predefined parameters to myMatlabFunction:
set(h_cw, 'KeyPressedCallback', {@myMatlabFunction,extraParam1,...});

% ...and now to clear the event trap:
set(h_cw, 'KeyPressedCallback', '');
```

A user on the StackOverflow forum asked[58] if it is possible to programmatically control the Command Window title, to enable easy window differentiation:

```
% For entire Desktop:
jDesktop.getMainFrame.setTitle('my new title');

% For Command Window only:
cmdWin = jDesktop.getClient('Command Window');
cmdWin.getTopLevelAncestor.setTitle('my new title');      % Matlab 7
%cmdWin.getTopLevelWindow.setTitle('my new title');       % Matlab 6
```

Another StackOverflow user asked[59] how to retrieve the Command Window's selected text:

```
text = char(jTextArea.getSelectedText);
```

8.4 Editor

The built-in MATLAB Editor is possibly the most complex Java-based component in MATLAB. In addition to standard editor functionalities, it also includes support for document docking, integrated debugging, code-folding and other nontrivial tasks. The editor itself has many possible customizations, once we get its Java handle:[60]

```
try
    % Matlab 7
    jDesktop = com.mathworks.mde.desk.MLDesktop.getInstance;
    jEditor = jDesktop.getFrameContainingGroup('Editor');†
    % => a com.mathworks.mde.desk.MLMultipleClientFrame object

catch

    % Matlab 6
    % Unfortunately Matlab 6 Desktop does not expose the Editor handle
    %jDesktop = com.mathworks.ide.desktop.MLDesktop.getMLDesktop;

    % So here is the workaround for Matlab 6:
    editorApp = com.mathworks.ide.editor.EditorApplication;
    openDocs = editorApp.getOpenDocuments;
```

---

† An alternative: jEditor = jDesktop.getGroupContainer('Editor').getTopLevelAncestor; we can also get jEditor by searching the Java Frames returned by java.awt.Frame.getFrames().

```
    % => a java.util.Vector

    firstDoc = openDocs.elementAt(0);
    % => a com.mathworks.ide.editor.EditorViewContainer object

    jEditor = firstDoc.getParent.getParent.getParent;
    % => a com.mathworks.mwt.MWTabPanel object or
    %    a com.mathworks.ide.desktop.DTContainer object
end
```

Note that we cannot use the shortcut presented for the Profiler in the previous section namely, jDesktop.getClient('Editor'). While this is syntactically correct in both MATLAB 6 and 7, it will normally return a null ([]) object because the Editor client constantly changes its name based on the currently presented filename. For example,

```
>> jDesktop.getClient('Editor')
ans =
     []
>> jDesktop.getClient('Untitled')
ans =
com.mathworks.ide.editor.EditorViewContainer[...]
```

8.4.1 *The EditorServices/matlab.desktop.editor Object*

Now that we have the Editor handle, let us retrieve its currently open (active) file name from the Editor's title (remember to strip away possible "dirty" indication — a prepended '*' character):

```
title = jEditor.getTitle;
currentFilename = char(title.replaceFirst('Editor - ',''));
currentFilename = strrep(currentFilename,'*','');
```

An alternative way to get the title of the currently active Editor document:[61]

```
% Get the handle to the currently-active document container
jDocContainer = jDesktop.getLastDocumentSelectedInGroup('Editor');

% Get this container's title from the Desktop
currentFilename = char(jDesktop.getTitle(jDocContainer));

% Strip away possible "dirty" indication (a prepended '*' character)
currentFilename = strrep(currentFilename,'*','');
```

The entire list of open file names can be retrieved in several ways (some may not work on some MATLAB releases):

```
% Alternative #1:
jEditorServices = com.mathworks.mlservices.MLEditorServices;
editorFilenames = char(jEditorServices.builtinGetOpenDocumentNames);
```

```
% Alternative #2:
openFiles = jDesktop.getWindowRegistry.getClosers.toArray.cell;
editorFilenames = cellfun(@(c)c.getTitle.char,openFiles,'uniform',false);

% Alternative #3:
jEditorServices = com.mathworks.mlservices.MLEditorServices;
openFiles = jEditorServices.getEditorApplication.getOpenEditors.toArray;
editorFilenames = arrayfun(@(c)c.getLongName.char,openFiles,'uniform',false);
```

In MATLAB R2009b to R2010b, we can also use the built-in ***editorservices*** package,[62] which is basically just a MATLAB wrapper for `MLEditorServices`:

```
% In Matlab R2009b, editorservices is a predefined built-in package
currentFile = editorservices.getActive;
openFiles = editorservices.getAll;

% The new editorservices package has a hidden undocumented property
% 'JavaEditor' which is a bridge to the Java side with all the editor
% panes, toolbars, etc.
jEditor = get(editorservices.EditorDocument,'JavaEditor');
edit('editorservices.EditorDocument'); % to see usage examples
```

editorservices was renamed ***matlab.desktop.editor*** in MATLAB 7.12 (R2011a).[63] The syntax for the corresponding code in R2011a is very similar:

```
currentFile = matlab.desktop.editor.getActive;
openFiles = matlab.desktop.editor.getAll;
jEditor = matlab.desktop.editor.getActive().JavaEditor;
edit('matlab.desktop.editor.Document'); % to see usage examples
```

Here are some interesting methods of the pre-R2009b `jEditorServices` handle[†] (note that in R2010b this interface has changed somewhat — see below):

■ *builtinAppendDocumentText(fileName, text)* — appends the specified text to the specified filename. Filename may be specified using the full pathname or just the simple file name. Note that if the file is not currently open in the Editor, then nothing happens, and no warning/error message is displayed.

■ *builtinGetActiveDocument()* — returns the full pathname of the currently active editor file.[‡]

■ *builtinGetDocumentText(fileName)* — returns the text of the specified Editor filename (which may be specified using its simple or full-path form). Note that the returned text is a `java.lang.String` object, which should be converted to MATLAB using the built-in ***char*** function. If the specified file is not currently open in the Editor, [] is returned.

[†] This refers to the `MLEditorServices` Java object, not the R2009b+***editorservices*** package, although many of the latter's methods are similar to those presented here. This is not a coincidence: they share a common origin.

[‡] The corresponding R2009b+***editorservices*** method name is *getActive()*.

- *builtinGetNumOpenDocuments()* — returns the number of open documents.[64]
- *builtinGetOpenDocumentNames()* — returns a Java array of java.lang.String objects,[†] of the full-path filenames of the files currently open in the Editor.[65] Use the built-in *cell* function to convert this array to a MATLAB string cell array:

```
>> openFiles = jEditorServices.builtinGetOpenDocumentNames.cell
openFiles =
    'C:\Yair\Utils\Matlab\EditorMacro\EditorMacro.m'
    'C:\Program Files\Matlab\R2008a\toolbox\matlab\uitools\uiundo.m'
    'C:\Yair\Undocumented Matlab\Untitled2'
```

- *closeAll()* — closes all open documents.[‡]
- *closeDocument(fileName)* — closes specified document, if currently loaded.[66]
- *isDocumentDirty(filename)* — returns a flag indicating whether or not the specified document is modified and unsaved compared to its disk image. Such documents are usually indicated with an asterix (*) in their title.
- *newDocument(text)* — creates a new (untitled) Editor document, with the optional specified text.[§]
- *openDocument(fileName)* — opens the specified document in the Editor and sets the cursor caret at line #1.[¶] If the file does not exist, a popup message will ask the user wheter to create a new file by this name or not.
- *openDocumentToFunction(fileName, functionName, string)* — opens the specified document at the specified function/sub-function. [††] If the filename does not exist, nothing happens (no error/warning is displayed); if the function does not exist in the document, then a new empty function having the specified name is added at the bottom of the document (a new line with the text: "function *functionName*"). I do not know what the third string is used for — as far as I could tell, it accepts any string value and has no visible effect.
- *openDocumentToLine(fileName, lineNum, focusFlag, highlightFlag)* — opens the specified document at the requested line number.[‡‡] The optional *focusFlag* (default=true) determines whether the Editor window should receive focus (be moved to the front); the accompanying *highlightFlag* (default=false)[§§] determines whether the entire line should be highlighted.

[†] The corresponding R2009b+*editorservices* method name is *getAll()*.

[‡] http://blogs.mathworks.com/desktop/2007/03/29/shortcuts-for-commonly-used-code/#comment-5753 (orhttp://bit.ly/aRvA9c). The corresponding R2009b+*editorservices* method name is *closeGroup()*. To leave the group up and just close all the open editors use: close(editorservices.getAll).

[§] The corresponding R2009b+*editorservices* method name is *new(text)*.

[¶] http://www.mathworks.com/matlabcentral/newsreader/view_thread/154050#386543 (or http://bit.ly/c775qm). The corresponding R2009b+*editorservices* method name is *open(fileName)*.

[††] The corresponding R2009b+*editorservices* method name is *openAndGoToFunction(fileName,functionName,...)*.

[‡‡] The corresponding R2009b+*editorservices* method name is *openAndGoToLine(fileName,lineNum,columnNum,...)*.

[§§] If *focusFlag* is specified, then so must *highlightFlag* — these two flags must either both be present or both be absent.

- *openDocumentToLineAndColumn(fileName, lineNum, colNum, focusFlag)* — opens the specified document at the requested line/column. *focusFlag* (default=true) determines whether the Editor window should receive focus.
- *reloadDocument(filename, onlyIfNonDirtyflag)* — reloads the specified document if it was modified. If *onlyIfNonDirtyflag* is true, then the document is not unloaded if it is "dirty" (i.e., modified and unsaved), causing data loss if the document were reloaded; if *onlyIfNonDirtyflag* is false, then the document is reloaded regardless of the document's "dirty" state (see *isDocumentDirty*).
- *saveDocument(fileName)* — saves the specified document.

The combination of *builtinGetOpenDocumentNames()* and *openDocument()* enables storing the current editor state in a disk file, for later reload.[67] This enables easy switching between projects, each having its own set of open Editor documents.

```
% Save the current Editor state:
jEditorServices = com.mathworks.mlservices.MLEditorServices;
editorState = jEditorServices.builtinGetOpenDocumentNames();
save('editorState.mat', 'editorState');

% Restore the Editor state:
jEditorServices = com.mathworks.mlservices.MLEditorServices;
load('editorState.mat');
for i = 1:length(editorState)
    jEditorServices.openDocument(editorState(i))
end
```

This functionality was encapsulated into a small but useful ***setEditorState*** utility:[68]

```
setEditorState('projectA','save');
setEditorState('projectB','load');
```

As an alternative, we can preserve and later load a copy of the current *MATLAB_Editor_State.xml* file in the ***prefdir*** folder (the editor will need to be closed and reopened for changes to take effect). This file stores both document names and their code-folding state (but not breakpoints or bookmarks).

As noted above, in R2010b (MATLAB 7.11), the Java class interface has changed. In R2010b, we get the editor handle as follows:[69]

```
jEditorServices = com.mathworks.mlservices.MLEditorServices;
jEditorApp = jEditorServices.getEditorApplication;
```

The method names and parameters have also changed from those presented above. In R2010b, interesting `jEditorApp` methods are (use the ***uiinspect*** utility for more info):

- *close(), closeNoPrompt()*
- *findEditor(com.mathworks.matlab.api.datamodel.StorageLocation)*
- *getActiveEditor()*
- *getOpenEditors()*

- *getEditor(com.mathworks.matlab.api.datamodel.StorageLocation)*
- *getLastActiveEditorViewClient()*
- *isEditorOpen(com.mathworks.matlab.api.datamodel.StorageLocation)*
- *isEditorOpenAndDirty(com.mathworks.matlab.api.datamodel.StorageLocation)*
- *newEditor(java.lang.String)*
- *openEditor(java.io.File)*
- *openEditorForDebug(java.io.File, int)*
- *openEditorForExistingFile(java.io.File)*

As can be seen from this new interface, each edited document has a separate `com.math-works.mde.editor.MatlabEditor` object. The *getActiveEditor()* method returns the `MatlabEditor` object for the currently edited document, and the others can be retrieved via the *getOpenEditors()*, which returns a `java.util.Collections.UnmodifiableList` of such `MatlabEditors`.

Each document's functionality can be accessed via its `MatlabEditor`'s methods (*bringToFront(), close(), goToLine(...), reload(), replaceText(...), setEditable(...)*, etc.) or properties (e.g. **CaretPosition, Document, Language, Length, LongName, ShortName, Selection, Text**, etc.):

- *appendText(java.lang.String)*
- *bringToFront()*
- *close(), closeNoPrompt(), dispose()*
- *fireEditorEvent(com.mathworks.matlab.api.editor.EditorEvent)*
- *firePropertyChange(java.lang.String, java.lang.Object, java.lang.Object)*
- *getCaretPosition()*
- *getComponent()*
- *getDocument()*
- *getLanguage()*
- *getLength()*
- *getShortName(), getLongName(), getStorageLocation()*
- *getProperty(java.lang.Object)*
- *getSelection()*
- *getText()*
- *getTextWithSystemLineEndings()*
- *goToFunction(java.lang.String, java.lang.String)*
- *goToLine(int, boolean), goToLine(int, int)*
- *insertTextAtCaret(java.lang.String)*
- *isBuffer()*
- *isDirty()*
- *isEditable()*
- *isMCode()*
- *isOpen()*

- *lockIfOpen(), unlock()*
- *negotiateSave()*
- *refreshMenus()*
- *reload()*
- *replaceText(java.lang.String, int, int)*
- *setCaretPosition(int)*
- *setEditable(boolean)*
- *setSelection(int, int)*
- *setStatusText(java.lang.String)*
- *smartIndentContents()*

8.4.2 The Editor Frame Object

The jEditor handle is actually a container for many internal panels (toolbars, etc.) and documents. The entire object hierarchy can be seen with the ***FindJObj*** utility:

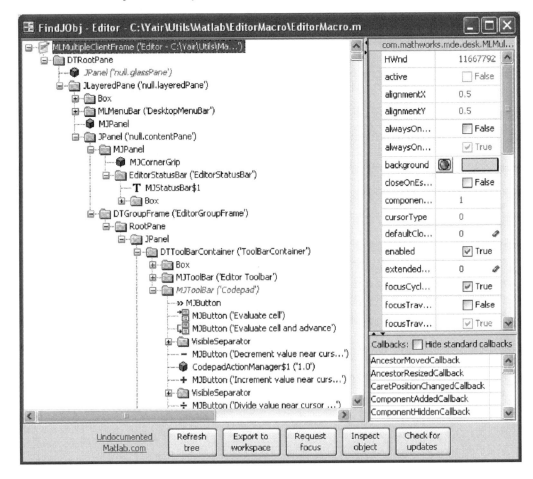

MATLAB Editor object hierarchy as seen by findjobj(jEditor) **or:** jEditor.findjobj

We can see the Editor hierarchy tree is rather complex. Here is a simplified version:

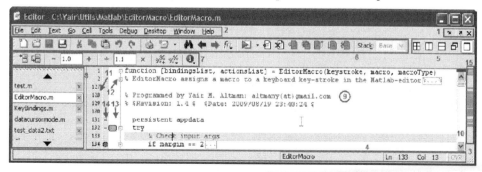

(See color insert.)

MLMultipleClientFrame (this is our top-level jEditor object)
  - DTMaximizedButtonPanel (1 — docking controls)
  - MLMenuBar (2 — main menu bar)
  - MJCornerGrip (3 — window lower-right corner grip)
  - EditorStatusBar (4 — status bar)[†]
  - DTGroupFrame
    - DTToolBarContainer
      - ArrangementControls (5 — document layout controls)
      - MJToolBar (6 — main Editor toolbar)
      - MJToolBar (7 — cell-mode or "codepad" toolbar)
    - DTDocumentContainer
      - DTDocumentBar (8 — document labels scroll-pane)
      - DTMaximizedPane
        - DTClientFrame (hidden — a nonactive document)
        - DTClientFrame (hidden — a nonactive document)
        - DTClientFrame (active document)
          - EditorSyntaxTextPane (9 — main editable text area)
          - ScrollBar (10 — vertical scrollbar)
          - ScrollBar (horizontal scrollbar)
          - GlyphGutter (11 — line numbers)
          - MWCodeFoldingSideBar (12 — code folding lines and icons)
          - BreakpointPanel (13 — breakpoints and executable lines icons)
          - ExecutionPanel (14 — bookmarks and execution arrow icons)
          - MessagePanel (15 — m-lint code-analysis icons)
        - (other hidden nonactive documents) ...

Note that this list was generated by MATLAB R2008a on Windows XP with a specific Editor document layout; other releases, platforms or layouts might vary.

[†] Statusbar customization was described in Section 4.7. jEditor.*setStatusText(...)* updates the statusbar text.

Navigating down the hierarchy tree is easily done using the *getComponent()* method, as the following example to retrieve the main toolbar object illustrates. However, in real-life programs, extra care should be taken to account for other MATLAB releases, platforms and Editor layouts, all of which may cause changes to the hierarchy tree:

```
jLayeredPane = jEditor.getComponent(0).getComponent(1);
dtGroupFrame = jLayeredPane.getComponent(3).getComponent(1);
jToolBarPanel = dtGroupFrame.getComponent(0).getComponent(0);
dtToolBarContainer = jToolBarPanel.getComponent(0);
mjToolBar = dtToolBarContainer.getComponent(1);
```

The `jEditor` handle to the `MLMultipleClientFrame` at the top of the hierarchy tree has over 300 invokable methods and close to 200 gettable/settable properties. My ***uiinspect*** utility, described above, facilitates the discovery of interesting things we can programmatically do with the Editor handle:

```
uiinspect(jEditor); % or: jEditor.uiinspect
```

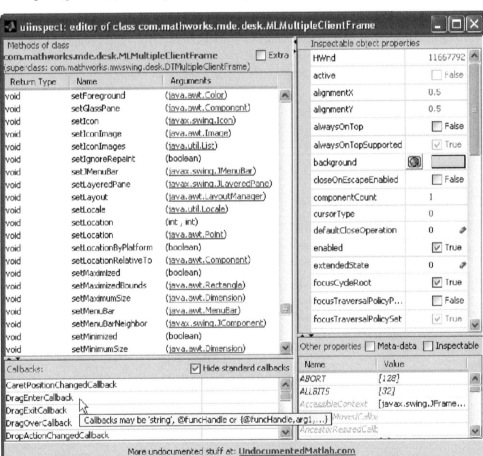

MATLAB Editor methods, callbacks and properties as seen by *uiinspect*

Using `jEditor` at the top-level Editor-window level, we can prevent the window's resizing, update its status bar, modify its toolbar/menu-bar, control docking/position/maximize/minimize and do other similar fun things:

```
% Actions via built-in methods:
jEditor.setResizable(false);
jEditor.setStatusText('testing 123...');
jEditor.setTitle('This is the Matlab Editor');

% Equivalent actions via properties:
set(jEditor, 'Resizable', 'off');
set(jEditor, 'StatusText', 'testing 123...');
set(jEditor, 'Title', 'This is the Matlab Editor');
```

Some other interesting methods and properties exposed by the `jEditor` handle:

- *getStatusBar/setStatusBar* (or **StatusBar** property) — returns or sets the handle to the `EditorStatusBar` panel at the Editor window's bottom (#4 in the Editor screenshot above). Updating the panel's main `javax.swing.JLabel` is easy:

    ```
    sbLabel = jEditor.getStatusBar.getComponent(0);
    set(sbLabel, 'Foreground',java.awt.Color.blue, 'Text','testing...');
    ```

    The entire status bar can be hidden by using *setStatusBarVisible(false)* (or: *set(`'StatusBarVisible`',0)*); to redisplay, simply set this property to true (or 1).

- *getWindowMenu, getDesktopMenu* (or corresponding read-only properties) — return a handle to the "Window" and "Desktop" children of the main `MLMenuBar` menu bar (#2 in the Editor screenshot above); *getJMenuBar/setJMenuBar* (or the **JMenuBar** property) returns or sets the handle to the `MLMenuBar` menu bar itself. We can customize menu items just like Figure menus (see Section 4.6).

While the `jEditor` and `jEditorServices` object reference handles enable access to many Editor niches, some aspects remain inaccessible. I discovered an example when a user requested[70] access to Editor bookmarks. Apparently, these are displayed in the `ExecutionPanel` panel (#14 in the Editor screenshot above), but I could not find a way to access them programmatically.†

8.5 Keyboard Bindings

Over the past years, there have been quite a few requests to enable keyboard macros and key-binding modifications in the MATLAB editor.[71] Some posters have even noted this lack as their main reason to use an external editor.

Based on the information presented in the previous section and some helpful feedback of early adopters, my ***EditorMacro*** utility on the MATLAB File Exchange[72] now provides a

---

† An ugly workaround: loop over all open files, invoke the editor's default action for <F2> ('next-bookmark' action reported by the ***EditorMacro*** utility) until detecting an earlier line number. To set bookmarks, use the 'toggle-book-mark' action.

solution for this need on all MATLAB releases since 6.0 (R12). Following ***EditorMacro***, another user has submitted a similar utility (***KeyBindings***[73]) that complements ***EditorMacro*** in several aspects. In this section, I will detail some of the inner workings, which heavily rely on undocumented MATLAB features.

MATLAB 7.9 (R2009b) introduced key-binding customization of the MATLAB Editor and Desktop as part of the Systems Preferences window.[74] The desktop design team appears to have done a good job of enabling easy keyboard shortcuts customization, saving/loading sets of shortcuts, and so on, using a new easy-to-use preference panel. However, there may still be reasons for using ***EditorMacro*** and/or ***KeyBindings***:

- **Earlier MATLAB versions** — those who have a MATLAB release earlier than R2009b have no option but to use these utilities. The first version of ***EditorMacro*** even supported MATLAB 6.0, which is a decade old!
- **Programmatic access** — some users may wish to have programmatic access to the keyboard bindings. For example, by saving sets of bindings in an m-file and accessing any of these sets via GUI or the desktop shortcuts toolbar. Note that in R2009b, these preferences can probably be accessed programmatically via the preferences interface (see Section 8.2).
- **Understanding the underlying workings** of the MATLAB desktop/editor, for those interested in exploring and using these undocumented subjects.

User-contributed utilities, especially those relying on undocumented features like ***EditorMacro*** and ***KeyBindings***, will always pale next to slick GUI preferences that are well-integrated by design. They do have a place in niche usages, as explained above. But the hope is that all these needs will eventually be addressed by well-documented integrated features, as has happened in this particular case.

8.5.1 *Inserting/Replacing Text*

In a nutshell, ***EditorMacro*** sets the **KeyPressedCallback** property (see Section 3.4) for each of the editor's document panes, to an internal function. This internal function then checks each keystroke against a list of registered keybindings. The list itself is persisted in the editor object's hidden **ApplicationData** property (accessible via the ***getappdata/setappdata*** built-in functions). If a match is found, then the associated macro is invoked.

Depending on the macro type, text can be inserted at the current editor caret position (or as a replacement of the currently selected text); or a nontext MATLAB function/command/action can be invoked. This enables ***EditorMacro*** to be used for quickly inserting code templates (header comments, ***try-catch*** blocks, etc.) or for automating MATLAB unit testing.

Here is a typical usage example: start by defining a simple function that returns a dynamic header comment:

```
function comment = createHeaderComment(hDocument, eventData)
  timestamp = datestr(now);
  username = getenv('username');
  %computer = getenv('computername');  % unused
  lineStr = repmat('%',1,35);
  comment = sprintf(...
      ['%s\n' ...
       '%% \n' ...
       '%% Name:    functionName\n' ...
       '%% \n' ...
       '%% Desc:    enter description here\n' ...
       '%% \n' ...
       '%% Inputs:  enter inputs here\n' ...
       '%% \n' ...
       '%% Outputs: enter outputs here\n' ...
       '%% \n' ...
       '%% Created: %s\n' ...
       '%% \n' ...
       '%% Author:  %s\n' ...
       '%% \n' ...
       '%s\n'], ...
      lineStr, timestamp, username, lineStr);
end  % createHeaderComment
```

Now define a macro to use this function, and another simple ***try-catch*** template:

```
>> EditorMacro('Alt-Control-h', @createHeaderComment);
>> macroStr = ['try\n  % Main code here\ncatch\n  ' ...
                '% Exception handling here\nend'];
>> bindings = EditorMacro('Ctrl alt T', macroStr)
bindings =
    'ctrl alt pressed H'    @createHeaderComment    'text'    'text'
    'ctrl alt pressed T'              [1x60 char]   'text'    'text'
```

Now start with a blank document and click <Ctrl>-<Alt>-H and <Ctrl>-<Alt>-T. This will automatically insert the following text into the document:

```
%%%%%%%%%%%%%%%%%%%%%%%%%%%%%%%%%%%%%
%
% Name:    functionName
%
% Desc:    enter description here
%
% Inputs:  enter inputs here
%
% Outputs: enter outputs here
%
% Created: 01-Jul-2009 23:31:46
%
```

```
% Author:  Yair Altman
%
%%%%%%%%%%%%%%%%%%%%%%%%%%%%%%%%%%%%
try
    % Main code here
catch
    % Exception handling here
end
```

Keybindings are normalized using Java's built-in `javax.swing.KeyStroke.`*getKey-Stroke()* method, to enable the user a very wide range of keystroke naming formats (e.g., 'Alt-Control-T' or 'ctrl alt t').

Text can also be computed dynamically by the called macro function. For example, this macro inserts the current timestamp:

```
EditorMacro('Alt control t', @(a,b)datestr(now), 'text');
```

Note the odd-looking definition of ***datestr***: This is because each ***EditorMacro*** macro function must accept at least two input arguments (see the following section for details). It is ok to ignore these input arguments, but they must be defined since they will be passed to the macro function in runtime, and unless defined a runtime error will be thrown.

I have taken great pains to make ***EditorMacro*** compatible with all MATLAB versions since 6.0 (R12). This was no easy feat: MATLAB 7 made some significant changes to the editor layout. Discovering how to get a handle to the MATLAB 6 editor object took some hours of trial-and-error — the result is listed at the beginning of Section 8.4. Once I had this handle, listing its display hierarchy was simple and the modifications were generally straightforward, although nontrivial: different quirks due to missing default type-castings, missing eventData in invoked callbacks, and so on. ***EditorMacro***'s source code now has clearly marked MATLAB 6 segments.

Another complication arose due to the different layout used for floating/maximized/tiled document layout in the editor. Yet another was due to the different behavior (at least on MATLAB 6) between a one-document and a multi-document editor view.

Due to the way keyboard events are processed by the MATLAB editor, **KeyPressedCallback** needs to be set separately for all the open document panes and split-panes. Since we wish newly opened documents to recognize the macro bindings, we also need to set the common container ancestor's **ComponentAddedCallback** to an internal function that will handle the **KeyPressedCallback** instrumentation for each newly opened document. Again, this is done differently for MATLAB 6 and 7.

Note that ***EditorMacro*** relies on the MATLAB Editor's and Command Window's internal display layout, which is very sensitive to modification between MATLAB releases (as it has between MATLAB 6 and 7, for example).

Here is a screenshot of ***EditorMacro***'s report of existing keybindings on my system:

```
'ctrl pressed Q'                  'Exit MATLAB'                        'run'    'cmdwin menu action'
'ctrl pressed Z'                  'undo'                               'run'    'cmdwin menu action'
'ctrl pressed W'                  'cut-to-clipboard'                   'run'    'cmdwin menu action'
'alt pressed W'                   'copy-to-clipboard'                  'run'    'cmdwin menu action'
'ctrl pressed Y'                  'paste-from-clipboard'               'run'    'cmdwin menu action'
'pressed F10'                     'debug-step'                         'run'    'cmdwin menu action'
'pressed F11'                     'debug-step-in'                      'run'    'cmdwin menu action'
'shift pressed F11'               'debug-step-out'                     'run'    'cmdwin menu action'
'pressed F5'                      'debug-continue'                     'run'    'cmdwin menu action'
'shift pressed F5'                'exit-debug'                         'run'    'cmdwin menu action'
'shift ctrl pressed U'            'Undock Command Window'              'run'    'cmdwin menu action'
'ctrl pressed 0'                  '0 Command Window'                   'run'    'cmdwin menu action'
'ctrl pressed 1'                  '1 Command History'                  'run'    'cmdwin menu action'
'ctrl pressed 2'                  '2 Current Directory'                'run'    'cmdwin menu action'
'ctrl pressed 3'                  '3 Workspace'                        'run'    'cmdwin menu action'
'ctrl pressed 4'                  '4 Help'                             'run'    'cmdwin menu action'
'ctrl pressed 5'                  '5 Profiler'                         'run'    'cmdwin menu action'
'ctrl pressed 6'                  '6 Figure Palette'                   'run'    'cmdwin menu action'
'ctrl pressed 7'                  '7 Plot Browser'                     'run'    'cmdwin menu action'
'ctrl pressed 8'                  '8 Property Editor'                  'run'    'cmdwin menu action'
'shift ctrl pressed 0'            'Editor'                             'run'    'cmdwin menu action'
'shift ctrl pressed 1'            'Figures'                            'run'    'cmdwin menu action'
'shift ctrl pressed 2'            'Web Browser'                        'run'    'cmdwin menu action'
'shift ctrl pressed 3'            'Variable Editor'                    'run'    'cmdwin menu action'
'shift ctrl pressed 4'            'File and Directory Comparisons'     'run'    'cmdwin menu action'
'shift ctrl pressed 5'            'special figures'                    'run'    'cmdwin menu action'
'pressed F1'                      'Product Help'                       'run'    'cmdwin menu action'
'ctrl pressed D'                  'open-selection'                     'run'    'cmdwin native action'
'ctrl pressed D'                  'open-selection'                     'run'    'editor menu action'
'ctrl alt pressed T'              @(a,b)datestr(now)                   'text'   'text'
'alt pressed B'                   'beep'                               'run'    'editor native action'
'alt pressed B'                   'beep'                               'run'    'cmdwin native action'
>> bindings = EditorMacro('ctrl r')
bindings =
'ctrl pressed R'     'comment'              'run'    'editor native action'
'ctrl pressed R'     'inc-search-backward'  'run'    'cmdwin native action'
'ctrl pressed R'     'comment'              'run'    'editor menu action'
>> newBindings = EditorMacro('alt b','beep','run');
```

Screenshot of currently-defined key bindings, as reported by *EditorMacro*

8.5.2 Running Action Macros

For running action macros, as opposed to text insertion, we must specify a callback function, which returns no output value and accepts at least two input arguments:[75]

- sourceObject — The macro target, which is the Editor document's text pane (an EditorSyntaxTextPane object, described in Section 8.4.2), or the Command Window text area (an XCmdWndView object, described in Section 8.3).
- eventData — a java.awt.event.KeyEvent[76] object that contains the key-stroke event data, including alt/ctrl/shift depression state and the clicked character.

For example, let us set the <Ctrl-E> combination to a macro moving to the end-of-line (unix-style — equivalent to <End> on Windows), and <Ctrl-Shift-E> to a similar macro doing the same while also selecting the text (like <Shift-End> on Windows). We shall even use the same macro code, by simply checking in the eventData whether the <Shift> key is depressed:

First, let us define our macro function and place it in EOL_Macro.m:

```
function EOL_Macro(hDocument,eventData)

    % Find the position of the next EOL mark
    currentPos = hDocument.getCaretPosition;
    docLength = hDocument.getLength;
```

```
    textToEOF = char(hDocument.getTextStartEnd(currentPos,docLength));
    nextEOLPos = currentPos+find(textToEOF<=13,1)-1;  % next CR/LF pos
    if isempty(nextEOLPos)
        % no EOL found (=> move to end-of-file)
        nextEOLPos = docLength;
    end

    % Do action based on whether <Shift> was pressed or not
    %get(eventData)    % for debugging purposes
    if eventData.isShiftDown
        % Select to EOL
        hDocument.moveCaretPosition(nextEOLPos);
    else
        % Move to EOL (without selection)
        hDocument.setCaretPosition(nextEOLPos);
    end
end  % EOL_Macro
```

...and now let us activate this macro in the MATLAB Command Window:

```
>> macros = EditorMacro('ctrl-e',@EOL_Macro,'run');
>> macros = EditorMacro('ctrl-shift-e',@EOL_Macro,'run')
macros =
    'ctrl pressed E'        @EOL_Macro    'run'  'user-defined macro'
    'shift ctrl pressed E'  @EOL_Macro    'run'  'user-defined macro'
```

For a full list of methods made available in the hDocument source object, I suggest using the **_uiinspect_** utility. We will discover, for example, the very handy method of *delete(startPos,endPos)*, which can be used for defining character/word/line/sentence deletion macros. Similarly, the *insert(startPos,string)* method can be used to insert text.[†] I have used this to answer a user request of binding (Emacs-style) <Ctrl-O> to an insertion of a new line beneath the current line without moving the caret:[77]

```
EditorMacro('ctrl-o', @(ed,evd)ed.insert(...
    ed.getLineEndFromPos(ed.getCaretPosition),sprintf('\n')), 'run');
```

I have used the long one-liner above to illustrate a point of the ease in which action macros can be defined. In real life, however, it would probably be easier to debug and maintain more verbose code such as this:

```
function NL_Macro(hDocument,eventData)
    currentPos = hDocument.getCaretPosition;
    eolPos = hDocument.getLineEndFromPos(currentPos);
    hDocument.insert(eolPos,sprintf('\n'));
end  % NL_Macro

EditorMacro('ctrl-o', @NL_Macro, 'run');
```

[†] This is correct for MATLAB 7 releases — in MATLAB 6 the input arguments order for *insert()* is reversed.

When running a macro, we might wish to save the document before (or possibly after) our modifications, as requested by a blog reader. Here is how to do so:

```
function my_Macro(hDocument,eventData)
    filename = jDocument.getFilename;

    % do some user-defined stuff

    % Now save the file:
    com.mathworks.mlservices.MLEditorServices.saveDocument(filename);
end % my_Macro
```

User data can also be passed to the macro functions, as the third and subsequent parameters, just as for regular MATLAB callbacks. For example,

```
EditorMacro('Shift-Control d', {@computeDiameter,3.14159}, 'run');
% In this case, 3.14159 is computeDiameter()'s 3rd input argument:
%     function computeDiameter(source,keyEvent,data)
```

8.5.3 Running Built-In Actions

Menus in MATLAB (and in Java applications in general) are connected to corresponding action methods that are invoked whenever the menu item is selected.[78] Actions are identified by name, which is a lowercase dash-separated description such as 'selection-up' (this format is familiar to Emacs users).

Naturally, there are many more possible actions than displayed menu items. In fact, it turns out that both the MATLAB Editor and the Command Window have some 200 built-in (native) actions, about half of them common, giving a total of some 300 unique native actions. Of these, only some 100 have default (pre-assigned) key-bindings in MATLAB. A few dozen actions even have multiple key-bindings. For example, the 'selection-up' action is assigned to both 'shift pressed UP' (=<shift>-<up>) and 'shift pressed KP_UP' (<shift>-<Keypad-up>):

```
>> [bindings, actions] = EditorMacro
actions =
...[snip]
 'selection-page-down'    'shift pressed PAGE_DOWN'    'editor native action'
 'selection-page-up'      'shift pressed PAGE_UP'      'editor native action'
 'selection-previous-word'          {2x1 cell}        'editor native action'
 'selection-up'                     {2x1 cell}        'editor native action'
 'set-read-only'                           []         'editor native action'
 'set-writable'                            []         'editor native action'
 'shift-insert-break'     'shift pressed ENTER'        'editor native action'
 'shift-line-left'        'ctrl pressed OPEN_BRACKET'  'editor native action'
 'shift-line-right'       'ctrl pressed CLOSE_BRACKET' 'editor native action'
 'shift-tab-pressed'      'shift pressed TAB'          'editor native action'
...[snip]...
```

Even more interestingly, apparently some 200 actions do not have any pre-assigned default key-bindings, such as 'set-read-only' and 'set-writable' in the snippet above. Let us take the 'match-brace' action for example. This sounded promising so I assigned it an unused key-binding and indeed found that it can be very useful: if the cursor is placed on a beginning or end of some code, then clicking the assigned key-binding will jump the cursor to the other end, and then back again. This works nicely for (..), [..], for..end, try..end, if..end, and so on.

```
>> % Ensure that <Alt>-M is unassigned
>> bindings = EditorMacro('alt m')
bindings =
    Empty cell array: 0-by-4

>> % Assign the key-binding and verify
>> EditorMacro('alt m','match-brace','run');
>> bindings = EditorMacro('alt m')
bindings =
    'alt pressed M'  'match-brace'  'run'  'editor native action'
```

Some action assignments that I have found very useful:[†]

```
EditorMacro('ctrl back_space', 'delete-previous-word', 'run');
EditorMacro('ctrl back_space', 'remove-previous-word', 'run');
EditorMacro('ctrl delete',     'delete-next-word',     'run');
EditorMacro('ctrl delete',     'remove-next-word',     'run');
```

Here is the code snippet that retrieves the actions from the Editor's Java Map object:

```
% Get all available actions even those without any key-binding
function actionNames = getNativeActions(hEditorPane)
   try
      actionNames = {};
      actionKeys = hEditorPane.getActionMap.allKeys;
      actionNames = cellfun(@char,cell(actionKeys),'Uniform',0);
      actionNames = sort(actionNames);
   catch
      % never mind...
   end
end  % getNativeActions

% Get all active native shortcuts (key-bindings)
function accelerators = getAccelerators(hEditorPane)
   try
      accelerators = cell(0,2);
      inputMap = hEditorPane.getInputMap;
      inputKeys = inputMap.allKeys;
      accelerators = cell(numel(inputKeys),2);
      for ii = 1 : numel(inputKeys)
```

[†] The reason for the apparent duplication here is that corresponding Editor and Command Window actions sometimes have slightly different names, as in this case.

```
          thisKey = inputKeys(ii);
          thisAction = inputMap.get(thisKey);
          accelerators(ii,:) = {char(thisKey), char(thisAction)};
      end
      accelerators = sortrows(accelerators,1);
   catch
      % never mind...
   end
end  % getAccelerators
```

Menu retrieval was more difficult: while it is possible to directly access the menubar reference (jMainPane.*getRootPane.getMenuBar*), the menu items themselves are not visible until their main menu item is clicked (displayed). The only way I know to access menu actions/keybindings is to read them from the individual menu items (if anyone knows a better way please tell me — perhaps some central key-listener repository?). Therefore, a simulation of the menu-click events is done† and the menu hierarchy is traveled recursively to collect all its actions and key-bindings.

Unfortunately, MATLAB menus are dynamically recreated whenever the Editor is docked/undocked, or a separate type of file is edited (e.g., switching from an m-file to a c-file). Similarly, whenever the active desktop window changes from the Command Window to another desktop-docked window (e.g., Command History). In all these cases, the dynamically recreated menus override any conflicting key-binding previously done with *EditorMacro*.

Another limitation of *EditorMacro* is that Multi-key bindings are still not reported properly, nor can they be assigned. For example, the editor menu action 'to-lower-case' has a pre-assigned default key-binding of <Alt>-<U>-<L>, but this is reported as unassigned. Of course, we can always add another (single-key) assignment for this action, for example, <Alt>-<Ctrl>-<L>.

Both of these limitations have a workaround in Perttu Ranta-aho's *KeyBindings* utility,[79] which nicely complements *EditorMacro*. In addition to *KeyBindings*, there have been other follow-on submissions on the File Exchange that provide sets of *EditorMacro* macros.[80]

EditorMacro still has some other unresolved limitations, which will hopefully be resolved in future releases of this utility:

- Key bindings are sometimes lost when switching between a one-document editor and a two-document one (i.e., adding/closing the second doc).
- Key bindings are not saved between editor sessions.
- In split-pane mode, when inserting a text macro on the secondary (right/bottom) editor pane, both panes (and the actual document) are updated but the secondary pane does not display the inserted macro (the primary pane looks ok).

Recall the standard warning about the use of the Event Dispatch Thread (EDT, see Section 3.2): Actions that affect the GUI need to be invoked asynchronously (via the EDT) rather than synchronously (on the main MATLAB thread). This is not a real problem in the editor, but it is

---

† Using menuItem.*doClick()*, described in Section 4.6.4. A similar trick is used by the *findjobj* utility (see Section 7.2.2).

indeed an acute issue in the Command Window: Unless we use EDT, we would get ugly red stack-trace exceptions thrown on the Command Window whenever we run our ***EditorMacro***-assigned macro. Here is the code snippet that solves this:

```
try
    % Matlab 7:
    % Note: it is better to use replaceSelection() than insert()
    %jEditorPane.insert(caretPosition, macro);
    try
        % Try to dispatch on EDT
        awtinvoke(jEditorPane.java, 'replaceSelection', macro);
    catch
        % no good - try direct invocation
        jEditorPane.replaceSelection(macro);
    end
catch
    % Matlab 6:'
    % Note: it is better to use replaceRange() than insert()
    %jEditorPane.insert(macro, caretPosition);'
    try
        % Try to dispatch on EDT
        awtinvoke(jEditorPane.java, 'replaceRange', macro, ...
                jEditorPane.getSelStart, jEditorPane.getSelEnd);

    catch
        % no good - try direct invocation
        jEditorPane.replaceRange(macro, jEditorPane.getSelStart, ...
                                    jEditorPane.getSelEnd);
    end
end   % try-catch block
```

One final note: ***EditorMacro*** uses the Java containers that underlie the MATLAB Editor and Command Window, to gain access to the list of key-bindings. Some time after I have created ***EditorMacro***, I discovered that this functionality can also be done directly: `com.mathworks.services.binding.MatlabKeyBindings.`*getManager()* returns a `com.mathworks.mwswing.binding.KeyBindingManager` object that contains all the keybindings for all contexts (Editor and Commend Window, in both Emacs & non-Emacs configurations). `KeyBindingManager.`*setCurrentKeyBindingSet(setName)* can be used to switch binding sets; *parseAndRegisterCustomKeyBindingSet(...)* can be used to load a new or modified set.

Key-bindings, both the default and the user-defined, are stored in XML files. The default files (for Emacs, Windows, and Macs) can be found in the *%matlabroot%/java/jar/services.jar* file (under com/mathworks/services/binding/resources/), that can be opened using any zip utility (e.g., WinRar, WinZip, or unzip). The user-defined XML files are stored in the user's ***prefdir*** folder, in files named something similar to *Pre2009bWindowsDefaults.xml*.

---

† Note that the MATLAB 6 method *replaceRange()* was renamed to *replaceSelection()* in MATLAB 7.

‡ Note the reverse order of the *insert()* input arguments in MATLAB 6 compared to MATLAB 7.

8.6 Workspace

The MATLAB Workspace pane is just another Desktop client, which hosts a table of variable names and attributes (class, value, size, etc.). A CSSM user asked[81] whether it is possible to modify the appearance of the Bytes column in the Workspace pane, so that it will present data in KBytes rather than in Bytes. Here is the solution:[†]

First, we need to retrieve the Workspace table's Java reference handle:

```
jDesktop = com.mathworks.mde.desk.MLDesktop.getInstance;
jWSBrowser = jDesktop.getClient('Workspace');
jWSTable = jWSBrowser.getComponent(0).getComponent(0).getComponent(0);
```

Next, we note that `jWSTable` is a simple Java Swing `JTable`, and as such we can easily modify its column header (assume that the Bytes column is the third column => column index #2 in Java):

```
jWSTable.getColumnModel.getColumn(2).setHeaderValue('KBytes');
jWSBrowser.repaint;
```

Modifying the column's behavior to display 1/1024 of the initial values is trickier. We can use a custom `TableCellRenderer`,[82] replacing WorkspaceTable's `DefaultTableCellRenderer`. Create the following *KBytesCellRenderer.java* file:[83]

```
import javax.swing.SwingConstants.*;
import javax.swing.table.*;

public class KBytesCellRenderer extends DefaultTableCellRenderer
                                implements TableCellRenderer
{
  public java.awt.Component getTableCellRendererComponent(
          javax.swing.JTable table, Object value, boolean isSelected,
          boolean hasFocus, int row, int column)
  {
    java.awt.Component cell = super.getTableCellRendererComponent(
          table, value, isSelected, hasFocus, row, column);
    int bytes = Integer.parseInt(value.toString());
    ((KBytesCellRenderer)cell).setHorizontalAlignment(TRAILING);
    ((KBytesCellRenderer)cell).setText(bytes/1024 + "");  //Bytes => KB
    return cell;
  }
  public KBytesCellRenderer() { super(); }
}
```

Next, compile this file[84] and place the generated *KBytesCellRenderer.class* file in MATLAB's Java classpath (see Section 1.6 for details) and restart MATLAB. All this is only a one-time operation. After restarting MATLAB, we can place the following code in our *startup.m* script:

```
jDesktop = com.mathworks.mde.desk.MLDesktop.getInstance;
```

[†] http://UndocumentedMatlab.com/blog/customizing-matlabs-workspace-table/ (or http://bit.ly/70a72C). In this section, I will assume MATLAB release R2008a (7.6) or higher — the adaptations for earlier releases should be minor.

```
jWSBrowser = jDesktop.getClient('Workspace');
jWSTable = jWSBrowser.getComponent(0).getComponent(0).getComponent(0);
jWSTable.getColumn('Bytes').setHeaderValue('KBytes');
jWSTable.getColumn('Bytes').setCellRenderer(KBytesCellRenderer);
jWSBrowser.repaint;
```

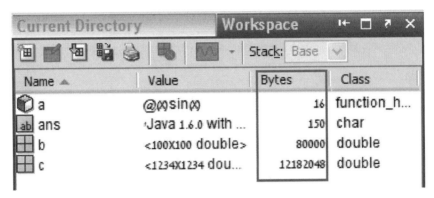

Before: Bytes

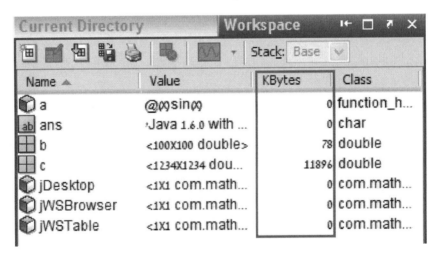

After: KBytes

An improved CellRenderer can highlight cells with illegal values or show a thousands (comma) separator. Just remember to restart MATLAB after each recompilation. Here is the version for the thousands separator:

```
import java.awt.Component;
import javax.swing.JTable;
import javax.swing.SwingConstants.*;
import javax.swing.table.*;
import java.text.NumberFormat;

public class KBytesCellRenderer extends DefaultTableCellRenderer
                            implements TableCellRenderer
```

```
{
  public Component getTableCellRendererComponent(
          JTable table, Object value, boolean isSelected,
          boolean hasFocus, int row, int column)
  {
    Component cell = super.getTableCellRendererComponent(
          table, value, isSelected, hasFocus, row, column);
    ((KBytesCellRenderer)cell).setHorizontalAlignment(TRAILING);
    int bytes = Integer.parseInt(value.toString());
    NumberFormat nf = NumberFormat.getInstance();
    ((KBytesCellRenderer)cell).setText(nf.format(bytes/1024));
    return cell;
  }
  public KBytesCellRenderer() { super(); }
}
```

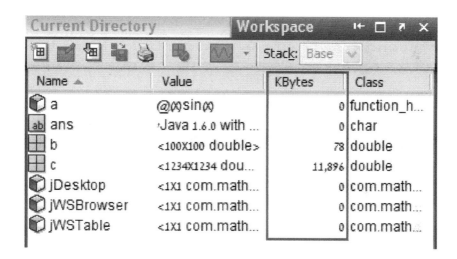

After: formatted Kbytes (US locale)

Even more control over the appearance of the numeric value can be achieved using Java's standard java.text.DecimalFormat[85] that enables using the computer's current Locale settings, which in turn determine the grouping character and location, the number of decimal digits, and other aspects of numeric value presentation.[86]

Note that Java objects appear as having 0 bytes. The reasons for this, and a workaround using Classmexer, were discussed in Section 1.1. Unfortunately, Classmexer cannot be used in our CellRenderer, since deep memory scans might take some time to execute whereas cell-renderers must be super-fast.

Another useful Workspace functionality is the ability to specify user-defined context menus for specific class types. This is done in several *prefspanel.m* files in the MATLAB codebase.†

---

† For example, \toolbox\matlab\audiovideo\prefspanel.m and \toolbox\signal\signal\prefspanel.m

It relies on the following Java-based mechanism, which is unsupported and undocumented, yet has existed in the present form for the past several releases:

```
classes = {'double', 'java.lang.Object'};
menuName = 'My context-menu';
menuItems = {'Inspect', 'Properties', '-', 'class name'};
menuActions = {'inspect($1)', 'get($1)', '', 'class($1)'};
com.mathworks.mlwidgets.workspace.MatlabCustomClassRegistry...
    .registerClassCallbacks(classes,menuName,menuItems,menuActions);
```

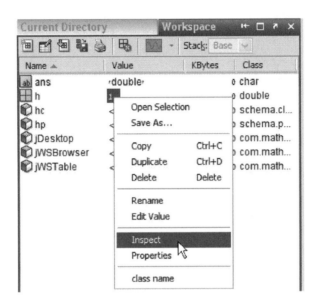

Customizable Workspace table context-menu

A few other supporting static methods are available in com.mathworks.mlwidgets. workspace.MatlabCustomClassRegistry class: *getClassCallbacksInformation (className), registerSimilarClassCallbacks(newClassNames,definedClassName)* and *unregisterClassCallbacks(definedClassName).* For example,

```
disp(com.mathworks.mlwidgets.workspace.MatlabCustomClassRegistry...
    .getClassCallbacksInformation('double'))

java.lang.Object[]:
    'My context-menu'
    [4 element array]
    [4 element array]
```

8.7 Other Desktop Tools

The Desktop contains several utilities and tools that can be accessed via buttons or menus but not programmatically.

One example of such a useful tool is the ***visdiff*** internal file comparison tool, which became documented[87] following a post I made in the CSSM newsgroup.[88]

Several such tools merit special attention and are detailed below: the Profiler, GUIDE, File-search utility, and the Variable Editor.

8.7.1 Profiler

The MATLAB profiler is an extremely helpful debugging tool, helping to diagnose code coverage and performance hotspot problems. It can be accessed via the following Desktop methods:

```
jDesktop = com.mathworks.mde.desk.MLDesktop.getInstance;

% The following methods show/hide the profiler window
jDesktop.showProfiler;      % or: showClient('Profiler')
jDesktop.hideProfiler;      % or: hideClient('Profiler')
jDesktop.closeProfiler;     % or: closeClient('Profiler')
jDesktop.toFront('Profiler');

% The following methods modify the profiler window appearance
jDesktop.setClientMaximized('Profiler',flag);   % flag = true/false
jDesktop.setClientMinimized('Profiler',flag);   % flag = true/false
jDesktop.setClientDocked('Profiler',flag);      % flag = true/false
jDesktop.setClientSelected('Profiler',flag);    % flag = true/false
jDesktop.setClientLocation('Profiler',location);

% The following methods return a status flag (true/false)
flag = jDesktop.isClientDocked('Profiler');
flag = jDesktop.isClientHidden('Profiler');
flag = jDesktop.isClientMaximized('Profiler');
flag = jDesktop.isClientMinimized('Profiler');
flag = jDesktop.isClientSelected('Profiler');
flag = jDesktop.isClientShowing('Profiler');
flag = jDesktop.isClientUnfurled('Profiler');

% The following methods also return true/false flags
flag = jDesktop.hadClient('Profiler');   % in the past
flag = jDesktop.hasClient('Profiler');   % currently

% This method returns the Profiler's Java object handle
>> jProfiler = jDesktop.getClient('Profiler')
jProfiler =
com.mathworks.mde.profiler.Profiler[…]
```

The profiler functionality (as opposed to the GUI) has programmatic access via the following (fully documented) built-in function: ***profile, profsave, profview,*** and ***coveragerpt*** (actually, ***coveragerpt*** is the undocumented function that launches the documented Coverage Report; I do not know why the report is documented but its gateway function is not).

Many of `jProfiler`'s methods have corresponding counterparts in the fully documented built-in ***profile*** function (*start()*, *stop()*, *clear()*, etc.). Some other possibly interesting methods include:

- *getHtmlText(), setHtmlText(string)* — control the displayed text in the Profiler window, which is HTML-rendered.[†]
- Other jProfiler methods: *getLazyProperty(obj), getSelectedLabsFromHtml(), getTempFilesManager(), setCommandText(string), setCurrentLocation(string), setCurrentLocationParallel(string), setNumLabs(int,int), setNumLabsParallel(int,int), setSelectedLab(int).*

To conclude this section, I would like to describe a couple of undocumented profiler features, which are unrelated to GUI or Java:[89]

To turn on memory stats in the profile report, run this (only once is necessary — it will be remembered for future profiling runs in the current MATLAB session):

```
profile -memory on;
profile('-memory','on'); % an alternative
```

To turn on JITC (Just-In-Time Java Compilation[‡]) information, run the following (again, only once is necessary, prior to profile report):

```
setpref('profiler','showJitLines',1);
```

> **Note:** JIT information has been removed in MATLAB 7.12 (R2011a). I assume that this was done so that programmers will not attempt to depend on (and code against) JITC functionality (see Scott Hirsh's comment below).

In R2010b and earlier MATLAB releases, we will now see additional JIT and memory (allocated, freed and peak) information displayed in the profile report, as well as the options to sort by allocated, freed and peak memory:

Profile Summary
Generated 03-Apr-2009 00:33:49 using cpu time.

Function Name	Calls	Total Time	Self Time*	Allocated Memory	Freed Memory	Self Memory	Peak Memory	Total Time Plot (dark band = self time)
uiinspect	1	0.904 s	0.001 s	4410.86 Kb	4372.22 Kb	-0.26 Kb	23.34 Kb	
uiinspect>displayObj	1	0.845 s	0.138 s	3563.64 Kb	3546.64 Kb	2.39 Kb	13.01 Kb	
com.jidesoft.grid.TableUtils (Java-method)	8	0.221 s	0.221 s	0.91 Kb	0.54 Kb	0.38 Kb	0.09 Kb	
uiinspect>getMethodsPane	1	0.221 s	0.036 s	153.38 Kb	151.03 Kb	-0.22 Kb	6.40 Kb	
uiinspect>getPropsPane	1	0.214 s	0.181 s	335.16 Kb	335.07 Kb	-3.69 Kb	9.01 Kb	

Profile summary report with additional memory usage information

---

[†] There are also *getHtmlTextParallel(), setHtmlTextParallel()* methods which I believe are related to the Distributed Computing Toolbox (DCT), but I do not know their exact use. Perhaps all the other *Parallel* methods are also DCT-related.

[‡] JITC can be statically and dynamically turned on/off in MATLAB — see Section 1.9 for details.

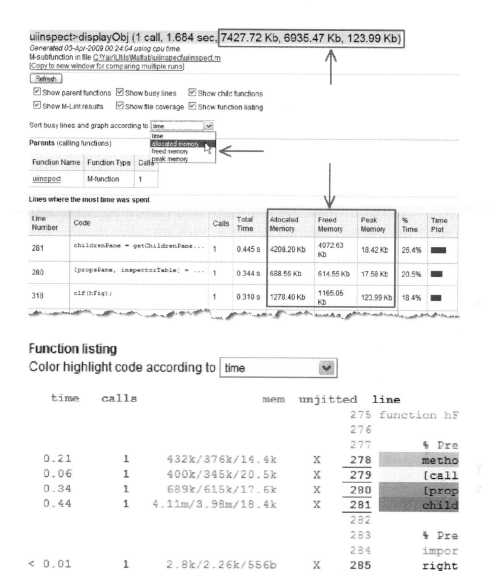

Profile drill-down report with additional memory-usage and JIT information (See color insert.)

For those interested, the references to these two options appear within the code of *profview.m* (line 1199 on R2007b), for the JIT option:

```
showJitLines = getpref('profiler','showJitLines',false);
```

...and in *profile.m* (lines 153-155 on R2011a), for the memory option:

```
if memory ~= -1
    callstats('memory', memory);
end
```

The JIT-profiling feature is permanent, since ***setpref*** is stored across MATLAB sessions; to make the memory-profiling feature permanent, we need to add the following line beneath the ***if memory — end*** block shown above:

```
callstats('memory',3);
```

Note that there appears to be two undocumented additional memory-related options in *profile.m* (lines 311–312):

```
options = {'detail','timer','history','nohistory','historysize',...
           'timestamp', 'memory', 'callmemory', 'nomemory' };
```

However, '-nomemory' appears to simply turn the memory stats collection off, and '-callmemory' is not recognized because of a bug in line 349, which looks for 'call**no**memory'...

```
case 'callnomemory' % should be 'callmemory'
     memory = 2;
```

When this bug is fixed, we see that we get only partial memory information, implying that the '-callmemory' option is not really useful — use '-memory' instead.

Note: When I published this information in the UndocumentedMatlab.com blog, Scott Hirsh, MATLAB's chief product manager, commented as follows:[90]

> *"We would love feedback on the memory profiling feature. We know there's a lot of interest in tools to help figure out where and how memory is chewed up in code, but it turns out to be not so obvious as to what information to present and how. I'll keep an eye out here for comments, or readers can feel free to email me directly. My address is on my MATLAB Central page.*
>
> *I would advise a bit of caution when working with the JIT profiler results. We turned them off after MATLAB 6.5 primarily because the JIT was such a moving target (changing every release) that we really didn't want to encourage users to code against it. That being said, your readers all know the standard warning that comes with your posts — so have fun poking around!!"*

8.7.2 *Find-Files Dialog*

The MATLAB file-finder dialog is used to find files containing some search (text) pattern. This is typically used in the MATLAB Editor (Edit/Find Files menu item) and so most users naturally assume that it is part of the Editor. However, the same menu option is also available in the main MATLAB desktop menu. In fact, this is a desktop tool, and just as other desktop tools, it can be accessed and invoked programmatically:

```
com.mathworks.mde.find.FindFiles.invoke('*.m','undocumented','C:\Yair')
```

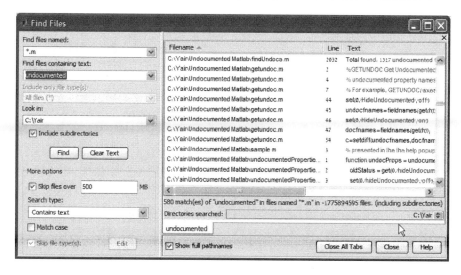

MATLAB's standard Find Files dialog window

8.7.3 GUIDE

GUIDE (MATLAB's **G**raphical **U**ser **I**nterface **D**esign **E**ditor) is very useful for designing simple GUI figures, although experience has shown that it has limitations for complex GUIs. Nevertheless, GUIDE is the tool used by most MATLAB developers when designing GUIs. In this section, I will show a few undocumented customizations that could help make GUIDE sessions more productive.[91]

The starting point is GUIDE's undocumented return value, which is a Java reference to the Layout Editor panel within the GUIDE figure Frame. This handle can be used to access GUIDE components and functionality. We can start by inspecting the interesting GUIDE layout hierarchy using my *FindJObj* utility (see Section 4.2.5), and the associated properties and method using my *UIInspect* utility (see Section 1.3):

```
>> h = guide
h =
Layout Document [untitled]

>> h.getClass
ans =
class com.mathworks.toolbox.matlab.guide.LayoutEditor

>> h.findjobj;
>> h.uiinspect;
```

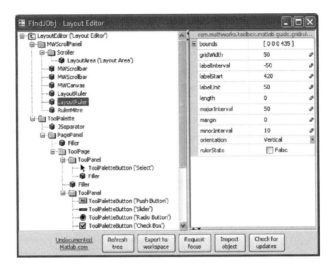

Hierarchy of Layout Editor within the GUIDE frame

To see the hierarchy of the entire GUIDE figure frame, simply run ***FindJObj*** on the frame reference, by either of the two following methods (and similarly for ***UIInspect***):

```
findjobj(h.getFrame);
findjobj(h.getTopLevelWindow);
```

We see that the Layout Editor contains, in addition to the expected `LayoutArea` and two `MWScrollbars`,[†] several ruler-related objects. These rulers can be activated via the GUIDE menu (Tools/Grid and Rulers), or the MATLAB Command Prompt, as follows:

Looking at the ruler properties in ***FindJObj*** or ***UIInspect***, we see a settable boolean property called **RulerState**. If we turn it on, then we see a very handy pixels-ruler. Once we set this property, it remains in effect for every future GUIDE session:

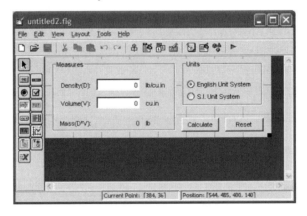

Before: GUIDE with no rulers

[†] These were changed in recent MATLAB releases to standard `MJScrollBar` objects.

```
h.getComponent(0).getComponent(4).setRulerState(true); % Horizontal
h.getComponent(0).getComponent(5).setRulerState(true); % Vertical
```

RulerState actually controls a system preference (**LayoutShowRulers**, a boolean flag) that controls the visibility of both rulers, and persists across MATLAB/GUIDE sessions. To change the visibility of only a single ruler for this GUIDE session only, or on old MATLAB versions (e.g. MATLAB 7.1 aka R14 SP3) that do not have the **RulerState** property, use the *hide()/show()/setVisible(flag)* methods, or set the **Visible** property:

```
% Equivalent ways to show horizontal ruler for this GUIDE session only
hRuler = h.getComponent(0).getComponent(4); % =top horizontal ruler
set(hRuler, 'Visible','on');
hRuler.setVisible(true); % or: hRuler.setVisible(1)
hRuler.show();
```

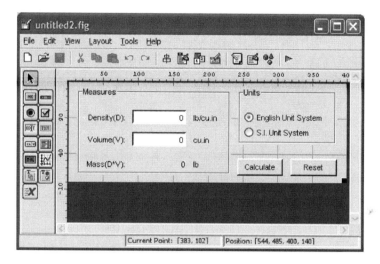

After: GUIDE with pixel rulers

Using `hRuler` properties, we can customize the rulers in manners that are unavailable using the standard GUIDE menu options: We can specify horizontal/vertical grid size, tick and label interval, and other similar ruler properties. For example, let us set a 5-pixel minor tick interval, 25-pixel major interval, labels every 50 pixels, starting offset of 40 pixels and a ruler size limited at 70 pixels:

```
hRuler = h.getComponent(0).getComponent(4); % =top horizontal ruler
set(hRuler, 'MinorInterval',5, 'MajorInterval',25);
set(hRuler, 'LabelInterval',50, 'LabelUnit',50);
set(hRuler, 'Margin',40, 'Length',260);
```

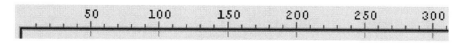

Default pixel ruler

Modified pixel ruler

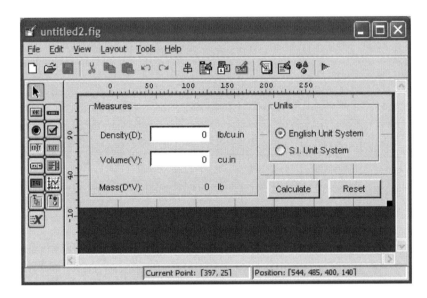

GUIDE with modified pixel rulers

Note that the vertical ruler's labels start (=**LabelStart** property) at the figure's height, and have a *decreasing* **LabelInterval** of –50. This is done because Java coordinates start counting from the top-left corner *downward*, whereas MATLAB counts from the bottom-left *upward* (see Section 3.1.1). In GUIDE, we naturally wish to display the MATLAB coordinates, hence the transformation.

Unfortunately, most properties have no corresponding system property, as far as I could tell. Here is a list of all the GUIDE-related system properties that I found:

- **LayoutActivate** — boolean, controls the ability to run (activate) unsaved figures without confirmation.
- **LayoutChangeDefaultCallback** — boolean, purpose unknown (I can see this preference in my MATLAB.prf file but I have no idea what it does or how it got there).
- **LayoutExport** — boolean, controls the ability to export unsaved figures without confirmation.
- **LayoutExtension** — boolean, controls display of file extension in the GUIDE window title.
- **LayoutFullPath** — boolean, controls display of file path in the GUIDE window title.

- **LayoutGridWidth** — integer, controls the size of the grid boxes.
- **LayoutMCodeComments** — boolean, controls generation of comments for m-file callbacks.
- **LayoutMRU1** to **LayoutMRU9** — string, contain the Most-Recently Used (MRU) full file paths.
- **LayoutQuickStartTab** — integer, controls whether GUIDE's Quick Start dialog (which is shown when GUIDE is launched with no input filename parameter) initially display the "Create New GUI" panel (=0) or the "Open Existing GUI" panel (=1).
- **LayoutShowGuides** — boolean, controls display of blue guidelines.
- **LayoutShowGrid** — boolean, controls display of gray gridlines.
- **LayoutShowRulers** — boolean, controls display of both rulers.
- **LayoutSnapToGrid** — boolean, controls snap-to-grid behavior.
- **LayoutToolBar** — boolean, controls display of the GUIDE widow toolbar.
- **LayoutToolNames** — boolean, controls display of tool names in the components palette.

In addition to these properties, GUIDE has some useful invokable methods. For example, the following method displays the Object Browser, which presents the hierarchical list of figure components:

```
h.showObjectBrowser;
```

And this displays the property inspector:

```
h.showInspector;
```

 Warning: These undocumented features are way deep in unsupported territory. They depend heavily on MATLAB's internal implementation, which may change without any prior notice between MATLAB releases. The very next MATLAB release might break these features, so beware.

8.7.4 Variable (Array) Editor

The Variable Editor (called the "Array Editor" prior to R2008a (MATLAB 7.6), and still called `ArrayEditor` in internal Java classes), is another Java-based Desktop tool. This tool is tightly coupled with the MATLAB Desktop. It can be invoked using the built-in ***openvar*** function, as follows:

```
data = magic(3);

% All the following are equivalent:
openvar('data');
com.mathworks.mlservices.MLArrayEditorServices.openVariable('data');
awtinvoke(com.mathworks.mlservices.MLArrayEditorServices, ...
          'openVariable(Ljava.lang.String;)', 'data');
```

The MATLAB Desktop's Variable Editor

The tight-coupling of the Variable Editor to the Desktop Workspace variables is unfortunate in some respects.[92] One would perhaps expect it to be a generic data editor that merely accepts the data when opened, but in fact it continues to be coupled even later. We can see this by modifying the variable in the Desktop Workspace and seeing the change automatically reflected in the Variable Editor.

An unfortunate side-effect of this design is that the Variable Editor needs to interact with the Desktop Workspace (probably via JMI). Since there is only one MATLAB computational thread, this interaction has to wait for the user's function to exit (the ">>" prompt returning) before it can query the variable data for display in the Editor.

A side-note about the internal MATLAB implementation is necessary to explain the preceding statement: Recall the explanation about MATLAB threads that was presented in Section 3.2: MATLAB only has a single computational thread, so MATLAB code has to finish before the JMI request can be handled. This means that the Variable Editor contents cannot be displayed synchronously by the MATLAB code that invokes it.

The bottom line is that we cannot wait (synchronously) on the Variable Editor in our code. In real-life, this is not a real nuisance, because invoking the Variable Editor from the Command Window immediately finishes the processing chunk, returning to the ">>" prompt and enabling the JMI chunk to be processed and the variable data to be displayed. However, if we wish to display the data from within our MATLAB code, we need to find another, a-synchronous, way.

We can make this work by modifying our request such that a callback function is called when the Variable Editor editing has ended (the window is closed) — this makes the design asynchronous, which enables the JMI interaction:

```
% Open the variable in the Variable Editor
openvar('data');
drawnow; pause(0.5); % wait for client to become visible

% Get handle of variable's client in the Variable Editor
jDesktop = com.mathworks.mde.desk.MLDesktop.getInstance;
jClient = jDesktop.getClient('data');
hjClient = handle(jClient,'CallbackProperties');
```

```
% Instrument the client to fire a callback when it is closed
set(hjClient,'ComponentRemovedCallback',{@my_callback, 'data'});

function my_callback(varEditorObj,eventData,varname)
   data = evalin('caller',varname);
   % do something with this modified data
end % my_callback
```

MLArrayEditorServices provides support services for the Variable Editor. This class's name reflects the tool's pre-R2008a name (Array Editor). Among the class's functions are *openVariable(varName), isEditable(varName)* and *setEditable(varName,flag)*. The *setEditable* method can be used to display non-modifiable (read-only) data:

```
m = magic(4);
com.mathworks.mlservices.MLArrayEditorServices.openVariable('m');
com.mathworks.mlservices.MLArrayEditorServices.setEditable('m',false);
```

References

1. http://www.mathworks.com/matlabcentral/newsreader/view_thread/146899#370458 (or http://tinyurl.com/yhj7kmu).
2. http://blogs.mathworks.com/desktop/2007/03/29/shortcuts-for-commonly-used-code/#comment-5753 (or http://bit.ly/aRvA9c).
3. http://java.sun.com/javase/6/docs/api/javax/swing/Action.html (or http://tinyurl.com/ybmpyxg).
4. http://java.sun.com/javase/6/docs/api/java/awt/Insets.html (or http://tinyurl.com/38calj).
5. http://www.mathworks.com/matlabcentral/newsreader/view_thread/155225#411327 (or http://bit.ly/6sjxUg).
6. http://blogs.mathworks.com/desktop/2007/05/18/do-you-dock-figure-windows-what-does-your-desktop-look-like/#comment-626 (or http://bit.ly/64pGv7) see comments #9 to #11.
7. http://www.mathworks.com/matlabcentral/newsreader/view_thread/128832 (or http://tinyurl.com/yjt6qz9).
8. http://stackoverflow.com/questions/826384/how-can-i-suppress-matlabs-command-window-when-calling-it-from-java#855923 (or http://bit.ly/cwpjLt); http://blogs.mathworks.com/desktop/2010/02/22/launching-matlab-without-the-desktop/#comment-6876 (or http://bit.ly/92Fux6); or this again: http://bit.ly/deRV1Z
9. http://www.mathworks.com/matlabcentral/newsreader/view_thread/303833#822981 (or: http://bit.ly/fxzceC).
10. http://www.mathworks.com/matlabcentral/newsreader/view_thread/285209 (or http://bit.ly/cRTs4g).

11. http://www.mathworks.com/matlabcentral/newsreader/view_thread/265337#693771   (or   http://bit.ly/aXFE17); also see Brad Phelan's earlier post: http://xtargets.com/snippets/posts/show/4 (currently offline — cached version: http://bit.ly/eWrHs3).

12. http://www.mathworks.com/matlabcentral/newsreader/view_thread/152888   (or   http://tinyurl.com/yl8ekyb).

13. http://UndocumentedMatlab.com/blog/setting-the-matlab-desktop-layout-programmatically/   (or   http://tinyurl.com/yd8pcl4).

14. http://blogs.mathworks.com/desktop/2007/08/29/i-came-i-saw-i-created-my-own-desktop-layout/   (or   http://bit.ly/acKIf8).

15. For example, http://www.mathworks.com/matlabcentral/newsreader/view_thread/146522   (or   http://tinyurl.com/yc44d4e).

16. http://blogs.mathworks.com/desktop/2007/08/29/i-came-i-saw-i-created-my-own-desktop-layout/#comment-159 (or http://tinyurl.com/yb93y4n).

17. http://UndocumentedMatlab.com/blog/modifying-default-toolbar-menubar-actions/#comment-49832 (or: http://bit.ly/j4gwja).

18. http://www.mathworks.com/matlabcentral/newsreader/view_thread/151033   (or   http://bit.ly/8atpPR); also see Section 5.4.

19. http://www.mathworks.com/matlabcentral/newsreader/view_thread/246731 (or http://bit.ly/bJTdI5).

20. http://stackoverflow.com/questions/2802233/programmatically-configure-matlab   (or   http://bit.ly/bGsT0l).

21. http://www.mathworks.com/matlabcentral/fileexchange/27567 (or http://bit.ly/aEYBJP).

22. http://4dpiecharts.com/2010/08/23/a-shortcut-to-success/ (or http://bit.ly/bCaKWt).

23. http://UndocumentedMatlab.com/blog/changing-system-preferences-programmatically   (or   http://bit.ly/gZpQZv).

24. http://blogs.mathworks.com/desktop/2010/01/11/the-preferences-folder-matlabprf/ (or http://bit.ly/7rf9t9).

25. http://blogs.mathworks.com/desktop/2009/12/07/the-preferences-directory/ (or http://bit.ly/6cqv5p).

26. http://www.mathworks.com/matlabcentral/newsreader/view_thread/100542 (or http://bit.ly/d8To1G).

27. http://www.mathworks.com/matlabcentral/newsreader/view_thread/154608   (or   http://tinyurl.com/ygg55y8).

28. http://www.mathworks.com/matlabcentral/newsreader/view_thread/151019#383576   (or   http://tinyurl.com/ybm4635).

29. http://java.sun.com/docs/books/tutorial/uiswing/components/textarea.html (or http://tinyurl.com/6o8n2); http://java.sun.com/javase/6/docs/api/javax/swing/JTextArea.html (or http://tinyurl.com/2db6j9).

30. Here's one of many: http://www.mathworks.com/matlabcentral/newsreader/view_thread/158678 (or http://bit.ly/98ht8T).

31. http://UndocumentedMatlab.com/blog/changing-matlab-command-window-colors/   (or   http://tinyurl.com/yjcazox).

32. http://java.sun.com/javase/6/docs/api/javax/swing/text/PlainDocument.html (or http://tinyurl.com/de6479).

33. http://java.sun.com/docs/books/tutorial/uiswing/components/text.html (or http://tinyurl.com/2xg2n).

34. http://UndocumentedMatlab.com/blog/cprintf-display-formatted-color-text-in-command-window/   (or   http://bit.ly/azTQt6).

35. http://www.mathworks.com/matlabcentral/fileexchange/24093 (or http://tinyurl.com/yaucfsk).

36. http://UndocumentedMatlab.com/blog/changing-matlab-command-window-colors-part2/#comment-539 (or http://bit.ly/bWRS58).

37. http://UndocumentedMatlab.com/blog/customizing-help-popup-contents (or http://bit.ly/5VuSAR).

38. http://www.mathworks.com/matlabcentral/newsreader/view_thread/155108 (or http://bit.ly/ahhpJM).

39. http://www.mathworks.com/help/techdoc/ref/doc.html (or http://bit.ly/8g8eUg).

40. http://java.sun.com/developer/technicalArticles/J2SE/Desktop/javase6/modality/ (or http://bit.ly/7dhU1c).

41. http://www.mathworks.com/matlabcentral/fileexchange/25975 (or http://bit.ly/7Smk3y).

42. http://UndocumentedMatlab.com/blog/gui-integrated-browser-control/ (or http://bit.ly/4HbOK0); http://www.mathworks.com/matlabcentral/newsreader/view_thread/243727 (or http://bit.ly/9me22R).

43. An example of this: http://stackoverflow.com/questions/1903516/matlab-displaying-markup-html-or-other-format/1903990#1903990 (or http://bit.ly/5aYj7d).

44. http://www.mathworks.com/matlabcentral/newsreader/view_thread/157032 (or http://bit.ly/alpy0g).

45. http://www.mathworks.com/support/solutions/en/data/1-91A14Y/ (or http://bit.ly/a7C9xS).

46. http://UndocumentedMatlab.com/blog/setprompt-setting-the-matlab-desktop-prompt/   (or   http://bit.ly/5N5Nnz).

47. Also see: http://undocumentedmatlab.com/blog/controlling-callback-re-entrancy/ (or http://bit.ly/qmFwYw).

48. http://www.mathworks.com/matlabcentral/fileexchange/26471 (or http://bit.ly/8TXCDa).

49. http://blogs.mathworks.com/desktop/2007/04/27/tab-completion-will-save-your-fingers/ (or http://bit.ly/cgmukU); http://blogs.mathworks.com/desktop/2009/08/10/tab-to-narrow-completions/ (or http://bit.ly/bxd6Rl).

50. http://www.mathworks.com/matlabcentral/newsreader/view_thread/264550 (or http://bit.ly/cn53PH).

51. http://stackoverflow.com/questions/1842804/tab-completion-of-filenames-as-arguments-for-matlab-scripts (or http://bit.ly/doyVDx).

52. http://www.mathworks.com/help/techdoc/ref/whos.html (or http://bit.ly/9qrlJm).

53. http://www.mathworks.com/help/techdoc/ref/openfig.html (or http://bit.ly/9DmvRB).

54. http://www.mathworks.com/matlabcentral/fileexchange/26830-tabcomplete (or http://bit.ly/caKwQS).

55. http://www.tim.cz/en/nfaq/matlab-simulink/tab-completion.php (or http://bit.ly/8XCCan).

56. http://www.mathworks.com/matlabcentral/newsreader/view_thread/30604 (or http://bit.ly/antBQ0).

57. http://www.mathworks.com/matlabcentral/newsreader/view_thread/257842 (or http://tinyurl.com/y93a9rh).

58. http://stackoverflow.com/questions/1924286/change-the-title-of-the-matlab-command-window (or http://bit.ly/8uND2K).

59. http://stackoverflow.com/questions/3533074/how-do-you-retrieve-the-selected-text-in-matlab (or http://bit.ly/c4Q2Wq).

60. http://UndocumentedMatlab.com/blog/accessing-the-matlab-editor/ (or http://tinyurl.com/yjel9kw).

61. http://www.mathworks.com/matlabcentral/newsreader/view_thread/135588#342288 (or http://tinyurl.com/ylnktyl).

62. http://blogs.mathworks.com/desktop/2009/10/26/the-matlab-editor-at-your-fingertips/ (or http://tinyurl.com/yaongph).

63. http://blogs.mathworks.com/desktop/2009/10/26/the-matlab-editor-at-your-fingertips/#comment-7532 (or http://bit.ly/gm6MHb); the code for this new object is in %matlabroot%\toolbox\matlab\codetools\++matlab\++desktop\++editor\

64. http://www.mathworks.com/matlabcentral/newsreader/view_thread/135588 (or http://tinyurl.com/ybb48zv).

65. http://www.mathworks.com/matlabcentral/newsreader/view_thread/127507 (or http://bit.ly/bChnMK); http://www.mathworks.com/matlabcentral/newsreader/view_thread/136808 (http://bit.ly/bSOSl9); http://www.mathworks.com/matlabcentral/newsreader/view_thread/159050#400521 (or http://bit.ly/boZuIY); http://blogs.mathworks.com/desktop/2009/10/26/the-matlab-editor-at-your-fingertips/#comment-6630 (or http://bit.ly/ct7cl9).

66. http://www.mathworks.com/matlabcentral/newsreader/view_thread/154050#386543 (or http://bit.ly/c775qm); http://www.mathworks.com/matlabcentral/newsreader/view_thread/285757#758517 (or http://bit.ly/bSmdSu).

67. http://blogs.mathworks.com/loren/2009/03/03/whats-in-your-startupm/#comment-30128   (or   http://tinyurl.com/y8wh7jw).

68. http://gkdot.blogspot.com/2009/04/matlab-gem.html (or http://bit.ly/bqjK3A).

69. http://www.mathworks.com/matlabcentral/newsreader/view_thread/291659#783173 (or http://bit.ly/cql8GK).

70. http://UndocumentedMatlab.com/blog/accessing-the-matlab-editor/comment-page-1/#comment-3433 (or http://bit.ly/9vnpfR); also http://www.mathworks.com/matlabcentral/newsreader/view_thread/288201 (or http://bit.ly/awtJ12).

71. For example, http://www.mathworks.com/matlabcentral/newsreader/view_thread/162069 (or http://tinyurl.com/ygfjg89).

72. http://www.mathworks.com/matlabcentral/fileexchange/24615 (or http://tinyurl.com/ydzznxb); http://UndocumentedMatlab.com/blog/EditorMacro (or http://bit.ly/cUJLJ6).

73. http://www.mathworks.com/matlabcentral/fileexchange/25089 (or http://tinyurl.com/yhd4go9).

74. http://blogs.mathworks.com/desktop/2009/09/04/r2009b-is-here/ (or http://tinyurl.com/ykjofmf); http://UndocumentedMatlab.com/blog/r2009b-keyboard-bindings/ (or http://bit.ly/d6VDU9).

75. http://UndocumentedMatlab.com/blog/non-textual-editor-actions/ (or http://tinyurl.com/ye62rjf).

76. http://java.sun.com/javase/6/docs/api/java/awt/event/KeyEvent.html (or http://tinyurl.com/2s63e9).

77. http://blogs.mathworks.com/desktop/2009/09/28/configurable-keyboard-shortcuts-have-arrived/#comment-6617 (or http://tinyurl.com/yfaw9gy).

78. http://UndocumentedMatlab.com/blog/editormacro-v2-setting-command-window-key-bindings/ (or http://bit.ly/9YOvPS).

79. http://www.mathworks.com/matlabcentral/fileexchange/25089 (or http://tinyurl.com/yhd4go9).

80. http://www.mathworks.com/matlabcentral/fileexchange/25122 (or http://tinyurl.com/ybkclvk), also by Perttu Ranta-aho; http://www.mathworks.com/matlabcentral/fileexchange/25217 (or http://tinyurl.com/y94d2e9) by Leif Persson.

81. http://www.mathworks.com/matlabcentral/newsreader/view_thread/269361 (or http://bit.ly/55EDCB)

82. http://java.sun.com/docs/books/tutorial/uiswing/components/table.html#renderer(orhttp://bit.ly/6g452w); also Section 4.1.1.

83. Downloadable from here: http://UndocumentedMatlab.com/files/KBytesCellRenderer.java (or http://bit.ly/7oRqls).

84. or download it directly from http://UndocumentedMatlab.com/files/KBytesCellRenderer.class (or http://bit.ly/6zxSQX).

85. http://download.oracle.com/javase/6/docs/api/java/text/DecimalFormat.html (or: http://bit.ly/nJmlml).

86. http://UndocumentedMatlab.com/blog/formatting-numbers/ (or http://bit.ly/qSvXy4).

87. http://www.mathworks.com/help/techdoc/ref/visdiff.html (or http://bit.ly/5faC1Z).

88. http://www.mathworks.com/matlabcentral/newsreader/view_thread/161225 (or http://bit.ly/5MgdBV).

89. http://UndocumentedMatlab.com/blog/undocumented-profiler-options/ (or http://tinyurl.com/yl7oenq).

90. http://UndocumentedMatlab.com/blog/undocumented-profiler-options/#comment-64 (or http://bit.ly/6PURhP).

91. http://UndocumentedMatlab.com/blog/guide-customization/ (or http://bit.ly/7jCpTz).

92. http://blogs.mathworks.com/desktop/2008/04/21/variable-editor/#comment-6790 (or http://bit.ly/8nqDR2).

Chapter 9

Using MATLAB® from within Java[†]

[†] A major portion of this chapter was contributed by Joshua Kaplan, who also authored the *matlabcontrol* package.

9.1 Approaches for Java Control of MATLAB

As explained in previous chapters, MATLAB can easily call Java. Unfortunately, the reverse, calling and controlling MATLAB from Java, is not so simple. MathWorks has taken a conscious decision not to support or document this functionality.

Note that this entire chapter is based on many hours of experimentation and online research and so while we (Joshua and Yair) are more or less certain of the following, it could be deficient or incorrect in many respects.

Generally speaking, seven technologies can be used to call MATLAB from within Java:[1]

- RMI/JMI (RMI for connection, JMI for the functionality)
- JNI (connecting to the supported C/C++ engine library using Java Native Interface)
- COM (Windows-only; connecting to MATLAB.Application ProgID or its like)
- Process-pipes (Unix/Macs-only)
- DDE (Windows-only; using Dynamic Data Exchange)
- Dedicated interface via — a Java/MEX communications wrapper
- Using MathWorks' commercial Java Builder toolbox

Most of this chapter will focus on the RMI/JMI approach. JMI (Java–MATLAB Interface) is widely used by MATLAB to interface the Java and the MATLAB codebases. RMI (Remote Method Invocation) can be used to run JMI on a different computer. In this manner, different computers can theoretically use a single MATLAB installation.

Programming JMI requires a medium-to-high level of Java experience. It is not a trivial task, and many complications need to be addressed, including threading, security, method completion blocking, virtual machine restrictions, and others. In practice, it is difficult to get all the pieces programmed correctly. This prevents using JMI directly from being practical and reliable. Several JMI wrappers were developed which attempt to handle these issues. The latest JMI wrapper package is called *matlabcontrol*.[2] JMI is described in Section 9.2, while *matlabcontrol* is described in Sections 9.3 and 9.4.

The JNI (Java Native Interface) approach uses the fact that MATLAB has a supported library for C/C++ integration with MATLAB.[3] The idea, described in Section 9.5, is to load this library in the Java code and invoke its entry-point methods using JNI.

A separate JNI-based approach uses the fact that R13 (MATLAB 6.5)'s compiler enabled creation of a shared library from the source MATLAB code. This shared library could then be imported and used directly in Java code.[4] Unfortunately, this approach is not possible with new MATLAB releases and/or compilers, which use MCR.†

MATLAB also has fully documented support for a COM (Windows Component Object Model) interface and process pipes (Unix/Mac) that allow remote communication from

---

† MCR (MATLAB Compiler Runtime) is a binary component that is called by MATLAB-compiled applications in order to execute core MATLAB functionality.

external applications.[5] In Windows, remember to register the MATLAB automation server before usage, using the *–regserver* startup option.[6] We may also need to modify the ProgID from "MATLAB.Application" to "MLApp.MLApp".[7]

Unfortunately, COM is a Windows-specific proprietary (Microsoft) technology that is not natively supported by Java. Interested readers can try using a Java/COM bridge using open-source JACOB[8] or JCOM,[9] or one of several commercial packages. This would have the benefits of enabling connection to an existing MATLAB session (a current *matlabcontrol* limitation), and of MathWorks' documented COM.

Unfortunately again, Windows-MATLAB does not appear to enable pipes, since it does not appear to use the Standard I/O framework,[10] so this approach can only be used on Unix/Mac.

DDE, an old inefficient ancestor of COM, can also be used on Windows to communicate between MATLAB and DDE clients.[11] Finally, dedicated interfaces can be implemented with other technologies (e.g., CORBA[12]).

STRONG WARNING: When implementing any of these approaches for interfacing Java to MATLAB, the information and warnings presented in Section 3.2 should be borne in mind. It is very easy to deadlock the MATLAB application by not paying attention to threading issues. Also, if the Java code is implemented in an applet, security issues need to be addressed.[13] In short, when connecting Java to MATLAB, extra attention and testing must be invested.

Quite a few Java-to-MATLAB interface implementations were developed over the years. Technology-wise, most implemented interfaces appear to use either RMI/JMI or JNI technologies. Functionality-wise, these solutions can roughly be grouped as described in the following subsections.

9.1.1 Controlling the MATLAB GUI

This approach is for those who want to control a MATLAB session (or multiple sessions) that will also be used by a user. All solutions in this approach rely on JMI:

- Kamin Whitehouse from the University of Virginia made the first known attempt to provide a JMI tutorial and wrapper.[14] His code enables controlling MATLAB from Java, but the Java program must be launched from MATLAB.
- Bowen Hui from the University of Toronto extended Whitehouse's code to control MATLAB from a Java program that is not launched from within MATLAB.[15] However, his code requires user configuration, and also writes the MATLAB results to a text file rather than returning those results as Java objects.
- Joshua Kaplan from Brown University wrote the ***matlabcontrol*** package that is also based on Whitehouse's work, and enables both local and remote JVM access. No user interaction or configuration is needed, and MATLAB results are returned as Java objects. *matlabcontrol* is described in Sections 9.3 and 9.4.

- Simon Caton from Cardiff University has a walk through on controlling MATLAB and the results it returns.[16]
- Debprakash Patnaik from Virginia Tech wrote a blog post[17] and some simple example codes[18] on how to call MATLAB from Java from within MATLAB. The code is much less sophisticated than the above solutions.
- Aguido Horatio Davis from Griffith University wrote a paper where he describes a framework for controlling MATLAB from Java.[19] Extensive code is provided in the paper beginning on page 117, although many previous pages describe the approach.
- David Allen from Virginia Polytechnic Institute addresses controlling MATLAB from Java in his thesis.[20] His description is on page 32 (page 38 of the PDF).
- Maksim Khadkevich created **JAMAL** (Java–MATLAB Library).[21] This project, like *matlabcontrol*, uses RMI to access the MATLAB process.

9.1.2 *Controlling the MATLAB Engine*

This approach uses MATLAB for non-interactive computation (an automation engine):

- Stefan Müller's **JMatLink** is a library that uses JNI to access MATLAB's C/C++ engine library.[22] JMatLink may require a fix of the environment PATH variable,[23] or to the Eclipse configuration.[24] Ying Bai wrote a book about JMatLink that can be previewed on Google Books (pages 76–286).[25] Note that an apparent JMatLink inconsistency prevents MATLAB Desktop access.[26]
- Andreas Klimke from the University of Stuttgart compared two different approaches for controlling the MATLAB engine: using JNI to MATLAB's engine library and using Java's `Runtime` class to start MATLAB and simulate Unix pipes using MATLAB's standard input and output streams (this fails on Windows).[27]
- Ma Li, Jiang Zhihong, Li Hao, and Wu Dan from Nanjing University wrote a paper entitled "The Combination of JAVA with MATLAB Apply to Meteorology" where they discuss using JNI calls to MATLAB's C++ API.[28]
- Erlangung der Würde from Technischen Universität Carolo-Wilhelmina zu Braunschweig describes another solution (in German)[29] using a combination of JNI calls to the C/C++ library, and applicative use of Java-sockets.
- There have been several reported adaptations of **SWIG**[30] (Simple Wrapper and Interface Generation) to generate a JNI wrapper for MATLAB.[31] I do not know the current status of the SWIG adaptations. As far as I could tell, they are not widely used. The official SWIG distribution once had a MATLAB adaptor and fully functional example,[32] but the current distribution does not (SWIG does support the MATLAB-compatible Octave).[33]
- Albert Strasheim has adapted the open-source **JNA** (Java Native Access) package[34] to connect to MATLAB's native C/C++ library.[35] JNA is basically a JNI replacement enabling simple native library integration. The adaptation is complete with some remaining quirks[36] and incorporated in **Array4J**.[37]

- Andrzej Karbowski from Warsaw's Instytut Automatyki i Informatyki Stosowanej created **jPar**,[38] a MATLAB computation-parallelization engine that connects remote MATLAB sessions using Java RMI.

- Chris Bunch from the University of California at Santa Barbara also posted a solution based on piping MATLAB's input/output streams[39] (plus a long rant about MathWorks not providing a supported solution …[40]).

- Pete Cappello and Andy Pippin, also from UCSB, created a package[41] that enables MATLAB control from their **JICOS**,[42] a Java-centric network framework.

- Markus Krätzig created **JMatlab/Link**,[43] which works as part of **JStatCom**.[44] His tutorial notes that it only works on Windows.

- Gianluca Magnani also used sockets for applicative interaction between the MATLAB server and Java clients, as a preferred way when compared with JMI.[45]

- A commercial approach is to use **JIntegra**.[46] They have a webpage explaining how to use their product with MATLAB.[47]

- Another, now defunct, commercial approach is called **MATLABServerAgent**. Their website no longer exists, but it can be found on the Internet Archive.[48]

9.1.3 Controlling a MATLAB Session from Another MATLAB Session

This approach links together multiple sessions of MATLAB, mostly using Java:

- A very simple non-Java approach is to use MATLAB's support for both client-side and server-side (automation) COM, to access another MATLAB using the built-in MATLAB command *actxserver*('MATLAB.application').

- Scott Gorlin from the Massachusetts Institute of Technology uses a different approach to control multiple sessions of MATLAB using Java, based on Kamin Whitehouse's code.[49] Some documentation is also available.[50]

- Gabor Cselle from the Swiss Federal Institute of Technology has a tutorial and source code on an approach to MATLAB distributed computing using Java.[51]

- Max Goldman and Da Guo from MIT created Java MPI in **MATLAB*P**, which enables parallel computing using MATLAB.[52]

- Brad Phelan, a MathWorks ex-employee, created a **Distributed Computing Toolbox** (DCT). His website is currently not functional, but a cached version can be found online.[53] He also has a simple example of using JMI.[54]

9.1.4 Running or Modifying MATLAB Code without MATLAB

This approach does not actually control MATLAB, but involves both MATLAB and Java:

- **MATLAB Builder JA**[55] is a MathWorks toolbox product that compiles MATLAB code into a Java wrapper that can be imported and used in Java code. The toolbox uses the MATLAB MCR, not interaction with the engine library.

■ Wojciech Gradkowski created **JMatIO**, a pure Java library for reading and writing to MATLAB's MAT-file format.[56] He also created a working example of JNI integration with MATLAB,[57] which is described in Section 9.5.

9.1.5 MATLAB Clones Written in Java

This approach does not use MATLAB at all, but implements a MATLAB-like package in Java that can easily be integrated in a Java application. However, these packages are significantly inferior to MATLAB in terms of their computational and graphic features.

■ **JMathLib** is an open-source clone of MATLAB written entirely in Java,[58] by Stefan Müller who also authored **JMatLink**.

■ **ARRAY4j** is another open-source package mentioned above (Albert Strasheim's adaptation of JNA — see Section 9.1.2).[59]

■ **jMatlab** is an Eclipse-based interpreter interface to linear algebra libraries.[60] It provides a MATLAB-like scripting language, GNUPLOT interface for plotting,[61] and the ability to write toolboxes using Java-based plugins.

9.2 JMI — Java-to-MATLAB Interface[62]

JMI takes the form of a *jmi.jar* file that comes with every copy of MATLAB released in the past decade. This jar file, located in the %matlabroot%/java/jar/ folder, is essentially a zip file containing Java packages and class files. This chapter will discuss com.mathworks.jmi. Matlab, the most important class in *jmi.jar*. Other JMI classes can be explored using the methods explained in Section 5.1.

JMI has been included with each MATLAB release since 5.3 (R11) when Java was first integrated into MATLAB. The oldest public reference to JMI comes from Aguido Horatio Davis, who posted about discovering and testing JMI on September 2000.[63] Since then, there have been only a handful of useful references posted online.[64] One of the notable posts has been Brad Phelan's **MatlabFunction** wrapper in 2005.[65]

In fact, JMI is so unsupported and undocumented that an article describing JMI written in 2002 by Peter Webb (a MathWorks employee) and published in the official newsletter[66] has been removed from their website in 2008, but can still be found in several online archives.[67] An attempt to document JMI (based on MATLAB 6.5) was made by Robert Fleming in March 2003,[68] and his javadocs could be a very good basic reference, taking into consideration they are incomplete and relatively old.

JMI easily allows calling two built-in MATLAB functions: *eval*[69] and *feval*.[70] Essentially, *eval* evaluates any string typed into MATLAB's Command Window, and *feval* allows calling any function by name and passing in arguments. For example,

```
>> sqrt(5)
ans =
    2.2361
```

```
>> eval('sqrt(5)')
ans =
    2.2361
>> feval('sqrt',5)
ans =
    2.2361
```

The first approach computes the square root of 5 by directly calling MATLAB's *sqrt* function. JMI does not enable this direct-invocation approach. Instead, JMI uses the second approach, where *eval* mimics the first call by evaluating the entire expression inside single quotes. The third option, which is also used by JMI, uses *feval* where the function and arguments are separated.

There are some differences between *eval* and *feval*. For example, assignment can be done using *eval*('x = 5') but cannot be done with *feval*.

The *eval* and *feval* functions have several relatives (e.g., *hgfeval*,[71] *evalc*, *evalin*), which will not be described here.

9.2.1 *com.mathworks.jmi.Matlab*

With all of that said, let us dive in! At the MATLAB command-prompt type:

```
>> methodsview('com.mathworks.jmi.Matlab')
```

to see the Matlab class's numerous methods. Many of these methods have very similar names; many others have the same names and just different parameters. In order to call *eval* and *feval*, we are going to use two of MATLAB's static methods:

```
public Object mtEval(String command, int returnCount)
public Object mtFeval(String funcName, Object[] args, int returnCount)
```

Since MATLAB can call Java, we can experiment with these methods from MATLAB. That is right, we are about to use MATLAB to call Java to call MATLAB! First, let us import the Java package that contains the MATLAB class, to reduce typing:

```
import com.mathworks.jmi.*
```

Now let us take the square root of 5 as we did above, but this time from Java. Using JMI's *eval*-equivalent:

```
>> Matlab.mtEval('sqrt(5)',1)
ans =
          2.23606797749979
```

Here, 'sqrt(5)' is passed to *eval*; 1 signifies that Matlab should expect a single return value. The return count is important: an empty string (") is returned if instead the call had used a return count of 0:

```
>> Matlab.mtEval('sqrt(5)',0)
ans =
     ' '
```

If instead, return count was set to 2 or higher, then a Java exception will occur, since the invoked function does not generate so many return values:

```
>> Matlab.mtEval('sqrt(5)',2)

??? Java exception occurred:
com.mathworks.jmi.MatlabException: Error using ==> sqrt
Too many output arguments.
  at com.mathworks.jmi.NativeMatlab.SendMatlabMessage(Native Method)
  at com.mathworks.jmi.NativeMatlab.sendMatlabMessage(NativeMatlab.java:212)
  at com.mathworks.jmi.MatlabLooper.sendMatlabMessage(MatlabLooper.java:121)
  at com.mathworks.jmi.Matlab.mtFeval(Matlab.java:1478)
  at com.mathworks.jmi.Matlab.mtEval(Matlab.java:1439)
```

This stack trace clearly shows how *mtEval()* is actually calling *mtFeval()* internally.

Now perform the square root using JMI's *feval*-equivalent:

```
>> Matlab.mtFeval('sqrt',5,1)
ans =
          2.23606797749979
```

Here, 'sqrt' is the name of the MATLAB function to be called, 5 is the argument to the function, and 1 is the expected return count. If the return count is set as 0 instead of 1, the function call will still succeed, but no results will be returned.

The second *mtFeval()* argument, which specifies the arguments to the invoked MATLAB function, can take any number of arguments as an array. So, the following is valid:

```
>> Matlab.mtFeval('sqrt',[5 3],1)
ans =
          2.23606797749979
          1.73205080756888
```

Note that although two values are returned (the square roots of both 5 and 3), they are considered as one, since it is a single array that is returned.

Multiple MATLAB arguments can be specified in *mtFeval()* using a cell array. For example, consider the following equivalent formats (note the array orientations):

```
>> min(1:4,2)
ans =
     1     2     2     2
```

```
>> Matlab.mtFeval('min',{1:4,2},1)
ans =
     1
     2
     2
     2
```

As we observed above, *mtEval()* is really just calling *mtFeval()*. This works because *eval* is a function, so *feval* can call it. An illustration:

```
Matlab.mtFeval('eval','sqrt(5)',1)
```

Both *mtFeval()* and *mtEval()* enable interaction with MATLAB, but the effects are not shown in the Command Window. Another static method will allow us to do this:

```
public static Object mtFevalConsoleOutput(String functionName,
                                Object[] args, int returnCount)
```

mtFevalConsoleOutput() is very similar to *mtFeval()*, except that *mtFeval()* suppresses any Command-Window output, whereas *mtFevalConsoleOutput()* does not. For instance,

```
>> Matlab.mtFeval('disp','hi',0);   % no visible output

>> Matlab.mtFevalConsoleOutput('disp','hi',0);
hi
```

There is no equivalent *mtEvalConsoleOutput()* method, but that is not a problem because we have seen that *eval* can be accomplished using *feval*:

```
>> Matlab.mtFevalConsoleOutput('eval','x = 5',0);
x =
     5
```

Finally, there is a large set of *eval** and *feval** methods that correspond to *mtEval** and allow setting a callback function that gets invoked when MATLAB finishes processing the JMI request, using a com.mathworks.jmi.CompletionObserver.[72]

There are many more *eval* and *feval* methods in the Matlab class. Most of these methods' names begin with *eval* or *feval* instead of *mtEval* and *mtFeval*. Many of these methods are asynchronous, which means that their effect on MATLAB can occur after the method call returns. This is often problematic because if one method call creates a variable which is then used by the next call, there is no guarantee that the first call has completed (or even begun) by the time the second call tries to use the new variable. Unlike *mtEval()* and *mtFeval()*, these methods are not static, meaning that we must have an instance of the Java class MATLAB:

```
>> proxy = Matlab
proxy =
com.mathworks.jmi.Matlab@1faf67f0
```

Using this instance, let us attempt to assign variable "a" and then store it into variable "b". This may cause an exception if the variable "a" does not yet exist by the time that it is needed to be stored in "b":

```
>> proxy.evalConsoleOutput('a = 5'); b = proxy.mtEval('a',1)
??? Java exception occurred:
com.mathworks.jmi.MatlabException: Error using ==> eval
Undefined function or variable 'a'.
    at com.mathworks.jmi.NativeMatlab.SendMatlabMessage(Native Method)
    at com.mathworks.jmi.NativeMatlab.sendMatlabMessage(NativeMatlab.java:212)
    at com.mathworks.jmi.MatlabLooper.sendMatlabMessage(MatlabLooper.java:121)
    at com.mathworks.jmi.Matlab.mtFeval(Matlab.java:1478)
    at com.mathworks.jmi.Matlab.mtEval(Matlab.java:1439)
a =
     5
```

If we run the above code, then we are not guaranteed to get that exception because of the nature of asynchronous method calls. However, this inherent unpredictability makes it difficult to perform almost any sequential action. It is therefore best to stick to *mtEval*, *mtFeval*, and *mtFevalConsoleOutput*, where such exceptions will be very rare. They can still occur, about 1 in 100 times, for an unknown reason. To prevent some problems, Java coders could use the *whenMatlabReady(*Runnable*)* method, which is executed on the MATLAB thread when it next becomes available.

Two potentially useful methods are *mtSet()* and *mtGet()*, which are the Java proxies for MATLAB's *set* and *get* functions, and similarly accept a MATLAB handle (a double value) and a property name (a string) or array of property names, and either set the value or return it. This can be used to update MATLAB HG handles from within Java code, without needing to pass through an intermediary MATLAB *eval* function:

```
>> Matlab.mtSet(gcf,'Color','b')

>> Matlab.mtGet(gcf,'Color')
ans =
     0
     0
     1

>> Matlab.mtGet(gcf,{'Color','Name'})
ans =
java.lang.Object[]:
    [3 x 1 double]
    'My figure'
```

In summary, using just *eval* and *feval*, an enormous amount of MATLAB's functionality can be accessed from Java. For instance, this enables creating sophisticated Java GUIs using Swing and then being able to call MATLAB code when the user clicks a button or moves a slider. Performance is not super fast, but it works splendidly.

9.2.2 Other Interesting JMI Classes

In addition to the important `com.mathworks.jmi.Matlab` class, a few other JMI classes merit at least a brief mention:

`com.mathworks.jmi.MatlabPath` handles changes to the MATLAB path or current folder (***pwd***). These can be modified from Java, or monitored asynchronously for changes.[†]

`com.mathworks.jmi.MLFileUtils` can be used to query file types (see ***mdbfileonpath***).

`com.mathworks.jmi.AWTUtilities` is a utility class that enables synchronous (*invoke-AndWait*) and asynchronous (*invokeLater*) invocation of functions on the main MATLAB execution thread, including the ability to *setTimeout* value (in milliseconds).

`com.mathworks.jmi.ClassLoaderManager` can be used as an alternative Java class-loader (CL), when the default MATLAB CL fails to load a class. This has been used in Section 2.2, in the ***checkClass*** and ***uiinspect*** utilities, that were discussed earlier in this book,[‡] and on the CSSM forum:[73]

```
try
    thisClass = java.lang.Class.forName(className);
catch
    classLoader =
        com.mathworks.jmi.ClassLoaderManager.getClassLoaderManager;
    thisClass = classLoader.loadClass(className);
end
```

A sibling method, `com.mathworks.jmi.ClassLoaderManager.`*findClass()*, is used to parse Java class names in the built-in ***awtcreate*** and ***awtinvoke*** functions.[§]

Alternatively, we could try to use `java.lang.ClassLoader.`*getSystemClassLoader. loadClass(className)*.[74] A different approach was used by Bred Phelan to load Groovy[75] classes in MATLAB, by registering Groovy's CL with the MATLAB CL.[76]

Finally, `com.mathworks.jmi.Support` is a utility class that reports the current MATLAB release's supported capabilities. For example,

```
>> cell(com.mathworks.jmi.Support.allMexExtensions)'
ans =
  '.mexglx'   '.mexa64'   '.mexmaci'   '.mexmaci64'   '.mexw32'   '.mexw64'
>> cell(com.mathworks.jmi.Support.allArches)'
ans =
  'glnx86'   'glnxa64'   'maci'   'maci64'   'win32'   'win64'
>> cell(com.mathworks.jmi.Support.allComputers)'
ans =
  'GLNX86'   'GLNXA64'   'MACI'   'MACI64'   'PCWIN'   'PCWIN64'
>> com.mathworks.jmi.Support.computer   % Equivalent function: computer
ans =
PCWIN
```

[†] A different mechanism for monitoring current folder changes was described in Section 8.1.4.

[‡] Sections 5.1.2 and 1.3, respectively.

[§] See %matlabroot%/toolbox/matlab/uitools/private/parseJavaSignature.m, which is used by both of these built-in functions.

com.mathworks.jmi.Support.*useDesktop(), useAWT(), useJVM(), useMWT()* and *useSwing()* are the equivalent of the corresponding built-in **usejava** function that was discussed in Section 1.1. For example:

```
>> com.mathworks.jmi.Support.useDesktop        % also: useAWT/JVM/MWT/Swing
ans =
      1      <= Equivalent function: usejava('desktop')
```

9.3 JMI Wrapper — Local MatlabControl[77]

9.3.1 *Local and Remote MatlabControl*

The previous section discussed how JMI enables calling MATLAB from Java. Joshua Kaplan has written an open-source wrapper for JMI called *matlabcontrol*, which is both documented and user-friendly.[78] *matlabcontrol* was originally created to enable a grading program written in Java used by teaching assistants at Brown University to programmatically control MATLAB in order to open, run, and close students' MATLAB code assignments,[79] and after several alternative approaches were analyzed as unsuitable for this need.[80]

matlabcontrol supports calling MATLAB in two different ways: *locally*, where the Java code is launched from MATLAB, and *remotely* where the Java code launches MATLAB. These are discussed in Sections 9.3 and 9.4, respectively.

matlabcontrol is a collection of Java classes, bundled together in a downloadable jar file.[81] As of this writing, the latest version is *matlabcontrol-3.1.0.jar*. Note down where we have downloaded the jar file — we will need to use this information shortly.

For local control, we shall use the LocalMatlabProxy and MatlabInvocationException classes. LocalMatlabProxy contains methods required for calling MATLAB; instances of MatlabInvocationException will be thrown if a problem occurs in the call.

To tell MATLAB where *matlabcontrol-3.1.0.jar* is, add the jar file path to MATLAB's dynamic (via the built-in **javaaddpath** function) or static (via **edit('classpath.txt')**) Java classpath. We will need to restart MATLAB if we modified the static classpath, but it has the benefit of working better in some situations (see Section 1.1).

MATLAB now knows where to find the Java class files in *matlabcontrol*. To save typing later on, run the following in MATLAB (or in the JMI-empowered MATLAB application):

```
import matlabcontrol.*
```

9.3.2 *LocalMatlabProxy*

LocalMatlabProxy is easy to use. All of its methods are static, meaning they can be called without needing to assign LocalMatlabProxy to a variable. The methods are:

```
void exit()
Object getVariable(String variableName)
```

```
void setVariable(String varName, Object value)
void eval(String command)
Object returningEval(String command, int returnCount)
void feval(String functionName, Object[] args)
Object returningFeval(String functionName, Object[] args)
Object returningFeval(String functionName, Object[] args, int returnCount)
void setEchoEval(boolean echoFlag)
```

matlabcontrol has a fair degree of documentation,[82] including detailed javadocs for all these methods.[83] Here is an overview:

■ *exit()* is as straightforward as it sounds: it will exit MATLAB. While it is possible to programmatically exit MATLAB by other means, they may be unreliable. So, to exit MATLAB from Java:

```
LocalMatlabProxy.exit();
```

■ Setting and getting variables can be done using the *getVariable(...)* and *setVariable(...)* methods. These methods will auto-convert between Java and MATLAB types where applicable.

Using *getVariable(...)*:
■ Java types in the MATLAB environment will be returned as Java types.
■ MATLAB types will be converted into Java types.[84]

Using *setVariable(...)*:
■ Java types will be converted into MATLAB types if they can, based on a set of rules.[85] Java Strings are converted to MATLAB char arrays. Additionally, arrays of one of those Java types are converted to arrays of the corresponding MATLAB type.

Using these methods is fairly intuitive:

```
>> LocalMatlabProxy.setVariable('x',5)
>> LocalMatlabProxy.getVariable('x')
 ans =
     5
```

Getting and setting basic types (numbers, strings, and Java objects) is quite reliable and consistent. However, it gets complicated when passing in an array (particularly multidimensional) from Java using *setVariable(...)*, or getting a MATLAB struct or cell array using *getVariable(...)*. The type conversion in such cases is unpredictable and may be inconsistent across MATLAB versions. In such cases, we are best off building a Java object with MATLAB code and then getting the Java object that we have created.

■ The *eval()* and *feval()* methods were described in detail in Section 9.2.1. These functions will return the result, if any, as a Java object.

Due to the way that the underlying JMI operates, it is necessary to know in advance the number of expected return arguments — we can use MATLAB's built-in **nargout** function for this. Note that some functions (e.g., **feval**) return a variable number of arguments, in which case **nargout** returns –1. For example,

```
>> nargout sqrt
ans =
     1
>> nargout feval
ans =
    -1
```

- `LocalMatlabProxy`'s *returningFeval(functionName, args)* method uses the **nargout** information to determine the number of returned arguments and provide it to JMI. It will likely not function as expected if the function specified by function-Name returns a variable number of arguments. In such a case, call *returningFeval(...)* with a third input argument that specifies the expected number of returned arguments.

  Since an *eval()* function can evaluate anything, just as if it were typed in the Command Window, there is no reliable way to determine what will be returned. All of this said, in most situations *returningEval(...)* can be used with a return count of 1, and the *returningFeval(...)* that automatically determines the return count will operate as expected.

9.3.3 Some Usage Examples

Let us perform some of the same simple square root operations that we performed in Section 9.2.1 with pure JMI, this time using *matlabcontrol*. First let us take the square root of 5, assigning the result to the MATLAB variable y (note that we are calling MATLAB from Java, which itself is called from within MATLAB):

```
>> LocalMatlabProxy.eval('sqrt(5)')
ans =
    2.2361
>> y = LocalMatlabProxy.returningEval('sqrt(5)',1)
y =
    2.2361
>> LocalMatlabProxy.feval('sqrt',5)  % no return value
>> y = LocalMatlabProxy.returningFeval('sqrt',5)
y =
    2.2361
>> y = LocalMatlabProxy.returningFeval('sqrt',5,1)
y =
    2.2361
```

There is no major difference between using *eval()* or *feval()* here. However, if instead of *sqrt(5)* we need the square root of a variable, then *eval()* is our only option:

```
>> a = 5
a =
     5
>> LocalMatlabProxy.eval('sqrt(a)')
ans =
    2.2361
>> y = LocalMatlabProxy.returningEval('sqrt(a)',1)
y =
    2.2361

>> LocalMatlabProxy.feval('sqrt','a')
??? Undefined function or method 'sqrt' for input arguments of type 'char'.
??? Java exception occurred:
matlabcontrol.MatlabInvocationException: Method could not return a value because
of an internal Matlab exception
   at matlabcontrol.JMIWrapper.returningFeval(JMIWrapper.java:256)
   at matlabcontrol.JMIWrapper.feval(JMIWrapper.java:210)
   at matlabcontrol.LocalMatlabProxy.feval(LocalMatlabProxy.java:132)
Caused by: com.mathworks.jmi.MatlabException: Undefined function or method
'sqrt' for input arguments of type 'char'.
   at com.mathworks.jmi.NativeMatlab.SendMatlabMessage(Native Method)
   at com.mathworks.jmi.NativeMatlab.sendMatlabMessage(NativeMatlab.ava:212)
   at com.mathworks.jmi.MatlabLooper.sendMatlabMessage(MatlabLooper.java:121)
   at com.mathworks.jmi.Matlab.mtFevalConsoleOutput(Matlab.java:1511)
   at matlabcontrol.JMIWrapper.returningFeval(JMIWrapper.java:252)
   ... 2 more
```

The automatic MATLAB/Java type conversions discussed above are equally applicable to *eval()* and *feval()*: *feval()* automatically converted the argument 'a' into a MATLAB char, instead of considering it as a MATLAB variable. As seen above, the *feval()* invocation fails with a Java `MatlabInvocationException`. So, the only way to interact with MATLAB variables is via *eval()* methods; *feval()* will not work.

Lastly, there is the *setEchoEval(echoFlag)* method: If this method is called with a *true* argument, then all Java to MATLAB calls will be logged in a dedicated window. This can be very helpful for debugging.

Let us now put *matlabcontrol* to work from Java. Below is a very simple Java class called LocalExample which uses Swing to create a window (`JFrame`), a text field (`JTextField`), and a button (`JButton`). When the button is pressed, the text in the field will be evaluated in MATLAB using `LocalMatlabProxy.`*eval(...)*, as explained above.

```java
import java.awt.event.*; //ActionEvent/Listener,WindowAdapter/Event

import javax.swing.*; //JButton,JFrame,JPanel,JTextField

import java.util.concurrent.Executors;
import java.util.concurrent.ExecutorService;
```

```
import matlabcontrol.LocalMatlabProxy;
import matlabcontrol.MatlabInvocationException;

/**
 * A simple demo of some of matlabcontrol's functionality.
 *
 * @author Joshua Kaplan
 */
public class LocalExample extends JFrame
{
  /**
   * Constructs this example and displays it.
   */
  public LocalExample() {
    //Window title
    super("Local Session Example");

    //Panel to hold field and button
    JPanel panel = new JPanel();
    this.add(panel);

    //Input field
    final JTextField inputField = new JTextField();
    inputField.setColumns(15);
    panel.add(inputField);

    //Eval button
    JButton evalButton = new JButton("eval");
    panel.add(evalButton);
    final Runnable evalRunnable = new Runnable()
    {
      public void run() {
        try { LocalMatlabProxy.eval(inputField.getText()); }
        catch (MatlabInvocationException exc) { }
      }
    };

    //Eval runnable, to execute in separate (non-EDT) thread
    final ExecutorService executor =
                Executors.newSingleThreadExecutor();

    //Eval action event for button and field
    ActionListener evalAction = new ActionListener()
    {
      public void actionPerformed(ActionEvent e) {
        //This uses the EDT thread, which may hang Matlab...
        //LocalMatlabProxy.eval(inputField.getText());

        //Execute runnable on a separate (non-EDT) thread
        executor.execute(evalRunnable);
      }
    };
    evalButton.addActionListener(evalAction);
    inputField.addActionListener(evalAction);
```

```
    //On closing, release resources of this frame
    this.addWindowListener(new WindowAdapter()
    {
      public void windowClosing(WindowEvent e) {
        LocalExample.this.dispose();
      }
    });

    //Display
    this.pack();
    this.setResizable(false);
    this.setVisible(true);
  }
}
```

We can either copy and compile the above code (remember to import the *matlabcontrol* JAR file[86] in the compiler; four class files will be created) or download the pre-compiled code.[87] Note that the pre-compiled classes were compiled using JDK 1.6, so if we have **MATLAB** R2007a (7.4) or earlier we will need to compile using an earlier Java compiler.

We now need to tell **MATLAB** where to find the *matlabcontrol* JAR and `LocalExample` class files, by adding them to **MATLAB**'s static or dynamic classpath (see Section 1.1):

```
% Add Java files to current MATLAB session's dynamic Java classpath
javaaddpath LocalExample.zip
javaaddpath matlabcontrol-3.1.0.jar
```

To run the Java program, simply type `LocalExample` in **MATLAB**'s Command Window. A small Java window will appear and JMI will evaluate any expression typed into it:

```
>> LocalExample
ans =
LocalExample[frame0,0,0,202x67,title=Local Session Example,...]
```

Java window calling MATLAB via JMI

```
a =
        1.77245385090552
```

When we called `LocalMatlabControl.eval(...)` from the Command Window earlier, what occurred behind the scenes was actually quite different from what happens when we press the "eval" button in the Java GUI: in the Command Window, everything executes in **MATLAB**'s single main thread; when we press "eval" in the GUI, it executes in the Event Dispatch Thread (EDT, see Section 3.2) which is a separate thread.

EDT is used extensively by **MATLAB** when accessing graphical components such as a figure window or plots. When calling a function from JMI, the calling thread always blocks (pauses) until MATLAB completes doing whatever it was asked. If we call JMI/*matlabcontrol* from EDT and MATLAB needed to use EDT everything will lock up.

To fix this potential EDT problem, when the "eval" button is pressed, the command is dispatched to a separate thread that can block without preventing MATLAB from doing its work.[88] Future versions of *matlabcontrol* will try to further simplify this process.

When using *matlabcontrol* over a remote connection, the topic of the following section, this EDT complication does not arise, since MATLAB and the Java program do not share the same EDT or even the same Java runtime process (JVM).

Similar solutions to the threading issue (i.e., using a separate thread or process) were suggested on CSSM.[89]

While the above Java example is quite simple, using a combination of all the methods described, a much more sophisticated program can be created. To explore the methods in more detail, use the downloadable demo.[90] The demo uses a remote connection to MATLAB, but the available methods are the same.

9.4 JMI Wrapper — Remote MatlabControl[91]

9.4.1 Remote Control of MATLAB

The previous section demonstrated using *matlabcontrol* to call MATLAB from Java from within the MATLAB application. This section will explain how to control MATLAB from a remote Java session.

We will create a small Java program that allows us to launch and connect to MATLAB, then send it ***eval*** commands and receive the results. While this example will involve creating a dedicated user interface, *matlabcontrol* can be integrated into any existing Java program without requiring any user interface.

matlabcontrol was originally created for controlling MATLAB, not for performing computations. If our exclusive concern is to perform MATLAB computations and use the results in Java, then check the MATLAB Builder JA toolbox,[92] which is made by MathWorks and is officially supported. Unfortunately, this toolbox is quite expensive and does not enable interaction with a running MATLAB session (it uses the non-GUI MATLAB engine, much as the compiler does). It is for this purpose that the open-source (free) *matlabcontrol* package was created.

Note that *matlabcontrol* opens a new running MATLAB session and does not connect to an already-running session. MATLAB commands can then be invoked either interactively (in MATLAB's Command Window) or remotely (from Java).

Debugging an already-open MATLAB session can be done with jdb over a dedicated port, using an altogether different mechanism than *matlabcontrol*, or alternatively using COM or process pipes. As noted at the beginning of this chapter, MATLAB has fully documented support for a COM interface (Windows) and process pipes (Unix/Mac) that allow remote communication from external applications. Using this approach would have the benefits of enabling communication with an existing MATLAB session (a current limitation of *matlabcontrol*) and of MathWorks' documented support.

9.4.2 A Simple RemoteExample

This section's RemoteExample demo is too long to paste here; instead we can download the source code,[93] or a jar file[94] that contains both the pre-compiled classes and *matlabcontrol*. If we wish to download and compile the source file, remember that we will need the *matlabcontrol* jar referenced in our Java classpath.

To run this jar, simply double click its file icon. Alternatively, in MATLAB:

```
>> javaaddpath RemoteExample.jar
>> RemoteExample.main('')
```

RemoteExample's user interface is built using standard Swing components — there is nothing special here, just some panels, panes, text fields, buttons, and so on.

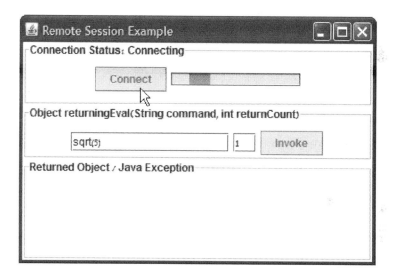

Connecting Java to MATLAB using RMI

The interesting part begins when the code creates a RemoteMatlabProxyFactory object:

```
RemoteMatlabProxyFactory factory = new RemoteMatlabProxyFactory();
```

This MATLAB-proxy factory object is used when the user clicks the "Connect" button:

```
factory.requestProxy();
```

This creates a RemoteMatlabProxy object. RemoteMatlabProxys must be created by a RemoteMatlabProxyFactory and cannot be directly constructed. When *request-*

Proxy() is called, *matlabcontrol* launches a new MATLAB process and connects to it using RMI.[95] When the connection is established, a `MatlabConnectionListener` added to the factory will be notified using its *connectionEstablished(*`RemoteMatlabProxy` *proxy)* callback method, passing in the now-connected *RemoteMatlabProxy* object.

While this example only deals with communicating with a single MATLAB session, *matlabcontrol* can handle multiple remote sessions. Whenever a new session is established, *connectionEstablished(*`RemoteMatlabProxy` *proxy)* is invoked on each connection listener. When a connection is lost due to MATLAB closing, or in very rare cases MATLAB encountering extremely severe errors, *connectionLost(*`RemoteMatlabProxy` *proxy)* is called. Calling methods on this proxy will lead to exceptions being thrown, as it can no longer communicate with MATLAB. The *proxy* is passed because this information is useful when controlling multiple sessions of MATLAB simultaneously.

When the "Invoke" button is pressed, the command and the number of expected return arguments are sent to MATLAB. If the number of return arguments is 0, the command will still execute but nothing will be returned. If the number is positive but less than the total number of return arguments, then only up to that number of arguments will be returned. If the number of return arguments specified exceeds the actual amount of arguments returned, a Java exception will be thrown. By default, the fields are populated to return the result of '*sqrt(5)*'. Press "Invoke" to see what happens. Change the number of return arguments to 0, and click "Invoke" again. Now change the number to 2 and try once more.

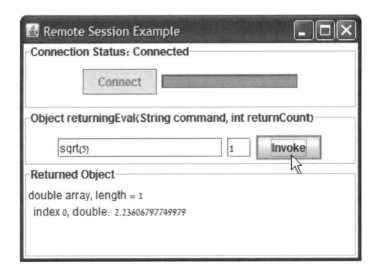

Invoking MATLAB commands from Java using JMI (See color insert.)

When the Java program closes, it also exits MATLAB. This is accomplished by adding a `WindowListener` to the program, which detects the Java closure event. It is important to directly invoke MATLAB's ***exit*** command as opposed to ***eval('exit')***, since all other *proxy* methods block (pause) until completion, but in the case of exiting MATLAB no signal will ever be sent by MATLAB to indicate that it has closed.

9.4.3 Parsing MATLAB's Return Values

Our ***eval*** commands are being sent using `RemoteMatlabProxy`'s *returningEval(*`String` *command,* `int` *returnCount)* method, whose return type is `Object`, because MATLAB can return multiple return types. For example, the expression "***sqrt(5)***" will return an array of doubles, "***pwd***" will return a `java.lang.String`, and "***whos***" will return a complicated array of arrays with a variety of base types and `Objects`.

This behavior may be different on different MATLAB releases. We will have to experiment to find out what is being returned on each particular platform that is being used. The demo can help as it lists everything returned, including array contents. The demo contains the *formatResult* method which recursively goes through the object returned from MATLAB and builds a description of what it contains. As discussed above, MATLAB functions may return a base type, an array of base types, a `String`. A Java object might also be returned — this might actually arise if we are trying to control the MATLAB GUI, as described in previous chapters of this book.

When returning Java objects from MATLAB, certain restrictions and limitations apply. First, the object must be `Serializable`[96] because of the underlying use of RMI. In practice, this is a minor issue because many built-in Java classes are `Serializable`, and making our own classes `Serializable` is usually trivial. Unfortunately, not all built-in MATLAB classes are `Serializable`, and so cannot be transferred from MATLAB.

Secondly, classes returned by MATLAB to Java must be defined in our Java program. For any standard built-in Java class, this is easy. However, if we send MATLAB classes, our Java program must have those classes in its classpath. In practice, this means to reference the jar file containing that class. For HG (or rather, UDD — see Appendix B) classes, we have the alternative of using the following MATLAB function to create a Java class interface, which can then be used in Java code to access the MATLAB object:

```
% This will create a figure.java file in the current folder:
myClassHandle = classhandle(handle(gcf));
myClassHandle.createJavaInterface(myClassHandle.name, pwd);


% alternately, use myClassHandle.JavaInterfaces{1}
% = 'com.mathworks.hg.Figure' in this particular case
% i.e., in your Java code import com.mathworks.hg.Figure
```

Finally, note that a copy of the Java object is transferred — not a reference to the original object. This means that if we modify the transferred object, then the original object in MATLAB will not be affected. However, we can then send the modified object back to MATLAB, so in practice this limitation can be bypassed.

None of these restrictions is applicable to *matlabcontrol* for local sessions.

9.5 Using JNI to Connect Java and MATLAB

JNI (Java Native Interface)[97] is a standard Java technology that enables access from Java to functionality exposed by native libraries. JNI was developed entirely unrelated to MATLAB. However, since MATLAB provides documented support to its native libraries for C and Fortran integration,[98] we can use the same native libraries for Java integration using JNI.

A major benefit of using JNI is that while using JNI is undocumented and not officially supported, the MATLAB functionality it uses is both documented and supported.[99] The official MATLAB support is only for C and Fortran interfaces, but the same MATLAB libraries that support C/Fortran are being loaded and used by JNI. This means that if we run into a problem in development, then we could indeed expect to get MathWorks support.[100] This contrasts with the JMI approaches, which rely on undocumented and unsupported technology.

JNI only works in a Java application that runs on the same computer as an installed MATLAB (not simply an installed MCR but a full MATLAB installation), since the native libraries are apparently not included in the MCR. Theoretically we could try loading native libraries remotely using direct UNC filepaths, but I never tried this, and I do not know whether it works. On the other hand, JMI can easily run on a remote computer which does not have MATLAB, and connect to the MATLAB-installed computer via RMI.

Unfortunately, Java can only access functions that have a very specific interface (*prototype*) declaration.[101] For example, a simple *print()* function should be declared as follows in the dynamic library:

```
JNIEXPORT void JNICALL Java_HelloWorld_print (JNIEnv *, jobject);
```

For example, the C-code which generates this function could be as follows:

```
#include <jni.h >
#include <stdio.h >

JNIEXPORT void JNICALL
Java_HelloWorld_print(JNIEnv *env, jobject obj)
{
   printf("Hello World!\n");
   return;
}
```

This *print()* function prototype informs the JVM it is meant for Java and is part of the HelloWorld class. After we compile this C-code into a dynamic library (e.g., *HelloWorld.dll*), the JVM expects to find the library in Java's *librarypath*. This *librarypath* is specified using the -Djava.library.path = directive in the JVM's command-line, or by setting the LD_LIBRARY_PATH environment variable.[102]

Once we have all this set up properly, we can now use our dynamic library in Java and access our native function using the regular Java syntax of HelloWorld.*print()*. The corresponding Java class would be as follows:

```
public class HelloWorld
{
    private native void print();
    public static void main(String[] args)
    {
        new HelloWorld().print();
    }
    static {
        try {
            System.loadLibrary("HelloWorld");
        } catch (UnsatisfiedLinkError error) {
            System.out.println("Error loading library!");
        }
    }
}
```

An UnsatisfiedLinkError exception will be thrown if JNI encounters any error when loading the file. This could range from file-not-found, to not being on the *librarypath*, to not being loadable as a dynamic library, and so on.[103] It is therefore always prudent to test for this exception when loading native libraries.

Admittedly, all this is far from trivial. It gets even more complicated when we try to access MATLAB native (dynamic library) functionality using JNI: The basic building block of JNI usage is the java.lang.System.*loadLibrary(libName)* method. Unfortunately, unlike almost any other Java method that can be tested from the MATLAB Command Prompt (as shown throughout this book), it appears that some internal bug or limitation in MATLAB's classloader prevents direct usage of System.*loadLibrary* from the Command Prompt.[104] Instead, it can only be used from within Java code (i.e., a user-created Java class).

Therefore, to test our dynamic library, we need to create a simple Java class that does the actual *loadLibrary*, and then call that class from MATLAB:

```
public class LoadLibrary
{
    public static void loadLibrary(String s)
    {
        try {
            System.loadLibrary(s);
        } catch (UnsatisfiedLinkError error) {
            System.out.println("Error loading library "+s);
        }
    }
}
```

We have several alternatives of specifying the *librarypath* in MATLAB: we can set the `LD_LIBRARY_PATH` environment variable, or add a corresponding `-Djava.library.path=` directive to our *java.opts* file (see Section 1.9).

Alternatively, we could add a line in the *librarypath.txt* file, which is located in the %matlabroot%/toolbox/local/ folder. Type ***edit('librarypath.txt')*** at the MATLAB Command Prompt to edit this file in the MATLAB Editor, or use any external text editor for this. If we do not have administrator access to this file, then we can also place a copy of this file in our user's MATLAB startup folder.

Once we have *librarypath.txt* set-up correctly and have restarted MATLAB, we can now load our library in MATLAB as follows:

```
javaaddpath('path-to-the-folder-that-contains-LoadLibrary.class');
LoadLibrary.loadLibrary('libMylib.so'); % or libMylib.dll in Windows
```

The *librarypath.txt* file should contain separate lines with the paths of each folder that contains loadable dynamic libraries. When MATLAB installs, *librarypath.txt* already includes a single line, with the path to MATLAB's internal dynamic libraries:

```
$matlabroot/bin/$arch
```

For example, for MATLAB R2011a running on Windows XP, this could translate into: C:\Program Files\MATLAB\R2011a\bin\win32\. Taking a look at this folder, we find some 200 dynamic libraries, depending on the MATLAB release and platform.

Some of these dynamic libraries, those that are named *native** (e.g., *nativecmdwin.dll*, *nativehg.dll*, *nativejava.dll*, etc.) as well as a few others (*JavaAccessBridge.dll*, *JAWTAccessBridge.dll*, *jogl.dll*, etc.), provide JNI-compliant functions that can be used in our Java programs. Unfortunately, we do not know the prototype interface for any of these functions (i.e., we do not know what input arguments they expect or what return value they provide). Without this information, we can only guess the actual prototype of these functions. For example, *nativejava.dll* defines the following function, as reported by the Dependency Walker utility:[105]

```
_Java_com_mathworks_util_NativeJava_getMenuBar@16
```

From this declaration, we know that the function is part of the `com.mathworks.util.NativeJava` class, and from the function name (`getMenuBar`) we can infer that it returns a reference to a figure window's menu-bar. However, which input argument does it expect? A MATLAB handle value? Or perhaps a `JFrame` reference? We cannot really know except by very extensive trial-and-error. Providing the wrong input arguments will cause errors, hangs, or crashes on our system.

Even more unfortunate is that most of MATLAB's dynamic libraries do not conform to the JNI prototype syntax. This means that we cannot load these libraries in Java without interface libraries that DO conform to the required JNI prototype syntax.

Theoretically, we could create a C-code wrapper for all these libraries, thereby creating JNI-compliant interface libraries. However, as noted above, MATLAB does not document the prototypes for the functions in its libraries. Without these prototypes, we cannot create the necessary JNI-compliant interface libraries. Luckily, there are two exceptions that we can use to our advantage:

Firstly, many of the MATLAB libraries rely on C++ rather than C code. C++ provides internal documentation of its prototype in the function name (its so-called *mangled* name). We can use these prototypes even if we do not really know what the functions actually do under their hood. I will not explore this option here.

An easier route is to use the few MATLAB's libraries that are meant for interfacing with external C/Fortran code, and which provide standard C header (*.h) files. These header files are located in the %matlabroot%/extern/include/ folder, and their associated [static] libraries are available under %matlabroot%/extern/lib/%arch% (e.g., C:\Program Files\MATLAB\R2011a\ extern\lib\win32\lcc).

The functions are exposed by four fully documented **static** libraries, which actually provide access to much of MATLAB's functionality.[†] These are the provided libraries:

- libmat.lib (header file: *mat.h*) — functionality for reading/writing MAT files.
- libmx.lib (*matrix.h*) — functionality relating to MATLAB data: creation/deletion, copying, testing, size, storage, casting/conversion, and so on.
- libmex.lib (*mex.h*) — functionality for executing MATLAB functions/commands, accessing MATLAB variables, and displaying messages.
- libeng.lib (*engine.h*) — functionality for controlling the MATLAB engine: starting and quiting the MATLAB engine,[‡] evaluating MATLAB expressions, and accessing/ updating MATLAB variables.

In addition to these documented functions, MATLAB provides a few additional libraries with header files in the same location as the documented libraries:

- libmwblas.lib (*blas.h*) — a library encapsulating the open-source BLAS (Basic Linear Algebra Subprograms) package,[106] auto-generated from Fortran.
- libmwblascompat32.lib (*blascompat32.h*) — a 32-bit compatible version of libmwblas, for bridging between 32-bit Embedded MATLAB and 64-bit BLAS.
- libmwlapack.lib (*lapack.h*) — a library encapsulating the open-source LAPACK (Linear Algebra) Package,[107] auto-generated from Fortran. LAPACK depends on BLAS, so we may possibly need to include BLAS when linking our library.

[†] http://www.mathworks.com/help/techdoc/apiref/bqoqnz0.html (or http://bit.ly/gZqD1A). Each of these **static** libraries also has a **dynamic** counterpart library, in the $matlabroot/bin/$arc/ folder that was mentioned above.

[‡] Note that starting the MATLAB engine (via *engOpen()*) uses csh on Unix/Linux, so you must have csh installed, otherwise the connection will fail: http://www.mathworks.com/matlabcentral/newsreader/view_thread/238828 (or http://bit. ly/i9Cc1H).

- libmwmathutil.lib — a library providing access to all the basic math functions (sin, log, sqrt, etc.). Unfortunately, there is no corresponding header file.
- libut.lib — a library with general-purpose functions, also without a header file.

There are a few header files that are not directly related to any particular library: *tmwtypes.h* provides definitions for data types; *fintrf.h* provides some definitions of pointer types and compatibility overrides of functions on old MATLAB releases; *io64.h* provides 64-bits I/O support (i.e., for files larger than 2 GB).

So here now is the roadmap for connecting Java to MATLAB via JNI:

- First, create a C-code file that wraps the functions in the aforementioned libraries (libmat, libmx, and so on), using the expected JNI notation.
- Link the C-code with the **static** MATLAB libraries, creating a **dynamic** library.
- Now, place this library in a folder that is specified in the LD_LIBRARY_PATH environment variable or JVM's -Djava.library.path= directive.
- Within the Java code, declare the dynamic library's functions as native. Ensure that the class name and declared function name match the JNI-compliant declaration in the library. For example, if the JNI declaration was for Java_HelloWorld_print, then ensure that the HelloWorld Java class declares a native print function.
- Finally, use java.lang.System.loadLibrary to load the library an access its internal functions within the Java code.

In practice, setting up JNI and all its plumbing can be quite tedious and error-prone. Several researchers reported their experience using JNI with MATLAB, as follows:

- Andreas Klimke from the University of Stuttgart wrote a detailed paper[108] (and source code[109]) about JNI access to MATLAB's engine library.
- Yousef Farschtschi from the University of Hamburg has posted a complete (although limited) example of setting up JNI for MATLAB based on Klimke's work. Yousef reported that the code does not work, but I suspect that this has something to do with his specific environment, rather than with his code.[110]
- Ma Li, Jiang Zhihong, Li Hao, and Wu Dan from Nanjing University wrote a paper entitled "The Combination of JAVA with MATLAB Apply to Meteorology" where they discuss using JNI calls to MATLAB's C++ API.[111]
- Erlangung der Würde from Technischen Universität Carolo-Wilhelmina zu Braunschweig described another solution (in German)[112] using a combination of JNI calls to the C/C++ library and applicative use of Java-sockets.

Still, I suggest using one of the following established wrapper packages, rather than pure JNI. These wrappers greatly ease the burden of using JNI with MATLAB:

- There have been several reported adaptations of **SWIG**[113] (*Simple Wrapper and Interface Generation*) to generate a JNI wrapper for MATLAB.[114] I do not know the current status of the SWIG adaptations. As far as I could tell, they are not widely used. The official SWIG distribution once had a MATLAB adaptor and fully functional example,[115] but the current distribution does not (SWIG does support the MATLAB-compatible Octave[116]).

- Stefan Müller's **JMatLink** is a library that uses JNI to access MATLAB's C/C++ engine library.[117] JMatLink may require a fix of the environment PATH variable,[118] or to the Eclipse configuration.[119] Ying Bai wrote a book about JMatLink that can be previewed on Google Books (pages 76–286).[120] Note that an apparent JMatLink inconsistency prevents MATLAB Desktop access.[121]

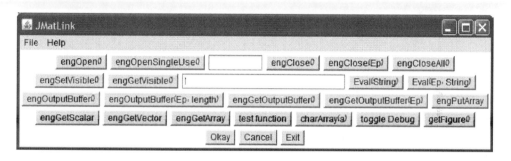

JMatLink's built-in testing GUI (See color insert.)

- Albert Strasheim has adapted the open-source **JNA** (*Java Native Access*) package[122] to connect to MATLAB's native C/C++ library.[123] JNA is basically a JNI replacement that enables simple native library integration, without the requirement for JNI-compliant prototypes. Using JNA, we do not need to write any C-code nor do we need to create any interfacing library (so-called "glue-code"). The JNA-MATLAB adaptation is complete with some remaining quirks[124] and is incorporated in the **Array4J** project.[125]

Here is a basic guide for integrating Strasheim's JNA adapter for MATLAB:

- First, download *jna.jar* from the JNA project archive.[126] This file contains the JNA project classes as well as dynamic libraries for Windows, Linux, SunOS, Darwin (Mac OS X), and Open/FreeBSD.
- Next, open *jna.jar* using WinZip or WinRar and extract the relevant dynamic library into a folder that is specified in the Java *librarypath*, as explained above. For example, on Windows x86, we would extract *com\sun\jna\win32-x86\ jnidispatch.dll*. This extraction step is not required, but it may help solve a minor problem of a leftover temporary file in our %TEMP% folder: JNA automatically extracts the relevant dynamic library from the JAR whenever it is activated, if it does not detect this dynamic library on its *librarypath*.[127]

- Next, download *jna4matlab.jar* from this book's website,[128] or download the contained classes directly from the ARRAY4J project's archives.[129] This contains MATLAB adapter classes and test classes (`MATLABTest`, `MXArrayTest`). *jna4matlab.jar* contains both the source (**.java*) and compiled (**.class*) files.
- Next, add both *jna.jar* and *jna4matlab.jar* (or the folder that contains the *jna4matlab* class files) to the static or dynamic Java classpath. If we are testing this from within MATLAB, then we can use the **javaaddpath** function.
- Within the Java code, use the following code skeleton (more complete examples are in the *MatlabTest.java* and *MXArrayTest.java* files):

```
import net.lunglet.matlab.*;      // Engine, MXArray
public void testFunction()
{
    // Open a visible Matlab engine instance
    // (will throw a RuntimeException upon failure)
    Engine engine = new Engine(true);     // false for non-visible

    // Eval MATLAB expression; get command prompt output as string
    String versionStr = engine.eval("version");

    // Do some MATLAB processing, read the results
    engine.eval("matlabVar = sqrt(pi);");
    MXArray pMatlabVar = engine.getVariable("matlabVar");
    double scalar = pMatlabVar.getScalar();                    // = 1.77245...

    engine.close(); // Close the MATLAB engine
}
```

Adapting the above code skeleton to MATLAB for testing purposes is straightforward:

```
engine = net.lunglet.matlab.Engine(true);       % an Engine object
versionStr = char(engine.eval('version'));
engine.eval('matlabVar = sqrt(pi);');
pMatlabVar = engine.getVariable('matlabVar');    % an MXArray object
scalar = pMatlabVar.getScalar();                 % = 1.77245...
engine.close();
```

Let us now turn to a more complex example, by converting MATLAB's own example for integrating an external C program with MATLAB. The original C-code is located in *%matlabroot%/extern/examples/eng_mat/engdemo.c*. Here is its Java variant, which is also included in *jna4matlab.jar*:

```
import net.lunglet.matlab.*;      // Engine, MXArray, MXLibrary
public class EngDemo
{
    /*
     * Start a non-visible MATLAB engine locally.
     * To start the session on a remote host, use the host name as
     * the startCmd: Engine(startCmd,visible,singleUse,bufSize)
     * For more complicated cases, use any string with whitespace,
```

```
 * and that string will be executed literally to start MATLAB.
 */
Engine ep = new Engine(false);  // may throw a RuntimeException

// Send data to MATLAB, analyze the data, and plot the result
public void testMatlabJNA() throws java.io.IOException
{
    final MXLibrary mx = MXLibrary.INSTANCE;
    MXArray result;
    double[] time = { 0.0, 1.0, 2.0, 3.0, 4.0, 5.0, 6.0, 7.0 };

    // Create a variable T for our data
    MXArray T = MXArray.createDoubleMatrix(1,10);

    // Alternative: MXArray T = mx.mxCreateDoubleMatrix(new NativeLong(1), new
NativeLong(10), MXConstants.REAL);

    // Populate the variable T with our time data
    com.sun.jna.Pointer prPtr = mx.mxGetPr(T);
    prPtr.write(0,time,0,time.length);

    // Place the variable T into the MATLAB workspace
    ep.putVariable("T", T);

    // Evaluate distance as a function of time = (1/2)*g*t.^2
    ep.eval("D = 0.5 * (-9.8) * T.^2;");

    // Plot the result (and pause a bit to ensure that we see it)
    ep.eval("plot(T,D);");
    ep.eval("title('Position vs. Time for a falling object');");
    ep.eval("xlabel('Time (seconds)');");
    ep.eval("ylabel('Position (meters)');");
    ep.eval("pause(10);");

    // Get the results of the MATLAB computation back into Java
    System.out.println("Retrieving D...");
    if ((result = ep.getVariable("D")) == null) {
        System.out.println("Oops! You did not create variable D");
    } else {
        System.out.println("D is class "+mx.mxGetClassName(result));
        double[] results = result.getPr();
        System.out.print("D values = [");
        for (double value : results)
            System.out.print(value + ", ");
        System.out.println("]");
    }

    // We're done! Free memory, close MATLAB engine and exit
    System.out.println("Done!");
    mx.mxDestroyArray(result);
    mx.mxDestroyArray(T);
    ep.close();
}
}
```

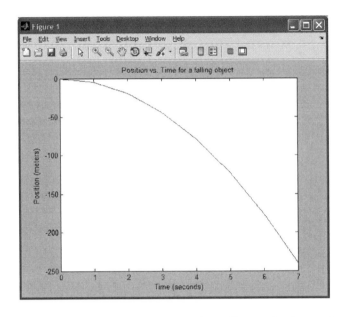

MATLAB plot generated via JNI/JNA (See color insert.)

In a related usage of JNI, Wojciech Gradkowski uploaded to the MATLAB File Exchange[130] a working example of using JNI to interface with user-generated dynamic libraries that were prepared using the MATLAB compiler. Note that this is somewhat different from the need to use JNI to connect to core MATLAB dynamic libraries, although the basic JNI principles are the same.

Important note: Testing JNI/JNA functionality may cause JVM to crash due to memory access errors if we pass incorrect values to MATLAB engine libraries. Do not be afraid of this: just remember to kill any leftover MATLAB process.

References

1. http://code.google.com/p/matlabcontrol/wiki/ApproachesToControl (or http://bit.ly/bnpVUR).
2. http://matlabcontrol.googlecode.com/ (or http://bit.ly/d5uggN).
3. http://www.mathworks.com/help/techdoc/mfatlab_external/f38569.html (or http://bit.ly/bpg7VF).
4. http://www.mathworks.com/matlabcentral/newsreader/view_thread/77971 (or http://bit.ly/cGHDGH).
5. http://www.mathworks.com/help/techdoc/matlab_external/brd4at8.html (or http://bit.ly/cVQ9s6).
6. http://www.mathworks.com/matlabcentral/newsreader/view_thread/278705 (or http://bit.ly/cEG3JQ).
7. http://www.mathworks.com/matlabcentral/newsreader/view_thread/245298 (or http://bit.ly/aIgwt5); http://www.mathworks.com/help/techdoc/matlab_external/f135590.html (or http://bit.ly/ahOnXi).
8. http://sourceforge.net/projects/jacob-project/ (or http://bit.ly/bbzakx).
9. http://sourceforge.net/projects/jcom/ (or http://bit.ly/bR21pq).
10. http://www.mathworks.com/matlabcentral/newsreader/view_thread/171255 (or http://bit.ly/bXcEjv).

11. An example for a .Net client: http://www.codeproject.com/KB/dotnet/matlabeng.aspx (or http://bit.ly/dwgVug).

12. http://www.mathworks.com/matlabcentral/newsreader/view_thread/36079#91688 (or http://bit.ly/g1CzCC).

13. http://www.mathworks.com/matlabcentral/newsreader/view_thread/160196 (or http://bit.ly/9irShd).

14. http://www.cs.virginia.edu/~whitehouse/matlab/JavaMatlab.html (or http://bit.ly/cscPuE).

15. http://www.cs.utoronto.ca/~bowen/code/code.html#matjav (or http://bit.ly/bjOAXq).

16. http://beryl.cs.cf.ac.uk/Web/Guides/Java%20Control%20of%20Matlab/1.php (or http://bit.ly/a5cIWI).

17. http://debprakash.blogspot.com/2007/01/java-matlab-connection-11-jan-2007-see.html (or http://bit.ly/b9Uu1w).

18. http://debprakash.googlepages.com/MatlabInterface.zip (or http://bit.ly/a7EMLe).

19. http://citeseerx.ist.psu.edu/viewdoc/download?doi=10.1.1.133.1715&rep=rep1&type=pdf (or http://bit.ly/cF7as8).

20. http://institutes.lanl.gov/ei/pdf_files/dallenThesis1.pdf (or http://bit.ly/asxQLF).

21. http://jamal.sourceforge.net/about.shtml (or http://bit.ly/d2u4zm).

22. http://jmatlink.sourceforge.net/ (or http://bit.ly/bwfLLv).

23. http://www.mathworks.com/matlabcentral/newsreader/view_thread/263275#710541 (or http://bit.ly/dAbH4B).

24. http://www.mathworks.com/matlabcentral/newsreader/view_thread/263275#731945 (or http://bit.ly/b4xk03).

25. http://books.google.com/books?id=l0wG3sV6UGkC&pg=PA266 (or http://bit.ly/aOHbN4).

26. http://www.mathworks.com/matlabcentral/newsreader/view_thread/260855#693135 (or http://bit.ly/cYjiz1); also see related http://www.mathworks.com/matlabcentral/newsreader/view_thread/286797#762248 (or http://bit.ly/ddrCAB).

27. http://mathforum.org/kb/message.jspa?messageID=884742 (or http://bit.ly/gGNybn); the source code location mentioned in this post is now defunct — here is a cached version: http://bit.ly/h3pdsP; the source code is extensively described by Klimke in the following document: http://preprints.ians.uni-stuttgart.de/downloads/2003/2003-005.pdf (or http://bit.ly/9S5kOs).

28. http://doi.ieeecomputersociety.org/10.1109/IFITA.2009.494 (or http://bit.ly/aa5vFo). Access to this paper is restricted by IEEE Xplore membership, or the article can be purchased.

29. http://bit.ly/99Ylet. The relevant pages appear to be 91 and 92.

30. http://swig.org

31. http://www.mathworks.com/matlabcentral/answers/618-calling-java-methods-using-jni (or http://bit.ly/gHghfF); http://alumni.media.mit.edu/~sbasu/code/swigmatlabplus/ (or http://bit.ly/fSBsuX); http://lnc.usc.edu/~holt/matwrap/; http://stackoverflow.com/questions/6983324/calling-java-from-matlab (or http://bit.ly/qFOD57).

32. An ancient SWIG MATLAB example is available here: http://fifi.org/doc/swig-examples/MATLAB/ (dated 1997) and referenced here: http://www.mathworks.com/matlabcentral/newsreader/view_thread/8451 (or http://bit.ly/fY6naj).

33. http://www.swig.org/Doc1.3/Octave.html (or http://bit.ly/fC2rw7).

34. https://jna.dev.java.net/

35. https://jna.dev.java.net/servlets/BrowseList?list=users&by=thread&from=935824 (or http://bit.ly/9zGSDm); http://www.mathworks.com/matlabcentral/newsreader/view_thread/154026 (or http://bit.ly/bHFoqD).

36. https://jna.dev.java.net/servlets/ReadMsg?list=users&msgNo=524 (or http://bit.ly/bjlEo5).

37. http://code.google.com/p/array4j/ (or http://bit.ly/djQTG9).

38. http://www.mathworks.com/matlabcentral/fileexchange/24924-jpar-parallelizing-matlab (or http://bit.ly/cOMVSP); described in this paper: http://www.ia.pw.edu.pl/~karbowsk/jpar/jpar-para08-abstract.pdf (or http://bit.ly/atiWw7).

39. http://shatterednirvana.wordpress.com/2007/07/12/howto-call-matlab-from-java-kinda/ (or http://bit.ly/935zSG).

40. http://shatterednirvana.wordpress.com/2007/06/06/calling-matlab-from-java-now-20-less-painful/ (or http://bit.ly/cZBVL3).

41. http://www.cs.ucsb.edu/projects/jicos/javadoc/edu/ucsb/cs/jicos/services/external/services/matlab/package-summary.html (or http://bit.ly/aguA39).

42. http://www.cs.ucsb.edu/projects/jicos/ (or http://bit.ly/dBxXlk).

43. http://www.jstatcom.com/jmatlab.html (or http://bit.ly/akj80x).

44. http://www.jstatcom.com/ (or http://bit.ly/9ha7Zs).
45. http://www.mathworks.com/matlabcentral/newsreader/view_thread/250598 (or http://bit.ly/aBdNMK).
46. http://j-integra.intrinsyc.com/ (or http://bit.ly/cDTVBT).
47. http://j-integra.intrinsyc.com/support/com/doc/other_examples/Matlab.htm (or http://bit.ly/b7ZbDF); also read http://www.mathworks.com/matlabcentral/newsreader/view_thread/100497 (or http://bit.ly/9sBiV2).
48. http://web.archive.org/web/20040318183559/http://www.matlabserveragent.com (or http://bit.ly/92wUpg).
49. http://www.scottgorlin.com/2007/07/matlabdispatch/ (or http://bit.ly/bKwavh); http://git.scottgorlin.com/?p=MatlabDispatch.git (or http://bit.ly/cZEpI7).
50. http://www.scottgorlin.com/wp-content/uploads/2008/01/day6.pdf (or http://bit.ly/bE7cTP).
51. http://www.gaborcselle.com/mdct/ (or http://bit.ly/9mRbTS).
52. http://beowulf.csail.mit.edu/18.337-2003/projects/web.mit.edu/maxg/www/18.337/ (or http://bit.ly/bNuotr).
53. For example, http://bit.ly/9sW6Rm; also read here: http://bit.ly/dsFyz1. Brad's DCT was possibly MATLAB's DCT ancestor.
54. Cached version again: http://bit.ly/c8Bkw3
55. http://www.mathworks.com/products/javabuilder/ (or http://bit.ly/bhZtT2).
56. http://www.mathworks.com/matlabcentral/fileexchange/10759 (or http://bit.ly/cRd7CA); http://www.mathworks.com/matlabcentral/newsreader/view_thread/254065 (or http://bit.ly/btlItE).
57. http://www.mathworks.com/matlabcentral/fileexchange/10463 (or http://bit.ly/9xsium).
58. http://sourceforge.net/projects/mathlib/ (or http://bit.ly/aScWPf); http://www.jmathlib.de/ (or http://bit.ly/dr9Cq6).
59. http://code.google.com/p/array4j/ (or http://bit.ly/djQTG9).
60. http://www.jmatlab.org/
61. http://www.mathworks.com/matlabcentral/newsreader/view_thread/163723#415046 (or http://bit.ly/ay0y30).
62. http://UndocumentedMatlab.com/blog/jmi-java-to-matlab-interface/ (or http://bit.ly/dBevUE).
63. http://www.mathforum.com/kb/message.jspa?messageID=851527&tstart=0 (or http://bit.ly/90oFqs).
64. For example, http://www.mathworks.com/matlabcentral/newsreader/view_thread/38383#97557 (or http://bit.ly/928Sja).
65. http://xtargets.com/snippets/posts/show/32 (currently offline — cached version: http://bit.ly/c8Bkw3).
66. http://www.mathworks.com/company/newsletters/news_notes/win02/patterns.html (or http://bit.ly/9U8BHq).
67. For example, http://bit.ly/d4Ox9w (with images). Another archive contains no images but includes a discussion in Chinese: http://bit.ly/assmCj (translated: http://bit.ly/atki6G).
68. http://inneoin.org/matlab/6.5.0.180913a/api/ (or http://bit.ly/be7JDz).
69. http://www.mathworks.com/help/techdoc/ref/eval.html (or http://bit.ly/bonUXT).
70. http://www.mathworks.com/help/techdoc/ref/feval.html (or http://bit.ly/9U8BHq).
71. http://UndocumentedMatlab.com/blog/hgfeval/ (or http://bit.ly/aIgaOa).
72. http://code.google.com/p/matlabcontrol/wiki/JMI (or http://bit.ly/ajstlA); http://www.mathworks.com/matlabcentral/newsreader/view_thread/97653#248522 (or http://bit.ly/a3rdYk).
73. http://www.mathworks.com/matlabcentral/newsreader/view_thread/282516 (or http://bit.ly/9N681K).
74. http://www.mathworks.com/matlabcentral/newsreader/view_thread/36994 (or http://bit.ly/aBhi0I).
75. http://groovy.codehaus.org/; http://en.wikipedia.org/wiki/Groovy_(programming_language) (or http://bit.ly/9kBeyP).
76. http://xtargets.com/snippets/posts/show/7 (currently offline — cached version: http://bit.ly/aOB0y1).
77. http://UndocumentedMatlab.com/blog/jmi-wrapper-local-matlabcontrol-part-1/ (or http://bit.ly/aAqOy0) and http://UndocumentedMatlab.com/blog/jmi-wrapper-local-matlabcontrol-part-2/ (or http://bit.ly/bE5BhP).
78. http://matlabcontrol.googlecode.com/ (or http://bit.ly/d5uggN).
79. http://code.google.com/p/matlabcontrol/wiki/FAQ (or http://bit.ly/chHXDb).
80. http://stackoverflow.com/questions/2047283/change-directory-in-matlab-from-terminal-java (or http://bit.ly/9NCu2M); note Joshua Kaplan's answer to the original posted question on that webpage.
81. http://code.google.com/p/matlabcontrol/downloads/list (or http://bit.ly/cyk1U6).
82. http://code.google.com/p/matlabcontrol/w/list (or http://bit.ly/csI7zb).
83. http://matlabcontrol.googlecode.com/svn/javadocs/doc/matlabcontrol/LocalMatlabProxy.html (or http://bit.ly/9aCFuE).
84. http://www.mathworks.com/help/techdoc/matlab_external/f6425.html#bq__5xw-1 (or http://bit.ly/cNXrVd).

85. http://www.mathworks.com/help/techdoc/matlab_external/f6671.html#bq__508-1 (or http://bit.ly/cBbYL8).

86. http://code.google.com/p/matlabcontrol/downloads/list (or http://bit.ly/cyk1U6).

87. http://UndocumentedMatlab.com/files/LocalExample.zip (or http://bit.ly/bWreyv).

88. http://java.sun.com/javase/6/docs/api/java/util/concurrent/Executors.html#newSingleThreadExecutor() (or http://bit.ly/8XTY0R).

89. http://www.mathworks.com/matlabcentral/newsreader/view_thread/171255 (or http://bit.ly/bXcEjv).

90. http://code.google.com/p/matlabcontrol/downloads/list (or http://bit.ly/cyk1U6).

91. http://UndocumentedMatlab.com/blog/jmi-wrapper-remote-matlabcontrol/ (or http://bit.ly/986irT).

92. http://www.mathworks.com/products/javabuilder/ (or http://bit.ly/bhZtT2).

93. http://UndocumentedMatlab.com/files/RemoteExample.java (or http://bit.ly/bXYcdy).

94. http://UndocumentedMatlab.com/files/RemoteExample.jar (or http://bit.ly/cmHy1E).

95. http://java.sun.com/javase/technologies/core/basic/rmi/index.jsp (or http://bit.ly/cJvo7D); http://java.sun.com/docs/books/tutorial/rmi/index.html (or http://bit.ly/bD47dg).

96. http://java.sun.com/javase/6/docs/api/java/io/Serializable.html (or http://bit.ly/b7Pc0l).

97. http://java.sun.com/docs/books/jni/

98. http://www.mathworks.com/help/techdoc/matlab_external/f38569.html (or http://bit.ly/bpg7VF).

99. http://www.mathworks.com/help/techdoc/apiref/ (or http://bit.ly/fxqxq4).

100. For example, http://www.mathworks.com/matlabcentral/newsreader/view_thread/298719#804190 (or http://bit.ly/ebS20D).

101. http://java.sun.com/docs/books/jni/html/start.html#1309 (or http://bit.ly/hokrtY).

102. http://java.sun.com/docs/books/jni/html/start.html#27157 (or http://bit.ly/eYSHUX).

103. http://www.mathworks.com/matlabcentral/answers/618-calling-java-methods-using-jni (or http://bit.ly/gHghfF); http://stackoverflow.com/questions/1168567/matlab-jni-error (or http://bit.ly/gk7Ra6); http://bit.ly/g2qmbr; http://www.mathworks.com/matlabcentral/newsreader/view_thread/116639 (or http://bit.ly/dNbqCV); and many others.

104. https://www.kitware.com/InfovisWiki/index.php/Matlab_Titan_Toolbox#Overcome_Matlab_loadLibrary_bug (or http://bit.ly/gU7Tpc). Note that the report mentioned above references an official Matlab bug report that for some unknown reason has since been removed: http://www.mathworks.com/support/solutions/data/1-1A2HO.html?solution=1-1A2HO.
Also see: http://www.mathworks.com/matlabcentral/answers/618-calling-java-methods-using-jni (or http://bit.ly/gHghfF).

105. http://dependencywalker.com/, http://en.wikipedia.org/wiki/Dependency_Walker (or http://bit.ly/gbD0L0); MathWorks recommends using this utility: http://www.mathworks.com/support/solutions/en/data/1-2RQL4L/ (or http://bit.ly/gmb8DU).

106. http://www.netlib.org/blas/, http://en.wikipedia.org/wiki/BLAS; BLAS has extensive documentation of its functionality, albeit in a simple textual format.

107. http://www.netlib.org/lapack/, http://en.wikipedia.org/wiki/LAPACK; LAPACK has even more extensive documentation.

108. http://preprints.ians.uni-stuttgart.de/downloads/2003/2003-005.pdf (or http://bit.ly/9S5kOs).

109. http://mathforum.org/kb/message.jspa?messageID=884742 (or http://bit.ly/gGNybn); the source code location mentioned in this post is now defunct — here is a cached version: http://bit.ly/h3pdsP

110. http://www.mathworks.com/matlabcentral/newsreader/view_thread/286771 (or http://bit.ly/he5UUW).

111. http://doi.ieeecomputersociety.org/10.1109/IFITA.2009.494 (or http://bit.ly/aa5vFo). Access to this paper is restricted by IEEE Xplore membership, or the article can be purchased.

112. http://bit.ly/99Ylet — the relevant pages appear to be 91 and 92.

113. http://swig.org

114. http://www.mathworks.com/matlabcentral/answers/618-calling-java-methods-using-jni (or http://bit.ly/gHghfF); http://alumni.media.mit.edu/~sbasu/code/swigmatlabplus/ (or http://bit.ly/fSBsuX); http://lnc.usc.edu/~holt/matwrap/; http://stackoverflow.com/questions/6983324/calling-java-from-matlab (or http://bit.ly/qFOD57).

115. An ancient SWIG MATLAB example is available here: http://fifi.org/doc/swig-examples/MATLAB/ (dated 1997) and referenced here: http://www.mathworks.com/matlabcentral/newsreader/view_thread/8451 (or http://bit.ly/fY6naj).

116. http://www.swig.org/Doc1.3/Octave.html (or http://bit.ly/fC2rw7).

117. http://jmatlink.sourceforge.net/ (or http://bit.ly/bwfLLv).

118. http://www.mathworks.com/matlabcentral/newsreader/view_thread/263275#710541   (or   http://bit.ly/dAbH4B).
119. http://www.mathworks.com/matlabcentral/newsreader/view_thread/263275#731945 (or http://bit.ly/b4xk03).
120. http://books.google.com/books?id=l0wG3sV6UGkC&pg=PA266 (or http://bit.ly/aOHbN4).
121. http://www.mathworks.com/matlabcentral/newsreader/view_thread/260855#693135 (or http://bit.ly/cYjiz1); also see related .http://www.mathworks.com/matlabcentral/newsreader/view_thread/286797#762248 (or http://bit.ly/ddrCAB).
122. http://jna.dev.java.net/; http://en.wikipedia.org/wiki/Java_Native_Access (or http://bit.ly/gQ4MzV); http://www.javaworld.com/javaworld/jw-02-2008/jw-02-opensourcejava-jna.html (or http://bit.ly/dHtcoS).
123. http://markmail.org/message/7b722ablic4j2bdx (or http://bit.ly/hNW67B); http://www.mathworks.com/matlabcentral/newsreader/view_thread/154026 (or http://bit.ly/bHFoqD).
124. http://jna.dev.java.net/servlets/ReadMsg?list=users&msgNo=524 (or http://bit.ly/bjlEo5).
125. http://code.google.com/p/array4j/ (or http://bit.ly/djQTG9).
126. http://java.net/projects/jna/sources/svn/content/trunk/jnalib/dist/jna.jar?rev=1182 (or http://bit.ly/hz37K2).
127. http://java.net/jira/browse/JNA-49(orhttp://bit.ly/gjUEnD);http://markmail.org/message/35ahe33bswag4qzf (or http://bit.ly/gJap7e); http://markmail.org/message/4uzk5ilwisa6rdvf (or http://bit.ly/h7orfq); http://markmail.org/message/23j76gf3gnx6sk6a (or http://bit.ly/epGKOQ).
128. http://UndocumentedMatlab.com/files/jna4matlab.jar (or http://bit.ly/g6sviR).
129. http://array4j.googlecode.com/svn/trunk/src/main/java/net/lunglet/matlab/ (or http://bit.ly/eWmMfo) for the basic MATLAB files; http://array4j.googlecode.com/svn/trunk/src/test/java/net/lunglet/matlab/ (or http://bit.ly/gnnd3j) for the test files.
130. http://www.mathworks.com/matlabcentral/fileexchange/10463 (or http://bit.ly/b69Tg0).

Chapter 10

Putting It All Together

10.1 UISplitPane[1]

An innovative[†] example of using many of the features discussed in this book is the ***UISplitPane*** utility,[2] which was chosen by the MATLAB Central team as "Pick of the Week" on March 27, 2009.[3] Split-pane functionality was always sorely missed in MATLAB GUI. Most other standard GUI controls have a MATLAB counterpart, but as of MATLAB 7.13 (R2011b) there is no split-pane control. The Swing class `JSplitPane` provides access to split-pane functionality, but `JSplitPane` cannot be used as is, because MATLAB axes and controls cannot be placed in its two Java subcontainers. Enter ***UISplitPane***:

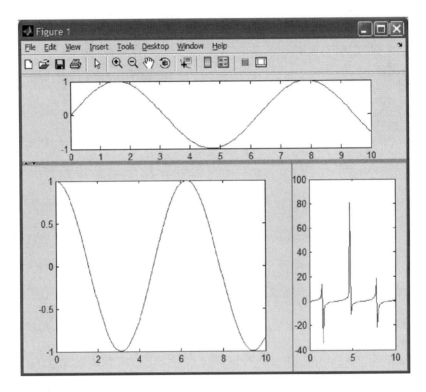

A horizontal *UISplitPane* contained within a vertical *UISplitPane* (See color insert.)

Recently, Malcolm Lidierth has posted a set of Java Swing-based utilities to the MATLAB File Exchange (collectively called *Project Waterloo*).[4] This set includes the GSplitPane and GSplitPaneDivider classes, that basically do what ***UISplitPane*** does. Readers are highly encouraged to investigate and use Malcolm's submission, which even includes an interactive demo.

[†] If I may be allowed a minor indulgence … I was quite satisfied with myself after coming up with the basic solution for *uisplitpane*, having spent several days investigating other potential routes which proved fruitless. Tying all the loose ends took another few days.

10.1.1 Technical Description

The technical idea behind *uisplitpane* is to create an invisible off-screen `JSplitPane`, extract only its narrow central divider subcomponent, and then place that divider onscreen using *javacomponent*. Pure-MATLAB code then attaches standard MATLAB *uipanel*s as subpanes on either side of the divider, and some smart property linkages (using *handle.listener*, Appendix B) ensure that whenever the divider is dragged, its two side-panes will be resized accordingly, together with all their content (axes and controls).

Since the two split panes are simple *uipanel*s, they can contain not only axes and controls, but also other *uisplitpane*s, creating a hierarchy of split panes. For example, the screenshot in the previous page displays a horizontal (left and right panes) *UISplitPane* that is contained within a vertical (up and down panes) *UISplitPane*.

The basic idea of tying a lean Swing component (in this case, the divider subcomponent) to standard MATLAB panels was taken from MATLAB's implementation of the semi-documented *uitab/uitabgroup* functions (Section 4.3). In that case, a lean Swing component (the tabs row) is displayed above several overlapping *uipanel*s, in which only one *uipanel* is visible at any one time. Selecting a Java tab fires a MATLAB callback that simply switches the current visible *uipanel* into view.

Unfortunately, *uitabgroup*'s solution cannot be applied as-is to the split-pane problem, since the `JSplitPane` containers overlap (hide) the axes-containing panes. This required using only the narrow divider subcomponent, giving an illusion of being connected to the split panes within a single component — a similar illusion as *uitabgroup* provides.

In addition to *javacomponent*, *uisplitpane* uses several other undocumented MATLAB features, which are all explained in other sections of this book, as well as on the http://www.UndocumentedMatlab.com website.[5]

The Java divider's reference is converted into a MATLAB *handle*, so that some extra properties can be added using *schema.prop* and will become visible when using regular *get*. *handle.listener* property linkages (Appendix B) ensure the above-mentioned linkage between the divider and its side panes.

Semi-documented internal function *hgfeval*[6] is used in mouse callbacks to chain the original **WindowButton** callback (if available), and *setptr* is used to set the mouse pointer (cursor). *UISplitPane* behaves nicely in the presence of Mode Managers (zoom, pan, etc.) by using the figure's undocumented **ModeManager** property and setting its **ButtonDownFilter** to bypass mode.

Finally, the figure's **JavaFrame** property (Chapter 7) is used to get the figure's AxisCanvas container (Section 7.3.1), which is needed for setting mouse callbacks that behave better than similar callbacks at the figure level.

The complete listing of *UISplitPane* can be downloaded from the MATLAB File Exchange[7] and is also presented here.

Note that the version on the File Exchange might be newer and contain additions or fixes that are not included in the code below. The objective of including the full listing here is therefore not for readers to copy the code as-is, but rather to study it in conjunction with the relevant sections presented earlier in this book.

10.1.2 *Source-Code Listing*

```
function [h1,h2,hDivider] = uisplitpane(varargin)
% uisplitpane Split a container (figure/frame/uipanel) into two resizable
  sub-containers
%
% Syntax:
%    [h1,h2,hDivider] = uisplitpane(hParent, 'propName',propVal,...)
%
% Description:
%    UISPLITPANE splits the specified container(s) (figure, panel or frame,
%    referenced by handle(s) hParent) into two distinct panes (panels)
%    separated by a movable divider. If no hParent container is specified,
%    then the current figure (gcf) is assumed. Matlab components may freely
%    be added to each of the panes. Pane sizes may be modified by dragging
%    or programmatically repositioning the movable divider.
%
%    UISPLITPANE returns the handles to the left/bottom sub-container h1,
%    right/top sub-container h2, and the split-pane divider hDivider.
%    If a vector of several hParents was specified, then h1, h2 & hDivider
%    will be corresponding vectors in the containing hParents. If the
%    hParents are found to be non-unique, then the returned handles will
%    correspond to the unique sorted vector of hParents, so that no hParent
%    will be split more than once.
%
%    The UISPLITPANE divider can be dragged to either side, up to the
%    specified DividerMinLocation to DividerMaxLocation property values
%    (defaults: 0.1 and 0.9, respectively, meaning between 10-90% of range).
%    In Matlab 7+, additional one-click buttons are added to the divider,
%    which enable easy flushing of the divider to either side, regardless
%    of DividerMinLocation & DividerMaxLocation property values.
%
%    Several case-insensitive properties may be specified as P-V pairs:
%       'Orientation':     'horizontal' (default) or 'vertical'
%                       Note: this specifies sub-pane alignment (R/L or U/D):
%                             divider direction is always perpendicular
%       'Parent':          Handle(s) of containing figure, panel or frame
%       'DividerWidth':    Divider width (1-25 [pixels], default=5)
%       'DividerColor':    Divider color (default = figure background color)
%                       Note: accepts both [r,g,b] and 'colorname' formats
%       'DividerLocation': Divider normalized initial location (.001-
%    .999,default=0.5)
%                       Note: 0 = far left/bottom, 1 = far right/top
%       'DividerMinLocation': Normalized minimal left/bottom pane size (0-1,
%    default=0.1)
%       'DividerMaxLocation': Normalized maximal left/bottom pane size (0-1,
%    default=0.9)
%
%    hDivider is a standard Matlab object handle possessing all these additional
%    properties. All these properties are gettable/settable via the hDivider
%    handle, except for the 'Orientation' & 'Parent' properties which become
%    read-only after the UISPLITPANE is constructed. hDivider also exposes
```

```
%   the following read-only properties:
%      'LeftOrBottomPaneHandle': the h1 value returned by this function
%      'RightOrTopPaneHandle':   the h2 value returned by this function
%      'DividerHandle':          the HG container handle (a numeric value)
%      'JavaComponent':          handle to the underlying java divider obj
%      'ContainerParentHandle':  handle to hParent container
%                         Note: this is important in Matlab 6 which does
%                         ^^^^  not allow hierarchical UI containers
%      'ContainerParentVarName': name of the hParent variable (if available)
%
% Example:
%    [hDown,hUp,hDiv1] = uisplitpane(gcf,'Orientation','ver','dividercolor',[0,1,
%  0]);
%    [hLeft,hRight,hDiv2] = uisplitpane(hDown,'dividercolor','r','divid
%  erwidth',1);
%    t=0:.1:10;
%    hax1=axes('Parent',hUp);   plot(t,sin(t));
%    hax2=axes('parent',hLeft); plot(t,cos(t));
%    hax3=axes('parent',hRight); plot(t,tan(t));
%    hDiv1.DividerLocation = 0.75;   % one way to modify divider properties...
%    set(hDiv2,'DividerColor','red'); % ...and this is another way...
%
% Bugs and suggestions:
%    Please send to Yair Altman (altmany at gmail dot com)
%
% Warning:
%    This code heavily relies on undocumented and unsupported Matlab
%    functionality. It works on Matlab 6+, but use at your own risk!
%    A detailed list of undocumented/unsupported functionality can
%    be found at: <a href="http://UndocumentedMatlab.com/blog/uisplitpane/">
% http://UndocumentedMatlab.com/blog/uisplitpane/</a>
%
% Change log:
%    2009-03-30: Fixed DividerColor parent's color based on Co Melissant's
% suggestion; re-fixed JavaFrame warning
%    2009-03-27: Fixed R2008b JavaFrame warning
%    2009-02-23: First version posted on <a href="http://www.mathworks.com/
% matlabcentral/fileexchange/authors/27420">MathWorks File Exchange</a>
%
% See also:
%    gcf, javax.swing.JSplitPane

% Technical implementation:
%    UISPLITPANE is a Matlab implementation of the Java-Swing
%    javax.swing.JSplitPane component. Since Matlab currently prevents
%    Matlab objects (axes etc.) to be placed within java containers (such as
%    those returned by JSplitPane), a pure-Matlab implementation was needed.
%    JSplitPane is actually used (if available) for the user-interface, but
%    hidden Matlab containers actually display the pane contents.
%
%    The basic idea was to take the platform-dependent divider sub-component
%    created by Java's JSplitPane, and place this divider in a stand-alone
```

```
%    Matlab container. Two sub-panes (uipanels or frames) are then placed
%    on either side of this divider. Property linking and divider callbacks
%    are then set in order to ensure that whenever the divider is dragged or
%    programmatically modified, the two sub-panes are updated accordingly.
%
%    Matlab 6 needs special treatment because in that version Java UI
%    components and uipanels were still unavailable. Therefore, standard
%    Matlab uicontrol buttons are used to represent the divider, and frames
%    (instead of uipanels) represent the sub-panes. Also, hierarchical UI
%    controls were not allowed - all controls and axes need to be direct
%    children of the containing figure frame, so special handling needs to
%    be done to correctly handle hierarchical dividers. Additional special
%    handling was also done to overcome bugs/limitations with mouse event
%    tracking in Matlab 6.

% On a personal note, this has been my most challenging project of all my
% submissions to the File Exchange. Ensuring backward compatibility all the
% way back to Matlab 6 proved extremely difficult.
% Programmed by Yair M. Altman: altmany(at)gmail.com

% $Revision: 1.2 $  $Date: 2009/03/30 22:07:23 $

    try
        %dbstop if error
        h1 = [];  %#ok
        h2 = [];  %#ok
        hDivider = handle([]);  %#ok

        % Process input arguments
        paramsStruct = processArgs(varargin{:});

        % Capture the parent var name, if supplied
        try
            paramsStruct.parentName = inputname(1);
        catch
            paramsStruct.parentName = '';
        end

        % Split the specified parent container(s)
        [h1, h2, hDivider] = splitPanes(paramsStruct);
        % TODO - setup hContainer return arg

        return;  % debug point
    % Error handling
    catch
        v = version;
        if v(1)<='6'
            err.message = lasterr;  % no lasterror function...
        else
            err = lasterror;
        end
        try
            err.message = regexprep(err.message,'Error using ==> [^\n]+\n','');
        catch
```

```
        try
            % Another approach, used in Matlab 6 (where regexprep is unavailable)
            startIdx = findstr(err.message,'Error using ==> ');
            stopIdx = findstr(err.message,char(10));
            for idx = length(startIdx) : -1 : 1
                idx2 = min(find(stopIdx > startIdx(idx)));   %#ok ML6
                err.message(startIdx(idx):stopIdx(idx2)) = [];
            end
        catch
            % never mind...
        end
    end
    if isempty(findstr(mfilename,err.message))
        % Indicate error origin, if not already stated within the error message
        err.message = [mfilename ': ' err.message];
    end
    if v(1)<='6'
        while err.message(end)==char(10)
            err.message(end) = [];  % strip excessive Matlab 6 newlines
        end
        error(err.message);
    else
        rethrow(err);
    end
  end

%% Internal error processing
function myError(id,msg)
    v = version;
    if (v(1) >= '7')
        error(id,msg);
    else
        % Old Matlab versions do not have the error(id,msg) syntax...
        error(msg);
    end
%end  % myError  %#ok for Matlab 6 compatibility

%% Process optional arguments
function paramsStruct = processArgs(varargin)
    % Get the properties in either direct or P-V format
    [parent, pvPairs] = parseparams(varargin);

    % Now process the optional P-V params
    try
        % Initialize
        paramName = [];
        paramsStruct = [];
        paramsStruct.dividercolor = '';
        supportedArgs = {'orientation','parent','tooltip',...
                         'dividerwidth','dividercolor','dividerlocation',...
                         'dividerminlocation','dividermaxlocation'};
        while ~isempty(pvPairs)
            % Ensure basic format is valid
```

```matlab
        paramName = '';
        if ~ischar(pvPairs{1})
            myError('YMA:uisplitpane:invalidProperty','Invalid uisplitpane prop');
        elseif length(pvPairs) == 1
            myError('YMA:uisplitpane:noPropertyValue', ...
                    ['No value specified for property ''' pvPairs{1} '''']);
        end

        % Process parameter values
        paramName  = pvPairs{1};
        paramValue = pvPairs{2};
        %paramsStruct.(lower(paramName)) = paramValue;  % no good on ML6...
        paramsStruct = setfield(paramsStruct, lower(paramName),
                                paramValue);  %#ok
        pvPairs(1:2) = [];
        if ~any(strcmpi(paramName,supportedArgs))
            url = ['matlab:help ' mfilename];
            urlStr = getHtmlText(['<a href="' url '">' strrep(url,'matlab:','')
                                 '</a>']);
            myError('YMA:uisplitpane:invalidProperty',...
                    ['Unsupported property - type "' urlStr ...
                     '" for a list of supported properties']);
        end
end  % loop pvPairs

% Process parent container property
if isfield(paramsStruct,'parent')
    % Parent property supplied as a P-V pair
    if ~all(ishandle(paramsStruct.parent))
        myError('YMA:uisplitpane:invalidProperty', ...
                'Parent must be a handle of a figure, panel or frame');
    end
elseif ~isempty(parent)
    % Parent container was supplied as a direct (first) parameter
    paramsStruct.parent = parent{1};
    if ~all(ishandle(paramsStruct.parent))
        myError('YMA:uisplitpane:invalidProperty', ...
                'Parent must be a handle of a figure, panel or frame');
    end
else
    % Default parent container = current figure (gcf)
    paramsStruct.parent = gcf;
end

% Ensure we do not split any parent container more than once...
if length(paramsStruct.parent) > length(unique(paramsStruct.parent))
    % Do not sort hParents (a side-effect of unique() function) unless we must...
    paramsStruct.parent = unique(paramsStruct.parent);
end

% Process DividerColor property
paramsStruct.dividercolor = processColor(paramsStruct.dividercolor, ...
                                         paramsStruct.parent);
```

```
    % Set default param values
    if ~isfield(paramsStruct,'orientation'),      paramsStruct.orientation =
'horizontal';  end
    if ~isfield(paramsStruct,'tooltip'),            paramsStruct.tooltip =
'';  end
    if ~isfield(paramsStruct,'dividerwidth'),       paramsStruct.
dividerwidth=5; end
    if ~isfield(paramsStruct,'dividerlocation'),    paramsStruct.
dividerlocation=0.5; end
    if ~isfield(paramsStruct,'dividerminlocation'), paramsStruct.
dividerminlocation=0.1;  end
    if ~isfield(paramsStruct,'dividermaxlocation'), paramsStruct.
dividermaxlocation=0.9;  end

    % Check min/max data
    checkNumericValue(paramsStruct.dividerminlocation,0,1,'DividerMinLocation');
    checkNumericValue(paramsStruct.dividermaxlocation,0,1,'DividerMaxLocation');
    if paramsStruct.dividermaxlocation <= paramsStruct.dividerminlocation
        myError('YMA:uisplitpane:invalidProperty', ...
              'DividerMaxLocation must be greater than DividerMinLocation');
    end

    % Check other properties
    checkNumericValue(paramsStruct.dividerlocation,0,1,'DividerLocation');
    checkNumericValue(paramsStruct.dividerwidth,1,25,'DividerWidth');
    if isfield(paramsStruct,'tooltip') & ~ischar(paramsStruct.tooltip)  %#ok
ML6
        myError('YMA:uisplitpane:invalidProperty','Tooltip must be a string');
    elseif isfield(paramsStruct,'orientation') &
          (~ischar(paramsStruct.orientation) | ...
          (~strncmpi(paramsStruct.orientation, 'horizontal', ...
                  length(paramsStruct.orientation)) & ...
           ~strncmpi(paramsStruct.orientation, 'vertical',  ...
                  length(paramsStruct.orientation)))) %#ok ML6
        myError('YMA:uisplitpane:invalidProperty', ...
              'Orientation must be ''horizontal'' or ''vertical''');
    elseif lower(paramsStruct.orientation(1)) == 'h'
        paramsStruct.orientation = 'horizontal';
    else
        paramsStruct.orientation = 'vertical';
    end
  catch
    if ~isempty(paramName),  paramName = [' ''' paramName ''''];  end
    myError('YMA:uisplitpane:invalidProperty', ...
          ['Error setting uisplitpane property' paramName ':' char(10)
           lasterr]);
  end
%end  % processArgs  %#ok for Matlab 6 compatibility

%% Check a property value for numeric boundaries
function checkNumericValue(value,minVal,maxVal,propName)
   errMsg = sprintf('number between %g - %g', minVal, maxVal);
   if ~isnumeric(value) | isempty(value)  %#ok ML6
```

```
              myError('YMA:uisplitpane:invalidProperty', ...
                   sprintf('%s must be a %s',propName,errMsg));
         elseif numel(value) ~= 1
             myError('YMA:uisplitpane:invalidProperty', ...
                   sprintf('%s must be a single %s',propName,errMsg));
         elseif value<minVal | value>maxVal   %#ok ML6
             myError('YMA:uisplitpane:invalidProperty', ...
                   sprintf('%s must be a %s',propName,errMsg));
         end
     %end   % checkNumericValue   %#ok for Matlab 6 compatibility


     %% Strip HTML tags for Matlab 6
     function txt = getHtmlText(txt)
         v = version;
         if v(1)<='6'
             leftIdx  = findstr(txt,'<');
             rightIdx = findstr(txt,'>');
             if length(leftIdx) ~= length(rightIdx)
                 newLength = min(length(leftIdx),length(rightIdx));
                 leftIdx  = leftIdx(1:newLength);
                 rightIdx = leftIdx(1:newLength);
             end
             for idx = length(leftIdx) : -1 : 1
                 txt(leftIdx(idx) : rightIdx(idx)) = [];
             end
         end
     %end   % getHtmlText   %#ok ML6


     %% Process color argument
     function color = processColor(color,hParent)
         try
             % Convert color names to RBG triple (0-1) if not already in that format
             if isempty(color)
                 % Get the parent's background color
                 if isprop(hParent,'Color')
                     color = get(hParent,'color');
                 elseif isprop(hParent,'BackgroundColor')
                     color = get(hParent,'BackgroundColor');
                 elseif isprop(hParent,'Background')
                     color = get(hParent,'Background');
                 else
                     color = get(gcf,'color');   % default = figure background color
                 end
             end
             if ischar(color)
                 switch lower(color)
                     case {'y','yellow'},   color = [1,1,0];
                     case {'m','magenta'},  color = [1,0,1];
                     case {'c','cyan'},     color = [0,1,1];
                     case {'r','red'},      color = [1,0,0];
                     case {'g','green'},    color = [0,1,0];
                     case {'b','blue'},     color = [0,0,1];
```

```
              case {'w','white'},    color = [1,1,1];
              case {'k','black'},    color = [0,0,0];
              otherwise,  myError('YMA:uisplitpane:invalidColor', ['''' color '''']);
          end
      elseif ~isnumeric(color) | length(color)~=3   %#ok ML6
          myError('YMA:uisplitpane:invalidColor', color);
      end

      % Convert decimal RGB format (0-255) to fractional format (0-1)
      if max(color) > 1
          color = color / 255;
      end
  catch
      myError('YMA:uisplitpane:invalidColor',['Invalid color specified: '
  lasterr]);
  end
%end  % processColor   %#ok ML6


%% Split the specified parent container(s)
function [h1, h2, hDivider] = splitPanes(paramsStruct)
  % Initialize
  h1 = [];
  h2 = [];
  hDivider = handle([]);

  % Loop over all specified parent containers
  for parentIdx = 1 : length(paramsStruct.parent)
     % Add the divider button to the parent container
     % Note: use temp vars a,b,c to bypass []-handle errors
     [a,b,c] = splitPane(paramsStruct.parent(parentIdx), paramsStruct);
     if ~isempty(a),  h1(parentIdx) = a;  end  %#ok grow
     if ~isempty(b),  h2(parentIdx) = b;  end  %#ok grow
     if ~isempty(c),  hDivider(parentIdx) = c;  end
  end

  % Clear any invalid handles
  if ~isempty(h1),       h1(h1==0) = [];  end
  if ~isempty(h2),       h2(h2==0) = [];  end
  if ~isempty(hDivider), hDivider(hDivider==0) = [];  end
%end  % splitPanes   %#ok ML6


%% Split a specific parent container
function [h1, h2, hDivider] = splitPane(hParent, paramsStruct)
  % Initialize
  h1 = [];  %#ok in case of premature exit
  h2 = [];  %#ok in case of premature exit

  % Matlab 6 has a bug that causes mouse movements to be ignored over Frames
  % The workaround is to leave a very small margin next to the divider
  dvMargin = 0;
  v = version;
  if v(1)<='6'
     dvMargin = 0.005;
  end
```

```
    % Get the container dimensions
    if strcmpi(paramsStruct.orientation(1),'v')
       % vertical
       dvPos = [0,paramsStruct.dividerlocation,1,paramsStruct.dividerwidth];
       h1Pos = [0,0,1,paramsStruct.dividerlocation-dvMargin];
    else
       % horizontal
       dvPos = [paramsStruct.dividerlocation,0,paramsStruct.dividerwidth,1];
       h1Pos = [0,0,paramsStruct.dividerlocation-dvMargin,1];
    end

    % Prepare the divider
    transformFlag = 0;
    originalParent = hParent;
    try
       hDivider = addDivider(hParent, paramsStruct, dvPos);
    catch
       % Matlab 6 required a uicontrol parent to be a figure, not a frame...
       % get the hParent position in containing figure coordinates
       T = getPos(hParent,'normalized');

       % Hide parent frames so mouse movements around the divider can be found &
    fired
       if isa(handle(hParent),'hg.uicontrol')
          set(hParent,'Visible','off');
          % TODO: link originalParent resizing events to this divider (listener?)
       end
       hParent = get(hParent,'parent');
       while ~isempty(hParent) & ishandle(hParent) & hParent~=0  %#ok for Matlab 6
          %if ~isa(handle(hParent),'figure')  % this is best but returns 0 in
          Matlab6!
          if ~strcmpi(get(hParent,'type'),'figure')
             parentPos = getPos(hParent,'normalized');
             T = transformParentChildCoords(parentPos, T);
             hParent = get(hParent,'parent');
          else
             break;
          end
       end

       % Reconfigure the split-pane positions in normalized figure coords
       dvPos = transformParentChildCoords(T, dvPos);
       h1Pos = transformParentChildCoords(T, h1Pos);
       %h2Pos = transformParentChildCoords(T, h2Pos);
       transformFlag = 1;

       % Now try again...
       hDivider = addDivider(hParent, paramsStruct, dvPos);
    end

    % Recompute the sub-containers dimensions now that the divider is displayed
    dvPos = get(hDivider,'pos');
    if strcmpi(paramsStruct.orientation(1),'v')
```

```matlab
        % vertical
        h2PosStart = paramsStruct.dividerlocation + dvPos(4) + dvMargin;
        h2Pos = [0,h2PosStart,1,1-h2PosStart];
    else
        % horizontal
        h2PosStart = paramsStruct.dividerlocation + dvPos(3) + dvMargin;
        h2Pos = [h2PosStart,0,1-h2PosStart,1];
    end
    if transformFlag
        h2Pos = transformParentChildCoords(T, h2Pos);
    end

    % Setup the mouse-click callback
    mouseDownSetup(hParent);

    % Help messages (right-click context menu)
    %hMenu = uicontextmenu;
    %set(hDivider, 'UIContextMenu',hMenu);
    %uimenu(hMenu,'Label','drag-able divider','Callback',@moveCursor,'UserData',h
Divider);

    % Set the mouse callbacks
    hFig = ancestor(hParent,'figure');
    winFcn = get(hFig,'WindowButtonMotionFcn');
    if ~isempty(winFcn) & ~isequal(winFcn,@mouseMoveCallback) & ...
        (~iscell(winFcn) | ~isequal(winFcn{1},@mouseMoveCallback))   %#ok for
    Matlab 6
        setappdata(hFig, 'uisplitpane_oldButtonMotionFcn',winFcn);
    end
    set(hFig,'WindowButtonMotionFcn',@mouseMoveCallback);

    % Prepare the sub-panes
    h1 = addSubPane(hParent,h1Pos);
    h2 = addSubPane(hParent,h2Pos);

    % Add extra props to hDivider
    addSpecialProps(hDivider, h1, h2, paramsStruct, originalParent);

    % Add listeners to hDivider props
    listenedPropNames = {'DividerColor', 'DividerWidth', 'DividerLocation', ...
                         'DividerMinLocation', 'DividerMaxLocation'};
    listeners = addPropListeners(hFig, hDivider, h1, h2, listenedPropNames);
    setappdata(hDivider, 'uisplitpane_listeners',listeners);
    % These will die with hDivider so no need to un-listen upon hDivider deletion
%end  % splitPane6  %#ok ML6

%% Add the divider button
function hDivider = addDivider(hParent,paramsStruct,position)
    try
        % Get a handle to a platform-specific Java divider object
        % by creating an invisible temporary javax.swing.JSplitPane container
        if lower(paramsStruct.orientation(1)) == 'h'
            jsp = javax.swing.JSplitPane(javax.swing.JSplitPane.HORIZONTAL_SPLIT);
        else  % =vertical
            jsp = javax.swing.JSplitPane(javax.swing.JSplitPane.VERTICAL_SPLIT);
```

```
   end
   jsp.setOneTouchExpandable(1);
   jdiv = jsp.getComponent(0);
   clear jsp; % release memory
   jpanel = javax.swing.JPanel;
   jpanel.add(jdiv);

   % Place onscreen at the correct position & size (still normalized to
container)
   [jdiv,hDivider] = javacomponent(jdiv, [], hParent);   %#ok jdiv is for
debugging
   %[jdiv2,hDivider] = javacomponent(jpanel, [], hParent);   %#ok jdiv is for
debug
   jdiv = handle(jdiv,'CallbackProperties');
   jdiv.Visible = 1;
   drawnow;
   %pause(0.03);
   jdiv.setLocation(java.awt.Point(0,0));
   set(hDivider, 'tag','uisplitpane divider', 'units','norm', 'pos',position);
   drawnow;
   dvPosPix = getPixelPos(hDivider);
   if lower(paramsStruct.orientation(1)) == 'h'
      newPixelPos = [dvPosPix(1:2) paramsStruct.dividerwidth dvPosPix(4)];
   else  % =vertical
      newPixelPos = [dvPosPix(1:3) paramsStruct.dividerwidth];
   end
   setPixelPos(hDivider,newPixelPos);
   jdiv.setSize(java.awt.Dimension(newPixelPos(3),newPixelPos(4)));
   jdiv.DividerSize = paramsStruct.dividerwidth;

   % Set the divider color
   color = mat2cell(paramsStruct.dividercolor,1,[1,1,1]);
   jdiv.setBackground(java.awt.Color(color{:}));

   % Add cross-referencing data
   storeHandles(handle(hDivider),jdiv,hDivider);
   addNewProp(jdiv,'Orientation',paramsStruct.orientation,1);

   % Add resizing & drag/click callbacks
   jdiv.ComponentResizedCallback = @dividerResizedCallback;
   jdiv.MouseDraggedCallback     = @dividerResizedCallback;
   import java.awt.*
   if paramsStruct.orientation(1)=='h'
      jLeft  = jdiv.getComponent(0);
      jRight = jdiv.getComponent(1);
      set(jLeft, 'ActionPerformedCallback', {@dividerActionCallback,
   handle(hDivider), jRight, 'left'}, 'ToolTipText','Click to hide left
   sub-pane');
      set(jRight,'ActionPerformedCallback', {@dividerActionCallback,
   handle(hDivider), jLeft, 'right'}, 'ToolTipText','Click to hide right
   sub-pane');
```

```
        % should be Cursor.W/E_RESIZE_CURSOR but problematic icon on JRE
    1.6=Matlab R2007b+
        jLeft.setCursor(Cursor(Cursor.HAND_CURSOR));   % should be
        Cursor.W_RESIZE_CURSOR
        jRight.setCursor(Cursor(Cursor.HAND_CURSOR)); % should be
        Cursor.E_RESIZE_CURSOR
    else
        jTop = jdiv.getComponent(0);
        jBot = jdiv.getComponent(1);
        set(jTop,'ActionPerformedCallback', {@dividerActionCallback,
    handle(hDivider),jBot,'top'},   'ToolTipText','Click to hide top sub-
    pane');
        set(jBot,'ActionPerformedCallback', {@dividerActionCallback, handle(hDivid
er),jTop,'bottom'},'ToolTipText','Click to hide bottom sub-pane');
        % should be Cursor.N/S_RESIZE_CURSOR but problematic icon on JRE
    1.6=Matlab R2007b+
        jTop.setCursor(Cursor(Cursor.HAND_CURSOR));   % should be
        Cursor.S_RESIZE_CURSOR
        jBot.setCursor(Cursor(Cursor.HAND_CURSOR));   % should be
        Cursor.N_RESIZE_CURSOR
    end

catch

    % Prepare & display the divider button
    hDivider = uicontrol('parent',hParent, 'style','togglebutton', ...
                         'tag','uisplitpane divider', ...
                         'background',paramsStruct.dividercolor, ...  %TODO
                         'tooltip',   paramsStruct.tooltip, ...
                         'enable', 'inactive', ...
                         ... %'callback',@mouseDownCallback, ...
                         ... %'ButtonDownFcn',@mouseDownCallback, ...
                         'units','norm', 'position',position);
    drawnow;
    dvPosPix = getPixelPos(hDivider);
    if lower(paramsStruct.orientation(1)) == 'h'
        newPixelPos = [dvPosPix(1:2) paramsStruct.dividerwidth dvPosPix(4)];
    else  % =vertical
        newPixelPos = [dvPosPix(1:3) paramsStruct.dividerwidth];
    end
    setPixelPos(hDivider,newPixelPos);
    try
        set(hParent,'ResizeFcn',@dividerResizedCallback);
    catch
        % never mind... :-(((
    end
    %jDivider = javax.swing.JButton;
    %set(jDivider,'parent',hParent, 'tag','uisplitpane divider', ...
    %            'background',paramsStruct.dividercolor, ...  %TODO
    %            'tooltip',   paramsStruct.tooltip, ...
    %            'ButtonDownFcn',@mouseDownCallback);
end
```

```
    % Transform HG double to handle obj, so extra props will become visible in
get()
    hDivider = handle(hDivider);
    set(double(hDivider), 'UserData', hDivider);
%end   % addDivider   %#ok ML6

%% Add a sub-pane to a parent container
function h = addSubPane(hParent,hPos)
    try
        % Try a uipanel first...
        h = uipanel('parent',hParent, 'units','norm', 'position',hPos, ...
                    'bordertype','none', 'tag','uisplitpane');
    catch
        % Error - probably Matlab 6... - try using a frame instead of a panel
        h = uicontrol('parent',hParent, 'style','frame', 'enable', 'inactive', ...
                    'units','norm', 'position',hPos, 'tag','uisplitpane');
    end
%end   % addSubPane   %#ok ML6

%% Divider one-click callback function
function dividerActionCallback(varargin)
    try
        jButton = varargin{2}.getSource;
        hDivider = varargin{3};
        jOther = varargin{4};
        str = varargin{5};
        dvPos = hDivider.DividerLocation;
        if any(strcmp(str,{'right','top'}))
            flag = (dvPos <= hDivider.DividerMinLocation);   % flushed left/bottom
            dvFlush = 0.99;
        else   % left/bottom
            flag = (dvPos >= hDivider.DividerMaxLocation);   % flushed right/top
            dvFlush = 0.001;
        end

        if flag   % flushed on the side => move back to center
            hDivider.DividerLocation = 0.5;
            jButton.setToolTipText(['Click to hide ' str ' sub-pane']);
            jOther.setVisible(1);
        else
            hDivider.DividerLocation = dvFlush;
            jOther.setToolTipText(['Click to restore ' str ' sub-pane']);
            jButton.setVisible(0);
        end
    catch
        % never mind...
        disp(lasterr);
    end
%end   % dividerActionCallback   %#ok for Matlab 6 compatibility

%% Divider property pre-change callback
function newValue = dividerPropChangedCallback(varargin)
        [prop,newValue,hFig,hDivider,h1,h2,propName] = deal(varargin{:});
```

```
try newValue = newValue.NewValue;  catch,   end   %#ok

% Note: Matlab 6 sends EventData obj, not scalar newValue
try jDivider = get(hDivider,'JavaComponent'); catch, end   %#ok
switch propName

   case 'DividerLocation'

      checkNumericValue(newValue,.001,.999,'DividerLocation');
      dvPos = get(hDivider,'pos');
      hParent1 = get(hDivider,'Parent');
      hParent2 = get(hDivider,'ContainerParentHandle');
      if ~isequal(hParent1,hParent2)
         % Matlab 6 required a uicontrol parent to be a figure, not
      frame...
         % get the hParent position in containing figure coordinates
         T = getPos(hParent2,'normalized');
         %variant of transformParentChildCoords(T, newValue*[1,1,0,0]);
         newVal2 = T(1:2) + T(3:4) .* newValue([1,1]);
         if lower(hDivider.Orientation(1))=='h'
            newValue = newVal2(1);
         else
            newValue = newVal2(2);
         end
      end
      if lower(hDivider.Orientation(1))=='h'
         set(hDivider,'position',[newValue dvPos(2:4)]);
      else
         set(hDivider,'position',[dvPos(1) newValue dvPos(3:4)]);
      end
      h1 = get(hDivider, 'LeftOrBottomPaneHandle');
      h2 = get(hDivider, 'RightOrTopPaneHandle');
      updateSubPaneSizes(h1,h2,hDivider,newValue);

      % Both flush buttons should now become visible,
      % since divider cannot be flushed
      try
         jDivider = get(hDivider,'JavaComponent');
         jDivider.getComponent(0).setVisible(1);
         jDivider.getComponent(1).setVisible(1);
      catch
         % never mind - probably Matlab 6 without jDivider...
      end

   case 'DividerColor'

      newValue = processColor(newValue,get(hDivider,'Parent'));  % =>
   [R,G,B]
      color = mat2cell(newValue,1,[1,1,1]);  % java-readable format
      try
         jDivider.setBackground(java.awt.Color(color{:}));
      catch
         % probably Matlab 6 without jDivider...
         set(hDivider, 'BackgroundColor', newValue);
```

```
            end
            jDivider.repaint;

        case 'DividerWidth'

            checkNumericValue(newValue,1,25,'DividerWidth');
            dvPos = getPixelPos(hDivider);
            if lower(hDivider.Orientation(1))=='h'
               setPixelPos(hDivider,[dvPos(1:2),newValue,dvPos(4)]);
            else
               setPixelPos(hDivider,[dvPos(1:3),newValue]);
            end
            updateSubPaneSizes(h1,h2,hDivider,hDivider.DividerLocation);
            try
               jDivider.setDividerSize(newValue);
            catch
               % never mind - probably Matlab 6 without jDivider...
            end
            jDivider.repaint;

        case 'DividerMinLocation'

            % nothing to do except check the value and store it for later use
            checkNumericValue(newValue,0,1,propName);
            if newValue >= hDivider.DividerMaxLocation
               myError('YMA:uisplitpane:invalidProperty', ...
                       'DividerMaxLocation must be > DividerMinLocation');
            end

        case 'DividerMaxLocation'

            % nothing to do except check the value and store it for later use
            checkNumericValue(newValue,0,1,propName);
            if newValue <= hDivider.DividerMinLocation
               myError('YMA:uisplitpane:invalidProperty', ...
                       'DividerMaxLocation must be > DividerMinLocation');
            end

        otherwise

            disp(['Unrecognized property: ' propName  ...
                  ' (new value: ' num2str(newValue) ')']);
    end
%end  % dividerPropChangedCallback  %#ok for Matlab 6 compatibility

%% Divider resizing callback function
function outsideLimitsFlag = dividerResizedCallback(varargin)
    try
        outsideLimitsFlag = 0;
        try
           hDivider = varargin{1}.MatlabHGContainer;
           hDivider = get(hDivider, 'UserData');
        catch
           try
              hDivider = varargin{2}.AffectedObject;
           catch
```

```
        hDivider = handle(findobj(gcbf,'tag','uisplitpane divider'));
    end
end

% exit if invalid handle or already in Callback
if ~ishandle(hDivider) | ~isempty(getappdata(hDivider(1),'inCallback'))
%#ok ML6
    % | length(dbstack)>1   %exit also if not called from user action
    return;
end
setappdata(hDivider(1),'inCallback',1);   % used to prevent endless
recursion

if isempty(varargin{1}) | (~isa(hDivider(1),'hg.uicontrol') &  ...
    varargin{2}.getID == java.awt.event.MouseEvent.MOUSE_DRAGGED)   %#ok ML6
    pixelPos = getPixelPos(hDivider);
    hParent = get(hDivider,'ContainerParentHandle');
    parentPixelPos = getPixelPos(hParent);
    if isequal(hParent, get(hDivider,'Parent'))
        parentPixelPos(1:2) = 0;
    end
    if hDivider.Orientation(1) == 'h'
        deltaX = varargin{2}.getX;
        newDvPos = (pixelPos(1)+deltaX-parentPixelPos(1)) /
    parentPixelPos(3);
    else  % vertical
        deltaY = -varargin{2}.getY;
        newDvPos = (pixelPos(2)+deltaY-parentPixelPos(2)) /
    parentPixelPos(4);
    end
    outsideLimitsFlag = (newDvPos > hDivider.DividerMaxLocation+.02) |  ...
                   (newDvPos < hDivider.DividerMinLocation-.02);
    newDvPos = max(hDivider.DividerMinLocation, newDvPos);
    newDvPos = min(hDivider.DividerMaxLocation, newDvPos);
    hDivider.DividerLocation = newDvPos;

else  % uicontrol - probably ML6

    for hIdx = 1 : length(hDivider)  % might be several if Frame resized in ML6
        pixelPos = getPixelPos(hDivider(hIdx));
        if lower(hDivider(hIdx).Orientation(1)) == 'h'
            newPixelPos = [pixelPos(1:2) hDivider(hIdx).DividerWidth  ...
                    pixelPos(4)];
        else  % =vertical
            newPixelPos = [pixelPos(1:3) hDivider(hIdx).DividerWidth];
        end
        if ~isequal(pixelPos,newPixelPos)
            setPixelPos(hDivider(hIdx),newPixelPos);
            hLeft  = get(hDivider(hIdx),'LeftOrBottomPaneHandle');
            hRight = get(hDivider(hIdx),'RightOrTopPaneHandle');
            updateSubPaneSizes(hLeft, hRight, hDivider(hIdx), ...
                        get(hDivider(hIdx),'DividerLocation'));
```

```
                end
            end
        end
    catch
        % never mind...
        disp(lasterr);
    end
    drawnow;
    pause(0.01);
    setappdata(hDivider(1),'inCallback',[]);  % used to prevent endless recursion
%end  % dividerResizedCallback  %#ok for Matlab 6 compatibility

%% Update sub-pane sizes after the divider has moved
function updateSubPaneSizes(h1,h2,hDivider,dvPos)
    try
        dvPixPos = getPixelPos(hDivider);
        hDivider = handle(hDivider);

        if lower(hDivider.Orientation(1))=='h' % horizontal

            if ~isa(hDivider,'hg.uicontrol')  % regular java obj
                % Left sub-pane
                set(h1,'position',[0,0,dvPos,1]);
                h1PixPos = getPixelPos(h1);
                setPixelPos(h1,[0,0,max(1,h1PixPos(3)-1),h1PixPos(4)+1]);
                % Right sub-pane
                set(h2,'position',[dvPos,0,1-dvPos,1]);
                h2PixPos = getPixelPos(h2);
                parentPixPos = getPixelPos(hDivider.Parent);
                h2Width = max(1, parentPixPos(3)-dvPixPos(1)-dvPixPos(3)+2);
                setPixelPos(h2,[dvPixPos(1)+dvPixPos(3)-1,0,h2Width,h2PixPos(4)+1]);
            else  % old ML6 uicontrol obj
                % Left sub-pane
                dvPos = hDivider.position;
                hParent = get(hDivider,'ContainerParentHandle');
                if ~isequal(hParent,hDivider.Parent)
                    h1Pos = get(hParent,'pos');
                else
                    h1Pos = get(h1,'pos');
                end
                newPos = [h1Pos(1), dvPos(2), ...
                         max(0.001,dvPos(1)-h1Pos(1)-0.005),dvPos(4)];
                % 0.5% margin due to ML6 frame bug: not firing mouse movement events
                set(h1,'pos',newPos);
                updateLogicalSubPane(h1);
                % Right sub-pane
                if ~isequal(hParent,hDivider.Parent)
                    h2Pos = get(hParent,'pos');
                else
                    h2Pos = get(h2,'pos');
                end
                newPos = dvPos(1)+dvPos(3)+0.005;
                % 0.5% margin due to ML6 frame bug: not firing mouse movement events
```

```
          newPos = [newPos,dvPos(2),max(0.001,h2Pos(1)+h2Pos(3)-
      newPos),dvPos(4)];
          set(h2,'pos',newPos);
          updateLogicalSubPane(h2);
      end


  else  % vertical

      if ~isa(hDivider,'hg.uicontrol')  % regular java obj
          % Bottom sub-pane
          set(h1,'position',[0,0,1,dvPos]);
          h1PixPos = getPixelPos(h1);
          setPixelPos(h1,[0,0,max(1,h1PixPos(3)),max(1,h1PixPos(4))]);
          % this is theoretically unneeded, used to align with pixel
      boundaries
          % Top sub-pane
          set(h2,'position',[0,dvPos,1,1-dvPos]);
          h2PixPos = getPixelPos(h2);
          parentPixPos = getPixelPos(hDivider.Parent);
          h2Height = max(1, parentPixPos(4)-dvPixPos(2)-dvPixPos(4));
          setPixelPos(h2, [0,dvPixPos(2)+dvPixPos(4),h2PixPos(3)+3,h2Height]);
      else  % old ML6 uicontrol obj
          % Bottom sub-pane
          dvPos = hDivider.position;
          hParent = get(hDivider,'ContainerParentHandle');
          if ~isequal(hParent,hDivider.Parent)
              h1Pos = get(hParent,'pos');
          else
              h1Pos = get(h1,'pos');
          end
          newPos = [h1Pos(1:2),dvPos(3),max(0.001,dvPos(2)-h1Pos(2)-0.005)];
          % 0.5% margin due to ML6 frame bug: not firing mouse movement events
          set(h1,'pos',newPos);
          updateLogicalSubPane(h1);
          % Top sub-pane
          if ~isequal(hParent,hDivider.Parent)
              h2Pos = get(hParent,'pos');
          else
              h2Pos = get(h2,'pos');
          end
          newPos = dvPos(2)+dvPos(4)+0.005;
          % 0.5% margin due to ML6 frame bug: not firing mouse movement events
          newPos = [h2Pos(1),newPos,dvPos(3),max(0.001,h2Pos(2)+h2Pos(4)-
      newPos)];
          set(h2,'pos',newPos);
          updateLogicalSubPane(h2);
      end
  end
  catch
      disp(lasterr);      % never mind...
  end
%end  % updateSubPaneSizes  %#ok for Matlab 6 compatibility
```

```
%% Update logical child sub-pane size (necessary in ML6 which requires all
%% frames to be children of the figure)
function updateLogicalSubPane(hPane)
    try
        hFig = gcbf;
        if isempty(hFig) %& isa(handle(hPane),'hg.uicontrol')
            hFig = ancestor(hPane,'figure');
        end
        hDivider = handle(findobj(hFig, 'ContainerParentHandle', hPane));
        for hIdx = 1 : length(hDivider)
            hParent = get(hDivider(hIdx),'Parent');
            if ~isequal(hPane,hParent)   % ML6
                dvLoc = get(hDivider(hIdx),'DividerLocation');
                pixelPos = getPixelPos(hDivider(hIdx));
                hPanePos = getPixelPos(hPane);
                orientation = get(hDivider(hIdx),'Orientation');
                if lower(orientation(1)) == 'h'
                    newDvPos = hPanePos(1) + hPanePos(3)*dvLoc;
                    newPixelPos = [newDvPos hPanePos(2) hDivider(hIdx).
                DividerWidth ...
                                    hPanePos(4)];
                else  % =vertical
                    newDvPos = hPanePos(2) + hPanePos(4)*dvLoc;
                    newPixelPos = [hPanePos(1) newDvPos hPanePos(3) ...
                                    hDivider(hIdx).DividerWidth];
                end
                if ~isequal(pixelPos,newPixelPos)
                    setPixelPos(hDivider(hIdx),newPixelPos);
                    hLeft  = get(hDivider(hIdx),'LeftOrBottomPaneHandle');
                    hRight = get(hDivider(hIdx),'RightOrTopPaneHandle');
                    updateSubPaneSizes(hLeft, hRight, hDivider(hIdx), dvLoc);
                end
            end
        end
    catch
        disp(lasterr);        % never mind...
    end
%end  % updateLogicalSubPane  %#ok for Matlab 6 compatibility

%% Get ancestor figure - used for old Matlab versions that don't have a built-in
ancestor()
function hObj = ancestor(hObj,type)
    if ~isempty(hObj) & ishandle(hObj)   %#ok for Matlab 6 compatibility
        try
            hObj = get(hObj,'Ancestor');
        catch
            % never mind...
        end
        try
            %if ~isa(handle(hObj),type)  % this is best but always returns 0 in
            Matlab6!
```

```
        %if ~isprop(hObj,'type') | ~strcmpi(get(hObj,'type'),type) % ML6 no
      isprop()
          objType=''; try objType=get(hObj,'type'); catch, end   %#ok
          if ~strcmpi(objType,type)
             try
                parent = get(handle(hObj),'parent');
             catch
                parent = hObj.getParent;   % some objs have no prop, just this
              method
             end
             if ~isempty(parent)  % empty parent means root ancestor, so exit
                hObj = ancestor(parent,type);
             end
          end
       catch
          % never mind...
       end
    end
%end  % ancestor  %#ok for Matlab 6 compatibility


%% Get position of an HG object in specified units
function pos = getPos(hObj,units)
    % Matlab 6 did not have hgconvertunits so use the old way...
    oldUnits = get(hObj,'units');
    if strcmpi(oldUnits,units)  % do not modify units unless we must!
       pos = get(hObj,'pos');
    else
       set(hObj,'units',units);
       pos = get(hObj,'pos');
       set(hObj,'units',oldUnits);
    end
%end  % getPos  %#ok for Matlab 6 compatibility


%% Get pixel position of an HG object - for Matlab 6 compatibility
function pos = getPixelPos(hObj)
    try
       % getpixelposition is unvectorized unfortunately!
       pos = getpixelposition(hObj);
    catch
       % Matlab6 did not have getpixelposition nor hgconvertunits so use the old
way...
       pos = getPos(hObj,'pixels');
    end
%end  % getPixelPos  %#ok for Matlab 6 compatibility


%% Set pixel position of an HG object - for Matlab 6 compatibility
function setPixelPos(hObj,pos)
    try
       % getpixelposition is unvectorized unfortunately!
       setpixelposition(hObj,pos);
    catch
```

```
    % Matlab6 did not have setpixelposition nor hgconvertunits so use the old
    way...
        old_u = get(hObj,'Units');
        set(hObj,'Units','pixels');
        set(hObj,'Position',pos);
        set(hObj,'Units',old_u);
    end
%end    % setPixelPos   %#ok for Matlab 6 compatibility

%% Transform parent=>child normalized coordinates
function normalizedChildCoords = ...
            transformParentChildCoords(normalizedParentCoords, ...
            normalizedChildCoords)
    normalizedChildCoords(1:2) =normalizedParentCoords(1:2) +  ...
                    normalizedParentCoords(3:4).*normalizedChildCoords(1:2);
    normalizedChildCoords(3:4) =normalizedParentCoords(3:4) .* ...
                    normalizedChildCoords(3:4);
%end    % transformParentChildCoords  %#ok for Matlab 6 compatibility

%% Store the container & component's handles in the component
function storeHandles(hcomp,jcomp,hcontainer)
    try
        % Matlab HG container handle
        sp(1) = schema.prop(jcomp,'MatlabHGContainer','mxArray');
        %sp(2) = schema.prop(hcomp,'MatlabHGContainer','mxArray');
        %set([hcomp,jcomp],'MatlabHGContainer',hcontainer);
        set(jcomp,'MatlabHGContainer',hcontainer);
        linkprops([hcomp,jcomp],'DividerHandle','MatlabHGContainer');

        % Java component handle (no need to store within jcomp - only in hcomp...)
        sp(end+1) = schema.prop(hcomp,'JavaComponent','mxArray');
        set(hcomp,'JavaComponent',jcomp);

        % Store the handle in the container's UserData
        % Note: javacomponent placed the jcomp classname in here, but the correct
    place
        % ^^^^  is really in the Tag property, and use UserData to store the
    handle ref
        set(hcontainer,'UserData',hcomp);

        % Disable public set of these handles - read only
        set(sp,'AccessFlags.PublicSet','off');
    catch
        disp(lasterr);      % never mind...
    end
%end    % storeHandles  %#ok for Matlab 6 compatibility

%% Add special properties to the hDivider handle
function addSpecialProps(hDivider, h1, h2, paramsStruct, hParent)
    try
        hhDivider = handle(hDivider);

        % Read-only props: handles & Orientation
        addNewProp(hhDivider,'Orientation',            paramsStruct.orientation,1);
        addNewProp(hhDivider,'LeftOrBottomPaneHandle', h1,1);
```

```
      addNewProp(hhDivider,'RightOrTopPaneHandle',   h2, 1);
      addNewProp(hhDivider,'DividerHandle',          double(hDivider),1);
      addNewProp(hhDivider,'ContainerParentHandle',  hParent,1); % necesary for
   ML6
      % Note: ML6 requires uicontrols to have a figure parent (not uipanel)
      addNewProp(hhDivider,'ContainerParentVarName', paramsStruct.parentName,1);

      % Read/write divider props:
      addNewProp(hhDivider,'DividerColor',          paramsStruct.dividercolor);
      addNewProp(hhDivider,'DividerWidth',          paramsStruct.dividerwidth);
      addNewProp(hhDivider,'DividerLocation',       paramsStruct.
   dividerlocation);
      addNewProp(hhDivider,'DividerMinLocation',    paramsStruct.
   dividerminlocation);
      addNewProp(hhDivider,'DividerMaxLocation',    paramsStruct.
   dividermaxlocation);

      % Note: setting the property's GetFunction is cleaner but doesn't work in
   ML6
   catch
      % Never mind...
   end
%end % addSpecialProps  %#ok for Matlab 6 compatibility


%% Add new property to supplied handle
function addNewProp(hndl,propName,initialValue,readOnlyFlag,getFunc,setFunc)
   sp = schema.prop(hndl,propName,'mxArray');
   set(hndl,propName,initialValue);
   if nargin>3 & ~isempty(readOnlyFlag) & readOnlyFlag  %#ok for Matlab 6
compatibility
      set(sp,'AccessFlags.PublicSet','off');  % default='on'
   end
   if nargin>4 & ~isempty(getFunc)  %#ok for Matlab 6 compatibility
      set(sp,'GetFunction',getFunc);  % unsupported in Matlab 6
   end
   if nargin>5 & ~isempty(setFunc)  %#ok for Matlab 6 compatibility
      set(sp,'SetFunction',setFunc);  % unsupported in Matlab 6
   end
%end % addNewProp  %#ok for Matlab 6 compatibility

%% Add divider property listeners
function listeners = addPropListeners(hFig, hDivider, h1, h2, propNames)
   hhDivider = handle(hDivider);  % ensure a handle obj
   listeners = handle([]);
   for propIdx = 1 : length(propNames)
      callback = {@dividerPropChangedCallback, hFig, hDivider, h1, h2, ...
               propNames{propIdx}};  %TODO
      prop = findprop(hhDivider, propNames{propIdx});
      try
         set(prop, 'SetFunction', callback);
         % Note: this fails in Matlab 6 so we do not have sanity checks revert in
      ML6
      catch
```

```
               listeners(propIdx) = handle.listener(hhDivider, prop, 'PropertyPreSet', ...
                                      callback);  %#ok mlint - preallocate
        end
     end
     listeners(end+1) = handle.listener(hhDivider, findprop(hhDivider,'Extent'), ...
                            'PropertyPostSet', @dividerResizedCallback);
%end  % addPropListeners  %#ok for Matlab 6 compatibility


%% Link property fields
function linkprops(handles,propName,h2PropName)
     if nargin < 3,  h2PropName = propName;  end
     msp = findprop(handles(1),propName);
     msp.GetFunction = {@localGetData,handles(2),h2PropName};
     msp.SetFunction = {@localSetData,handles(2),h2PropName};
%end  % linkprop  %#ok for Matlab 6 compatibility


%% Get the relevant property value from jcomp
function propValue = localGetData(object,propValue,jcomp,propName)   %#ok
     propValue = get(jcomp,propName);
%end  % localGetData  %#ok for Matlab 6 compatibility


%% Set the relevant property value in jcomp
function propValue = localSetData(object,propValue,jcomp,propName)   %#ok
     set(jcomp,propName,propValue);
%end  % localSetData  %#ok for Matlab 6 compatibility


%% Setup the mouse-click callback
function mouseDownSetup(hParent)
     % Matlab 6 has several bugs/problems/limitations with buttonDownFcn, so use
figure callback
     try
        v = version;
        if v(1)<='6'
           axisComponent = getAxisComponent(hParent);
           if ~isempty(axisComponent)
              winDownFcn = get(axisComponent,'MouseClickedCallback');
           else
              winDownFcn = get(hParent,'WindowButtonDownFcn');
           end
           if isempty(winDownFcn) | (~isequal(winDownFcn,@mouseDownCallback) &  ...
              (~iscell(winDownFcn) | ~isequal(winDownFcn{1},@
           mouseDownCallback)))   %#ok


              % Set the ButtonDownFcn callbacks
              if ~isempty(winDownFcn)
                 setappdata(hParent, 'uisplitpane_oldButtonUpFcn',winDownFcn);
                  setappdata(hParent, 'uisplitpane_oldButtonUpObj',axisComponent);
              end
              if ~isempty(axisComponent)
                 set(axisComponent, 'MouseClickedCallback', {@
              mouseDownCallback,hParent});
```

```
                     % remember ancestor HG handle...
                     addNewProp(axisComponent,'Ancestor',hParent,1);
                 else
                     set(hParent, 'WindowButtonDownFcn',@mouseDownCallback);
                 end
             end
             % TODO: chain winDownFcn
         end
     catch
         disp(lasterr);
     end
%end   % mouseDownSetup   %#ok ML6

%% Mouse click down callback function
function mouseDownCallback(varargin)
     try
         % Modify the cursor shape (close hand)
         hFig = gcbf;   %varargin{3};
         if isempty(hFig) & ~isempty(varargin)   %#ok for Matlab 6 compatibility
             hFig = ancestor(varargin{1},'figure');
         end
         if isempty(hFig) | ~ishandle(hFig),  return;  end  %#ok just in case..
         setappdata(hFig, 'uisplitpane_mouseUpPointer',getptr(hFig));
         newPtr = getappdata(hFig, 'uisplitpane_mouseDownPointer');
         if ~isempty(newPtr)
             setptr(hFig, newPtr);
         end

         % Determine the clicked divider
         hDivider = getCurrentDivider(hFig);
         if isempty(hDivider),  return;  end

         % Store divider handle for later use (mouse move/up)
         setappdata(hFig, 'uisplitpane_clickedDivider', hDivider);
     catch
         disp(lasterr);        % Never mind...
     end
%end   % mouseDownCallback   %#ok for Matlab 6 compatibility

%% Mouse movement callback function
function mouseMoveCallback(varargin)   %#ok varargin used for debug only
     try
         % Get the figure's current cursor location & check if it is over any
divider
         hFig = gcbf;
         if isempty(hFig) | ~ishandle(hFig),  return;  end  %#ok just in case..
         inDragMode = isappdata(hFig, 'uisplitpane_clickedDivider');
         % Exit if already in progress - do not want to mess everything...
         if isappdata(hFig,'uisplitpane_inProgress'),  return;  end

         % Fix case of Mode Managers (pan, zoom, ...)
         try
             modeMgr = get(hFig,'ModeManager');
             hMode = modeMgr.CurrentMode;
```

```
            set(hMode,'ButtonDownFilter',@shouldModeBeInactiveFcn);
        catch
            % Never mind - either an old Matlab (no mode managers) or no active mode
        end

        % If in drag mode, mode the divider to the new cursor's position
        if inDragMode
            hDivider = getappdata(hFig, 'uisplitpane_clickedDivider');
            event.AffectedObject = hDivider;
            event.getID = java.awt.event.MouseEvent.MOUSE_DRAGGED;
            cp = get(hFig,'CurrentPoint');   % TODO: convert from pixels => norm
            orientation = get(hDivider, 'orientation');
            pixelPos = getPixelPos(hDivider);
            if lower(orientation(1))=='h'   % horizontal
                event.getX = cp(1,1) - pixelPos(1);   % x location
                event.getY = 0;
            else   % vertical
                event.getX = 0;
                event.getY = pixelPos(2) - cp(1,2);   % y location
                % Note: Y has a negative value to simulate Java behavior
            end
            if (event.getX == 0) & (event.getY == 0)   %#ok ML6
                return;
            elseif dividerResizedCallback([],event)
                mouseUpCallback([],[],hFig);
            end
        else   % regular (non-drag) mouse movement
            % If mouse pointer is not currently over any divider
            hDivider = getCurrentDivider(hFig);
            if isempty(hDivider) %&& ~inDragMode   %#ok for Matlab 6 compatibility
                % Perform cleanup
                mouseOutsideDivider(hFig,inDragMode,hDivider);
            else
                % From this moment on, do not allow any interruptions
                setappdata(hFig,'uisplitpane_inProgress',1);
                mouseOverDivider(hFig,inDragMode,hDivider);
            end
        end

        % Try to chain the original WindowButtonMotionFcn (if available)
        try
            hgfeval(getappdata(hFig, 'uisplitpane_oldButtonMotionFcn'));
        catch
            % Never mind...
        end
    catch
        disp(lasterr);        % Never mind...
    end
    rmappdataIfExists(hFig,'uisplitpane_inProgress');
%end  % mouseMoveCallback  %#ok for Matlab 6 compatibility

%% Mouse click up callback function
function mouseUpCallback(varargin)
```

```
    try
        % Restore the previous (pre-click) cursor shape
        hFig = gcbf;   %varargin{3};
        if isempty(hFig) & ~isempty(varargin)   %#ok for Matlab 6 compatibility
            hFig = varargin{3};
            if isempty(hFig)
                hFig = ancestor(varargin{1},'figure');
            end
        end
        if isempty(hFig) | ~ishandle(hFig),  return;  end  %#ok just in case..
        if isappdata(hFig, 'uisplitpane_mouseUpPointer')
            mouseUpPointer = getappdata(hFig, 'uisplitpane_mouseUpPointer');
            set(hFig,mouseUpPointer{:});
            rmappdata(hFig, 'uisplitpane_mouseUpPointer');
        end


        % Cleanup data no longer needed
        rmappdataIfExists(hFig, 'uisplitpane_clickedDivider');

        % Try to chain the original WindowButtonUpFcn (if available)
        oldFcn = getappdata(hFig, 'uisplitpane_oldButtonUpFcn');
        if ~isempty(oldFcn) & ~isequal(oldFcn,@mouseUpCallback) & (~iscell(oldFcn)
    | ...
            ~isequal(oldFcn{1},@mouseUpCallback))   %#ok for Matlab 6 compatibility
            hgfeval(oldFcn);
        end
    catch
        disp(lasterr);       % Never mind...
    end
%end  % mouseUpCallback   %#ok for Matlab 6 compatibility


%% Mouse movement outside the divider area
function mouseOutsideDivider(hFig,inDragMode,hDivider)   %#ok hDivider is unused
    try
        % Restore the original figure pointer (probably 'arrow', but not
    necessarily)
        % On second thought, it should always be 'arrow' since zoom/pan etc. are
        % disabled within hDivider
        %if ~isempty(hDivider)
            % Note: Only modify this within hDivider (outside the patch area)
            % - not in other axes - TODO!!!
            set(hFig, 'Pointer','arrow');
        %end
        oldPointer = getappdata(hFig, 'uisplitpane_oldPointer');
        if ~isempty(oldPointer)
            %set(hFig, oldPointer{:});  % see comment above
            drawnow;
            rmappdataIfExists(hFig, 'uisplitpane_oldPointer');
            if isappdata(hFig, 'uisplitpane_mouseUpPointer')
                setappdata(hFig, 'uisplitpane_mouseUpPointer',oldPointer);
            end
        end
```

```
        % Restore the original ButtonUpFcn callback
        if isappdata(hFig, 'uisplitpane_oldButtonUpFcn')
            oldButtonUpFcn = getappdata(hFig, 'uisplitpane_oldButtonUpFcn');
            axisComponent  = getappdata(hFig, 'uisplitpane_oldButtonUpObj');
            if ~isempty(axisComponent)
                set(axisComponent, 'MouseReleasedCallback',oldButtonUpFcn);
            else
                set(hFig, 'WindowButtonUpFcn',oldButtonUpFcn);
            end
            rmappdataIfExists(hFig, 'uisplitpane_oldButtonUpFcn');
        end

        % Additional cleanup
        rmappdataIfExists(hFig, 'uisplitpane_mouseDownPointer');
        drawnow;
    catch
        disp(lasterr);      % never mind...
    end
%end  % mouseOutsideDivider  %#ok for Matlab 6 compatibility

%% Mouse movement within the divider area
function mouseOverDivider(hFig,inDragMode,hDivider)
    try
        % Separate actions for H/V
        orientation = get(hDivider, 'orientation');
        if lower(orientation(1))=='h'  % horizontal
            shapeStr = 'lrdrag';
        else  % vertical
            shapeStr = 'uddrag';
        end

        % If we have entered the divider area for the first time
        axisComponent = getAxisComponent(hFig);
        if ~isempty(axisComponent)
            winUpFcn = get(axisComponent,'MouseReleasedCallback');
        else
            winUpFcn = get(hFig,'WindowButtonUpFcn');
        end
        if isempty(winUpFcn) | (~isequal(winUpFcn,@mouseUpCallback) &  ...
           (~iscell(winUpFcn) | ~isequal(winUpFcn{1},@mouseUpCallback)))  %#ok ML6
            % Set the ButtonUpFcn callbacks
            if ~isempty(winUpFcn)
                setappdata(hFig, 'uisplitpane_oldButtonUpFcn',winUpFcn);
                setappdata(hFig, 'uisplitpane_oldButtonUpObj',axisComponent);
            end
            if ~isempty(axisComponent)
                set(axisComponent, 'MouseReleasedCallback',{@mouseUpCallback,hFig});
            else
                set(hFig, 'WindowButtonUpFcn',@mouseUpCallback);
            end
            % Clear up potential junk that might confuse us later
            rmappdataIfExists(hFig, 'uisplitpane_clickedBarIdx');
        end
```

```
      % If this is a drag movement (i.e., mouse button is clicked)
      if inDragMode
         % Act according to the dragged object
         dvLimits = get(hDivider, {'dividerMinLocation','dividerMinLocation'});
         cp = get(hFig,'CurrentPoint');  % TODO: convert from pixels => norm
         if strcmpi(orientation,'horizontal')
            dvLocation = cp(1,1);  % x location
         else  % vertical
            dvLocation = cp(1,2);  % y location
         end
         dvLocation = min(max(dvLocation,dvLimits{1}),dvLimits{2});
         set(hDivider,'DividerLocation',dvLocation);
         % Mode managers (zoom/pan etc.) modify cursor shape so we need
         to force ours
         newPtr = getappdata(hFig, 'uisplitpane_mouseDownPointer');
         if ~isempty(newPtr)
            setptr(hFig, newPtr);
         end
      else  % Normal mouse movement (no drag)
         % Modify the cursor shape
         oldPointer = getappdata(hFig, 'uisplitpane_oldPointer');
         if isempty(oldPointer)
            % Preserve original pointer shape for future use
            setappdata(hFig, 'uisplitpane_oldPointer',getptr(hFig));
         end
         setptr(hFig, shapeStr);
         setappdata(hFig, 'uisplitpane_mouseDownPointer',shapeStr);
      end
      drawnow;
   catch
      disp(lasterr);       % never mind...
   end
%end  % mouseOverDivider  %#ok for Matlab 6 compatibility

%% Remove appdata if available
function rmappdataIfExists(handle, name)
   if isappdata(handle, name)
      rmappdata(handle, name)
   end
%end  % rmappdataIfExists  %#ok for Matlab 6 compatibility

%% Get the figure's java axis component
function axisComponent = getAxisComponent(hFig)
   try
      if isappdata(hFig, 'uisplitpane_axisComponent')
         axisComponent = getappdata(hFig, 'uisplitpane_axisComponent');
      else
         axisComponent = [];
         oldJFWarning=warning('off','MATLAB:HandleGraphics:ObsoletedProperty:Jav
      aFrame');
         javaFrame = get(hFig,'JavaFrame');
         warning(oldJFWarning.state,'MATLAB:HandleGraphics:ObsoletedProperty:Jav
      aFrame');
```

```
            axisComponent = get(javaFrame,'AxisComponent');
            axisComponent = handle(axisComponent, 'CallbackProperties');
            if ~isprop(axisComponent,'MouseReleasedCallback')
                axisComponent = [];  % wrong axisComponent...
            else
                setappdata(hFig, 'uisplitpane_axisComponent',axisComponent);
            end
        end
    catch
        % never mind...
    end
%end  % getAxisComponent  %#ok for Matlab 6 compatibility

%% Get the divider (if any) that the mouse is currently over
function hDivider = getCurrentDivider(hFig)
    try
        hDivider = handle([]);
        hDividers = findall(hFig, 'tag','uisplitpane divider');
        if isempty(hDividers), return; end  % should never happen...
        for dvIdx = 1 : length(hDividers)
            dvPos(dvIdx,:) = getPixelPos(hDividers(dvIdx));  %#ok mlint -
        preallocate
        end
        cp = get(hFig, 'CurrentPoint');  % in Matlab pixels
        inXTest = (dvPos(:,1) <= cp(1)) & (cp(1) <= dvPos(:,1)+dvPos(:,3));
        inYTest = (dvPos(:,2) <= cp(2)) & (cp(2) <= dvPos(:,2)+dvPos(:,4));
        hDivider = hDividers(inXTest & inYTest);
        hDivider = hDivider(min(1:end));  % return no more than a single hDivider!
        hDivider = handle(hDivider);  % transform into a handle object
    catch
        disp(lasterr);      % never mind...
    end
%end  % getCurrentDivider  %#ok for Matlab 6 compatibility

%% Determine whether a current mode manager should be active or not (filtered)
function shouldModeBeInactive = shouldModeBeInactiveFcn(hObj, eventData)  %#ok
- eventData
    try
        shouldModeBeInactive = 0;
        hFig = ancestor(hObj,'figure');
        hDivider = getCurrentDivider(hFig);
        shouldModeBeInactive = ~isempty(hDivider);
    catch
        disp(lasterr);      % never mind...
    end
%end  % shouldModeBeActiveFcn  %#ok for Matlab 6 compatibility

%% hgfeval replacement for Matlab 6 compatibility
function hgfeval(fcn,varargin)
    if isempty(fcn), return; end
    if iscell(fcn)
        feval(fcn{1},varargin{:},fcn{2:end});
    elseif ischar(fcn)
```

```
        evalin('base', fcn);
    else
        feval(fcn,varargin{:});
    end
%end  % hgfeval  %#ok for Matlab 6 compatibility
```

10.2 Integration Debriefing System

As another example that illustrates the use of some of the features presented in this book, I would like to present the IDS (Integration Debriefing System) application, which I developed for a large industrial client.

Integration engineers are often at the bottom of the "food chain" when it comes to development resource allocation. They often need to rely on simple tools such as Excel. In cases of large data sets, numerous input files, and multiple data formats, using Excel becomes a very painful experience. Work efficiency and productivity suffer greatly, and integration work often needs to wait for support from external programmers (e.g., modifying VB code or plugins).

IDS was designed to solve this problem, enabling integration engineers to easily load and analyze large amounts of data, originating from multiple systems, using different storage formats. Moreover, it enables engineers who do not necessarily have any programming experience to utilize the full set of mathematical, statistical, and other engineering functions available in MATLAB, writing analysis functions using a GUI.

In practice, IDS achieved a 10-fold improvement in processed data size, 80-fold improvement in analysis time, and a 5-fold improvement in overall integration productivity, compared with the old Excel-based system. Because of IDS's focus on analysis result anomalies, some of the issues discovered using IDS would never have been detected in the older system. Altogether, IDS was a major success for the client. I will be happy to provide additional details to readers who may be interested in a similar application for their own needs.

For such a generic large-scale application, MATLAB may seem to be an odd choice. However, MATLAB's powerful analysis and graphing functionality, coupled with the power of Java GUI, resulted in a very usable application. Here are some of the features that made this possible:

10.2.1 *Data Setup*

In the main application window, summary information is displayed in Java tables that are contained within Java tabbed panes (refer to Sections 4.1 and 4.3):

Java tables contained within a JTabbedPane

The panel-level checkbox that appears next to the "Objects" label uses a nice trick: It relies on the undocumented fact that the panel's title label is a simple hidden ***uicontrol*** child of the MATLAB panel handle (see Section 6.11). This ***uicontrol*** handle can be found and simply transformed from a 'style' = 'text' control into a 'style' = 'checkbox' control, as follows:[8]

```
hPanel = uipanel('position',[0.2,0.2,0.4,0.4], 'title','Objects');
hTitle = setdiff(findall(hPanel),hPanel); % retrieve title handle
hTitle = get(hPanel,'TitleHandle'); % alternative, uses hidden prop

% Modify the uicontrol style; add 20 pixel space for the checkbox
newPos = get(hTitle,'position') + [0,0,20,0]; % in pixels
set(hTitle, 'style','checkbox', 'value',1, 'pos',newPos);
```

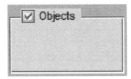

MATLAB *uipanel* with a checkbox title

Within the tables, I have used a custom CellRenderer (see Section 4.1.1) in order to automatically align the data based on its type (strings are left-aligned and numbers are right-aligned), display cell-specific tooltips (that provide cell-specific information that cannot be displayed due to space and usability considerations), and set cell-specific foreground and background colors (to highlight modified cells).

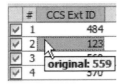

Data table with a non-standard cell-renderer

The table headers enable sorting by attaching a `TableSorter` class, as explained in Section 4.1.4. The headers tooltip implement multi-line HTML formatting (as explained in Section 3.3.3 and shown in Section 4.1.7) to explain the sorting functionality:

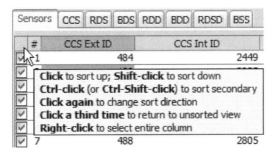

Data table with sortable columns

10.2.2 *Defining Data Items and Events*

In the application's data definition screen, I use a ***uitree*** (Section 4.2) to display the defined data items. In the events definition screen, I use a Java table with dynamic cell drop-down selection values, based on each event's context.

Here is how it works: Events are defined as some condition that occurs in some data file. For example, a value of "2" that occurs for the first time in data field *System_Status* of data file *System_Operability_Status.dat* indicates some sort of failure event. The *System_Status* data field only occurs in the *System_Operability_Status.dat* data file, and so the field-selection drop-down should only present the fields that are relevant for the selected data file (Section 4.1.1):

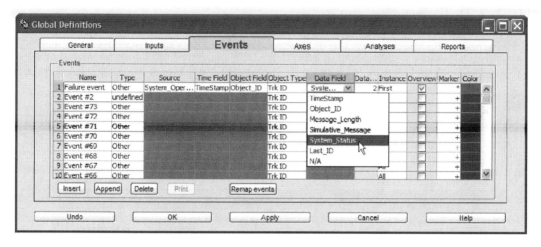

uitable with a non-standard column cell editor that supports dynamic data values

Since the event we have just defined is a failure event, we need to give it a prominent marker. Let us choose a red * marker, using the cell-attached color drop-down control (Section 4.1.1 again):

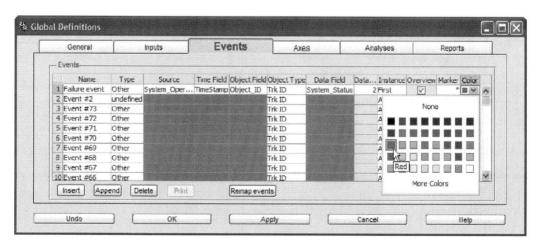

uitable with custom CellRenderer and CellEditor

New events are added to the events table with empty (undefined) data, prominently marked with a red background, until being populated. The non-mandatory fields are similarly marked with a gray background and editing-disabled (Section 4.1.1). Both these features are done using a custom CellRenderer for the relevant columns (mandatory and non-mandatory data columns have slightly different variants).

These CellRenderers are also used in other definition tables (Inputs, Axes, Analyses). For example, the Axes definition table contains some calculated data fields — these are presented as read-only disabled fields, with a gray background color.

10.2.3 Defining Analyses

The analyses definition GUI window is really where the engineering know-how is entered by the user. Each defined analysis can be edited in a table and saved separately. Analyses can use each other as building blocks, as well as use any available built-in or user-defined MATLAB function that is available on the MATLAB path.

When selecting from the list of available analyses building blocks, the listbox implements a dynamic tooltip, such that moving the mouse pointer over any list item presents relevant formatted (color/font-styled) information in a tooltip (Section 6.6.3):

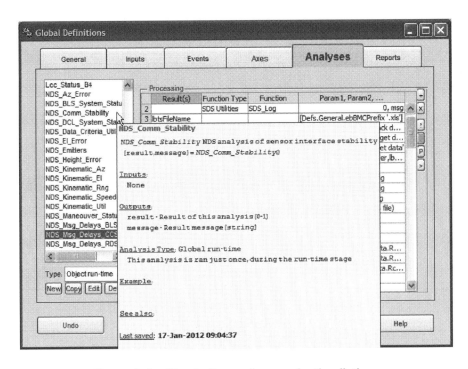

Dynamic tooltips in the analyses selection listbox

In order to assist integration engineers, who may have little programming experience, often-used functions are grouped in categories, so that novice users can simply select the category and then the required function from the drop-down list. This uses the same dynamic-data drop-downs described above for the Events table:

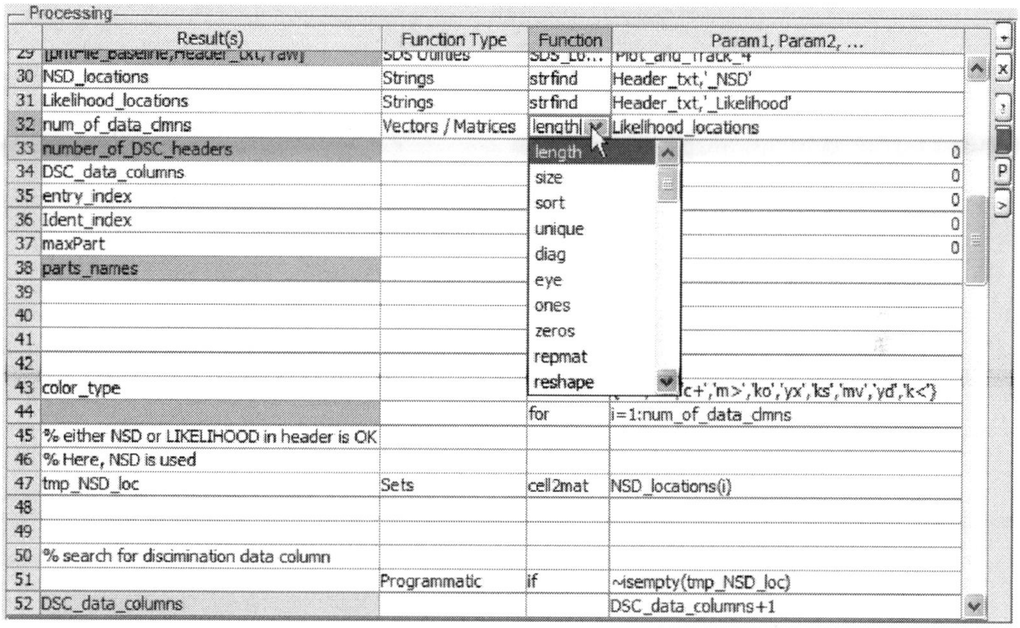

Analyses definition table

Note in the screenshot above how the drop-down only presents functions relevant to the selected function category ("Vectors/Matrices" in this case — category labels are themselves selected from a drop-down). The drop-down cell is fully editable. This means that users can select a drop-down value or type any other function name.

Also note that some cells (in the bottom table row in the screenshot above) have background highlights. This is done by automatically passing the generated analysis script through the built-in *mlint* preprocessor using *mlint's* undocumented interface[9] (*mlint* was not described in this book since it is not Java-based; I will possibly detail it in a future book about undocumented aspects of pure MATLAB).

A tooltip over the affected cell displays the specific warning or error. This was done by using the custom CellRenderer's ability to specify cell-specific tooltips as well as a cell-specific background color (pink/orange for warnings; red for errors):

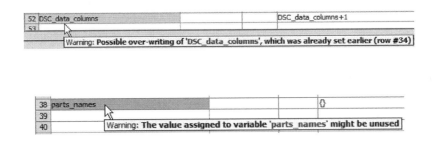

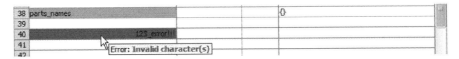

Analysis dynamic (cell-specific) alert tooltips: notice, warning, and error

In some cases, users may wish to view the auto-generated analysis code in a syntax-high-lighted code pane (mini-editor), rather than in a data table. This is done using a `SyntaxText-Pane` panel that has a MATLAB MIME type (Section 5.5.1):

```
% GUI Processing line #24
deltaTimes(1:amdrLength) = Defs.Consts.anal.NDS_Msg_Delays.infDelay;

% GUI Processing line #26
lastUpdTime = -Inf;

% GUI Processing line #27
for amdrIdx = 1 : amdrLength

    % GUI Processing line #28
    rdsIdx = find (rdsUpdTimes==amdrUpdTimes(amdrIdx), 1);

    % GUI Processing line #29
    if (~isempty(rdsIdx) && lastUpdTime<amdrUpdTimes(amdrIdx))

        % GUI Processing line #30
        deltaTimes(amdrIdx) = minus (rdsTimes(rdsIdx), amdrTimes(amdrIdx));

    % GUI Processing line #31
    end    % (if matching CCS data entry was found)
```

Auto-generated analysis code in a syntax-highlighted code pane

10.2.4 Defining Reports

In the final setup tab panel, *Reports*, the user can create and customize the appearance of defined reports. The reports can then be used as building blocks within any of the analyses.

In the Reports panel, I have integrated the plot-selection drop-down control (`PlotTypeCombo`) and plot-catalog dialog window (`PlotCatalog`) that were discussed in Section 5.4.2:

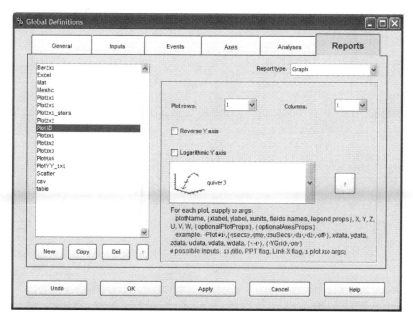

Plot-selection control embedded within the reports-definition panel

Users can select any of the recently used or predefined plot types specified in the drop-down list. In addition, users can always select the "More Plot Types …" option (at the bottom of the drop-down list) to display a dedicated plot catalog selection window.

After any plot is selected, it is automatically added to the drop-down list for possible future reuse.

Once a plot type is selected, its meta-data is examined in order to extract the plot's expected input arguments. These are then displayed immediately beneath the plot selection drop-down, for user reference.

The reader is referred to Section 5.4.2 for a description of all these features.

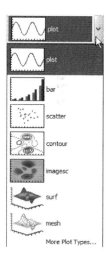

10.2.5 *Displaying Analysis Results*

Following automated running of the selected analyses on the tested data objects, a color-coded matrix of results is created and presented to the user. The brighter the color of a specific cell (a data-object vs. analysis combination), the more problematic was the corresponding combination (i.e., the specific analysis ran on the specific object). This enables immediate and intuitive drill-down into problematic cases, rather than sifting through hundreds of irrelevant graphs.

This not only significantly improves the data analysis time, but also improves accuracy, since users are not made careless by endless acceptable reports, before they encounter the infrequent problematic report.

In order to facilitate user decision on which of the cells to drill-down, I implemented dynamic data cursors: moving the mouse pointer over any of the matrix cells presents a tooltip box with information about the executed analysis and the reason for the unacceptable result:

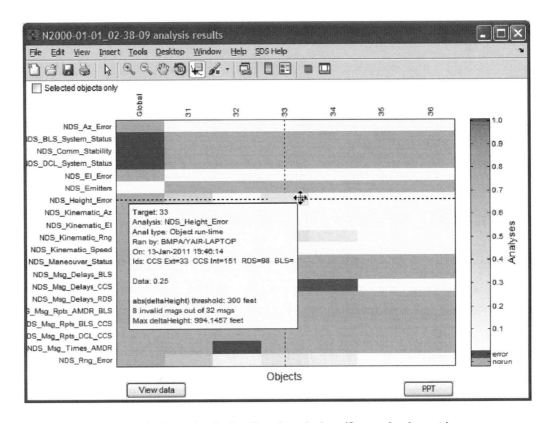

Dynamic data-tips in the Results window (See color insert.)

When clicking any of the result matrix's cells, the user is presented with the reports prepared by that particular analysis for the selected object. As shown above, these user-defined reports can be Excel tables, data tables, text files, or graphs.

A particularly useful and non-obtrusive addition I have used for all graphs is the addition of a small "+" button next to the graph origin.[10] Clicking on the button presents a small dialog window that enables controlling the graph properties:[11]

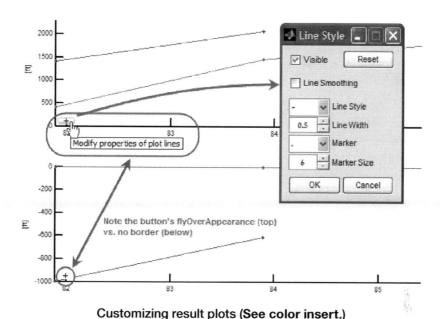

Customizing result plots (See color insert.)

The "+" button is actually a com.mathworks.mwswing.MJButton Java object (Section 6.1) that is displayed using *javacomponent*. We recall from Section 6.1 that MJButton added a **FlyOverAppearance** property to the standard Java Swing's JButton. It is this property that I now use, using the following code snippet:

```
btLinePropsCbStr = ['uisetlineprops(findall(' num2str(hAxes,99) ...
                 ',''type'',''line''))'];

btLinePropsPos = [axesPos(1:2) + 0.003, 0.02, 0.02];

%uicontrol('string','+', 'Units','Norm', 'BackgroundColor','white',
'Position',btLinePropsPos, 'tag',['bt' axName '_LineProps'], 'userdata',hAxes,
'callback',btLinePropsCbStr);


% Note: all the following code is just to have a specific cursor
% ^^^ (HAND_CURSOR) when hovering over the button...
btLineprops = com.mathworks.mwswing.MJButton('+');
btLineprops.setBorder([]);
btLineprops.setBackground(java.awt.Color.white);
btLineprops.setCursor(java.awt.Cursor(java.awt.Cursor.HAND_CURSOR));
btLineprops.setFlyOverAppearance(true);
btLineprops.setToolTipText('Modify properties of plot lines');
[dummy,btContainer] = javacomponent(btLineprops,[0 0 1 1],hFig); %#ok
set(btLineprops, 'ActionPerformedCallback',btLinePropsCbStr);
set(btContainer, 'Units','Norm', 'Position',btLinePropsPos);
```

10.3 Concluding Exercise: UIMultiListbox

As a suggested concluding exercise, implement the following ***UIMultiListbox*** utility, which covers many of the topics discussed in this book:

The ***UIMultiListbox*** utility will place two adjacent listboxes in the specified MATLAB container handle (frame, ***uipanel***, or ***gcf*** if none was specified). The listboxes will be correlated so that selecting an item in any of the listboxes will automatically select the corresponding item in the other listbox — use ***handle.listener***[†] to synchronize the selection/movement of the listboxes.

Update the figure status-bar[‡] and present a systray icon message[§] whenever the listbox selection is modified. Since the listboxes are correlated, there is no need for one of the listboxes to have scrollbars — hide them.[¶]

Set a specialized border[††] around the listboxes and update the listbox background color upon MouseEnter/MouseExit/FocusGained/FocusLost event callbacks.[‡‡]

Enable display of list-item icons (e.g., state flags).[§§]

Use ***uimenu***[¶¶] and ***uiundo***[†††] to undo listbox changes.

Enable Drag-and-Drop of listbox items.[‡‡‡]

Use HTML tooltips and labels.[§§§]

Set item-specific tooltips and context (right-click) menus.[¶¶¶]

Finally, use ***schema.prop***[††††] to return a single handle object with all the information combined (both listbox handles, etc.) upon creation of the ***UIMultiListbox*** object.

Here are several links that may provide ideas on how to use a JSlider to synchronize two listboxes:

- http://www.mathworks.com/matlabcentral/newsreader/view_thread/235484
- http://www.jroller.com/santhosh/entry/synchronize_scrolling_views
- http://www.jidesoft.com/javadoc/com/jidesoft/list/DualList.html[‡‡‡‡]

[†]   See Appendix B.

[‡]   See Section 4.7.

[§]   See Section 3.6.

[¶]   See Sections 6.5.2 and 6.6.

[††]   See several Border examples in Chapter 6.

[‡‡]   See Section 3.4.

[§§]   See Section 6.6.1.

[¶¶]   See Section 4.6.

[†††] See Section 4.4.

[‡‡‡] See Section 3.7.

[§§§] See Section 3.3.3.

[¶¶¶] See Section 6.6.2.

[††††] See Appendix B.

[‡‡‡‡] com.jidesoft.list.DualList is included in the JIDE Grids package that is included in MATLAB since release 7.7 (R2008b) — see Section 5.7.2.

References

1. See http://UndocumentedMatlab.com/blog/uisplitpane/ (or http://bit.ly/9iQoPJ).
2. http://www.mathworks.com/matlabcentral/fileexchange/23073 (or http://bit.ly/n73ucR).
3. http://blogs.mathworks.com/pick/2009/03/27/gui-layout-part-6 (or http://bit.ly/qEdlQV).
4. http://www.mathworks.com/matlabcentral/fileexchange/32697-making-matlab-swing (or http://bit.ly/reUuPk); http://sourceforge.net/projects/waterloo/ (or http://bit.ly/oiWP2q).
5. http://UndocumentedMatlab.com/blog/uisplitpane (or http://bit.ly/9iQoPJ).
6. http://UndocumentedMatlab.com/blog/hgfeval/ (or http://bit.ly/aIgaOa).
7. http://www.mathworks.com/matlabcentral/fileexchange/23073 (or http://tinyurl.com/ckbbo4).
8. http://UndocumentedMatlab.com/blog/panel-level-uicontrols/ (or http://bit.ly/9qbn3E).
9. http://www.mathworks.com/matlabcentral/newsreader/view_thread/255839#755939 (or http://bit.ly/e6TawD).
10. http://UndocumentedMatlab.com/blog/borderless-button-used-for-plot-properties/ (or http://bit.ly/rusMp1).
11. Adapted from http://www.mathworks.cn/matlabcentral/fileexchange/14143-uisetlineprops (or http://bit.ly/f0yT6S).

Appendix A: What Is Java?

This appendix presents an overview of Java. An experienced Java programmer can safely skip this section. For readers who are inexperienced with Java, reading this section should help to understand the Java concepts and code snippets used throughout this book, as well as to make simple adaptations.

Java is an object-oriented programming language, introduced in 1995. Its main strength when compared with other object-oriented languages of its time (C++ being the most important) was its portability: Java was designed to be architecture-neutral so that Java programs written on a Mac, for example, would look and behave exactly in the same way as on Windows, Linux, and any other platform that supports the Java specification. This design, coupled with modern object-oriented programming features, built-in security measures, and easily accessible GUI and I/O, significantly reduced development work and made Java a favorite among programmers worldwide.

All Java programs are object-oriented: whereas MATLAB functions can be created as standalone functions or scripts, Java code must be enclosed in a containing *class*. The Java class determines which *properties* belong to the class and which functions (*methods*) are available to act on these properties.

Classes are often grouped into *packages* and sub-packages based on semantic and functional relationships. For example, MATLAB has a `com.mathworks.mwswing` package that contains the `MJButton`, `MJLabel`, and `MJPanel` classes. A class's Fully Qualified Class Name (FQCN) includes the package name along with all its parent packages (e.g., `com.mathworks.mwswing.MJButton`). Sibling classes of the same package, or classes that explicitly import definition of another package, do not need to use FQCN, and can use just the short classname (e.g., `MJButton`). Classes whose package is not explicitly defined are still packaged, in a global default package. It is customary, although not mandatory, for package names to be lower-cased and for class names to be camel-cased (e.g., ClassName).[1]

Java classes, properties, methods, and so on, all have *accessibility attributes* that control whether these elements are accessible to external (non-class) code. A *public* attribute means that the element is usable by any external class (including our MATLAB code); *protected* and *private* attributes (and the default package visibility) reduce the element visibility/accessibility to Java subclasses, the same class and the same package, respectively. Only public elements are visible from our MATLAB code.

Like MATLAB, Java is *case-sensitive*. This means that ClassName is different from class-name, className, and Classname. Using a keyword or element name with incorrect capitalization will result in a compilation error.

Unlike MATLAB, all Java code statements must either conclude with a semicolon (;) or be encased within braces: {...}. Failing to do so causes a compilation error. Java is less forgiving than MATLAB to such seemingly trivial syntax errors.

Java comments can either extend to the end of the line (following a double-slash //, like MATLAB's %), or have a limited extent (between /*...*/, like MATLAB's %{...%} construct).

Unlike MATLAB, Java arrays use square, not round, brackets ([]). Java arrays start all index values at 0, and multi-dimensional arrays use multiple brackets. So, MATLAB's a(2,3) would be represented in Java as a[1][2].

Classes and methods can contain local properties. These properties have accessibility attributes as well as a type. All Java properties and methods must be declared to have a specific type (aka *strong-typing*).[†] The type can either be a primitive built-in type (boolean, char, byte, short, int, long, float, or double) or a class name (e.g., String or java.lang. String). Properties can be declared to be *static* (a class property, as opposed to an object instance property) and/or *final* (unchangeable, constant).[‡] All newly created properties get a default value, unless they were initialized during creation. Property values can be initialized upon creation, or set later on in the code:

```
public boolean myFlag1;                          // default value = false
private boolean myFlag2 = true;                  // override default value
protected double number = java.lang.Math.PI;     // override default = 0
private final double PI_2 = java.lang.Math.PI / 2.0;   // constant
public static String message = "Hello World!";   // override default = ""
private MyClass myObject = new MyClass();         // default constructor
```

Java elements (properties and methods) are either *static* or objectal. Static elements belong to their class, so only a single instance exists, referred-to by its class name (e.g., HelloWorld. *main()* or Math.PI). Non-static (object instance) elements are created by *instantiating* one or more *instances* (*objects*) of their class. The elements are then accessed using the object's *reference* handle (e.g., myObject.data or message.*length()*).

"Hello World" is a classic sample program that is typically used when learning new programming languages. Here is the Java version of this program:

```
public class HelloWorld
{
    public static void main(String[] args)
    {
```

---

† Notwithstanding Generics (http://java.sun.com/j2se/1.5.0/docs/guide/language/generics.html or http://tinyurl.com/6nfhp), which is an advanced Java programming topic outside the scope of this book (also see section 1.7).

‡ Additional modifiers that are encountered less frequently are native, synchronized, transient, and volatile.

```
        // Display a message on the console
        System.out.println("Hello World!");
    }
}
```

This simple program contains the public class `HelloWorld`, which contains only a single method — a public *main()*. This method is declared `static`, so it can be called without requiring object instantiation. Such a static *main()* method is the standard execution entry point of all Java programs. It accepts an array of `String` values (which are the command-line arguments supplied to the program), and returns nothing (`void`).

Within the *main()* method, we call the `System` class's[†] static `out` object's *println()* method to display a "Hello World" message (a `String`) to the console, which is automatically redirected to MATLAB's Command Window in MATLAB.

Classes can have multiple versions of the same method, which receive different arguments and/or return different result types. Java knows in run-time (and sometimes even in compilation-time) which of the methods to use, based on the types of the supplied and returned arguments (this is often called *method overloading*).

Classes can have special methods, called *constructors*, that have the same name as their class. They are invoked whenever a new class object is created (instantiated). Like other methods, constructors can be overloaded with different input arguments. Unlike regular methods, constructors have no return type and return no value.

For example, assume we declare the following class:

```
public class MyClass
{
    public MyClass()    // default constructor
    {
        // do something useful here...
    }

    /* Non-default (overloaded) constructor */
    public MyClass(boolean flag, double number)
    {
        // do something useful here...
    }
}
```

We could then use the following object instantiation code:

```
private MyClass myObject = new MyClass();    // use default constructor
private MyClass myObject = new MyClass(myFlag, PI_2);  // non-default
```

Java classes support a single *inheritance*, meaning they can inherit (*extend*) another class (the *superclass*). The subclass inherits all the methods and properties of its superclass and can

---

[†] `System` is actually `java.lang.System` — the `java.lang` package is automatically imported in Java and does not require explicit import. `String` is also such a class (`java.lang.String`).

override any of them. In addition, subclasses often add new properties and methods to support a specialized implementation. For example, MATLAB's com.mathworks.mwswing. MJButton class extends the standard Swing javax.swing.JButton class, in order to provide additional capabilities to MATLAB button controls that are unavailable in the standard Java Swing button implementation (many of these extensions are detailed in Chapters 5 and 6).

While supporting only a single inheritance, Java classes can support multiple *interfaces*. An interface (typically named, I ..., e.g., IGeometry) is a declaration of a set of methods that each supporting class must implement. For example, geometry interface IGeometry might require implementing the *area()*, *circumference()*, *maxWidth()*, and *maxHeight()* methods. Then, the Circle, Rectangle, and Square classes (that all extend the Geometry super-class) could all declare themselves as implementing IGeometry, if they indeed implement all the relevant methods:

```
public class Circle extends Geometry implements IGeometry, IPaintable
```

Now, if a method somewhere receives an input argument of type Geometry, it could invoke any IGeometry method on the input, without needing to know the input object's exact sub-type (shape). As noted, classes can implement multiple interfaces.

Java supports most of the MATLAB operators and quite a few others. This includes the standard arithmetic operators + (which can also be used to concatenate strings), -, *, /, and = (which is also used for property assignments), comparison operators (<, <=, ==, >=, >, and !=), and logical operators (&, |, &&, and ||). For example,

```
double PI_2 = java.lang.Math.PI / 2.0;        // arithmetic operation
boolean myFlag = condition1 && (PI_2 >= 3);   // logical operation
String message = "Hello "."World!";           // string operation
```

In addition, Java has autoincrement (++) and autodecrement (--), % (the remainder function), left and right shifts (<<<, <<, >>, >>>), logical complement (!, similar to MATLAB's ~), ternary conditional (? :), and the instanceof operator:

```
number++;                           // same as: number = number + 1;
number = Math.PI % 2;               // remainder function
myFlag = condition1 & !condition2;  // logical condition
myFlag = (a > b ? true : myFlag2);  // ternary condition
```

Finally, many simple operations on the properties can be simplified using the <op>= syntax. For example,

```
number -= 1.2;           // same as: number = number - 1.2;
message += "again!";     // same as: message = message + "again!";
```

The instanceof operator is special: it returns true if the object on its left-hand side is an instance of the class (or implements the interface) on its right-hand side. For example,

```
myFlag = (shape instanceof Geometry);
```

All the Java operators have predefined precedence. Therefore, a + b*c would give a different result than (a + b)*c. For this reason, it is always advisable to add parentheses () in order to explicitly determine the computation order.

Java's code-flow statements are similar to MATLAB, using the following constructs:

```
if (a > b)        { ... }
else if (c > d)  { ... }
else             { ... }
while (myFlag)   { ... }
do { ... } while (number-- > 0)
switch (value)  {
    case 1:
        // do something useful here...
        break;       // without this break, execution will fall-through
    case 2:
        // do something useful here...
        break;       // without this break, execution will fall-through
    default:
        // do something useful here...
}
for (int i=0;              // initial loop value
     i < maxIValue;        // loop continues while condition is true
     i++)                  // loop round post-processing
{
        // do something useful here...
}
```

To exit loops and functions, use one of the following keywords, similar to MATLAB: break, continue, and return. break and continue also support a labeled target, although this is often discouraged as bad programming style:

```
boolean foundFlag = false;
search:
  for (int i=0; i < maxIValue; i++)  {
    for (int j=0; j < maxJValue; j++)  {
      if (someCondition)  {
        foundFlag = true;
        break search;  //regular break would only exit the inner loop
      }
    }
  }
```

Exceptions can be *caught* and processed in Java as in MATLAB, using a try/catch mechanism. An optional finally section enables code execution following the exception-handled block, regardless of whether or not any exception was *raised*. This is similar to MATLAB's *onCleanup* function that was added in MATLAB 7.6 (R2008a).

```
try
{
```

```
    // do something useful here...
}
catch (AnException ex)
{
    // do some error-processing here...
}
catch (AnotherException ex)   { ... }
finally
{
    // do some cleanup processing here whether or not exception raised
}
```

Java enforces stricter exception fore-thought by the programmer: uncaught exceptions must be declared at the top method definition line:

```
public void readFile(String fileName) throws IOException { ... }
```

Unlike MATLAB, which directly interprets source code, Java requires source code to be compiled into binary *byte-code* (aka *class-files*). This compilation can be done using the command-line *javac* compiler, available from the main Java site as part of the free Java Development Kit (JDK).[2] To use javac, place the class code in a similarly named **.java* file (*HelloWorld.java* in our case), then run "javac HelloWorld.java" from the Operating System's command line. If the source code has errors, they will be reported in the console. Otherwise, we will get a file called *HelloWorld.class*. To run HelloWorld, run "java HelloWorld.class" in the operating system's command line. Sections 1.1 and 1.6 explain how to run this program from within MATLAB.

A much preferred alternative to javac is to use a Java IDE (integrated development environment), which enables easier code development and debugging. There are several excellent free and commercial Java IDEs. Two widely-used free IDEs are Eclipse[3] and NetBeans[4] (see Section 1.6). In both IDEs, compilation is done on the fly, and errors are visually displayed next to their offending source.

Similar to MATLAB's **path**, Java uses a *classpath* to locate external classes. If the class uses other classes, then we may need to indicate their classpath location during compilation using javac's–cp command-line switch or the IDE's project preferences.

References

1. http://en.wikipedia.org/wiki/CamelCase
2. http://java.sun.com/javase/downloads/previous.jsp
3 http://www.eclipse.org/downloads/ (or http://bit.ly/95wdV4).
4. http://netbeans.org/downloads/ (or http://bit.ly/cISU1K).

Appendix B: UDD[†]

In several places within this book, I referred to a few undocumented built-in MATLAB functions that relate to *UDD*, including ***handle.listener*** and ***schema.prop***. These functions are not documented anywhere outside the UndocumentedMatlab.com website, so this short appendix that describes them may be helpful.

This functionality, while related to Java objects as mentioned elsewhere in this book, is not in itself Java-based. As far as I know, UDD (Unified Data Definition?) is based on C++ code. Only one of the UDD roles is to provide wrapper functionality for Java objects within MATLAB. In this appendix, I will only describe issues that directly relate to using Java objects — other aspects are documented on the website.

UDD objects, also referred to as schema objects, were introduced with MATLAB 6.0 (R12) back in 2000. UDD is a foundation technology for using handle graphics, Simulink®, Java, and COM within MATLAB. MathWorks has consistently refused to document UDD, stating that UDD objects are for internal use by MathWorks developers. Although there is no formal documentation, there are plenty of examples and tools to help us learn about these classes.

Since UDD classes are a foundation technology, they appear both as built-in classes and as classes defined in m-code. We can tell whether a particular object is a UDD object by using the undocumented built-in function ***classhandle***, which takes one input argument (an object handle) and returns a ***schema.class*** object that describes the input object. If the input object is not a UDD object, then ***classhandle*** raises an exception. Using ***classhandle*** in code that cannot guarantee that the input argument is a UDD object requires placing it in a ***try-catch*** block, to handle such exceptions.

UDD and the newer, well-documented *MCOS* classes share many similarities. In many respects, ***classhandle*** is analogous to the MCOS ***metaclass*** function; property and method attributes are similar, and so on. This is not surprising, considering MCOS was developed based on the UDD experience.

MCOS classes can either be defined as a standalone class or scoped by placing it in a package or package hierarchy. With UDD, all classes must be defined in a package. UDD packages are not hierarchical, meaning that a UDD package may not contain other packages. UDD classes are instantiated as *packageName.className*. UDD classes whose **Global** flag is defined as 'yes' only report their *className*, not their package (instantiation must still be done using

---

[†] This appendix is the product of joint work with Donn Shull of http://aetoolbox.com. Additional details on UDD can be found at http://undocumentedmatlab.com/?s = UDD (or http://bit.ly/dShAuG).

the package name). As far as I know, handle graphics objects are the only built-in UDD objects that have the **Global** flag set to on, so they have global scope rather than package scope.

```
>> hFig = hg.figure    % note: global hg package name is not reported
hFig =
        figure

>> which hg.figure
hg.figure is a built-in method % hg.figure constructor
```

MCOS classes are *value classes* by default, but we can subclass the ***handle*** class to create *handle classes*. In contrast, UDD classes are handle classes by default, but we can create value classes by setting the **Handle** property to 'off'.

The m-files that create the UDD classes supplied with MATLAB are easy to find by searching for directories containing a schema.m file. These classes include the timeseries UDD package ***tsdata***, the user interface package ***uitools***, and many others.

Built-in UDD classes are harder to identify, and are hidden in normal usage. For example, in the current handle graphics version,[†] all HG elements are UDD objects from the ***hg*** package that are wrapped in normal usage by a numeric handle. We can get the underlying UDD object using the undocumented built-in ***handle()*** function:

```
>> hFig = handle(gcf)
hFig =
        figure

>> hLine = handle(plot(1:5))
hLine =
        graph2d.lineseries
```

We can pass any UDD object to Java classes in MATLAB. MATLAB automatically creates a bean adapter for the object, enabling access to the object's methods and properties.

Any Java object created in MATLAB has an associated UDD wrapper object that is either from the ***javahandle*** or the ***javahandle_withcallbacks*** package (the latter is invoked with the 'CallbackProperties' parameter, as described in Section 3.4):

```
>> hjButton = handle(javax.swing.JButton)
hjButton =
        javahandle.javax.swing.JButton

>> hjButton = handle(javax.swing.JButton, 'CallbackProperties')
hjButton =
        javahandle_withcallbacks.javax.swing.JButton
```

Java objects' properties and methods can be inspected using Reflection. MATLAB has a corresponding mechanism, which is accessible via the UDD ***classhandle***:

```
>> chFig = classhandle(handle(gcf))
chFig =
        schema.class
```

---

† This may change in the future HG2: http://UndocumentedMatlab.com/blog/matlab-hg2/ (or http://bit.ly/chL9iK).

```
>> get(chFig)                          % or: chFig.get
                Name: 'figure'
             Package: [1x1 schema.package]
         Description: ''
         AccessFlags: {0x1 cell}
              Global: 'on'
              Handle: 'on'
        Superclasses: [1x1 schema.class]
      SuperiorClasses: {0x1 cell}
      InferiorClasses: {0x1 cell}
             Methods: [2x1 schema.method]
          Properties: [90x1 schema.prop]
              Events: [10x1 schema.event]
      JavaInterfaces: {'com.mathworks.hg.Figure'}
>> chFig.Package.get               % or: get(chFig.Package)
                Name: 'hg'
     DefaultDatabase: [1x1 handle.Database]
             Classes: [42x1 schema.class]
           Functions: [0x1 handle]
         JavaPackage: 'com.mathworks.hg'
          Documented: 'on'
>> chFig.Superclasses.Name
ans =
GObject
```

As can be seen, UDD figure objects are global handles that belong to the hg package and inherit from the base GObject class. UDD figures have two methods, 90 properties and 10 events (callbacks). Let us inspect them, starting with the two methods:

```
>> cmhFig = chFig.Methods
cmhFig =
        schema.method: 2-by-1

>> cmhFig(1).get
                Name: 'setlayoutdirty'
         Description: 'setlayoutdirty'
           Signature: [1x1 schema.signature]
              Static: 'off'
    FirstArgDispatch: 'on'

>> cmhFig(1).Signature.get
     InputTypes: {'handle'}
    OutputTypes: {0x1 cell}
      Varargout: 'off'
       Varargin: 'off'

>> cmhFig(2).get
                Name: 'setDefaultButton'
         Description: 'setDefaultButton'
           Signature: [1x1 schema.signature]
              Static: 'off'
```

```
       FirstArgDispatch: 'on'
>> cmhFig(2).Signature.get
       InputTypes: {2x1 cell}
      OutputTypes: {0x1 cell}
        Varargout: 'off'
         Varargin: 'off'
>> cmhFig(2).Signature.InputTypes'
ans =
     'handle'      'handle'
```

As can be seen, two-figure methods are defined: *setlayoutdirty(handle)* and *setDefaultButton (handle,handle)*. Both methods do not return any output argument.

When MATLAB creates a UDD handle for a Java object, the Java object is inspected using Reflection and the corresponding UDD attributes are set. So, we can also use **classhandle** and the mechanism shown above for Java objects. hjButton in the example above exposes no less than 340 methods, 153 properties, and 31 events.

Events (callbacks) are invoked (*sent* or *raised*) when a specific condition occurs in the underlying object (the relationship with Java events was discussed in Section 1.4):

```
>> cehFig = chFig.Events
cehFig =
        schema.event: 10-by-1
>> reshape(get(cehFig,'Name'),5,2)
ans =
     'SerializeEvent'           'WindowButtonUpEvent'
     'FigureUpdateEvent'        'WindowButtonDownEvent'
     'ResizeEvent'              'WindowButtonMotionEvent'
     'WindowKeyReleaseEvent'    'WindowPostChangeEvent'
     'WindowKeyPressEvent'      'WindowPreChangeEvent''
>> cehFig(5).get
                  Name: 'WindowKeyPressEvent'
    EventDataDescription: 'Window KeyPress Event'
```

We can monitor these events using the undocumented built-in **handle.listener** function. The basic syntax is **handle.listener**(*hObject, eventName, callback*), where *hObject* is the handle of the requested object, *eventName* is a string that contains the case-insensitive possibly partial (if unambiguous) event name, and *callback* is a callback definition (function handle, string, or cell array):‡

```
hListener = handle.listener(gcf,'ResizeEvent','disp 123');
hListener = handle.listener(gcf,'ResizeEvent',@myCallbackFunc);
hListener = handle.listener(gcf,'ResizeEvent',{@myCallbackFunc,data});
hListener = handle.listener(gcf,'ResizeEvent',@(h,e) myFunc(h));
```

---

† WindowPreChangeEvent was removed starting in R2008a.
‡ http://UndocumentedMatlab.com/blog/continuous-slider-callback/#Event_Listener (or http://bit.ly/a6gtaX). In R2009b+, we can also use the **addlistener** function, but in earlier MATLAB releases **addlistener** only worked for Java objects. Starting with R2009b, **addlistener** accepts non-Java handles. See Section 6.12.4 for additional details about **addlistener** and its limitation.

The callback'ed function has the following interface:

```
function myCallbackFcn(hObject,hEventData,varargin)
```

where `hObject` contains the carrier object's handle, `hEventData` contains the event-specific data, which normally contains the **Type** and **Source** properties, as well as other event-specific information, and `varargin` contains optional extra data that is specified during listener setup via cell array (e.g., `{@myCallbackFunc,data}`), or anonymous function (`@(h,e)` `myFunc(h,e,data1,data2)`) declaration formats.

In many cases, although not all, the event names are simply the object's callback name without the 'Callback' suffix. This is the case for Java and COM objects (see Section 1.4). Thus, each *Event* has a corresponding **EventCallback** property.

Unfortunately, this does not always apply in Handle-Graphics objects. Some events have corresponding callbacks (e.g., `ResizeEvent` ⟺ **ResizeFcn**), but some events (e.g., `SerializeEvent`) and some callbacks (e.g., **CloseRequestFcn**) have no counterparts.

Note that MATLAB only keeps the callback alive as long as both the object and the listener handle are accessible somewhere. For this reason, we must store `hListener` somewhere persistent. A good place for this is in the source object's **ApplicationData**, since that property lives just long enough as the object itself:

```
setappdata(gcf, 'ResizeEventListener', hListener);
```

We can get the handle for a UDD event using the ***findevent**(classHandle, eventName)* function, which tries to match partial case-insensitive event names (if unambiguous).

Let us now inspect the UDD figure properties. If we ***get(gcf)***, we only see 62 properties, so why does the figure ***classhandle*** report 90? The answer is that 28 properties (e.g., **JavaFrame**, **UseHG2**, **ApplicationData**) are defined as hidden (their meta-property **Visible** = 'off'). We can access them just like the regular properties (e.g., ***get(gcf,'JavaFrame')***), but we cannot see them in ***get(gcf)*** or ***set(gcf)***:

```
>> cphFig = chFig.Properties
cphFig =
      schema.prop: 90-by-1

>> cphFig(1).get
           Name: 'Alphamap'
    Description: ''
       DataType: 'figureAlphamapType'
   FactoryValue: [1x64 double]
    AccessFlags: [1x1 struct]
        Visible: 'on'
    GetFunction: []
    SetFunction: []

>> reshape(get(cphFig,'Name'),30,3)   % hidden props in *bold italic*
ans =
    'Alphamap'              'PaperOrientation'           '*UseHG2*'
    '*BackingStore*'        'PaperPosition'              'BeingDeleted'
```

'CloseRequestFcn'	'PaperPositionMode'	'*PixelBounds*'
'Color'	'PaperSize'	'ButtonDownFcn'
'Colormap'	'PaperType'	'Clipping'
'CurrentAxes'	'Pointer'	'CreateFcn'
'CurrentCharacter'	'PointerShapeCData'	'DeleteFcn'
'*CurrentKey*'	'PointerShapeHotSpot'	'BusyAction'
'*CurrentModifier*'	'Position'	'HandleVisibility'
'CurrentObject'	'*OuterPosition*'	'*HelpTopicKey*'
'CurrentPoint'	'*ActivePositionProperty*'	'HitTest'
'*Dithermap*'	'*PrintTemplate*'	'Interruptible'
'*DithermapMode*'	'*ExportTemplate*'	'Selected'
'DockControls'	'Renderer'	'SelectionHighlight'
'*DoubleBuffer*'	'RendererMode'	'*Serializable*'
'FileName'	'Resize'	'Tag'
'*FixedColors*'	'ResizeFcn'	'Type'
'*HelpFcn*'	'SelectionType'	'UIContextMenu'
'*HelpTopicMap*'	'ToolBar'	'UserData'
'IntegerHandle'	'Units'	'*ApplicationData*'
'InvertHardcopy'	'*WaitStatus*'	'*Behavior*'
'KeyPressFcn'	'WindowButtonDownFcn'	'*XLimInclude*'
'KeyReleaseFcn'	'WindowButtonMotionFcn'	'*YLimInclude*'
'MenuBar'	'WindowButtonUpFcn'	'*ZLimInclude*'
'*MinColormap*'	'WindowKeyPressFcn'	'*CLimInclude*'
'Name'	'WindowKeyReleaseFcn'	'*ALimInclude*'
'*JavaFrame*'	'WindowScrollWheelFcn'	'*IncludeRenderer*'
'NextPlot'	'WindowStyle'	'Children'
'NumberTitle'	'WVisual'	'Parent'
'PaperUnits'	'WVisualMode'	'Visible'

```
>> cphFig(27).get
          Name: 'JavaFrame'
   Description: ''
      DataType: 'figureJavaFrameType'
  FactoryValue: []
   AccessFlags: [1x1 struct]
       Visible: 'off'
   GetFunction: []
   SetFunction: []

>> cphFig(90).get
          Name: 'Visible'
   Description: ''
      DataType: 'figureVisibleType'
  FactoryValue: 'on'
   AccessFlags: [1x1 struct]
       Visible: 'on'
   GetFunction: []
   SetFunction: []
```

The following trick uses a hidden property of the root (desktop, 0) handle, in order to automatically display hidden properties in *get()* and *set()*:[1]

```
set(0,'HideUndocumented','off'); % default = 'on'
```

Each property has a separate list of **AccessFlags**. The flags most useful to users are **PublicSet** and **PublicGet**, indicating whether users can *set* or *get* the property value:

```
>> cphFig(27).AccessFlags    % JavaFrame
ans =
       PublicSet: 'off'        <= this indicates a read-only property
       PublicGet: 'on'
      PrivateSet: 'on'
      PrivateGet: 'on'
            Init: 'off'
         Default: 'off'
           Reset: 'off'
       Serialize: 'off'
            Copy: 'off'
        Listener: 'off'
        AbortSet: 'off'

>> cphFig(90).AccessFlags    % Visible
ans =
       PublicSet: 'on'
       PublicGet: 'on'
      PrivateSet: 'on'
      PrivateGet: 'on'
            Init: 'off'
         Default: 'on'
           Reset: 'on'
       Serialize: 'on'
            Copy: 'on'
        Listener: 'on'
        AbortSet: 'off'
```

As noted above, UDD handles of Java objects include all object properties, methods, and events. For example, let us inspect JButton's read-only **UIClassID** property:

```
>> chjButton = classhandle(hjButton);
>> chjButton.Properties(2).get
            Name: 'UIClassID'
     Description: ''
        DataType: 'jstring'
    FactoryValue: ''
     AccessFlags: [1x1 struct]
         Visible: 'on'
     GetFunction: []
     SetFunction: []

>> chjButton.Properties(2).AccessFlags
ans =
      PublicSet: 'off'         <= this indicates a read-only property
      PublicGet: 'on'
     PrivateSet: 'off'         <= unsettable even within the class itself
     PrivateGet: 'on'
```

```
       Init: 'off'
    Default: 'off'
      Reset: 'off'
  Serialize: 'off'
       Copy: 'off'
   Listener: 'off'
   AbortSet: 'off'
```

Specific property handles can be found using the *findprop(uddHandle,propertyName)* function. Like *findevent*, *findprop* also tries to match partial case-insensitive property names
(if unambiguous), and returns [] if the property was not found:

```
>> get(chFig.findprop('resize'))    % or: findprop(chhFig,'resize')
           Name: 'Resize'
    Description: ''
       DataType: 'figureResizeType'
   FactoryValue: 'on'
    AccessFlags: [1x1 struct]
        Visible: 'on'
    GetFunction: []
    SetFunction: []

>> get(findprop(chFig,'Resize'));  % equivalent alternative format
>> get(chFig.findprop('Window'))    % or: findprop(chhFig,'Window')
??? The 'Window' property name is ambiguous in the 'figure' class.
>> hProp = findprop(chFig,'NoSuchProperty')
hProp =
     []              <= indicates the specified property was not found
```

Note that *findprop* acts on (accepts) either UDD object handle or the *classHandle*, whereas
findevent can only act on *classHandle*. This is due to *findevent* being a *schema.class* method,
whereas *findprop* belongs to many classes (*which findprop -all*).

Callback monitors can be placed on property access events using *handle.listener*:[2]

```
>> hFig = handle(gcf);
>> hProp = findprop(hFig, 'position');
>> hListener = handle.listener(hFig,hProp,'PropertyPreSet','123');

% Now resize or move the figure and see the callback being invoked:
ans =
    123
```

 Note: Unfortunately, callbacks do not work for some properties (e.g., **Children**).[3]

Using this format of *handle.listener*, four separate property access events can be monitored:
PropertyPreSet, PropertyPostSet, PropertyPreGet, and PropertyPostGet. The callback's event-
Data (second input argument) will contain information about the affected object and property,
and (for Set events) the new value:

```
>> eventData.get
            Type:  'PropertyPreSet'
```

```
    Source:  [1x1 schema.prop]
AffectedObject:  [1x1 figure]
      NewValue:  [734 361 477 325]      <= PropertyPre/PostSet only!
```

The fully documented built-in function *linkprop* uses this *handle.listener* mechanism to keep separate objects synchronized with respect to a specified property: whenever one of the objects modifies the monitored property, a callback is invoked that updates the other object's corresponding property. You may find the source code for this illuminating.[†] I uploaded a utility called *PropListener* to the MATLAB File Exchange that enables bulk setting of monitors on a set of properties and objects.[4]

handle.listener monitors property value changes, but cannot intercept these changes. Sometimes we need the ability to prevent illegal values (*data validation*). For this, we can use the property's **SetFunction** and **GetFunction** meta-properties:

```
>> hProp.SetFunction = {@myCallbackFunc,extraData1,extraData2,...};
```

The **SetFunction** and **GetFunction** callback functions expect two mandatory input arguments (in addition to any extra data that you may add, as in the example snippet above): the affected object (the figure's UDD handle in this case) and the expected value(s). The callback function returns a single output argument, which is then set in the property (**SetFunction**) or returned to the caller (**GetFunction**). Within the callback function, we can trap applicative errors and modify the set/retrieved value. The default behavior is to simply pass the value as-is:

```
function value = myCallbackFcn(hObject,value,varargin)
```

Unfortunately, **SetFunction** and **GetFunction** cannot be modified for existing Java object properties — only for newly added properties (see below). For existing Java properties, we can only use a *handle.listener* monitor.[5] Moreover, for Java properties, *handle.listener* callbacks are only invoked when the property is modified using MATLAB's *set()* function or = operator, not when using the Java setter method:

```
set(hjButton,'Text','test #1');    % listener callback invoked
hjButton.Text = 'test #2';         % listener callback invoked
hjButton.setText('test #3');       % listener callback NOT invoked
```

We can usually add properties to handle objects in run-time. This is done, as in UDD schema.m constructors, using the *schema.prop* function (actually, this is a class constructor, since *schema.prop* is a UDD class):[‡]

```
>> hProp = schema.prop(hFig,'MyName', 'string');
>> hProp = schema.prop(hFig,'MyValue','double');
```

[†] %matlabroot%/toolbox/matlab/graphics/@graphics/@linkprop\linkprop.m. Note that graphics.linkprop is itself a UDD class.

[‡] Note the similarities between UDD's undocumented *schema.prop* and MCOS's fully documented *addprop*.

```
>> hProp = schema.prop(hFig,'AnyData','mxArray'); % mxArray = anything

>> get(hFig)
              . . .
          UserData: []
          Children: [0x1 double]
            Parent: 0
           Visible: 'on'
            MyName: ''         <= new 'string' property, default = ''
           MyValue: 0          <= new 'double' property, default = 0
           AnyData: []         <= new 'mxArray' property, default = []
```

These new properties are displayed when we use *get(uddHandle)* or *set(uddHandle)* on our UDD handle, as shown above. Of course, both the property's corresponding **AcccessFlag** and **Visible** flags must be 'on', which is the default value for newly created properties. Unfortunately, new properties are sometimes not visible when retrieving the full list of object properties using *get()* or *set()* on the base (non-UDD) object, although both **AcccessFlag** and **Visible** flags are 'on'. Despite this minor issue, as with hidden properties, our new properties can always be accessed directly, as long as we know their name, regardless of whether or not they are visible:

```
>> set(gcf,'AnyData',magic(3));
>> get(gcf,'AnyData')
ans =
      8    1    6
      3    5    7
      4    9    2

% Set the public accessibility and visibility flags
>> hProp.Visible = 'off';     % set property as hidden
>> hProp.AccessFlags.PublicSet = 'off';     % set property as read-only

% Retry to set the property, now that it is read-only
>> set(gcf,'AnyData',{1,2,3})
??? Error using ==> set
Changing the 'AnyData' property of instance is not allowed.
```

Finally, to obtain a Java interface source file that represents our class to compile with our Java code, use *classhandle* once again: the *schema.class* object that is returned has a method called *createJavaInterface(interfaceName, folderPath)*. For example:

```
chFig.createJavaInterface('IFigure', pwd);
```

This will create an *IFigure.java* file in the current working directory. We can inspect this file to see the prototypes for accessing the object's properties/methods from Java. Note that read-only properties will not have a setter method in the interface; that set-only properties will have no getter method; and that private properties will not appear at all. Also note that object arrays are indicated using JNI notation[6] rather than the regular []-notation (Ljava.lang.String;

rather than `java.lang.String[]`). Finally, note that logical properties (flags) are converted into Java `int` values, not `boolean`:

```
package com.mathworks.hg;
public interface IFigure extends com.mathworks.jmi.bean.TreeObject
{
  /* Properties */
  public double[] getAlphamap();
  public void setAlphamap(double[] value);

  public com.mathworks.hg.types.HGCallback getCloseRequestFcn();
  public void setCloseRequestFcn(com.mathworks.hg.types.HGCallback value);

  public com.mathworks.hg.types.HGColor getColor();
  public void setColor(com.mathworks.hg.types.HGColor value);

  ...

  public int getType();      <= this is a read-only property

  public com.mathworks.jmi.types.MLArrayRef getUserData();
  public void setUserData(com.mathworks.jmi.types.MLArrayRef value);

  public com.mathworks.hg.types.HGHandle getParent();
  public void setParent(com.mathworks.hg.types.HGHandle value);

  public int getVisible();
  public void setVisible(int value);

  /* Methods */
  public void setlayoutdirty();
  public void setDefaultButton(com.mathworks.jmi.bean.UDDObject arg1);
}
```

References

1. http://UndocumentedMatlab.com/blog/displaying-hidden-handle-properties/ (or http://bit.ly/d25kyG); http://UndocumentedMatlab.com/blog/getundoc-get-undocumented-object-properties/ (or: http://bit.ly/ns3Cog).
2. http://UndocumentedMatlab.com/blog/continuous-slider-callback/#Property_Listener (or http://bit.ly/b49NwE).
3. See http://www.mathworks.com/matlabcentral/newsreader/view_thread/292813#800994 (or http://bit.ly/fPBfHU) for details.
4. http://www.mathworks.com/matlabcentral/fileexchange/18301-proplistener (or http://bit.ly/aDIcav).
5. http://www.mathworks.com/matlabcentral/newsreader/view_thread/157947 (or http://bit.ly/bTpl96).
6. http://java.sun.com/docs/books/jni/ or http://java.sun.com/j2se/1.5.0/docs/guide/jni/spec/types.html#wp276 (or http://bit.ly/cuQAiV).

Appendix C: Open Questions

Being investigative and unsupported in nature, some questions appeared during the preparation of this work, which are still unanswered. Perhaps answers to some of these will be discovered or reported as time passes. If so, I will report them on the http://www.UndocumentedMatlab.com/books/Java/ website and in future book editions.

1. What is the purpose of ***edtObject*** (Section 1.1)?
2. Are Java EDT exceptions trappable or suppressible in MATLAB (Section 1.3)?
3. Is it possible to automatically serialize non-primitive MATLAB constructs (e.g., structs and class objects) into Java objects that can be stored in Java Collections, without requiring use of a conversion function (Section 2.1.1)?
4. Is it possible to know in advance whether a Java logical flag property expects scalar (true/false) or string ('on'/'off') values (Section 3.3.1)?
5. How can the `com.jidesoft.plaf.LookAndFeelFactory` class be used to modify the Look & Feel (Section 3.3.2)?
6. Why does System-Tray throw errors on some MATLAB versions (Section 3.6)?
7. Is it possible to transfer/incorporate MATLAB plots into a pure-Java GUI container (Section 3.8)?
8. Is it possible to customize the new ***uitable*** without needing to replace its existing table **Model** (Section 4.1.1)?
9. Why does tree node collapsing/expansion sometimes fail (Section 4.2.3)?
10. Why do some standard `JTabbedPane` methods fail for ***uitabgroup*** (Section 4.3.1)?
11. Can the empty space left of menu-item icons be removed (Section 4.6.4)?
12. Why are *guide.jar* and *audiovideo.jar* placed in a separate folder than the rest of MATLAB's standard JAR files (Section 5.1.3)?
13. Why does Java place the external JAR files of org.netbeans and org.openide in the MathWorks /jar/ folder rather than the /jarext/ folder (Section 5.1.3)?
14. How can the *BDE.jar* package be used (Sections 5.1.3 and 5.8.3)?
15. Can the `com.mathworks.widgets.find.FindClientRegistry` class be used to specify user-defined Find/Replace functionality (Section 5.5.5)?

16. Can the `com.jidesoft.docking` package be used to customize MATLAB's docking (Section 5.7.1)?

17. What is the benefit of MATLAB's two custom `Timer` classes versus Swing's standard `javax.swing.Timer` (Section 5.8)?

18. How can the `com.mathworks.ide.filtermgr.FilterEditor` class be used (Section 5.8.3)?

19. Is it possible to access a ***uicontrol***'s underlying Java peer's reference without needing to scan the Java frame hierarchy using ***FindJObj*** (Chapter 6)?

20. Is it possible to set an editbox's **Document** property to a `StyledDocument` in order to implement HTML markup or syntax highlighting (Section 6.5.1)?

21. Why does MATLAB insist on calling its scrollbar ***uicontrol*** "slider" instead of "scrollbar" and why does it not have an actual slider control (Section 6.8)?

22. Is there an accessible underlying Java peer object for ***uipanel*** (Section 6.11)?

23. How can the Java frame's native HWND values be used to modify the frame's appearance (Sections 7.1.4 and 7.3.7)?

24. Are axes Handle-Graphics underlying objects accessible (via the JAWT canvas or elsewhere) (Section 7.3.1)?

25. Is it possible to set a transparent Swing component such that the axes behind it would appear (Section 7.3.3)?[1]

26. Is it possible to synchronize help browser's TOC with the displayed page?[2]

27. Is it possible to undecorate a MATLAB figure window (Section 7.3.7)?

28. What is the purpose of the second flag argument in `jDesktop.`*closeGroup()* (Section 8.1.1)?

29. What is the purpose of `jDesktop.`*getDocumentContainment()* (Section 8.1.1)?

30. Is it possible to reload tab-completion definitions in the current MATLAB session, without being required to restart MATLAB (Section 8.3.4)?

31. What is the purpose of the third string argument in `editorservices.` *openDocumentToFunction()* (Section 8.4.1)?

32. Why is ***coveragerpt*** undocumented (having a help section but no doc page or online help), while the Coverage Report that it launches is fully documented (Section 8.7.1)?

33. What is the purpose of `jProfiler`'s **Parallel()* methods (Section 8.7.1)?

34. What is the purpose of GUIDE's **LayoutChangeDefaultCallback** and **LayoutQuickStartTab** preference properties (Section 8.7.3)?

35. Is it possible to modify existing UDD objects' meta-properties, for example, **SetFunction** (Appendix B)?

References

1. http://www.mathworks.com/matlabcentral/newsreader/view_thread/248051 (or http://bit.ly/cEjXRQ).
2. http://www.mathworks.com/matlabcentral/newsreader/view_thread/265127 (or http://bit.ly/deYzhK).

Index